# The Science of Materials

# The Science of Materials

Edited by **Andrew Green**

**NY** RESEARCH
P R E S S

New York

Published by NY Research Press,
23 West, 55th Street, Suite 816,
New York, NY 10019, USA
www.nyresearchpress.com

**The Science of Materials**
Edited by Andrew Green

International Standard Book Number: 978-1-63238-466-9 (Hardback)

Printed in the United States of America.

# Contents

# Preface

Every book is initially just a concept; it takes months of research and hard work to give it the final shape in which the readers receive it. In its early stages, this book also went through rigorous reviewing. The notable contributions made by experts from across the globe were first molded into patterned chapters and then arranged in a sensibly sequential manner to bring out the best results.

This book covers in detail some of the existent theories and innovative concepts revolving around materials science. The ever growing need of advanced technology is the reason that has fueled the research in this field in recent times. This subject involves study of structure and properties of matter and design of new materials. It is an interdisciplinary field that combines chemistry, physics, mining, mineralogy and engineering. Different approaches, evaluations, methodologies and researches have been included in this text. As this discipline is emerging at a fast pace, this book will help the readers to better understand the concepts of materials science and keep pace with the latest advancements in the field.

It has been my immense pleasure to be a part of this project and to contribute my years of learning in such a meaningful form. I would like to take this opportunity to thank all the people who have been associated with the completion of this book at any step.

**Editor**

# Plasmonic modes in thin films: quo vadis?

*Antonio Politano\* and Gennaro Chiarello*

*Dipartimento di Fisica, University of Calabria, Cosenza, Italy*

**Edited by:**
Muhammad Rizwan Saleem,
University of Eastern Finland, Finland

**Reviewed by:**
Xiaofeng Li, Soochow University,
China
Yuehui Lu, Chinese Academy of
Sciences, China

**\*Correspondence:**
Antonio Politano, Dipartimento di
Fisica, University of Calabria, Rende,
Cosenza 87036, Italy
e-mail: antonio.politano@fis.unical.it

Herein, we discuss the status and the prospect of plasmonic modes in thin films. Plasmons are collective longitudinal modes of charge fluctuation in metal samples excited by an external electric field. Surface plasmons (SPs) are waves that propagate along the surface of a conductor with applications in magneto-optic data storage, optics, microscopy, and catalysis. In thin films, the electronic response is influenced by electron quantum confinement. Confined electrons modify the dynamical screening processes at the film/substrate interface by introducing novel properties with potential applications and, moreover, they affect both the dispersion relation of SP frequency and the damping processes of the SP. Recent calculations indicate the emergence of acoustic surface plasmons (ASPs) in Ag thin films exhibiting quantum well states and in graphene films. The slope of the dispersion of ASP decreases with film thickness. We also discuss open issues in research on plasmonic modes in graphene/metal interfaces.

**Keywords: thin films, plasmons, plasmonics, silver, gold, graphene, magnetoplasmonics**

Plasmons in low-dimensional systems never cease to amaze with new astonishing findings, although it has quite a long history, started with the discovery of surface plasmons (SPs) in thin films by Ritchie (1957).

Recently, novel modes, such as sheet (Langer et al., 2011; Politano et al., 2012a), Dirac (Fei et al., 2011; Stauber, 2014), and acoustic surface plasmons (ASPs) (Politano et al., 2011; Yuan et al., 2011) and, moreover, plasmarons (Krstajic and Peeters, 2013), have been observed in low-dimensional systems. Such excitations are supported by the two-dimensional electron gas (2DEG). The great interest toward plasmons arises from the exceptional range of the possible applications of plasmonics.

To date, plasmonic devices based on noble metals (Ag and Au) are widely diffused (Nyga et al., 2008; Pala et al., 2009). Nevertheless, current research is oriented toward the realization of graphene-based plasmonic devices. In fact, plasmons in graphene offer promising prospect of applications covering a wide frequency range, going from terahertz up to the visible (Vicarelli et al., 2012; García de Abajo, 2014).

Nanoscale thin films are an ideal playground for manipulating plasmon properties by peculiar phenomena occurring in thin films, such as quantum size effects (Hamawi et al., 1991; Wei and Chou, 2002) and quantum electron confinement (Ogando et al., 2005; Politano and Chiarello, 2010). Film morphology may originate plasmon confinement within disordered grains (Moresco et al., 1999) or periodic nanodomes (Politano et al., 2013a). Herein, the open challenges regarding plasmons modes in thin films will be presented to the reader, with a particular attention for the cases with higher prospect for plasmonic applications, i.e., noble metal (Ag and Au) and graphene films.

As a general rule, the electromagnetic fields of both sides forming an interface interact in such a way that the SP splits into two plasmonic excitations in which electron may oscillate in phase or not. For a Drude thin slab in vacuum of thickness $a$ (Pitarke et al.,

2007), the dispersion relations of these modes can be obtained by applying appropriate boundary conditions and solving Maxwell's equations (Raether, 1980):

$$\omega = \frac{\omega_p}{\sqrt{2}}\left(1 \pm e^{-qa}\right)^{1/2} \qquad (1)$$

The high energy plasmon in the **Figure 1A** has anti-symmetric field distribution, whereas the low-energy one has symmetric field distribution.

At short wavelengths ($qa \gg 1$), the surface waves become decoupled and each surface sustains independent oscillations at the reduced frequency $\omega_s = \omega_p/\sqrt{2}$ characteristic of a semi-infinite electron gas with a single plane boundary. At long wavelengths ($qa \ll 1$), there are normal oscillations at $\omega_p$ and tangential 2D oscillations at:

$$\omega_{2D} = \left(2\pi naq\right)^{1/2} \qquad (2)$$

which were later discussed by Stern (1967) and observed in artificially structured semiconductors (Allen et al., 1977) and, more recently, in a metallic surface-state band on a silicon surface (Nagao et al., 2001a,b).

The plasmon dispersion in Eq. 1 is modified by the interaction with phonons. Plasmon–phonon coupling is a striking manifestation of the breakdown of the Born–Oppenheimer approximation (Jablan et al., 2011), with consequences on transport (Tediosi et al., 2007) properties. The plasmon–phonon coupling phenomenon implies the hybridization of the plasmon modes of the 2DEG with the optical phonon modes, giving rise to the coupled plasmon–phonon modes (shown in **Figure 1B** for the sample case of graphene/SiO$_2$).

Concerning interfaces, different authors have invoked the existence of *interface plasmons* (Layet et al., 1986). Ahlqvist et al. (1982) have studied the electrodynamics of the interface between two

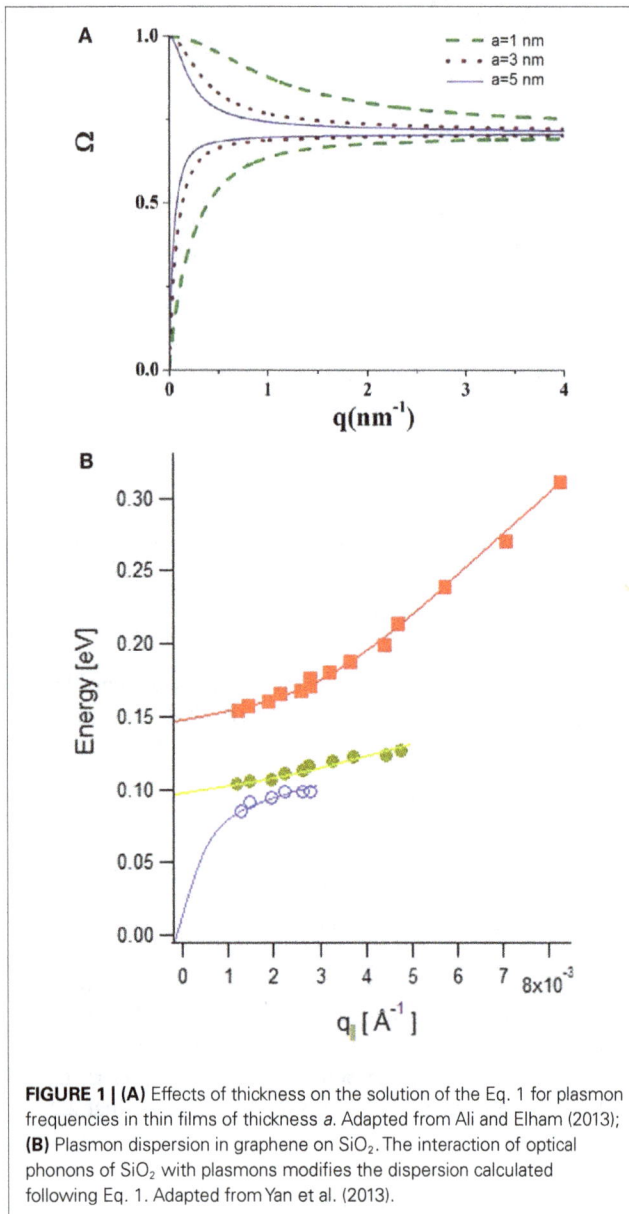

**FIGURE 1 | (A)** Effects of thickness on the solution of the Eq. 1 for plasmon frequencies in thin films of thickness $a$. Adapted from Ali and Elham (2013); **(B)** Plasmon dispersion in graphene on $SiO_2$. The interaction of optical phonons of $SiO_2$ with plasmons modifies the dispersion calculated following Eq. 1. Adapted from Yan et al. (2013).

semi-infinite electron gases, finding that the interface plasmon is characterized by

$$\omega_i^2 = \left(\omega_1^2 + \omega_2^2\right)/2 \qquad (3)$$

where $\omega_i$, $\omega_1$, and $\omega_2$ are the frequencies of the interface plasmon and of the two semi-infinite electron gases, respectively. Jewsbury and Summerside (1980) have suggested that an "interface plasmon" is not a pure mode but arises from the electronic band structure at the interface.

However, the traditional theoretical approach used to describe plasmons in thin films, based on Eqs 1 and 2 and on interface plasmons (Eq. 3) is inadequate to describe the extraordinary complexity of plasmon modes at interfaces. Thus, the overall

encouraging viewpoint for plasmonic applications is also accompanied by the possibility to carry out many other fascinating fundamental studies.

As an example, the strain resulting from the lattice mismatch between adlayer and substrate (Schell-Sorokin and Tromp, 1990; Sander et al., 1998) may further affect plasmonic excitations. Additional collective electronic modes may be induced by strain, as found by Pellegrino et al. (2010) for the case of graphene. However, experimental studies are still lacking due to the difficulties in following strain effects on plasmonic excitations.

Moreover, the influence of electron quantum confinement (presence of quantum well states, QWS) on the SP is still not clearly established. Theoreticians (Yuan and Gao, 2008) and experimentalists (Yu et al., 2005; Politano et al., 2009) have put in evidence the influence of QWS on the plasmon lifetime in films. Due to the opening of the decay channel of the SP into electron-hole pairs via interband transitions involving QWS, the line-width of the SP assumes an unusual dispersion relation as a function of the momentum transfer, as compared with the case of bulk samples. The effects of QWS on plasmon dispersion have been studied only for a few systems. In Politano et al. (2008) and Politano and Chiarello (2009), it has been shown that the screening properties are influenced by the presence of the modified electron distribution in the presence of QWS. However, rigorous and satisfactory theoretical description is still missing.

The presence of QWS and the subsequent enhanced SP density of states around the Fermi level in thin films may also increase the cross section for the excitation of intrinsically free-electron plasmons, such as the multipole surface plasmon (MP) (Liebsch, 1998). The nature of MP has been understood for alkali metals (Tsuei et al., 1990, 1991; Sprunger et al., 1992; Zielasek et al., 2006), alkaline-earth metals (Sprunger et al., 1992), and aluminum (Chiarello et al., 2000). Unfortunately, contradictory results are reported for the most popular plasmonic systems (Ag and Au). Calculations based on a s–d polarization model by Liebsch (1998) predicted the existence of the Ag and Au MP near $\omega_m = 0.8 \cdot \omega_p = 7.2\,eV$ ($\omega_p = 9.0\,eV$ is the s-electron bulk plasmon energy for both Ag and Au) as the density profile at the surface has predominantly s-electron character. Experiments on bulk Ag have not found this mode (Moresco et al., 1996; Barman et al., 2004a,b). In contrast, the Ag MP has been recently measured in Ag films on Ni(111). However, such excitation is revealed only in experimental conditions enhancing the surface sensitivity, i.e., at low impinging energies and grazing incidence (Politano et al., 2013b,d), in agreement with Liebsch's prediction (Liebsch, 1998). Therefore, electron quantum confinement in Ag 5 sp-derived QWS (Miller et al., 1994) enhances the cross section for Ag MP excitation in thin films compared with semi-infinite media (bulk samples).

Another open issue is related to the possible existence of acoustic plasmon modes in thin films. Unfortunately, to date no experimental works exist on this topic, while from the theoretical side Silkin et al. (2011) have shown that ASP emerge in the electronic response of thin Ag films. The presence of Ag QWS in

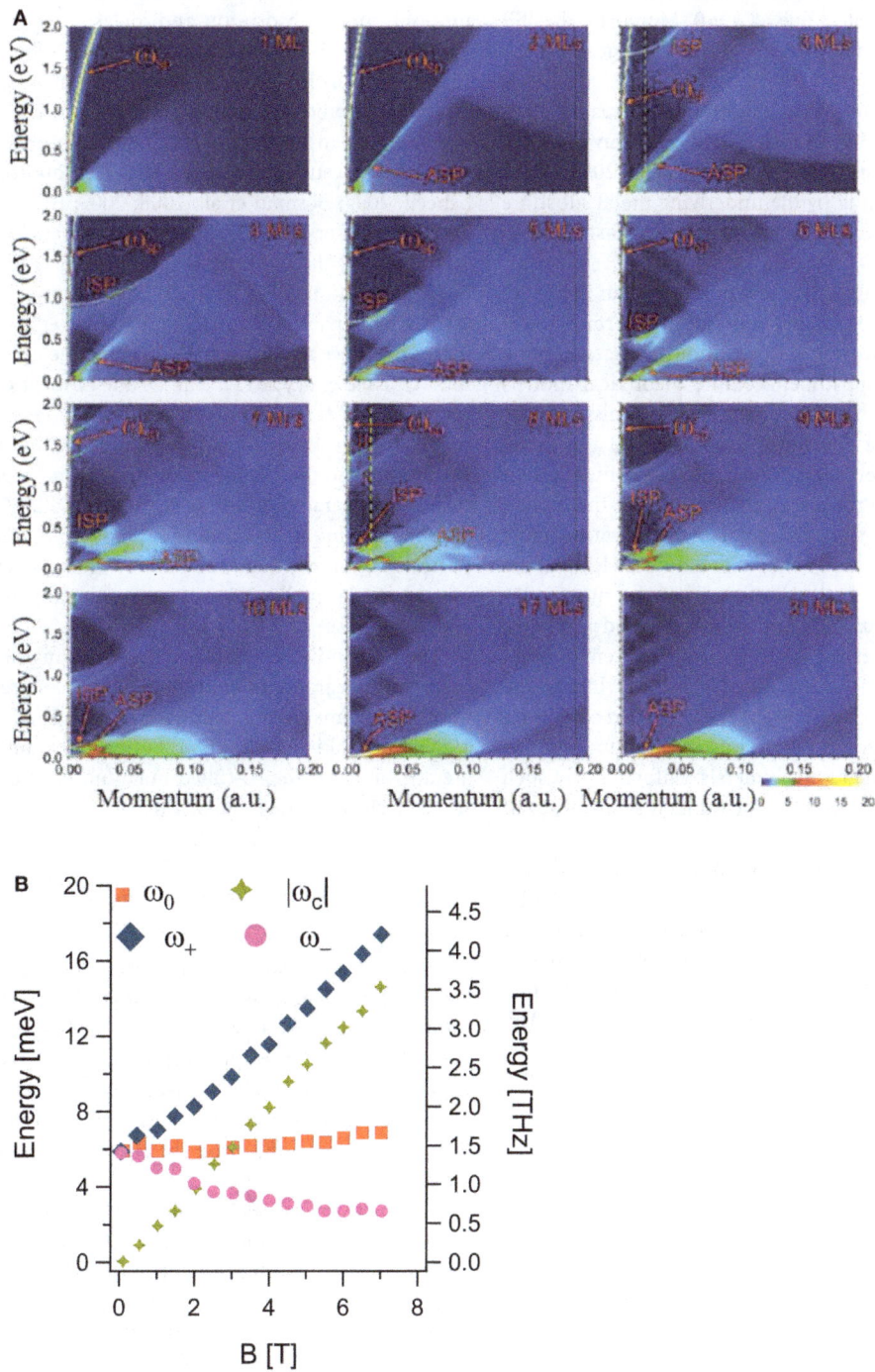

**FIGURE 2 | (A)** Normalized surface loss function Im[$g(q,\omega)$]/$q\omega$ for Ag(111) films with thickness ranging from 1 to 31 monolayers (ML) evaluated by using realistic effective masses in energy band dispersions. Note the strongly dispersing mode corresponding to a conventional $\omega_{SP}$ mode of a thin film. Peaks denoted with "ISP" are originated from the interband transition between the energy-split quantum states. Adapted from Silkin et al. (2011). **(B)** Dependence of the plasmon energy $\omega_0$, the cyclotron resonance energy $\omega_c$ and the magnetoplasmon energies $\omega_\pm$ on the magnetic field B. Adapted from Crassee et al. (2012).

ultrathin films induces the appearance of ASP, whose dispersion is determined by the QWS band. The slope of the dispersion relation decreases with film thickness.

The surface response function (**Figure 2A**) for film thickness higher than three layers shows an additional feature at about 2 eV, which correspond to interband transitions between energy-split

$SS^+$ and $SS^-$ electronic states (interband SP, ISP). In contrast with ASP, the ISP energy has finite value at $q = 0$. Moreover, the ISP energy decreases with increasing thickness and it merges with the ASP at higher thickness.

Acoustic surface plasmon owes its existence to the spatial coexistence of a 2DEG with a 3D electron gas. It has been also predicted to exist at the K/Be interface (Echeverry et al., 2010; Silkin et al., 2010a,b). The screening by the underlying metal substrate change the square-root-like dispersion of the 2D plasmon into linear.

Concerning graphene films, the most puzzling open issues are related to plasmonic modes in graphene/metal interfaces. Due to the difficulty in the theoretical description of the screening by the underlying metal substrate, accurate theoretical models for plasmons in graphene/metal interfaces are still missing. The out-of-plane charge transfer between graphene and the metal is determined by the difference between the work function of graphene and the metal surface and, in addition, by the metal–graphene chemical interaction that creates an interface dipole lowering the metal work function. The induced electrostatic potential decays weakly with the distance from the metal contact as $V(x) \approx x^{-1/2}$ and $\approx x^{-1}$ for undoped and doped graphene, respectively (Khomyakov et al., 2010). Instead, current models overestimate the screening by the metal substrate. Likely, the experimental study of plasmons in graphene deposited on jellium surfaces (Al) could help theoreticians to improve our understanding of screening processes at graphene/metals. Unfortunately, such experimental study is complicated by the difficult preparation of graphene on jellium surfaces.

Low-energy intraband plasmon in graphene is currently well understood (Shin et al., 2011; Stauber and Gómez-Santos, 2012b; Stauber, 2014). In contrast, theoretical models hitherto fail to describe the nature of a non-linear mode observed at ~0.5 eV (Politano and Chiarello, 2014) and, moreover, the quadratic dispersion of interband plasmon (Generalov and Dedkov, 2012; Politano et al., 2012b) in graphene/metal interfaces. The dispersion of the interband plasmon is instead linear in both freestanding graphene (Kramberger et al., 2008) and Cs-decoupled graphene/Ni(111) (Cupolillo et al., 2013b; Ligato et al., 2013). The observation of the change of the interband plasmon from linear to quadratic as a function of the number of graphene layers on silicon carbide (Lu et al., 2009) may in principle afford important information for shedding light on the still confusing state-of-the-art of plasmon modes in epitaxial graphene. However, theoretical models describing the increasing wealth of experimental results on interband plasmons are yet missing.

Moreover, experimental studies on plasmons in bilayer graphene grown on metals would be essential to verify and improve current theoretical models for both plasmon dispersion(Wang and Chakraborty, 2007; Sensarma et al., 2010; Stauber and Gómez-Santos, 2012a; Roldán and Brey, 2013) and plasmaron formation (Van-Nham and Holger, 2012; Krstajic and Peeters, 2013).

Finally, another intriguing topic is magnetoplasmonics, which recently is attracting huge interest for its potential applications in technology (Belotelov et al., 2011; Bonanni et al., 2011). The 2D magnetoplasmons are collective excitations between Landau levels (Lozovik and Sokolik, 2012). They can be observed through infrared optical absorption and inelastic light scattering (Kallin and Halperin, 1984; Oji and MacDonald, 1986; Cinà et al., 1999; Eriksson et al., 1999; Bychkov and Martinez, 2002; Li and Zhai, 2011). In layered and doped graphene structures, the instability and unusual dispersion of magnetoplasmon modes have been studied in recent years, within different approaches (Tahir and Sabeeh, 2007; Berman et al., 2008, 2009; Bychkov and Martinez, 2008a,b; Fischer et al., 2009, 2010; Roldán et al., 2009; Tahir et al., 2011; Wu et al., 2011; Bisti and Kirova, 2012; Ferreira et al., 2012; Lozovik and Sokolik, 2012; Wang et al., 2012; Yan et al., 2012; Chamanara et al., 2013a,b; Petkovic et al., 2013). Magnetoplasmons have been observed in graphene epitaxially grown on $SiC$ (Crassee et al., 2012). The Drude absorption is transformed into a strong terahertz plasmonic peak due to nanoscale inhomogeneities, such as substrate terraces and wrinkles. Plasmonic excitations also modify the magneto-optical response and, in particular, the Faraday rotation (Crassee et al., 2012). This makes graphene a unique playground for plasmon-controlled magneto-optical phenomena thanks to a cyclotron mass, which is two orders of magnitude smaller than in conventional plasmonic materials, such as noble metals.

The field-induced splitting of the plasmon peak resembles strikingly the appearance of collective resonances previously observed in other systems (Allen et al., 1983; Glattli et al., 1985; Mast et al., 1985; Kukushkin et al., 2003). The upper and lower branches are attributed to the so-called bulk and edge magnetoplasmons, respectively, with the frequencies

$$\omega_\pm = \sqrt{\frac{\omega_c^2}{4} + \omega_0^2} \pm \frac{|\omega_c|}{2}$$

where $\omega_0$ is the plasmon frequency at 0 field, $\omega_c = \pm eB/mc$ is the cyclotron frequency, defined as positive for electrons and negative for holes, $m$ is the cyclotron mass, and $c$ the speed of light. At high fields ($|\omega_c| \gg \omega_0$), the upper branch becomes essentially the usual cyclotron resonance with a linear dependence on magnetic field, while the lower branch represents a collective mode confined to the edges (Fetter, 1985) with the energy inversely proportional to the field (**Figure 2B**).

In conclusion, issues discussed herein provide the grounds for theoretical studies aimed at characterizing in more details how growth mode, quantum size effects, and the electron quantum confinement within the adlayer influence the dispersion and the lifetime of collective excitations in nanoscale thin films.

The comprehension of plasmonic excitations in thin films (Chiarello et al., 1997a,b), especially of noble metals (Politano and Chiarello, 2009; Politano, 2012a,b, 2013) and graphene (Politano et al., 2011, 2012a,b, 2013a,c; Cupolillo et al., 2012, 2013a,b; Politano and Chiarello, 2013a,b, 2014), could keep active researchers for a long time.

## REFERENCES

Ahlqvist, P., Monreal, R., Flores, F., and Garcia-Moliner, F. (1982). Interface plasmons at the boundary of two semi-infinite electron gases. *Phys. Scripta* 26, 35. doi:10.1088/0031-8949/26/1/006

Ali, B., and Elham, A. (2013). Effect of shell thickness on propagation of surface hybrid modes in metallic cylindrical nanoshells. *Phys. Scripta* 88, 035707. doi:10.1088/0031-8949/88/03/035707

Allen, S. J., Störmer, H. L., and Hwang, J. C. M. (1983). Dimensional resonance of the two-dimensional electron gas in selectively doped GaAs/AlGaAs heterostructures. *Phys. Rev. B* 28, 4875–4877. doi:10.1103/PhysRevB.28.4875

Allen, S. J., Tsui, D. C., and Logan, R. A. (1977). Observation of the two-dimensional plasmon in silicon inversion layers. *Phys. Rev. Lett.* 38, 980. doi:10.1103/PhysRevLett.38.980

Barman, S. R., Biswas, C., and Horn, K. (2004a). Collective excitations on silver surfaces studied by photoyield. *Surf. Sci.* 566-568, 538–543. doi:10.1016/j.susc.2004.06.059

Barman, S. R., Biswas, C., and Horn, K. (2004b). Electronic excitations on silver surfaces. *Phys. Rev. B* 69, 454131–454139. doi:10.1103/PhysRevB.69.045413

Belotelov, V. I., Akimov, I. A., Pohlm, M., Kotov, V. A., Kastures, S., Vengurlekar, A. S., et al. (2011). Enhanced magneto-optical effects in magnetoplasmonic crystals. *Nat. Nanotechnol.* 6, 370–376. doi:10.1038/nnano.2011.54

Berman, O. L., Gumbs, G., and Echenique, P. M. (2009). Quasiparticles for a quantum dot array in graphene and the associated magnetoplasmons. *Phys. Rev. B* 79, 075418. doi:10.1103/PhysRevB.79.075418

Berman, O. L., Gumbs, G., and Lozovik, Y. E. (2008). Magnetoplasmons in layered graphene structures. *Phys. Rev. B* 78, 085401. doi:10.1103/PhysRevB.78.085401

Bisti, V. E., and Kirova, N. N. (2012). Cyclotron excitations in pure bilayer graphene: electron-hole asymmetry and coulomb interaction. *Physica B Condens. Matter* 407, 1923–1926. doi:10.1016/j.physb.2012.01.065

Bonanni, V., Bonetti, S., Pakizeh, T., Pirzadeh, Z., Chen, J., Nogués, J., et al. (2011). Designer magnetoplasmonics with nickel nanoferromagnets. *Nano Lett.* 11, 5333–5338. doi:10.1021/nl2028443

Bychkov, Y. A., and Martinez, G. (2002). Magnetoplasmons and band nonparabolicity in two-dimensional electron gas. *Phys. Rev. B* 66, 193312. doi:10.1103/PhysRevB.66.193312

Bychkov, Y. A., and Martinez, G. (2008a). Magnetoplasmon excitations in graphene. *Physica E Low Dimens. Syst. Nanostruct.* 40, 1410–1411. doi:10.1016/j.physe.2007.09.026

Bychkov, Y. A., and Martinez, G. (2008b). Magnetoplasmon excitations in graphene for filling factors $\nu \leq 6$. *Phys. Rev. B* 77, 125417. doi:10.1103/PhysRevB.77.125417

Chamanara, N., Sounas, D., and Caloz, C. (2013a). Non-reciprocal magnetoplasmon graphene coupler. *Opt. Express* 21, 11248–11256. doi:10.1364/OE.21.011248

Chamanara, N., Sounas, D., Szkopek, T., and Caloz, C. (2013b). Terahertz magnetoplasmon energy concentration and splitting in graphene PN junctions. *Opt. Express* 21, 25356–25363. doi:10.1364/OE.21.025356

Chiarello, G., Cupolillo, A., Amoddeo, A., Caputi, L. S., Papagno, L., and Colavita, E. (1997a). Collective excitations of two layers of K on Ni(111). *Phys. Rev. B* 55, 1376–1379. doi:10.1103/PhysRevB.55.1376

Chiarello, G., Cupolillo, A., Caputi, L. S., Papagno, L., and Colavita, E. (1997b). Collective and single-particle excitations in thin layers of K on Ni(111). *Surf. Sci.* 377, 365–370. doi:10.1016/S0039-6028(96)01419-7

Chiarello, G., Formoso, V., Santaniello, A., Colavita, E., and Papagno, L. (2000). Surface-plasmon dispersion and multipole surface plasmons in Al(111). *Phys. Rev. B* 62, 12676–12679. doi:10.1103/PhysRevB.62.12676

Cinà, S., Whittaker, D. M., Arnone, D. D., Burke, T., Hughes, H. P., Leadbeater, M., et al. (1999). Magnetoplasmons in a tunable periodically modulated magnetic field. *Phys. Rev. Lett.* 83, 4425–4428. doi:10.1103/PhysRevLett.83.4425

Crassee, I., Orlita, M., Potemski, M., Walter, A. L., Ostler, M., Seyller, T., et al. (2012). Intrinsic terahertz plasmons and magnetoplasmons in large scale monolayer graphene. *Nano Lett.* 12, 2470–2474. doi:10.1021/nl300572y

Cupolillo, A., Ligato, N., and Caputi, L. (2013a). Low energy two-dimensional plasmon in epitaxial graphene on Ni (111). *Surf. Sci.* 608, 88–91. doi:10.1016/j.susc.2012.09.018

Cupolillo, A., Ligato, N., and Caputi, L. S. (2013b). Plasmon dispersion in quasi-freestanding graphene on Ni(111). *Appl. Phys. Lett.* 102, 111609. doi:10.1063/1.4798331

Cupolillo, A., Ligato, N., and Caputi, L. S. (2012). Two-dimensional character of the interface-π plasmon in epitaxial graphene on Ni(111). *Carbon N. Y.* 50, 2588–2591. doi:10.1016/j.carbon.2012.02.017

Echeverry, J. P., Chulkov, E. V., and Silkin, V. M. (2010). Collective electronic excitations in a potassium-covered BE surface. *Phys. Status Solidi C* 7, 2640–2643. doi:10.1002/pssc.200983842

Eriksson, M. A., Pinczuk, A., Dennis, B. S., Simon, S. H., Pfeiffer, L. N., and West, K. W. (1999). Collective excitations in the dilute 2D electron system. *Phys. Rev. Lett.* 82, 2163–2166. doi:10.1103/PhysRevLett.82.2163

Fei, Z., Andreev, G. O., Bao, W., Zhang, L. M., Mcleod, S., Wang, C., et al. (2011). Infrared nanoscopy of dirac plasmons at the graphene-$SiO_2$ interface. *Nano Lett.* 11, 4701–4705. doi:10.1021/nl202362d

Ferreira, A., Peres, N. M. R., and Castro Neto, A. H. (2012). Confined magneto-optical waves in graphene. *Phys. Rev. B* 85, 205426. doi:10.1103/PhysRevB.85.205426

Fetter, A. L. (1985). Edge magnetoplasmons in a bounded two-dimensional electron fluid. *Phys. Rev. B* 32, 7676–7684. doi:10.1103/PhysRevB.32.7676

Fischer, A. M., Dzyubenko, A. B., and Römer, R. A. (2009). Localized collective excitations in doped graphene in strong magnetic fields. *Phys. Rev. B* 80, 165410. doi:10.1103/PhysRevB.80.165410

Fischer, A. M., Römer, R. A., and Dzyubenko, A. B. (2010). Symmetry content and spectral properties of charged collective excitations for graphene in strong magnetic fields. *Europhys. Lett.* 92, 37003. doi:10.1209/0295-5075/92/37003

García de Abajo, F. J. (2014). Graphene plasmonics: challenges and opportunities. *ACS Photonics* 1, 135–152. doi:10.1021/ph400147y

Generalov, A. V., and Dedkov, Y. S. (2012). EELS study of the epitaxial graphene/Ni(111) and graphene/Au/Ni(111) systems. *Carbon N. Y.* 50, 183–191. doi:10.1016/j.carbon.2011.08.018

Glattli, D. C., Andrei, E. Y., Deville, G., Poitrenaud, J., and Williams, F. I. B. (1985). Dynamical hall effect in a two-dimensional classical plasma. *Phys. Rev. Lett.* 54, 1710–1713. doi:10.1103/PhysRevLett.54.1710

Hamawi, A., Lindgren, S. A., and Walldén, L. (1991). Quantum size effects in thin metal overlayers. *Phys. Scripta* T39, 339–345. doi:10.1088/0031-8949/1991/T39/053

Jablan, M., Soljacic, M., and Buljan, H. (2011). Unconventional plasmon-phonon coupling in graphene. *Phys. Rev. B* 83, 161409. doi:10.1103/PhysRevB.83.161409

Jewsbury, P., and Summerside, P. (1980). The nature of interface plasmon modes at bimetallic junctions. *J. Phys. F Met. Phys.* 10, 645. doi:10.1088/0305-4608/10/4/015

Kallin, C., and Halperin, B. I. (1984). Excitations from a filled Landau level in the two-dimensional electron gas. *Phys. Rev. B* 30, 5655–5668. doi:10.1103/PhysRevB.30.5655

Khomyakov, P. A., Starikov, A. A., Brocks, G., and Kelly, P. J. (2010). Nonlinear screening of charges induced in graphene by metal contacts. *Phys. Rev. B* 82, 115437. doi:10.1103/PhysRevB.82.115437

Kramberger, C., Hambach, R., Giorgetti, C., Rümmeli, M. H., Knupfer, M., Fink, J., et al. (2008). Linear plasmon dispersion in single-wall carbon nanotubes and the collective excitation spectrum of graphene. *Phys. Rev. Lett.* 100, 196803. doi:10.1103/PhysRevLett.100.196803

Krstajic, P. M., and Peeters, F. M. (2013). Energy-momentum dispersion relation of plasmarons in bilayer graphene. *Phys. Rev. B* 88, 165420. doi:10.1103/PhysRevB.88.165420

Kukushkin, I. V., Smet, J. H., Mikhailov, S. A., Kulakovskii, D. V., Von Klitzing, K., and Wegscheider, W. (2003). Observation of retardation effects in the spectrum of two-dimensional plasmons. *Phys. Rev. Lett.* 90, 156801. doi:10.1103/PhysRevLett.90.156801

Langer, T., Förster, D. F., Busse, C., Michely, T., Pfnür, H., and Tegenkamp, C. (2011). Sheet plasmons in modulated graphene on Ir(111). *New J. Phys.* 13, 053006. doi:10.1088/1367-2630/13/5/053006

Layet, J. M., Contini, R., Derrien, J., and Lüth, H. (1986). Coupled interface plasmons of the Ag-Si(111) system as investigated with high-resolution electron energy-loss spectroscopy. *Surf. Sci.* 168, 142–148. doi:10.1016/0039-6028(86)90844-7

Li, C., and Zhai, F. (2011). Anisotropic magnetoplasmon spectrum of two-dimensional electron gas systems with the Rashba and Dresselhaus spin-orbit interactions. *J. Appl. Phys.* 109, 093306. doi:10.1063/1.3583651

Liebsch, A. (1998). Prediction of a Ag multipole surface plasmon. *Phys. Rev. B* 57, 3803–3806. doi:10.1103/PhysRevB.57.3803

Ligato, N., Cupolillo, A., and Caputi, L. S. (2013). Study of the intercalation of graphene on Ni(111) with Cs atoms: towards the quasi-free graphene. *Thin Solid Films* 543, 59–62. doi:10.1016/j.tsf.2013.02.121

Lozovik, Y. E., and Sokolik, A. A. (2012). Influence of Landau level mixing on the properties of elementary excitations in graphene in strong magnetic field. *Nanoscale Res. Lett.* 7, 1–19. doi:10.1186/1556-276X-7-134

Lu, J., Loh, K. P., Huang, H., Chen, W., and Wee, A. T. S. (2009). Plasmon dispersion on epitaxial graphene studied using high-resolution electron energy-loss spectroscopy. *Phys. Rev. B* 80, 113410. doi:10.1088/0953-8984/23/1/012001

Mast, D. B., Dahm, A. J., and Fetter, A. L. (1985). Observation of bulk and edge magnetoplasmons in a two-dimensional electron fluid. *Phys. Rev. Lett.* 54, 1706–1709. doi:10.1103/PhysRevLett.54.1706

Miller, T., Samsavar, A., and Chiang, T. C. (1994). Photoexcitation of resonances in Ag films on Ni(111). *Phys. Rev. B* 50, 17686. doi:10.1103/PhysRevB.50.17686

Moresco, F., Rocca, M., Hildebrandt, T., and Henzler, M. (1999). Plasmon confinement in ultrathin continuous Ag films. *Phys. Rev. Lett.* 83, 2238–2241. doi:10.1103/PhysRevLett.83.2238

Moresco, F., Rocca, M., Zielasek, V., Hildebrandt, T., and Henzler, M. (1996). Evidence for the presence of the multipole plasmon mode on Ag surfaces. *Phys. Rev. B* 54, 14333–14336. doi:10.1103/PhysRevB.54.R14333

Nagao, T., Hildebrandt, T., Henzler, M., and Hasegawa, S. (2001a). Dispersion and damping of a two-dimensional plasmon in a metallic surface-state band. *Phys. Rev. Lett.* 86, 5747–5750. doi:10.1103/PhysRevLett.86.5747

Nagao, T., Hildebrandt, T., Henzler, M., and Hasegawa, S. (2001b). Two-dimensional plasmon in a surface-state band. *Surf. Sci.* 493, 680–686. doi:10.1016/S0039-6028(01)01282-1

Nyga, P., Drachev, V. P., Thoreson, M. D., and Shalaev, V. M. (2008). Mid-IR plasmonics and photomodification with Ag films. *Appl. Phys. B* 93, 59–68. doi:10.1007/s00340-008-3145-9

Ogando, E., Zabala, N., Chulkov, E. V., and Puska, M. J. (2005). Self-consistent study of electron confinement to metallic thin films on solid surfaces. *Phys. Rev. B* 71, 205401. doi:10.1103/PhysRevB.71.205401

Oji, H. C. A., and MacDonald, A. H. (1986). Magnetoplasma modes of the two-dimensional electron gas at nonintegral filling factors. *Phys. Rev. B* 33, 3810–3818. doi:10.1103/PhysRevB.33.3810

Pala, R. A., White, J., Barnard, E., Liu, J., and Brongersma, M. L. (2009). Design of plasmonic thin-film solar cells with broadband absorption enhancements. *Adv. Mater. Weinheim* 21, 3504–3509. doi:10.1002/adma.200900331

Pellegrino, F. M. D., Angilella, G. G. N., and Pucci, R. (2010). Dynamical polarization of graphene under strain. *Phys. Rev. B* 82, 115434. doi:10.1103/PhysRevB.82.115434

Petkovic, I., Williams, F. I. B., Bennaceur, K., Portier, F., Roche, P., and Glattli, D. C. (2013). Carrier drift velocity and edge magnetoplasmons in graphene. *Phys. Rev. Lett.* 110, 016801. doi:10.1103/PhysRevLett.110.016801

Pitarke, J. M., Silkin, V. M., Chulkov, E. V., and Echenique, P. M. (2007). Theory of surface plasmons and surface-plasmon polaritons. *Rep. Prog. Phys.* 70, 1–87. doi:10.1088/0034-4885/70/1/R01

Politano, A. (2012a). Influence of structural and electronic properties on the collective excitations of Ag/Cu(111). *Plasmonics* 7, 131–136. doi:10.1007/s11468-011-9285-5

Politano, A. (2012b). Interplay of structural and temperature effects on plasmonic excitations at noble-metal interfaces. *Philos. Mag.* 92, 768–778. doi:10.1080/14786435.2011.634846

Politano, A. (2013). Low-energy collective electronic mode at a noble metal interface. *Plasmonics* 8, 357–360. doi:10.1007/s11468-012-9397-6

Politano, A., Agostino, R. G., Colavita, E., Formoso, V., and Chiarello, G. (2008). Purely quadratic dispersion of surface plasmon in Ag/Ni(111): the influence of electron confinement. *Phys. Status Solidi Rapid Res. Lett.* 2, 86–88. doi:10.1002/pssr.200701307

Politano, A., Campi, D., Formoso, V., and Chiarello, G. (2013a). Evidence of confinement of the π plasmon in periodically rippled graphene on Ru(0001). *Phys. Chem. Chem. Phys.* 15, 11356–11361. doi:10.1039/c3cp51954f

Politano, A., Formoso, V., and Chiarello, G. (2013b). Collective electronic excitations in thin Ag films on Ni(111). *Plasmonics* 8, 1683–1690. doi:10.1007/s11468-11013-19587-x

Politano, A., Formoso, V., and Chiarello, G. (2013c). Evidence of composite plasmon-phonon modes in the electronic response of epitaxial graphene. *J. Phys. Condens. Matter* 25, 345303. doi:10.1088/0953-8984/25/34/345303

Politano, A., Formoso, V., and Chiarello, G. (2013d). Interplay between single-particle and plasmonic excitations in the electronic response of thin Ag films. *J. Phys. Condens. Matter* 25, 305001. doi:10.1088/0953-8984/25/30/305001

Politano, A., and Chiarello, G. (2009). Collective electronic excitations in systems exhibiting quantum well states. *Surf. Rev. Lett.* 16, 171–190. doi:10.1142/S0218625X09012482

Politano, A., and Chiarello, G. (2010). Enhancement of hydrolysis in alkali ultrathin layers on metal substrates in the presence of electron confinement. *Chem. Phys. Lett.* 494, 84–87. doi:10.1016/j.cplett.2010.05.089

Politano, A., and Chiarello, G. (2013a). Quenching of plasmons modes in air-exposed graphene-Ru contacts for plasmonic devices. *Appl. Phys. Lett.* 102, 201608. doi:10.1039/c3nr02027d

Politano, A., and Chiarello, G. (2013b). Unravelling suitable graphene-metal contacts for graphene-based plasmonic devices. *Nanoscale* 5, 8215–8220. doi:10.1039/c3nr02027d

Politano, A., and Chiarello, G. (2014). Emergence of a nonlinear plasmon in the electronic response of doped graphene. *Carbon N. Y.* 71, 176–180. doi:10.1016/j.carbon.2014.01.026

Politano, A., Formoso, V., and Chiarello, G. (2009). Damping of the surface plasmon in clean and K-modified Ag thin films. *J. Electron Spectros. Relat. Phenomena* 173, 12–17. doi:10.1016/j.elspec.2009.03.003

Politano, A., Marino, A. R., and Chiarello, G. (2012a). Effects of a humid environment on the sheet plasmon resonance in epitaxial graphene. *Phys. Rev. B* 86, 085420. doi:10.1103/PhysRevB.86.085420

Politano, A., Marino, A. R., Formoso, V., Farías, D., Miranda, R., and Chiarello, G. (2012b). Quadratic dispersion and damping processes of π plasmon in monolayer graphene on Pt(111). *Plasmonics* 7, 369–376. doi:10.1007/s11468-011-9317-1

Politano, A., Marino, A. R., Formoso, V., Farías, D., Miranda, R., and Chiarello, G. (2011). Evidence for acoustic-like plasmons on epitaxial graphene on Pt(111). *Phys. Rev. B* 84, 033401. doi:10.1103/PhysRevB.84.033401

Raether, H. (1980). *Excitation of Plasmons and Interband Transitions by Electrons*. Berlin: Springer-Verlag.

Ritchie, R. H. (1957). Plasma losses by fast electrons in thin films. *Phys. Rev. B* 106, 874–881. doi:10.1103/PhysRev.106.874

Roldán, R., and Brey, L. (2013). Dielectric screening and plasmons in AA-stacked bilayer graphene. *Phys. Rev. B* 88, 115420. doi:10.1103/PhysRevB.88.115420

Roldán, R., Fuchs, J. N., and Goerbig, M. O. (2009). Collective modes of doped graphene and a standard two-dimensional electron gas in a strong magnetic field: linear magnetoplasmons versus magnetoexcitons. *Phys. Rev. B* 80, 085408. doi:10.1103/PhysRevB.80.085408

Sander, D., Schmidthals, C., Enders, A., and Kirschner, J. (1998). Stress and structure of Ni monolayers on W(110): the importance of lattice mismatch. *Phys. Rev. B* 57, 1406–1409. doi:10.1103/PhysRevB.57.1406

Schell-Sorokin, A. J., and Tromp, R. M. (1990). Mechanical stresses in (sub)monolayer epitaxial films. *Phys. Rev. Lett.* 64, 1039–1042. doi:10.1103/PhysRevLett.64.1039

Sensarma, R., Hwang, E. H., and Das Sarma, S. (2010). Dynamic screening and low-energy collective modes in bilayer graphene. *Phys. Rev. B* 82, 195428. doi:10.1088/0957-4484/23/50/505204

Shin, S. Y., Hwang, C. G., Sung, S. J., Kim, N. D., Kim, H. S., and Chung, J. W. (2011). Observation of intrinsic intraband π-plasmon excitation of a single-layer graphene. *Phys. Rev. B* 83, 161403. doi:10.1103/PhysRevB.83.161403

Silkin, V. M., Chulkov, E. V., Echeverry, J. P., and Echenique, P. M. (2010a). Modification of response properties of the Be(0001) surface upon adsorption of a potassium monolayer: an Ab initio calculation. *Phys. Status Solidi B* 247, 1849–1857. doi:10.1002/pssb.200983843

Silkin, V. M., Hellsing, B., Walldén, L., Echenique, P. M., and Chulkov, E. V. (2010b). Photoelectron driven acoustic surface plasmons in p(2 × 2)K/Be(0001): Ab initio calculations. *Phys. Rev. B* 81, 113406. doi:10.1103/PhysRevB.81.113406

Silkin, V. M., Nagao, T., Despoja, V., Echeverry, J. P., Eremeev, S. V., Chulkov, E. V., et al. (2011). Low-energy plasmons in quantum-well and surface states of metallic thin films. *Phys. Rev. B* 84, 165416. doi:10.1103/PhysRevB.84.165416

Sprunger, P. T., Watson, G. M., and Plummer, E. W. (1992). The normal modes at the surface of Li and Mg. *Surf. Sci.* 269-270, 551–555. doi:10.1016/0039-6028(92)91307-W

Stauber, T. (2014). Plasmonics in Dirac systems: from graphene to topological insulators. *J. Phys. Condens. Matter* 26, 123201. doi:10.1088/0953-8984/26/12/123201

Stauber, T., and Gómez-Santos, G. (2012a). Plasmons and near-field amplification in double-layer graphene. *Phys. Rev. B* 85, 075410. doi:10.1103/PhysRevB.85.075410

Stauber, T., and Gómez-Santos, G. (2012b). Plasmons in layered structures including graphene. *New J. Phys.* 14, 105018. doi:10.1088/1367-2630/14/10/105018

Stern, F. (1967). Polarizability of a two-dimensional electron gas. *Phys. Rev. Lett.* 18, 546–548. doi:10.1103/PhysRevLett.18.546

Tahir, M., and Sabeeh, K. (2007). Theory of Weiss oscillations in the magneto-plasmon spectrum of Dirac electrons in graphene. *Phys. Rev. B* 76, 195416. doi:10.1103/PhysRevB.76.195416

Tahir, M., Sabeeh, K., and Mackinnon, A. (2011). Temperature effects on the magnetoplasmon spectrum of a weakly modulated graphene monolayer. *J. Phys. Condens. Matter* 23, 425304. doi:10.1088/0953-8984/23/42/425304

Tediosi, R., Armitage, N. P., Giannini, E., and Van Der Marel, D. (2007). Charge carrier interaction with a purely electronic collective mode: plasmarons and the infrared response of elemental bismuth. *Phys. Rev. Lett.* 99, 016406. doi:10.1103/PhysRevLett.99.016406

Tsuei, K. D., Plummer, E. W., Liebsch, A., Kempa, K., and Bakshi, P. (1990). Multipole plasmon modes at a metal surface. *Phys. Rev. Lett.* 64, 44–47. doi:10.1103/PhysRevLett.64.44

Tsuei, K. D., Plummer, E. W., Liebsch, A., Pehlke, E., Kempa, K., and Bakshi, P. (1991). The normal modes at the surface of simple metals. *Surf. Sci.* 247, 302–326. doi:10.1016/0039-6028(91)90142-F

Van-Nham, P., and Holger, F. (2012). Coulomb interaction effects in graphene bilayers: electron-hole pairing and plasmaron formation. *New J. Phys.* 14, 075007. doi:10.1088/1367-2630/14/7/075007

Vicarelli, L., Vitiello, M., Coquillat, D., Lombardo, A., Ferrari, A., Knap, W., et al. (2012). Graphene field-effect transistors as room-temperature terahertz detectors. *Nat. Mater.* 11, 865–871. doi:10.1038/nmat3417

Wang, W., Apell, S. P., and Kinaret, J. M. (2012). Edge magnetoplasmons and the optical excitations in graphene disks. *Phys. Rev. B* 86, 125450. doi:10.1021/nl3016335

Wang, X.-F., and Chakraborty, T. (2007). Coulomb screening and collective excitations in a graphene bilayer. *Phys. Rev. B* 75, 041404. doi:10.1103/PhysRevB.75.041404

Wei, C. M., and Chou, M. Y. (2002). Theory of quantum size effects in thin Pb(111) films. *Phys. Rev. B* 66, 233408. doi:10.1103/PhysRevB.66.233408

Wu, J. Y., Chen, S. C., Roslyak, O., Gumbs, G., and Lin, M. F. (2011). Plasma excitations in graphene: their spectral intensity and temperature dependence in magnetic field. *ACS Nano* 5, 1026–1032. doi:10.1021/nn1024847

Yan, H., Li, Z., Li, X., Zhu, W., Avouris, P., and Xia, F. (2012). Infrared spectroscopy of tunable Dirac terahertz magneto-plasmons in graphene. *Nano Lett.* 12, 3766–3771. doi:10.1021/nl3016335

Yan, H., Low, T., Zhu, W., Wu, Y., Freitag, M., Li, X., et al. (2013). Damping pathways of mid-infrared plasmons in graphene nanostructures. *Nat. Photonics* 7, 394–399. doi:10.1038/nphoton.2013.57

Yu, Y. H., Jiang, Y., Tang, Z., Guo, Q. L., Jia, J. F., Xue, Q. K., et al. (2005). Thickness dependence of surface plasmon damping and dispersion in ultrathin Ag films. *Phys. Rev. B* 72, 205405. doi:10.1103/PhysRevB.72.205405

Yuan, Z., and Gao, S. (2008). Landau damping and lifetime oscillation of surface plasmons in metallic thin films studied in a jellium slab model. *Surf. Sci.* 602, 460–464. doi:10.1016/j.susc.2007.10.040

Yuan, Z., Jiang, Y., Gao, Y., Käll, M., and Gao, S. (2011). Symmetry-dependent screening of surface plasmons in ultrathin supported films: the case of Al/Si(111). *Phys. Rev. B* 83, 165452. doi:10.1103/PhysRevB.83.165452

Zielasek, V., Ronitz, N., Henzler, M., and Pfnür, H. (2006). Crossover between monopole and multipole plasmon of Cs monolayers on Si(111) individually resolved in energy and momentum. *Phys. Rev. Lett.* 96, 196801. doi:10.1103/PhysRevLett.96.196801

**Conflict of Interest Statement:** The authors declare that the research was conducted in the absence of any commercial or financial relationships that could be construed as a potential conflict of interest.

# 2

# Group IV light sources to enable the convergence of photonics and electronics

*Shinichi Saito[1]\*, Frederic Yannick Gardes[1], Abdelrahman Zaher Al-Attili[1], Kazuki Tani[2,3,4], Katsuya Oda[2,3,4], Yuji Suwa[2,3,4], Tatemi Ido[2,3,4], Yasuhiko Ishikawa[5], Satoshi Kako[3,6], Satoshi Iwamoto[3,6] and Yasuhiko Arakawa[3,6]*

[1] Faculty of Physical Sciences and Engineering, University of Southampton, Southampton, UK
[2] Photonics Electronics Technology Research Association (PETRA), Tokyo, Japan
[3] Institute for Photonics-Electronics Convergence System Technology (PECST), Tokyo, Japan
[4] Central Research Laboratory, Hitachi Ltd., Tokyo, Japan
[5] Department of Materials Engineering, Graduate School of Engineering, The University of Tokyo, Tokyo, Japan
[6] Institute of Industrial Science, The University of Tokyo, Tokyo, Japan

**Edited by:**
*Jifeng Liu, Dartmouth College, USA*

**Reviewed by:**
*Androula Galiouna Nassiopoulou, National Centre for Scientific Research Demokritos, Greece*
*Raul J. Martin-Palma, Universidad Autonoma de Madrid, Spain*
*Jifeng Liu, Dartmouth College, USA*

**\*Correspondence:**
*Shinichi Saito, Nano Research Group, Electronics and Computer Science, Faculty of Physical Sciences and Engineering, Highfield Campus, University of Southampton, Southampton SO17 1BJ, UK*
*e-mail: s.saito@soton.ac.uk*

Group IV lasers are expected to revolutionize chip-to-chip optical communications in terms of cost, scalability, yield, and compatibility to the existing infrastructure of silicon industries for mass production. Here, we review the current state-of-the-art developments of silicon and germanium light sources toward monolithic integration. Quantum confinement of electrons and holes in nanostructures has been the primary route for light emission from silicon, and we can use advanced silicon technologies using top-down patterning processes to fabricate these nanostructures, including fin-type vertical multiple-quantum-wells. Moreover, the electromagnetic environment can also be manipulated in a photonic crystal nanocavity to enhance the efficiency of light extraction and emission by the Purcell effect. Germanium is also widely investigated as an active material in Group IV photonics, and novel epitaxial growth technologies are being developed to make a high quality germanium layer on a silicon substrate. To develop a practical germanium laser, various technologies are employed for tensile-stress engineering and high electron doping to compensate the indirect valleys in the conduction band. These challenges are aiming to contribute toward the convergence of electronics and photonics on a silicon chip.

Keywords: silicon, photonics, CMOS, germanium, epitaxy, luminescence, quantum, strain

## 1. INTRODUCTION

As the integration of transistors in a chip increases, the demands of the interconnections are expanding, since more information will be transferred between chips optically (Miller, 2009). The advantage of optical interconnection over electrical wiring is fundamentally coming from the elementary particles, photons, used for signal transmission. We can transmit photons without an electrical connection throughout an optical fiber, since photons do not have charge. Of course, optical loss exists, but still the total energy consumption of the optical interconnection can be much lower than that of the electrical connection, especially for the long-distance communications at higher data rate, even including the energy required to convert electrons to photons and vice versa (Miller, 2009). Si photonics is revolutionizing optical interconnections in terms of cost, power, bandwidth, and scalability (Zimmermann, 2000; Pavesi and Lockwood, 2004; Reed and Knights, 2004; Pavesi and Guillot, 2006; Reed, 2008; Deen and Basu, 2012; Fathpour and Jalali, 2012; Vivien and Pavesi, 2013). III-V (Wale, 2008; Evans et al., 2011) and Si-based platform technologies (Reed and Knights, 2004; Gunn, 2006; Rylyakov et al., 2011; Arakawa et al., 2013; Urino et al., 2013) are competing for the next generation of optical interconnections. The critical missing component for Si photonics is a monolithic light source compatible with the existing infrastructure of complementary-metal-oxide-semiconductor (CMOS) technologies for fabrication. The hybrid integration of III-V devices on an Si substrate (Fang et al., 2006) or feeding of an optical fiber to an Si waveguide coupled with a grating from a III-V laser diode (Gunn, 2006) would be the near-term solution, but it is desirable to realize monolithic light sources for the long term. Comprehensive reviews on developing practical lasers on Si have been published by various authors (Cullis et al., 1997; Ossicini et al., 2006; Daldosso and Pavesi, 2009; Liang and Bowers, 2010; Steger et al., 2011; Liu et al., 2012; Michel and Romagnoli, 2012; Boucaud et al., 2013; Shakoor et al., 2013; Liu, 2014). Here, we review this active field focusing on the progress of Si and germanium (Ge) light sources fabricated by standard CMOS processes.

Photoluminescence (PL) (Canham, 1990) and electroluminescence (EL) (Koshida and Koyama, 1992) from porous-Si are the most famous achievements to overcome the fundamental limitations of the indirect band-gap character of Si. The maximum PL (Gelloz and Koshida, 2000) and EL (Gelloz et al., 2005) quantum efficiency exceeded 23 and 1%, respectively. The mechanism of light emission from porous-Si is considered to originate from quantum confinement effects (Canham, 1990; Koshida and Koyama, 1992; Cullis et al., 1997; Nassiopoulou, 2004; Ossicini et al., 2006; Daldosso and Pavesi, 2009) in the self-organized nanostructure. The typical length scale to expect quantum confinement would be comparable to the exciton Bohr radius, which is about 5 nm for Si and 18 nm for Ge (Cullis et al., 1997). On the other hand, the gate length fabricated by CMOS technologies is comparable to the exciton Bohr radius so that we can fabricate various quantum structures, including quantum dots (Arakawa and Sakaki, 1982), nano-wires, and quantum-wells, by lithographically controlled top-down processes. In addition, novel cavity structures (Iwamoto and Arakawa, 2012) can be fabricated to enhance the internal quantum efficiency by the Purcell effect (Purcell, 1946) as well as the extraction efficiency by improved coupling to a lens. Ge is also intensively studied, since the direct band-gap energy is closer to the indirect transition energy than that of Si. Highly, $n$-type doping and strain engineering are effective to enhance the light emissions from Ge (Liu et al., 2012; Michel and Romagnoli, 2012; Boucaud et al., 2013; Liu, 2014), and some of these recent advances are reviewed in this paper.

## 2. STRATEGIES TO ENHANCE LIGHT EMISSION FROM GROUP IV MATERIALS

### 2.1. THEORETICAL STUDY OF LIGHT EMISSION FROM SILICON

Both Si and Ge are known to be poor light emitters because of their indirect band-gap structures. Even so, there are some methods for making direct transitions to occur in these materials. These possibilities were examined theoretically by first-principles calculations based on density functional theory using plane-wave-based ultra-soft pseudo-potentials (Vanderbilt, 1990; Laasonen et al., 1993). Generalized gradient approximation (Perdew et al., 1996) is used for the calculation of Si, and hybrid functional (Perdew et al., 1996) is used for Ge. The optical matrix elements are calculated with the aid of core-repair terms (Kageshima and Shiraishi, 1997).

The lowest conduction band (LCB) of bulk Si has a minimum near the $X$-point, and six electron valleys exist near $X$-points. Two valleys among the six are projected onto $\Gamma$-point in two-dimensional momentum ($k$)-space when an Si quantum-well (QW) with (001) surfaces is fabricated; this is called a valley-projection. Because the top of the valence band is also projected onto $\Gamma$, direct transitions are possible in an Si(001) QW.

Optical gain of Si(001) QWs is shown in **Figure 1** as a function of the thickness (Suwa and Saito, 2009). Here, losses due to transitions within conduction bands and those within valence bands are not taken into account. The thinner QW shows the larger gain, since the surface of the QW plays an important role in this direct transition and it dominates if the QW is thin. **Figure 1** also shows that the surface structure of the QW affects the efficiency of light emission strongly.

Experimentally, optical gain from the Si quantum dots (Pavesi et al., 2000) embedded in an insulating matrix (Pavesi et al., 2000; Nassiopoulou, 2004; Ossicini et al., 2006; Pavesi and Guillot, 2006) has been reported. It was confirmed that the interface states associated with oxygen atoms were important to explain the positive optical gain (Pavesi et al., 2000; Nassiopoulou, 2004; Ossicini et al., 2006; Pavesi and Guillot, 2006). It will be interesting to make these structures by top-down CMOS processes.

### 2.2. THEORETICAL STUDY OF LIGHT EMISSION FROM GERMANIUM

Ge has two important differences from Si. One is that Ge has the minimum of the LCB at the $L$-point ($L$-valley), while Si has it

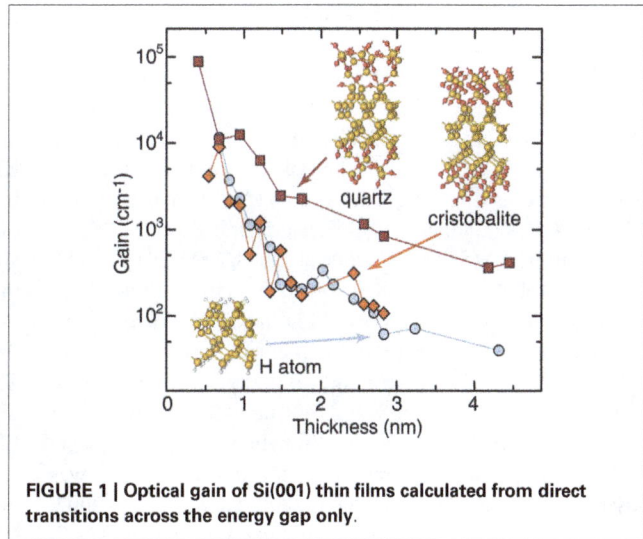

FIGURE 1 | Optical gain of Si(001) thin films calculated from direct transitions across the energy gap only.

near the $X$-point. The other is that Ge has a local minimum of the LCB at the $\Gamma$-point ($\Gamma$-valley), while Si does not. An $L$-valley is projected onto the $\Gamma$-point in the two-dimensional $k$-space for Ge(111) QW. For Ge(001) QW, no $L$-valley is projected onto the $\Gamma$-point. The small $\Gamma$-valley, which is not occupied unless a large number of electrons are injected, is always projected onto the $\Gamma$-point independently of the direction of the QW. While there are two approaches to obtain efficient light emission from Ge by direct transitions, using $L$-valleys of a Ge(111) QW or using the $\Gamma$-valley of bulk Ge, we think the latter is more promising. This is due to the fact that the calculated optical matrix element for the $\Gamma$-valley is very large compared to that for the $L$-valleys.

To enhance light emission from bulk Ge, applying tensile strain is known to be effective (Liu et al., 2010). Tensile strain makes the energy difference between the $\Gamma$ and $L$-valleys small, and that makes electron injection into the $\Gamma$-valley easier. Also heavy $n$-type doping is known to be effective, because electrons can be injected into the $\Gamma$-valley if the $L$-valleys are already occupied by doped electrons.

In order to predict required strength of strain and amount of doping, we calculated optical gains of bulk Ge with and without strain. **Figure 2** shows calculated optical gain as functions of injected electron density and hole densities. Here, the applied strain is assumed to be 0.25% biaxial tensile strain parallel to (001) surface and optical losses due to free carrier absorptions (Wang et al., 2013) are taken into account. This result shows that even bulk Ge without strain can have a positive optical gain, but number of electrons required for that is very large ($10^{20}$ cm$^{-3}$). Despite the relatively small amount of the strain (0.25%), the impact on the gain is clear. Owing to this enhancement, only half the electron density ($5 \times 10^{19}$ cm$^{-3}$) is needed to have positive gain. In experiment, applying 0.25% strain is rather easy, and making higher strain will be possible, as we see in the following sections. Therefore, Ge lasers will be realized when an appropriate strain and carrier injection are achieved.

## 3. ELECTRO-LUMINESCENCE FROM SILICON QUANTUM-WELL

As we reviewed in Section 1, it is well established that efficient recombination is observed in Si nanostructures by quantum confinement effects (Canham, 1990; Koshida and Koyama, 1992; Cullis et al., 1997; Ossicini et al., 2006; Daldosso and Pavesi, 2009). The nanostructures include quantum dots (Arakawa and Sakaki, 1982), nano-wires (Canham, 1990; Koshida and Koyama, 1992), quantum-well (QW) (Saito et al., 2006a,b, 2008, 2009; Saito, 2011), and fins (Saito et al., 2011a,b). One of the difficulties in developing an efficient light-emitting diode (LED) made of Si comes from the trade-off between quantum confinement and carrier injection. The surface of these Si nanostructures is easily oxidized to $SiO_2$, and the band offsets between Si and $SiO_2$ are too high to expect efficient current injection except for tunneling. In order to overcome this trade-off, lateral carrier injection into the Si QW was proposed (Saito et al., 2006a,b, 2008, 2009; Hoang et al., 2007; Noborisaka et al., 2011; Saito, 2011). As shown in **Figures 3A–C**, the Si QW LEDs were fabricated by local thinning of a silicon-on-insulator (SOI) substrate, and the Si QW was directly connected to the thick Si diffusion electrodes (Saito et al., 2006a,b). Both electrons and holes are laterally injected to the Si QW in these planar

$p$-$i$-$n$ diodes (Saito et al., 2006a,b, 2008, 2009; Noborisaka et al., 2011; Saito, 2011). Another advantage of these device structures is the fabrication of the Si QW through the LOCal-Oxidation of Si (LOCOS) process. The LOCOS process was originally developed for isolation of CMOS transistors (Sze and Lee, 2012; Taur and Ning, 2013). It was also used to evaluate the carrier mobility in the ultra-thin Si QW (Uchida and Takagi, 2003). Oxidation is one of the most precisely controlled processes in CMOS technologies, and we can routinely oxidize a large Si wafer (typically 8–10″ in diameter) within the local variation of <0.1 nm. Besides, the interface between Si and $SiO_2$ is excellent with low interface trap density ($<10^{11}$ cm$^{-2}$) (Sze and Lee, 2012; Taur and Ning, 2013). The excellent interfacial quality and strong quantum confinement in Si nanostructures are critical to ensure high quantum efficiency (Gelloz et al., 2005). As shown in **Figure 3E**, EL is observed exclusively from the thin Si QW and EL from thick Si electrodes is negligible (Saito et al., 2006a). This supports the mechanism of EL based on quantum confinement (Ossicini et al., 2006; Suwa and Saito, 2009). The high carrier density in the thin Si QW also contributes to enhance the emissions (Saito et al., 2006a). By applying the back gate to the Si substrate, we can modulate the intensities of light emission, and the device can be called as an Si light-emitting transistor (Saito et al., 2006b).

The next step toward the practical light source for Si photonics is to couple the light from Si to a cavity and a waveguide (WG). An Si-based WG cannot be used for emission from Si QW due to the absorption. An $Si_3N_4$ WG was fabricated on top of the Si QW by conventional lithography and dry etching (Saito et al., 2008, 2009). To enhance the optical confinement in the WG of the Si Resonant Cavity LED (RCLED), part of the supporting substrate was removed by using double sided aligner and anisotropic wet etching (Saito et al., 2008, 2009), as shown in **Figure 3B**. Evanescent coupling between the propagating optical mode and Si QW was expected, and the enhanced EL from the edge of the waveguide was observed (**Figure 3F**). More recently, SOI substrates with superior uniformities with thick Buried-OXide (BOX) ($>2\,\mu$m) became available, and by using these wafers, strong optical confinement within the $Si_3N_4$ WG was ensured without removing the supporting Si substrate (Saito, 2011), as shown in **Figure 3C**. In fact, the near-field image of the propagating optical mode was taken at the edge of the WG (**Figure 3G**).

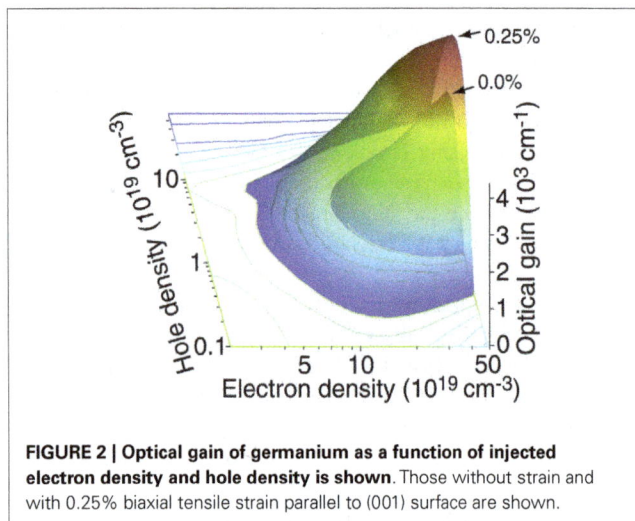

**FIGURE 2 | Optical gain of germanium as a function of injected electron density and hole density is shown**. Those without strain and with 0.25% biaxial tensile strain parallel to (001) surface are shown.

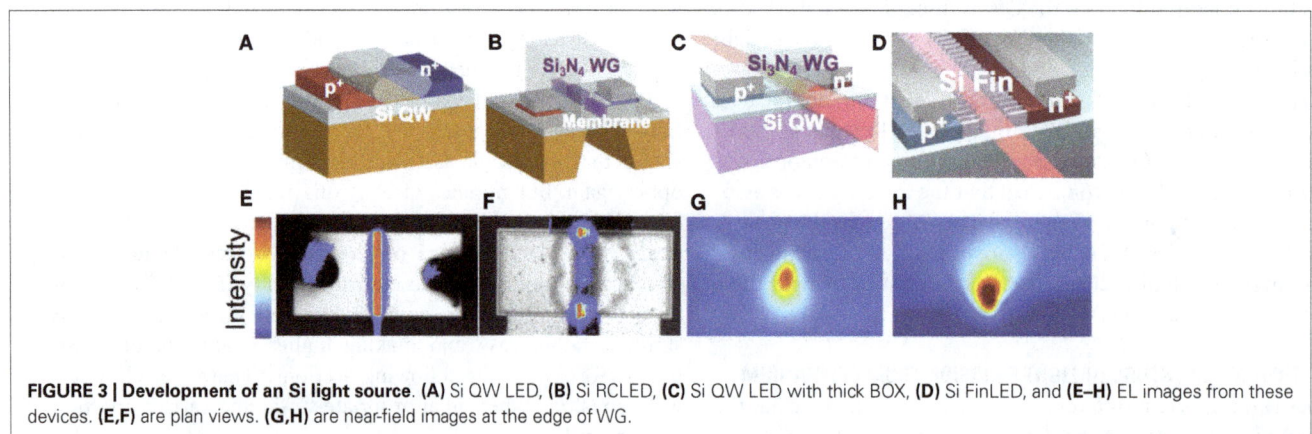

**FIGURE 3 | Development of an Si light source. (A)** Si QW LED, **(B)** Si RCLED, **(C)** Si QW LED with thick BOX, **(D)** Si FinLED, and **(E–H)** EL images from these devices. **(E,F)** are plan views. **(G,H)** are near-field images at the edge of WG.

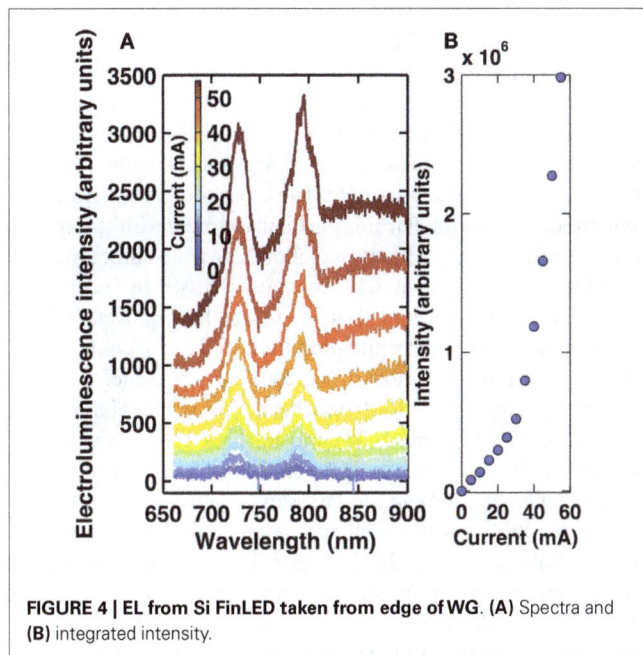

**FIGURE 4 | EL from Si FinLED taken from edge of WG. (A)** Spectra and **(B)** integrated intensity.

The obvious disadvantage of using the planar Si QW is the small confinement factor of the optical mode in the Si QW due to the thin single QW layer. It is not straightforward to make Si Multiple QWs (MQWs) (Fukatsu et al., 1992), if the surface of the Si QW is covered with the amorphous $SiO_2$. As an alternative to the stacking of the Si MQWs, the Si FinLED has been proposed (Saito et al., 2011b), as shown in **Figure 3D**. Si fin is a vertical QW located perpendicular to an Si substrate, and it was proposed for a self-aligned double-gate CMOS field-effect-transistor, called a FinFET (Hisamoto et al., 2000). FinFETs are already used for mass production and more than one billion of FinFETs are integrated in the most recent MPU (INTEL, 2013[1]; ITRS, 2012[2]). Therefore, we can fabricate thousands of Si fins as MQWs at the same time simply by conventional photolithography and dry etching (Saito et al., 2011b). By applying forward bias to the Si FinLED, we can observe edge emission from the $Si_3N_4$ WG (**Figure 3H**). The EL spectra from the edge of the Si FinLED are shown in **Figure 4A**. The enhanced peaks from the edge of the stop band were observed due to the distributed-feedback structure of the periodic fins (Saito et al., 2011b). The non-linear increase of the EL intensity against the current is considered to come from stimulated emission (**Figure 4B**), but the estimated gain of $<1$ cm$^{-1}$ was too low to overcome the threshold for a laser operation (Saito et al., 2011b).

## 4. APPLICATION OF PHOTONIC NANOSTRUCTURES TO GROUP IV MATERIALS

### 4.1. CONTROL OF LIGHT EMISSION BY PHOTONIC CRYSTALS

The light emission properties of materials depend not only on material characteristics such as the dipole moment and the

[1]http://www.intel.com/content/www/us/en/history/museum-transistors-to-transformations-brochure.html
[2]http://www.itrs.net

refractive index but also on the electromagnetic environment surrounding the material. In the previous sections, engineering group IV materials themselves such as quantum confinement, doping, and strain engineering have been discussed. Here, we discuss another approach, i.e., tailoring the electromagnetic environment by photonic nanostructures for improving light emission properties. The total efficiency of light-emitting devices can be expressed as a product of three factors: light emission efficiency $\eta_{emission}$, extraction efficiency $\eta_{extraction}$, and collection efficiency $\eta_{collection}$. $\eta_{emission}$ denotes how efficiently injected carriers recombine by emitting photons. $\eta_{extraction}$ takes into account the fact that only a part of emitted photons can be extracted from the material. $\eta_{collection}$ expresses how much extracted photons can be collected by the first lens of the setup. All of them can be improved by photonic nanostructures. Photonic crystal (PhC) (Jannopoulos and Winn, 1995), which has a wavelength-scale periodic variation of refractive index, is an important photonic nanostructure for this application (see discussions in Iwamoto and Arakawa, 2012). **Figure 5A** shows a scanning electron microscope (SEM) image of a two-dimensional (2D) PhC slab, which is the most widely studied PhC structure. The structure can be fabricated by forming air holes in a thin semiconductor plate using conventional lithography and etching processes. In the structure, owing to the periodic modulation in refractive index, in-plane light propagation is governed by the photonic band structure. Strikingly, propagation is forbidden in photonic bandgaps (PBGs). Photonic band structures and PBGs can play roles to improve mainly $\eta_{extraction}$ and $\eta_{collection}$. Another important structure is the PhC nanocavity (**Figure 5B**), which is created by omitting air holes from the regular array. Photons are confined in in-plane and out-of-plane directions due to the PBG effect and total internal reflection, respectively. PhC nanocavities have a high quality factor $Q$ and small mode volume $V_c$ ($\sim 1$ cubic wavelength or less). These two quantities are key parameters to enhance the spontaneous emission rate through the Purcell effect (Purcell, 1946) and improve $\eta_{emission}$. Particularly, for light emitters with broad linewidth such as bulk Si, $V_c$ has a stronger impact (Ujihara, 1995). Such high-$Q$ PhC nanocavities can uncover the quantum nature of light-matter interaction. Cavity quantum electrodynamics in a high-$Q$ PhC nanocavity coupled with a single semiconductor quantum dot is a hot topic in the field (see, for example, Arakawa et al., 2012). Purcell enhancement factors of as large as 12 (Lo Savio et al., 2011) and 30 (Sumikura et al., 2014) were reported, which would be limited by the emission linewidth and the $Q$ factor, respectively.

### 4.2. ENHANCED LIGHT EMISSION FROM SILICON PHOTONIC CRYSTAL STRUCTURES

PhC structures without cavities have been firstly applied to control the light emission from crystalline Si. In 2003, Zelsmann et al. (2003) reported enhanced PL extraction from a 2D PhC slab fabricated into the top Si layer of a SOI substrate at low temperature (Zelsmann et al., 2003). Similar enhancements at room temperature have been observed from arrays of Si nanoboxes (Cluzel et al., 2006a) and rods (Cluzel et al., 2006b) formed on SOI substrates. Strong light emission was observed at wavelengths corresponding to photonic band edges at the $\Gamma$ point. Increasing the number of band edge within the emission spectrum of Si can lead to higher luminescence intensity. This is experimentally verified by increasing the lattice constant of PhC so that normalized frequencies corresponding to the Si emission wavelengths are increased (Fujita et al., 2008). Si light-emitting diodes (LEDs) with PhC patterns have also been demonstrated (Nakayama et al., 2010a; Iwamoto and Arakawa, 2012). The device schematically shown in **Figure 6A** was fabricated using a SOI substrate. Firstly, a lateral *p-i-n* junction was formed into the top 200-nm-thick Si layer by area-selective implantations of boron and phosphorous ions. Then, a PhC structure was patterned. To keep mechanical stability and better thermal conductivity, the buried-oxide (BOX) layer was not removed. An SEM image of the central part of a device is shown in **Figure 6A**. The *i*-region is 5 $\mu$m in length and 250 $\mu$m in width. EL spectra from devices with different PhC periods and from a device without PhC are shown in the inset of **Figure 6B**. EL emission increased as the period *a* increased. **Figure 6B** shows the integrated intensities from these devices as a function of injected current. The integrated intensity from the device with *a* = 750 nm is ~14 times stronger than that from an unpatterned LED. This enhancement is mainly caused by the improvement of $\eta_{extraction}$ and $\eta_{collection}$ due to the photonic band structures as discussed above. $\eta_{emission}$ is also expected to be enhanced in PhC nanocavities. **Figure 7** shows room-temperature $\mu$-PL spectra measured at the center of

an L3-type PhC nanocavity compared to a non-patterned region (see the inset). The L3 PhC nanocavity was also fabricated into an SOI substrate. In this sample, the BOX layer was etched out in order to confine the photons strongly in the vertical direction. The PL intensity from the cavity was much larger than that from the non-patterned region. In addition, sharp peaks are observed only in the spectrum from the cavity. These peaks originate from the cavity resonant modes. For this particular sample with the air hole radius $r = 0.37a$, large enhancement of PL over 300 times was obtained for a cavity mode at 1,191 nm. As discussed in Section 1, this enhancement can be attributed to three factors. Detailed analysis including numerical simulation indicated that $\eta_{emission}$ is improved by ~5 times (Iwamoto et al., 2007). The enhancement factors in $\eta_{emission}$ ranging ~5−10 have been reported for Si interband transition (Fujita et al., 2008) and for light emission from optically active defects in Si (Lo Savio et al., 2011). The temperature dependence of cavity mode emission (Hauke et al., 2010; Lo Savio et al., 2011) and the dependence of PL on cavity mode volume $V_c$ (Nakayama et al., 2012) suggest that the Purcell effect plays a role in this enhancement. The enhancement in $\eta_{emission}$ reported so far is still too small for practical applications. However, this research would provide important insights for further development of light-emitting devices using group IV materials. Indeed, these pioneering works have stimulated theoretical investigations, which discuss the possibility of lasing oscillation in Si (Escalante and Martínez, 2012, 2013). Recent advances in this field are developments of Si LEDs with PhC nanocavities (Nakayama et al., 2011; Shakoor et al., 2013). Shakoor et al. (2013) recently reported Si LEDs using L3-type nanocavity structure, in which optically active defects created by hydrogen bombardment are used as light emission centers. They carefully designed the cavity structure to improving $\eta_{collection}$ and obtained sharp light emission at around 1.5 $\mu$m with a power density of 0.4 mW/cm$^2$. The strain-induced dislocations (Ng et al., 2001; Kittler et al., 2013) will also be compatible to PhC nanocavities, since the emission energies are smaller than the band gap of Si. The combination of PhC nanocavities and defect engineering is very promising, and a wall plug efficiency of $0.7 \times 10^{-8}$ was reported (Shakoor et al., 2013).

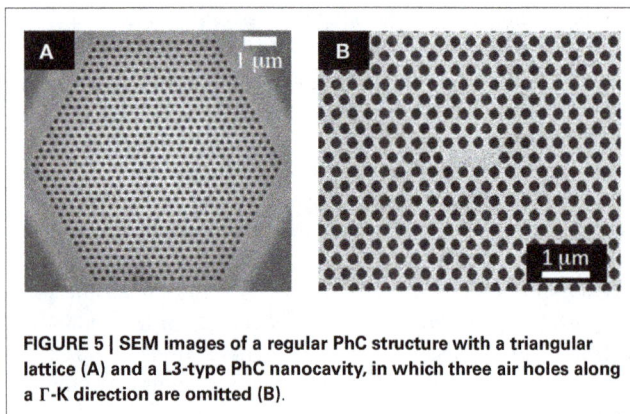

**FIGURE 5 | SEM images of a regular PhC structure with a triangular lattice (A) and a L3-type PhC nanocavity, in which three air holes along a Γ-K direction are omitted (B).**

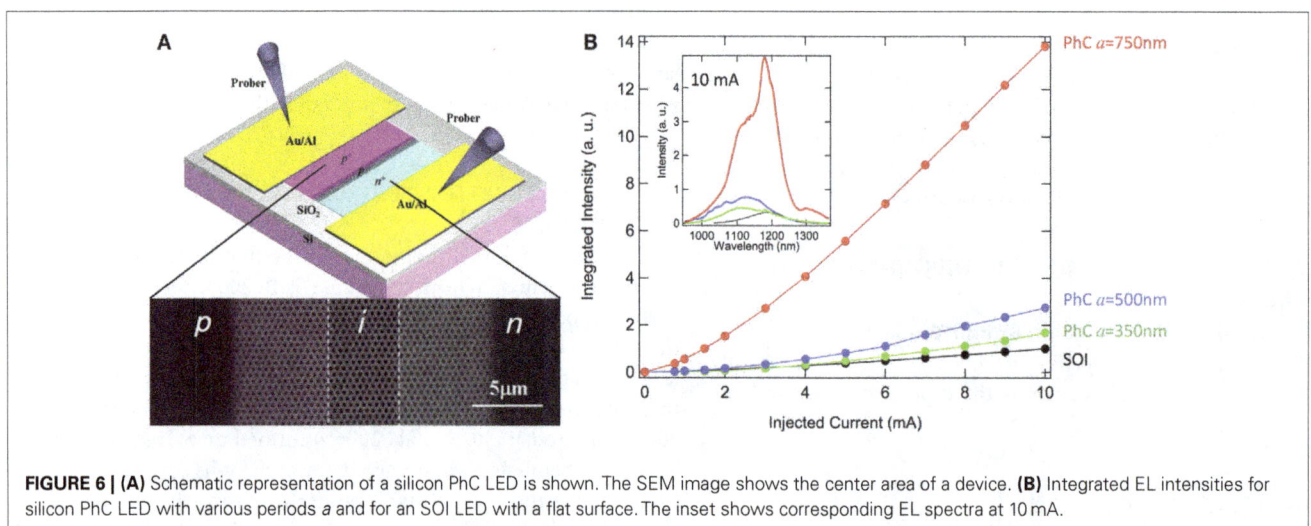

**FIGURE 6 | (A)** Schematic representation of a silicon PhC LED is shown. The SEM image shows the center area of a device. **(B)** Integrated EL intensities for silicon PhC LED with various periods *a* and for an SOI LED with a flat surface. The inset shows corresponding EL spectra at 10 mA.

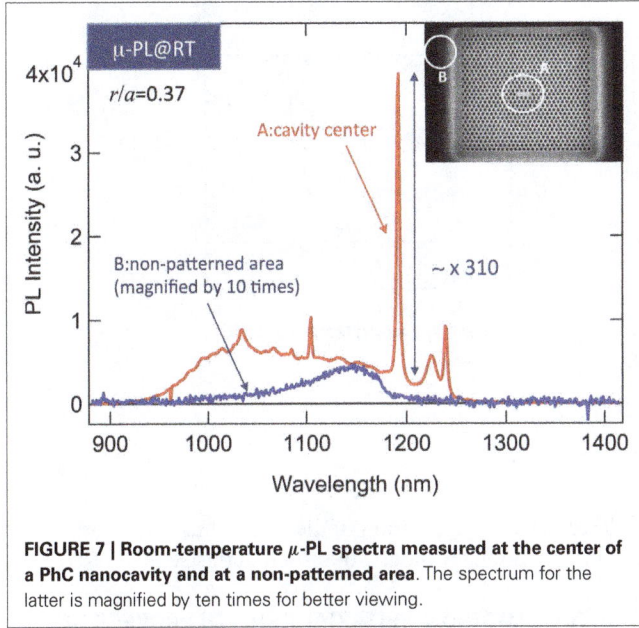

**FIGURE 7 | Room-temperature $\mu$-PL spectra measured at the center of a PhC nanocavity and at a non-patterned area.** The spectrum for the latter is magnified by ten times for better viewing.

### 4.3. APPLICATION OF PHOTONIC CRYSTAL STRUCTURES TO OTHER EMITTERS IN GROUP IV MATERIALS

Erbium ions have been investigated as one of the promising light emitters in Si. PhC nanocavities have been also applied to enhance the light emission from Er ions (Wang et al., 2012; Savio et al., 2013). Narrowing the cavity linewidth in Er-doped silicon nitride PhC nanocavities has been also demonstrated under optical pumping condition (Gong et al., 2010). As discussed in the previous sections, Ge is, at present, the most important material for future light-emitting devices in Si photonics. PhC (Nakayama et al., 2010b) and PhC nanocavities (Kurdi et al., 2008; Ngo et al., 2008) have been applied to increase the light emission from bulk Ge. Applying advanced strain/doping engineering technologies to photonic nanostructures would open a new route for boosting the light emission efficiency of Ge.

### 5. GENERATION OF TENSILE STRAIN IN Ge LAYERS EPITAXIALLY GROWN ON Si SUBSTRATE

In epitaxial growth of Ge on an Si substrate, a compressive strain in Ge, derived from the 4.2% lattice mismatch with Si, should be relaxed after growth beyond the critical thickness, while it has been reported by one of the authors that, during the cooling from the growth temperature to room temperature, a biaxial tensile strain as large as 0.2% is built-in due to the thermal expansion mismatch (Ishikawa et al., 2003, 2005; Cannon et al., 2004; Liu et al., 2005). It is known that the strain in semiconductors causes shifts in band edge energies, e.g., de Walle, 1989, modifying the gap energies, i.e., properties of optical transitions. The 0.2% tensile strain in Ge reduces the direct bandgap energy from 0.80 to ~0.77 eV, and as a result, the optical absorption edge (or the longer limit of detection wavelength) shifts from 1.55 to >1.60 $\mu$m, causing the increase of optical absorption coefficient at 1.55 $\mu$m (Ishikawa et al., 2003, 2005; Cannon et al., 2004; Liu et al., 2005). This property is effective for the detection of near-infrared (NIR) light

used in the optical fiber communications (1.3–1.6 $\mu$m). A further attractive feature of the tensile strain in Ge is the reduction of energy difference in the conduction band between the direct $\Gamma$ valley and indirect $L$-valley (e.g., Fischetti and Laux, 1996; Wada et al., 2006; Camacho-Aguilera et al., 2012; Nama et al., 2013; Süess et al., 2013). This feature stimulates researchers to obtain efficient NIR light emission from tensile-strained Ge due to the enhanced direct transition around the $\Gamma$ point (e.g., Liu et al., 2007; Lim et al., 2009). In this section, the grown-in tensile strain in Ge on Si, generated due to the thermal expansion mismatch, is described. **Figure 8A** shows typical $\omega - 2\theta$ x-ray diffraction (XRD) curves taken for 0.6-$\mu$m-thick Ge grown on a 525-$\mu$m-thick Si(001) substrate with the Cu K$\alpha$ radiation as the x-ray source (0.15406 nm in wavelength). The samples were grown by ultrahigh-vacuum chemical vapor deposition with a source gas of GeH$_4$ (9%) diluted in Ar. The growth temperature was 600°C, while a lower temperature of 370°C was used at the initial stage of Ge growth (~50 nm) in order to prevent the islanding, leading to Ge layers uniform in thickness (Luan et al., 1999; Ishikawa and Wada, 2010). After the growth, high-temperature annealing was carried out for one of the samples at 800°C for 20 min. Such annealing is often performed in order to reduce the threading-dislocation density (Luan et al., 1999). In our case, the density was reduced from $1 \times 10^9$ to $1 - 2 \times 10^8$ cm$^{-2}$. In **Figure 8A**, the peaks due to the (004) diffraction are clearly seen at around $2\theta \sim 66°$ for both of the as-grown and annealed samples. It is important that the peaks were located at larger diffraction angles than that for unstrained Ge, indicating the reduction of out-of-plane lattice constant, i.e., the increase of in-plane lattice constant due to the generation of tensile strain. According to the peak positions, the in-plane biaxial tensile strain was estimated to be 0.11 and 0.22% for the as-grown and annealed samples, respectively.

As mentioned above, such a tensile strain is generated in Ge due to the mismatch of thermal expansion coefficient with Si. As schematically shown in **Figure 8B**, the compressive strain in Ge due to the 4.2% lattice mismatch should be relaxed at the growth/annealing temperature, while the shrinkage in the Ge lattice during the cooling should be prevented by the thick Si substrate, since Si has a smaller thermal expansion coefficient than that of Ge. This means that a tensile (compressive) stress/strain is generated in Ge (Si), as in the bottom of **Figure 8B**. Taking into account the balance of forces together with the balance of moments in the stacked structure of Ge and Si, the tensile (compressive) strain in Ge (Si) is theoretically expressed as:

$$\in_{||}(Ge) = \frac{1}{R}\left[\frac{Y_1 t_1^3 + Y_2 t_2^3}{6Y_1 t_1 (t_1 + t_2)} + \left(\frac{t_1}{2} - z_1\right)\right] \quad (1)$$

$$\in_{||}(Si) = \& - \frac{1}{R}\left[\frac{Y_1 t_1^3 + Y_2 t_2^3}{6Y_2 t_2 (t_1 + t_2)} - \left(\frac{t_2}{2} - z_2\right)\right], \quad (2)$$

where, $\alpha_i$, $Y_i$, $t_i$, and $z_i$ represent the thermal expansion coefficient, the Young's modulus, the layer thickness, and the location in the layer measured from the bottom of the layer for the $i$-th layer (1 for Ge and 2 for Si), respectively. The radius of curvature $R$ is

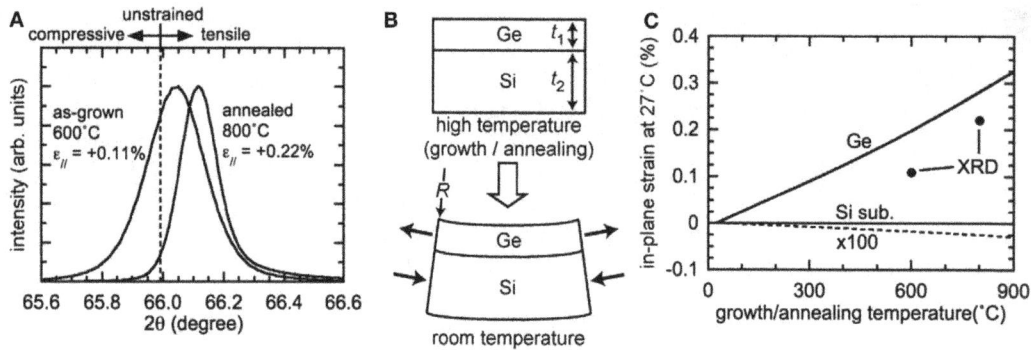

**FIGURE 8 | (A)** $\omega - 2\theta$ XRD curves for 0.6-$\mu$m-thick Ge on Si (001) substrate, **(B)** schematic illustration showing the generation of tensile stress/strain in Ge, and **(C)** theoretical curves and experimental data for biaxial tensile strain in Ge.

represented from

$$\frac{1}{R} = \frac{6\,(t_1 + t_2)\,Y_1 Y_2 t_1 t_2 \int_{T_{\mathrm{GR/AN}}}^{T_{\mathrm{RT}}} (\alpha_1 - \alpha_2)\,dT}{3(t_1 + t_2)^2\,Y_1 Y_2 t_1 t_2 + \left(Y_1 t_1^3 + Y_2 t_2^3\right)(Y_1 t_1 + Y_2 t_2)}, \quad (3)$$

where $T_{\mathrm{GR/AN}}$ and $T_{\mathrm{RT}}$ represent the growth/annealing temperature before cooling and the room temperature (after the cooling), respectively. Since the first term is dominant in the right side of equation (1), the strains are almost independent of $z_i$, the location within the layer. Therefore, equations (1) and (2) are simplified to:

$$\in_{\parallel} (\mathrm{Ge}) \sim \frac{1}{R}\frac{Y_1 t_1^3 + Y_2 t_2^3}{6 Y_1 t_1\,(t_1 + t_2)} \quad (4)$$

$$\in_{\parallel} (\mathrm{Si}) \sim -\frac{1}{R}\frac{Y_1 t_1^3 + Y_2 t_2^3}{6 Y_2 t_2\,(t_1 + t_2)}. \quad (5)$$

The lines in **Figure 8C** represent the strains calculated for the Ge thickness of 0.6 $\mu$m and the Si thickness of 525 $\mu$m. Note that almost identical results can be obtained when the thickness of Si substrate $t_2$ is much larger (more than ~100 times) than the Ge thickness $t_1$. The parameters used in the calculation can be found in Ishikawa et al. (2005). It is found that a tensile strain on the order of 0.1% is generated in Ge at room temperature, while the compressive strain in Si is negligible. It is also found that higher growth/annealing temperature generates larger tensile strain after the cooling. These properties are qualitatively in good agreement with the XRD results in **Figure 8A**. However, quantitatively, the tensile strain observed by XRD was smaller than the theoretical one. This is probably ascribed to the residual compressive strain in Ge at the growth/annealing temperature (Ishikawa et al., 2005). From the viewpoint of optoelectronic integration of Ge devices on an Si platform, Si-on-insulator (SOI) wafers have been widely used. For Ge layers grown on SOI wafers, a similar amount of tensile strain should be generated, since the elastic deformation, derived from the thermal expansion mismatch, is governed by the thick Si substrate, rather than the buried SiO$_2$ and the top Si layers with the thicknesses on the order of 1 $\mu$m or below. Patterning of

the Ge layer as well as deposition of dielectric films embedding a strain could intentionally modify the strain in the Ge.

## 6.  DIRECT GERMANIUM EPITAXIAL GROWTH PROCESS ON SILICON

The Ge was epitaxially grown by using a cold-wall rapid thermal chemical vapor deposition system. Germane (GeH$_4$) was used as a source gas, which was supplied with H$_2$ carrier gas. As the starting point of improving the crystallinity and controlling the lattice strain, Ge layers with good surface morphology were grown at 420°C under relatively high pressure of 7,000 Pa. Then, the Ge layers were annealed in the same H$_2$ atmosphere to improve the crystallinity. **Figure 9** shows a reciprocal space map (RSM) of XRD (XRD-RSM) from the 130-nm-thick Ge layer directly grown on the Si substrate before and after H$_2$ annealing. An intense Si (-1-13) peak was observed, which represented the diffraction from the Si substrate under the Ge layer. Since the XRD-RSM was measured by using semiconductor array detectors, errors in the counts occur if the diffraction intensity is very high; therefore, the streak line observed around the Si (-1-13) peak does not represent any actual diffraction. Since a Ge (-1-13) diffraction peak was observed from the Ge layer without annealing (**Figure 9A**), it could be confirmed that a single crystalline Ge layer was obtained by using low-temperature epitaxial growth. The displacement of the diffraction peak shows that the as-grown Ge layer still contained a compressive strain just after the low-temperature epitaxial growth at 420°C due to the larger lattice constant of Ge compared to that of the Si substrate. It has been reported that cyclic annealing at a relatively higher temperature can reduce the threading-dislocation density (Luan et al., 1999) in Ge layers. This has led to studies on the effect of annealing on the crystallinity and lattice strain of Ge layers. After low-temperature epitaxial growth of Ge layers at 420°C, the temperature was increased to the annealing temperature in the same H$_2$ atmosphere as that during the epitaxial growth, and the Ge layers were then annealed at various temperatures for 10 min. XRD-RSMs of Ge layers annealed at a temperature ($T_{\mathrm{GR/AN}}$) of 700°C after the low-temperature epitaxial growth are shown in **Figure 9B**. The Ge (-1-13) diffraction peaks became much steeper and the peak intensity increased when the annealing temperature

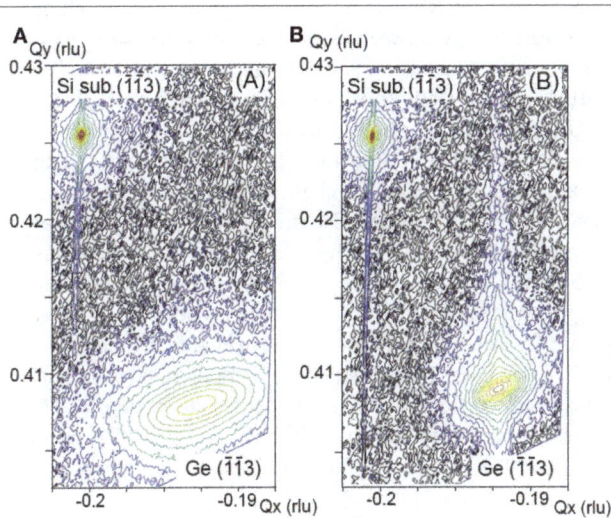

**FIGURE 9 | XRD-RSM of (-1-13) diffraction from Ge layer grown on Si substrate, (A) after low-temperature epitaxial growth and (B) after post-annealing.**

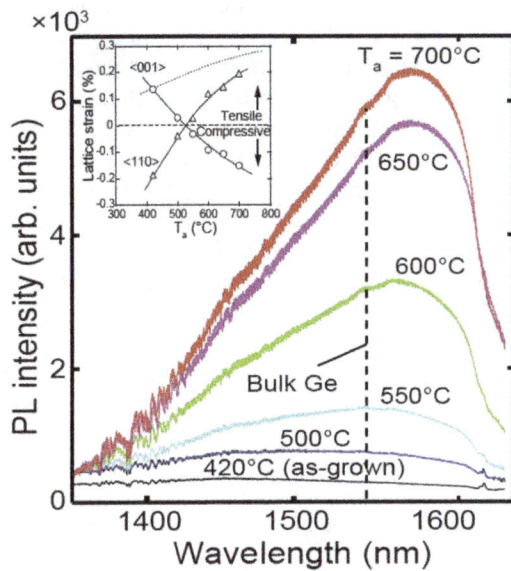

**FIGURE 10 | Photoluminescence spectra from Ge layers annealed with different temperatures are shown**. Peak wavelength of photoluminescence from Ge layers red-shifted as annealing temperature increased, consistent with temperature induced tensile strain. Inset shows lattice strain of Ge layers grown on Si substrate along <001> and <110> crystal orientations as a function of annealing temperature. Dotted line indicates lattice strain calculated with difference between thermal expansion coefficients of Si and Ge.

for this experiment, so that <001> is perpendicular to the surface of the Ge film, while <110> is in the plane of Ge film. The lattice strain in the <110> crystal orientation increased as $T_{GR/AN}$ increased, and the strain in the <001> crystal orientation showed an opposite dependence. Although the Ge layer contained a compressive strain in the <110> crystal orientation at $T_{GR/AN} = 420°C$, i.e., without annealing, this strain started decreasing when $T_{GR/AN}$ was increased, and the Ge was completely un-strained at $T_{GR/AN} = 530°C$. Furthermore, the sign of the lattice strain changed from compressive to tensile after annealing at $T_{GR/AN} > 530°C$, and the tensile strain at $T_{GR/AN} = 700°C$ reached 0.19%. This result is consistent with previous studies (Cannon et al., 2004). Normally, a grown layer with a larger lattice constant compared to a substrate contains a compressive strain within the growth plane. However, since the Ge layers grown on the Si substrate were almost completely relaxed even after low-temperature growth, the Ge lattice could be dislocated at the Ge/Si interface by post-annealing, and the lattice strain of the Ge layer was relaxed during annealing at the relatively higher temperature with the volumes of Ge and Si determined by the thermal expansion coefficients (Singh, 1968; Okada and Tokumaru, 1984). After annealing, the volume of the Ge layer and the Si substrate both shrunk as the temperature decreased, and there was barely any change to the lattice alignment at low-temperature. The volume of the Si substrate returned to its original value because it was thick enough. However, the volume of the Ge layer could not return due to its larger thermal expansion coefficients. Therefore, the tensile lattice strain remained only in the Ge layers after cooling (Cerdeira et al., 1972). The ideal lattice strain in <110> crystal orientation was also plotted in the inset of **Figure 10**, which was calculated with only the difference of the thermal expansion coefficients between Si and Ge, so these values indicate the maximum lattice strain. Since there are large discrepancies between calculation and measured values, it seems that relaxation ratio has a large effect on the lattice strain even at the lower temperatures. PL spectra from the post-annealed Ge layers with various annealing temperatures are shown in **Figure 10**. Although Ge is an indirect bandgap material and the L-valley has the lowest energy level in the conduction band, we were able to observe recombination between electrons and holes at the Γ-valley as luminescence at a wavelength of 1,550 nm, even from the bulk Ge (dashed line in **Figure 10**). A comparison with the post-annealed Ge layers shows that although the spectrum was very weak and broad for the as-grown Ge layer, an obvious peak could be observed from annealed samples at $T_{GR/AN} > 530°C$. Moreover, the PL intensity increased and the peak shape became sharper as the annealing temperature was increased. The PL spectrum is strongly affected by crystallinity, because non-radiative recombination was significantly increased with defects such as dislocation and stacking faults. Therefore, these results suggest that the crystallinity of the Ge layers was improved by the post-annealing. The peak was observed at a shorter wavelength from the Ge layer annealed at 500°C compared with that from bulk Ge, and a red shift of the PL peaks occurred after post-annealing at a higher temperature. In addition, the peak wavelength from the unstrained Ge was 1,550 nm, which is almost the same value as that of the bulk Ge. These results show that the bandgap energy at the Γ-point was varied by the lattice strain in the Ge layers (Cerdeira

was increased, indicating that the crystallinity of the Ge layers was increased by the post-annealing.

The inset of **Figure 10** shows the lattice strain in the Ge layers in the <001> and <110> crystal orientations as a function of the annealing temperature. We used standard Si wafers et al., 1972; de Walle and Martin, 1986; de Walle, 1989), which is consistent with the XRD measurements. These results indicate that, in the range of this study, the most favorable PL characteristic can be obtained from the Ge layer after post-annealing at higher temperatures.

## 7. GERMANIUM LIQUID-PHASE EPITAXY AND DEVICES FOR PHOTONIC APPLICATION

Liquid-phase epitaxy (LPE) is a technique that was invented in the 1960s (Nelson, 1963) and developed in the 1970s (Wieder et al., 1977) for the fabrication of detectors, solar cells, LEDs (Saul and Roccasecca, 1973), and laser diodes (Panish et al., 1970). Originally used for III-V crystal growth, it has been adapted for SiGe-on-insulator (SGOI) and Ge-on-Insulator (GOI) growth by various groups (Liu et al., 2004; Tweet et al., 2005; Feng et al., 2008; Hashimoto et al., 2009; Miyao et al., 2009; Ohta et al., 2011) and is also referred to as rapid melt growth (RMG). The GOI technique was pioneered by Liu et al. (2004) for Ge-on-insulator fabrication. In this technique, a thin insulating layer is deposited on an Si substrate and patterned to open up seed windows. The target material, in this case Ge, is deposited using a non-selective method and patterned to form the desired features. This is then encapsulated using an insulating layer and heated up in a rapid-thermal-annealer (RTA) in order to melt the Ge. The micro-crucible holds the melt in place until the liquid epitaxial growth is complete. Upon cooling, liquid-phase epitaxial growth starts from the seed and propagates to the extremities of the strip structure. For the realization of single crystal Ge, epitaxial growth must proceed faster than unseeded random nucleation, so that the crystal regrowth starting from the seed is uninterrupted. Misfit dislocations arising at the SiGe interface in the seed area are necked down to the seed window as shown in **Figure 11**. The RMG is limited to the growth of structures of the order of around 3 $\mu$m in width and with a length of above 100 $\mu$m. The limitation is largely due to the surface tension of the insulator causing the Ge to form ball shapes while in the liquid phase.

RMG is very attractive for the heterogeneous integration of Ge-based devices on insulator for electronics and photonics and has been demonstrated for Gate all around P-MOSFET (Feng et al., 2008), P-Channel FinFET (Feng et al., 2007), waveguide integrated Ge/Si heterojunction photodiodes (Tseng et al., 2013), or Ge Gate PhotoMOSFET (Going et al., 2014). These devices demonstrate the possibility of using RMG to obtain high quality Ge crystalline layers to create a bridge between electronic components and photonic components. This vision is clearly demonstrated by Going et al. (2014) in a Ge Gate PhotoMOSFET (Carroll et al., 2012) where a Ge-gated NMOS phototransistor is integrated on an Si photonics platform on SOI substrate. The resulting device, with 1-$\mu$m channel length, and 8-$\mu$m channel width, demonstrates a responsivity of over 18 A/W at 1550 nm with 583 nW of incident light. By increasing the incident power to 912 $\mu$W, the device operates at 2.5 GHz. Ge RMG or LPE on Si is therefore a promising technology for the fabrication of heterogeneous devices requiring high quality Ge layers such as MOSFETs, near-infrared detectors but also Ge-based lasers that are still to be demonstrated using this specific process technique. In fact, a highly tensile strain of 0.4% has successfully been applied to a Ge film grown by RMG process (Matsue et al., 2014), which is quite promising for light emission.

## 8. TIME-RESOLVED PHOTOLUMINESCENCE STUDY OF GERMANIUM ON SILICON

The use of $n$-type tensile-strained Ge grown on Si substrates is one promising way to realize an efficient light source for Si photonics through the enhanced direct recombination from the $\Gamma$ valley. However, the large lattice mismatch between Ge and Si inherently causes misfit dislocations at the interface, and threading dislocations during the growth. Besides, epitaxially grown Ge is usually a thin layer, so that both the interface and the surface become important. Therefore, investigation of the excess carrier lifetime is crucial for the realization of efficient light-emitting devices. Recently, the excess carrier dynamics of thin Ge film grown on either Si or SOI substrates have been investigated by time-resolved photoluminescence (Kako et al., 2012), microwave photoconductive decay (Sheng et al., 2013), and pump-probe transmission (Geiger et al., 2014) methods. Here, we present the time-resolved photoluminescence study of both non-doped and $n$-type Ge samples grown on Si.

The Ge samples were epitaxially grown on (100) Si substrates by using a cold-wall rapid thermal chemical vapor deposition system (Oda et al., 2014). There were two primary growth steps. The first step was the growth of an intrinsic Ge thin layer ($\approx$100 nm) at low temperature followed by an annealing process. The second step was the regrowth of Ge on the first layer with another annealing

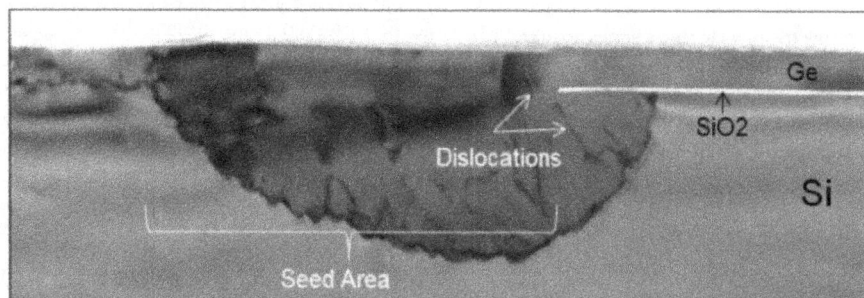

**FIGURE 11 | Transmission-electron-microscope (TEM) image of a high quality single crystalline Ge-on-insulator obtained using RMG**. It can clearly be seen that the misfit dislocations from the lattice mismatch are confined to the seed region and that the crystalline Germanium lateral overgrowth is free from defects.

process. *In situ* *n*-type doping was carried out during the second growth step by supplying phosphine. The Ge becomes biaxially strained ($\approx 0.15\%$) due to the difference of the thermal expansion coefficients between Si and Ge. Time-resolved photoluminescence measurements were performed using a time-correlated single-photon counting method employing a superconducting single-photon detector (SSPD) with a time resolution of about 50 ps. A Ti:Sapphire pulsed laser was used as the excitation source (wavelength 710 nm, repetition rate 80 MHz, and pulse-duration 100 fs). The laser beam was focused on the sample surface using an objective lens. The photoluminescence from the samples was collected by the same objective and focused on to an optical fiber connected to the SSPD. Photoluminescence ranging from 1.2 to 1.8 $\mu$m was detected.

**Figure 12A** shows a time-resolved photoluminescence decay curve measured from a nominally undoped Ge sample (thickness 500 nm). In order to limit the effects of lateral diffusion, the laser spot size was set to $\approx 10\ \mu$m. The decay is a single exponential with a lifetime of 1 ns, which corresponds to the excess carrier lifetime of 2 ns. Germanium has an indirect bandgap, and as such, its excess carrier dynamics are determined by non-radiative recombination processes, such as Shockley-Read Hall (SRH) recombination and surface recombination processes. The photoluminescence decay lifetime, $\tau_{PL}$, of undoped Ge is then related to the excess carrier lifetime, $\tau_{ex}$, as $2\tau_{PL} = \tau_{ex}$. The lifetime of excess carriers $\tau_{ex}$ of an indirect semiconductor film depends on the thickness and can be represented by Sproul (1994) and Gaubas and Vanhellemont (2006) as:

$$\frac{1}{\tau_{ex}} = \frac{1}{\tau_B} + \frac{1}{\frac{d}{2S} + \frac{d^2}{\pi^2 D}}, \tag{6}$$

where $\tau_B$ is the bulk lifetime, $S$ is the surface recombination velocity, $D$ is the ambipolar diffusion constant, and $d$ is the layer

thickness. The excess carrier lifetimes obtained for undoped Ge layers with different thicknesses (filled black circles) are shown in the inset of **Figure 12A** together with the black curve, which is a fit to the data using equation (6) with parameters of $\tau_B = 3.5$ ns, $S = 5.5 \times 10^3$ cm/s, and $D = 30$ cm$^2$/s (The ambipolar diffusion constant $D_a$ could be estimated by changing the spot size and measuring the photoluminescence decay time). Both SRH bulk recombination and the surface recombination processes determine the excess carrier dynamics in our undoped Ge samples.

**Figure 12B** shows time-resolved photoluminescence decay curves measured at two different excitation power densities from an *n*-type Ge sample (thickness 500 nm, doping concentration $7 \times 10^{19}$ cm$^{-3}$). The measured decay depends on both the excitation power density and time (in contrast to those measured from undoped samples, which are independent of the excitation power). The instantaneous lifetime (that measured at a particular point during the decay) depends on the photoluminescence intensity, and thus the excess carrier density. Based on the SRH non-radiative recombination model, the lifetime of excess carriers depends on their density (Linnros, 1998). This dependence can be simplified to $\tau_{hl} = \tau_n + \tau_p$ ($\tau_{ll} = \tau_p$ for *n*-type doping) in the two extreme conditions where the carrier density is high (low) when compared to the doping concentration ($\tau_n$ and $\tau_p$ are the inverse capture rates of the electrons and holes, respectively). The photoluminescence lifetime can be expressed as $2\tau_{PL} = \tau_{hl}$ (high excess carrier density) and $\tau_{PL} = \tau_{ll}$ (low excess carrier density). Therefore, from our measurements, we estimate $\tau_{ll} = 0.14$ ns, $\tau_{hl} = 0.8$ ns based on SRH theory. The estimated $\tau_{hl}$ value is shorter than those found from the undoped samples. This difference might be attributed to an increased dislocation density introduced by the doping, but the estimation of $\tau_{hl}$ could be underestimation because the Auger process becomes important for doped samples (Gaubas and Vanhellemont, 2006). Further investigation is needed in order to obtain a better understanding.

## 9. ELECTRO-LUMINESCENCE FROM GERMANIUM

Realization of monolithic light sources compatible with the existing Si photonics platform is one of the most difficult challenges. Ge has attracted much attention as for possible future monolithic light sources owing to its emission wavelengths of ~1.6 $\mu$m suitable for an Si-based WG, in addition to the CMOS compatibility and the pseudo-direct band-gap character (Menéndez and Kouvetakis, 2004; Liu et al., 2007, 2012; Liang and Bowers, 2010; Michel et al., 2010; Boucaud et al., 2013; Liu, 2014). Recently, laser operation from Ge pumped optically (Liu et al., 2010) and electrically (Cheng et al., 2007; Camacho-Aguilera et al., 2012) has been reported. However, there is no report so far to reproduce their results. The optical gain from Ge is also achieved by the tensile-stress engineering (de Kersauson et al., 2011). The precise nature of the optical gain in Ge is still controversial (Carroll et al., 2012), but the high crystalline quality of Ge is one of the most critical factor to avoid non-radiative recombinations at dislocations. It is confirmed by several groups (Michel et al., 2010; Liu et al., 2012; Boucaud et al., 2013; Liu, 2014) that the primary challenges for engineering Ge as an active layer are: (i) crystallinity, (ii) high *n*-type doping, (iii) tensile strain, as confirmed theoretically (Suwa and Saito, 2010,

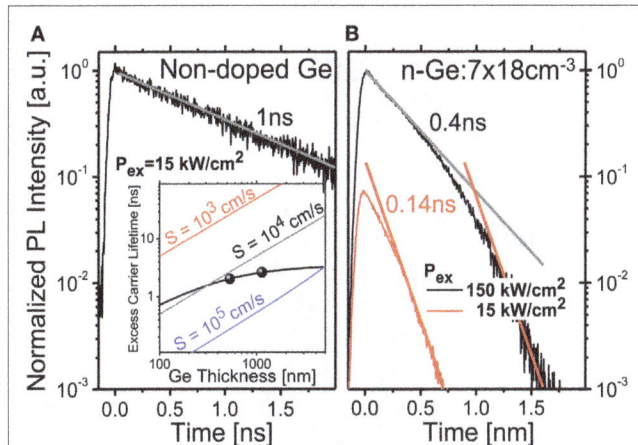

**FIGURE 12 | (A)** Time-resolved PL curve for an undoped Ge sample. The inset shows the measured excess carrier lifetimes for two Ge thicknesses with simulated lifetimes using the equations shown in the text.
**(B)** Time-resolved PL curves of an *n*-type sample for two excitation powers, 150 kW/cm$^2$ (black line) and 15 kW/cm$^2$ (red line).

2011; Virgilio et al., 2013a,b). Here, we review some of the Ge light sources developed on SOI substrates.

## 9.1. DEVICE STRUCTURE AND FABRICATION PROCESS

As we discussed in section for Si light sources, lateral carrier injection is a natural choice for electrical pumping, since fabrication processes are based on planar CMOS technologies. We show several candidates for Ge light sources suitable for lateral carrier injection in **Figure 13**.

**Figures 13A,E,I** show schematic views and a transmission-electron-microscope (TEM) image of a Ge FinLED (Saito et al., 2011a), which uses Ge fins as MQWs embedded in $Si_3N_4$ WG. Ge fins were fabricated by the oxidation condensation technique (Tezuka et al., 2009) applied to SiGe fins (Saito et al., 2011a). Relatively, high crystallinity is expected in Ge fins, since the lattice mismatch between Si and Ge would be relaxed by stretching the fins during the oxidation (Saito et al., 2011a). In fact, the low dark current density of $1.86 \times 10^{-5}$ A/cm$^{-2}$ at a reverse bias of $1 - V$ and the strong breakdown current density of $>1$ MA/cm$^{-2}$ were confirmed (Saito et al., 2011a).

In order to enhance the overlap between an optical mode and fins, Ge fins with (111) orientation at the sidewall were also developed (Tani et al., 2012), as shown in **Figures 13B,F,J**. To improve the patterning accuracy, Si (111) fins were fabricated by anisotropic wet etching, and $n$-Ge was re-grown after the condensation oxidation of SiGe fins (Tani et al., 2011).

Further increase of the coupling is realized by using a bulk Ge WG (Liu et al., 2007; Camacho-Aguilera et al., 2012; Tani et al., 2013a,b), as shown in **Figures 13C,G,K** for schematic views and the scanning electron microscope (SEM) image, rather than using Ge QW or Ge fins. The $p$- and $n$-type diffusion regions were formed in the 40 nm-thick SOI layer, and the Ge waveguide with 500-nm width and 500-$\mu$m length was directly grown on the SOI diode. The SOI thickness was designed to minimize the optical

loss due to free carrier absorption in the diffusion electrodes. The Ge waveguide was doped with $1 \times 10^{19}$ cm$^{-3}$ of phosphorus, and the surface of the Ge waveguide was then passivated with GeO$_2$ formed by low-temperature oxidation to reduce interfacial traps (Tani et al., 2012, 2013a). Then, metal electrodes were made on both diffusion regions.

To enhance light emission efficiency from Ge by tensile stress, several techniques have been developed, e.g., the use of the thermal expansion of relaxed Ge grown on Si (Ishikawa et al., 2003), the growth on buffer layers with larger lattice parameter (Huo et al., 2011), the mechanical deformation using membrane structures (Kurdi et al., 2010), the stress concentration in a membrane structure (Nama et al., 2013), and using external stressors (Ortolland et al., 2009; Ghrib et al., 2013). Considering the process compatibility to the lateral carrier injection, the $Si_3N_4$ film with the tensile stress of 250-MPa was employed (Tani et al., 2013a), as shown in **Figures 13D,H,L**.

## 9.2. IMPACT OF STRESS ENGINEERING FOR LATERAL GERMANIUM ON SILICON DIODE

**Figure 14A** shows EL spectra of the Ge waveguide with 500-nm width and 500-$\mu$m length taken from the top of the substrate under continuous current injection of 60-mA. EL peak wavelength of the device with an SiN stressor is slightly longer than that without the SiN stressor due to the tensile strain-induced band-gap shrinkage, although the exact band-gap energy cannot be quantitatively estimated due to the additional peak shifts caused by heating under high currents. Moreover, as shown in **Figure 14B**, the peak intensity of the EL of the device with SiN stressor is 1.65 times larger than that without SiN stressors. **Figure 14C** shows two-dimensional stress mapping calculated by a finite element modeling of the Ge waveguide on the Si substrate covered by $Si_3N_4$ stressor. The tensile stress of 100 MPa

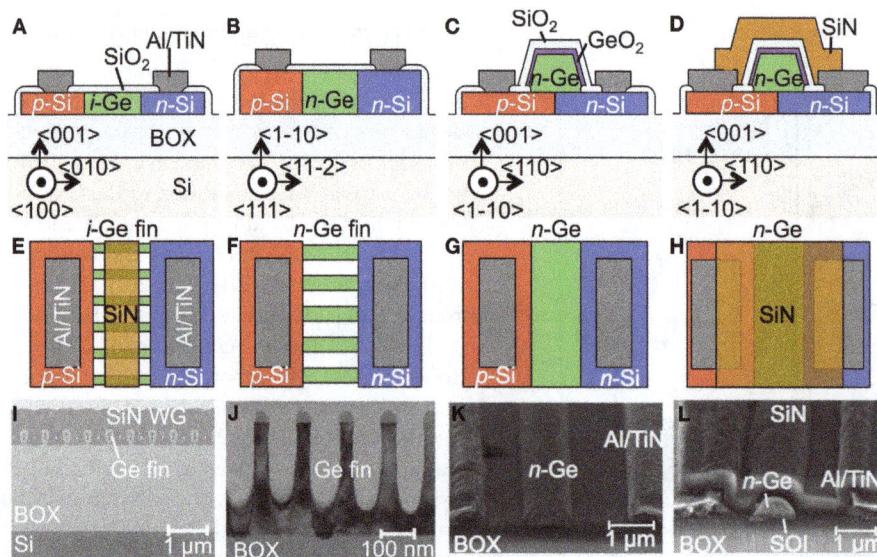

**FIGURE 13 | Development of a Ge light source. (A)** $i$-Ge FinLED, **(B)** $n$-Ge FinLED, **(C)** $n$-Ge-WG-on-Si LED without SiN, and **(D)** $n$-Ge-WG-on-Si LED with SiN. **(A–D)** Cross section, **(E–H)** plan views, and **(I–L)** microscope images.

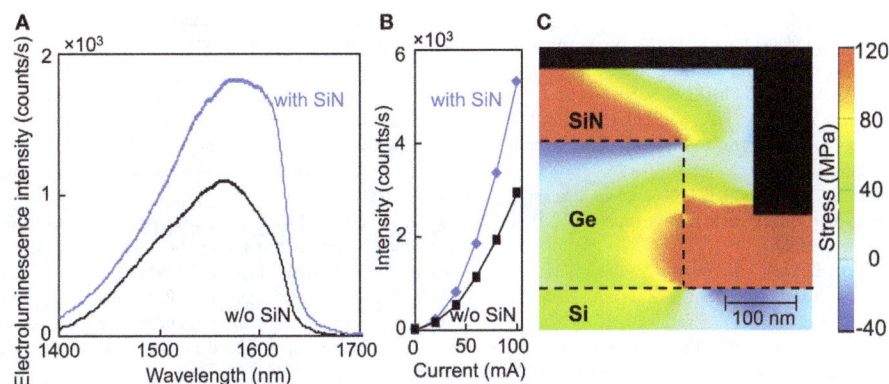

**FIGURE 14 | Strain engineering for _n_-Ge-WG-on-Si LED. (A)** Spectra and **(B)** integrated intensity from experiments. **(C)** Stress mapping simulation.

is localized on the side wall of the Ge waveguide, while the in-plane compressive stress of 40 MPa exists on the top part of the Ge waveguide. The increase of the light emission efficiency was 22% caused by the tensile stress, after subtracting of the additional increase of 35% caused by the light extraction efficiency due to the reduced reflectance at the surface of the Ge waveguide by the 500 nm-thick $Si_3N_4$ layer (Tani et al., 2013a). Therefore, the stress engineering by $Si_3N_4$ is an appropriate option to improve the performance of Ge light sources. Recently, there are significant advances in stress engineering by manipulating free-standing Ge structures (Jain et al., 2012; Boztug et al., 2013; Süess et al., 2013; Sukhdeo et al., 2014), and enhanced direct recombination has been achieved.

## 10. CONCLUSION AND FUTURE OUTLOOK

In this paper, we reviewed the recent progress on the developments of silicon and germanium light sources. There are many process options to fabricate silicon- and germanium-based nanostructures by using modern silicon technologies. For active materials, planar silicon single-quantum-well (Saito et al., 2006a,b, 2008, 2009; Hoang et al., 2007; Noborisaka et al., 2011; Saito, 2011) or multiple-quantum-wells made of silicon or germanium fins (Saito et al., 2011a,b) can be used. To enhance the recombination rates and the extraction efficiencies, photonic crystal structures have been introduced (Fujita et al., 2008; Nakayama et al., 2010a; Iwamoto and Arakawa, 2012). The further increase of the efficiency can be achieved by introducing tensile strain and _n_-type doping of the germanium (Ishikawa et al., 2003; Menéndez and Kouvetakis, 2004; Liu et al., 2007, 2012; Kurdi et al., 2010; Michel et al., 2010; Huo et al., 2011; Boucaud et al., 2013; Ghrib et al., 2013; Nama et al., 2013; Tani et al., 2013a; Liu, 2014).

Considering the success of the laser operation using the bulk germanium waveguides (Liu et al., 2010; Camacho-Aguilera et al., 2012), the next step will be to reduce the threshold current for pumping. It is critical to develop a process technology to fabricate a high crystalline quality germanium quantum-well compatible with the silicon photonics platform. If practical silicon or germanium laser diodes are available in the future, these group IV lasers will realize the convergence of electronics and photonics on a silicon chip.

## ACKNOWLEDGMENTS

We would like to thank research collaborators, engineers, and line managers in Hitachi, the University of Tokyo, and University of Southampton for supporting this project. We are also grateful to Prof. H. N. Rutt for his careful reading of the manuscript and constructive comments. Funding: parts of the studied discussed here was supported by Japan Society for the Promotion of Science (JSPS) through its "Funding Program for World-Leading Innovation R&D on Science and Technology (FIRST Program)," the Project for Developing Innovation Systems, and Kakenhi 216860312, MEXT, Japan. This work is also supported by EU, FP7, Marie-Curie, Carrier Integration Grant (CIG), PCIG13-GA-2013-618116, and University of Southampton, Zepler Institute, Research Collaboration Stimulus Fund.

## REFERENCES

Arakawa, Y., Iwamoto, S., Nomura, M., Tandaechanurat, A., and Ota, Y. (2012). Cavity quantum electrodynamics and lasing oscillation in single quantum dot-photonic crystal nanocavity coupled systems. _IEEE J. Selec. Top. Quant. Elec._ 18, 1818–1829. doi:10.1109/JSTQE.2012.2199088

Arakawa, Y., Nakamura, T., and Urino, Y. (2013). Silicon photonics for next generation system integration platform. _Commun. Mag. IEEE_ 51, 72–77. doi:10.1109/MCOM.2013.6476868

Arakawa, Y., and Sakaki, H. (1982). Multidimensional quantum well laser and temperature dependence of its threshold current. _Appl. Phys. Lett._ 40, 939–941. doi:10.1063/1.92959

Boucaud, P., Kurdi, M. E., Ghrib, A., Prost, M., de Kersauson, M., Sauvage, S., et al. (2013). Recent advances in germanium emission. _Photon. Res._ 1, 102–109. doi:10.1364/PRJ.1.000102

Boztug, C., Sanchez-Perez, J. R., Yin, J., Lagally, M. G., and Paiella, R. (2013). Graiting-coupled mid-infrared light emission from tensile strained germanium nanomembranes. _Appl. Phys. Lett._ 103, 201114. doi:10.1002/smll.201201090

Camacho-Aguilera, R. E., Cai, Y., Patel, N., Bessette, J. T., Romagnoli, M., Kimerling, L. C., et al. (2012). An electrically pumped germanium laser. _Opt. Express_ 20, 11316–11320. doi:10.1364/OE.20.011316

Canham, L. T. (1990). Silicon quantum wire array fabrication by electrochemical and chemical dissolution of wafers. _Appl. Phys. Lett._ 57, 1046–1048. doi:10.1063/1.103561

Cannon, D. D., Liu, J., Ishikawa, Y., Wada, K., Danielson, D. T., Jongthammanu-rak, S., et al. (2004). Tensile strained epitaxial Ge film on Si(100) substrate with potential application to L-band telecommunications. _Appl. Phys. Lett._ 84, 906. doi:10.1063/1.1645677

Carroll, L., Friedli, P., Neunschwander, S., Sigg, H., Cecchi, S., Isa, F., et al. (2012). Direct-gap gain and optical absorption in germanium correlated to the density of photoexcited carriers, doping, and strain. *Phys. Rev. Lett.* 109, 057402. doi:10.1103/PhysRevLett.109.057402

Cerdeira, F., Buchenauer, C. J., Pollak, F. H., and Cardona, M. (1972). Stress-induced shifts of first-order Raman frequencies of diamond- and zinc-blende-type semiconductors. *Phys. Rev. B* 5, 580–593. doi:10.1103/PhysRevB.5.580

Cheng, T. H., Kuo, P. S., Lee, C. T., Liao, M., Hung, T. A., and Liu, C. W. (2007). "Electrically pumped Ge laser at room temperature," in *IEEE Int. Conf. Electron Devices Meeting (IEDM)*, Washington, 659–662.

Cluzel, B., Pauc, N., Calvo, V., Charvolin, T., and Hadji, E. (2006a). Nanobox array for silicon-on-insulator luminescence enhancement at room temperature. *Appl. Phys. Lett.* 88, 133120. doi:10.1063/1.2191089

Cluzel, B., Calvo, V., Charvolin, T., Picard, E., Noé, P., and Hadji, E. (2006b). Single-mode room-temperature emission with a silicon rod lattice. *Appl. Phys. Lett.* 89, 201111. doi:10.1063/1.2364876

Cullis, A. G., Canham, L. T., and Calcott, P. D. J. (1997). The structural and luminescence properties of porous silicon. *J. Appl. Phys.* 82, 909–965. doi:10.1063/1.366536

Daldosso, N., and Pavesi, L. (2009). Nanosilicon photonics. *Laser Photon. Rev.* 3, 508–534. doi:10.1002/lpor.200810045

de Kersauson, M., Kurdi, M. E., David, S., Checoury, X., Fishman, G., Sauvage, S., et al. (2011). Optical gain in single tensile-strained germanium photonic wire. *Opt. Express* 19, 17925–17934. doi:10.1364/OE.19.017925

de Walle, C. G. V. (1989). Band lineups and deformation potentials in the model-solid theory. *Phys. Rev. B* 39, 1871. doi:10.1103/PhysRevB.39.1871

de Walle, C. G. V., and Martin, R. M. (1986). Theoretical calculations of heterojunction discontinuities in the Si/Ge system. *Phys. Rev. B* 34, 5621–5634. doi:10.1103/PhysRevB.34.5621

Deen, M. J., and Basu, P. K. (2012). *Silicon Photonics Fundamentals and Devices.* West Sussex: Wiley.

Escalante, J. M., and Martínez, A. (2012). Theoretical study about the gain in indirect bandgap semiconductor optical cavities. *Physica B Condens. Matter* 407, 2044–2049. doi:10.1016/j.physb.2012.02.002

Escalante, J. M., and Martínez, A. (2013). Optical gain by simultaneous photon and phonon confinement in indirect bandgap semiconductor acousto-optical cavities. *Opt. Quant. Electron.* 45, 1045–1056. doi:10.1007/s11082-013-9715-z

Evans, P., Fisher, M., Malendevich, R., James, A., Studenkov, P., Goldfarb, G., et al. (2011). "Multi-channel coherent PM-QPSK InP transmitter photonic integrated circuit (PIC) operating at 112 gb/s per wavelength," in *Optical Fiber Communication Conference (OFC)* (Los Angeles: (IEEE) PDPC, OSA).

Fang, A. W., Park, H., Cohen, O., Jones, R., Paniccia, M. J., and Bowers, J. E. (2006). Electrically pumped hybrid AlGaInAs-silicon evanescent laser. *Opt. Express* 14, 9203–9210. doi:10.1364/OE.14.009203

Fathpour, S., and Jalali, B. (eds) (2012). *Silicon Photonics for Telecommunications and Biomedicine.* Boca Raton: CRC Press.

Feng, J., Thareja, G., Kobayashi, M., Chen, S., Poon, A., Bai, Y., et al. (2008). High-performance gate-all-around GeOI p-MOSFETs fabricated by rapid melt growth using plasma nitridation and ALD $Al_2O_3$ gate dielectric and self-aligned NiGe contacts. *IEEE Electron Device Lett.* 29, 805–807. doi:10.1109/LED.2008.2000613

Feng, J., Woo, R., Chen, S., Liu, Y., Griffin, P. B., and Plummer, J. D. (2007). P-channel germanium FinFET based on rapid melt growth. *IEEE Electron Device Lett.* 28, 637–639. doi:10.1109/LED.2007.899329

Fischetti, M. V., and Laux, S. E. (1996). Band structure, deformation potentials, and carrier mobility in strained Si, Ge, and SiGe alloys. *J. Appl. Phys.* 80, 2234. doi:10.1063/1.363052

Fujita, M., Tanaka, Y., and Noda, S. (2008). Light emission from silicon in photonic crystal nanocavity. *IEEE J. Selec. Top. Quant. Elec.* 14, 1090–1097. doi:10.1109/JSTQE.2008.918941

Fukatsu, S., Usami, N., Chinzei, T., Shiraki, Y., Nishida, A., and Nakagawa, K. (1992). Electroluminescence from strained SiGe/Si quantum well structures grown by solid source si molecular beam epitaxy. *Jpn. J. Appl. Phys.* 31, L1015–L1017. doi:10.1143/JJAP.31.L1015

Gaubas, E., and Vanhellemont, J. (2006). Dependence of carrier lifetime in germanium on resistivity and carrier injection level. *Appl. Phys. Lett.* 89, 142106. doi:10.1063/1.2358967

Geiger, R., Frigerio, J., Süess, M. J., Chrastina, D., Isella, G., Spolenak, R., et al. (2014). Excess carrier lifetimes in Ge layers on Si. *Appl. Phys. Lett.* 104, 062106. doi:10.1063/1.4865237

Gelloz, B., Kojima, A., and Koshida, N. (2005). Highly efficient and stable luminescence of nanocrystalline porous silicon treated by high-pressure water vapor annealing. *Appl. Phys. Lett.* 87, 031107. doi:10.1063/1.2001136

Gelloz, B., and Koshida, N. (2000). Electroluminescence with high and stable quantum efficiency and low threshold voltage from anodically oxidized thin porous silicon diode. *J. Appl. Phys.* 88, 4319–4324. doi:10.1063/1.1290458

Ghrib, A., Kurdi, M. E., de Kersauson, M., Prost, M., Sauvage, S., Checoury, X., et al. (2013). Tensile-strained germanium microdisks. *Appl. Phys. Lett.* 102, 221112. doi:10.1063/1.4809832

Going, R. W., Loo, J., Liu, T.-J. K., and Wu, M. C. (2014). Germanium gate PhotoMOSFET integrated to silicon photonics. *IEEE J. Selec. Top. Quant. Elec.* 20, 8201607. doi:10.1109/JSTQE.2013.2294470

Gong, Y., Makarova, M., Selçuk Yerci, R. L., Stevens, M. J., Baek, B., Nam, S. W., et al. (2010). Linewidth narrowing and Purcell enhancement in photonic crystal cavities on an Er-doped silicon nitride platform. *Opt. Express* 18, 2601–2612. doi:10.1364/OE.18.002601

Gunn, C. (2006). Cmos photonics for high-speed interconnects. *Micro IEEE* 26, 58–66. doi:10.1109/MM.2006.32

Hashimoto, T., Yoshimoto, C., Hosoi, T., Shimura, T., and Watanabe, H. (2009). Fabrication of local Ge-on-insulator structures by lateral liquid-phase epitaxy: effect of controlling interface energy between Ge and insulators on lateral epitaxial growth. *Appl. Phys. Express* 2, 066502. doi:10.1143/APEX.2.066502

Hauke, N., Zabel, T., Müller, K., Kaniber, M., Laucht, A., Bougeard, D., et al. (2010). Enhanced photoluminescence emission from two-dimensional silicon photonic crystal nanocavities. *New J. Phys.* 98, 053005. doi:10.1088/1367-2630/12/5/053005

Hisamoto, D., Lee, W., Kedzierski, J., Takeuchi, H., Asano, K., Kuo, C., et al. (2000). FinFET-a self-aligned double-gate MOSFET scalable to 20 nm. *IEEE Trans. Electron Devices* 47, 2320–2325. doi:10.1109/16.887014

Hoang, T., LeMinh, P., Holleman, J., and Schmitz, J. (2007). Strong efficiency improvement of SOI-LEDs through carrier confinement. *IEEE Electron Device Lett.* 28, 383–385. doi:10.1109/LED.2007.895415

Huo, Y., Lin, H., Chen, R., Makarova, M., Rong, Y., Li, M., et al. (2011). Strong enhancement of direct transition photoluminescence with highly tensile-strained Ge grown by molecular beam epitaxy. *Appl. Phys. Lett.* 98, 011111. doi:10.1063/1.3534785

Ishikawa, Y., and Wada, K. (2010). Germanium for silicon photonics. *Thin Solid Films* 518, S83–S87. doi:10.1016/j.tsf.2009.10.062

Ishikawa, Y., Wada, K., Cannon, D. D., Liu, J., Luan, H.-C., and Kimerling, L. C. (2003). Strain-induced band gap shrinkage in Ge grown on Si substrate. *Appl. Phys. Lett.* 82, 2044. doi:10.1063/1.1564868

Ishikawa, Y., Wada, K., Liu, J., Cannon, D. D., Luan, H.-C., Michel, J., et al. (2005). Strain-induced enhancement of near-infrared absorption in Ge epitaxial layers grown on Si substrate. *J. Appl. Phys.* 98, 013501. doi:10.1063/1.1943507

Iwamoto, S., and Arakawa, Y. (2012). Enhancement of light emission from silicon by utilizing photonic nanostructures. *IEICE Trans. Electron* E95-C, 206–212. doi:10.1587/transele.E95.C.206

Iwamoto, S., Arakawa, Y., and Gomyo, A. (2007). Observation of enhanced photoluminescence from silicon photonic crystal nanocavity at room temperature. *Appl. Phys. Lett.* 91, 211104. doi:10.1063/1.2816892

Jain, J. R., Hryciw, A., Baer, T. M., Miller, D. A. B., Brongersma, M. L., and Howe, R. T. (2012). A micromachining-based technology for enhancing germanium light-emission via tensile strain. *Nat. Photonics* 6, 398–405. doi:10.1038/nphoton.2012.111

Jannopoulos, J. D., and Winn, R. D. M. J. N. (1995). *Photonic Crystals.* Princeton, NJ: Princeton University Press.

Kageshima, H., and Shiraishi, K. (1997). Momentum-matrix-element calculation using pseudopotentials. *Phys. Rev. B* 56, 14985. doi:10.1103/PhysRevB.56.14985

Kako, S., Okumura, T., Oda, K., Suwa, Y., Saito, S., Ido, T., et al. (2012). "Time-resolved photoluminescence study of highly n-doped germanium grown on silicon," in *The 9th International Conference on Group IV Photonics (GFP)*, Vol. 7, San Diego, 340–342.

Kittler, M., Reiche, M., and Arguirov, T. (2013). "1.55 $\mu$m light emitter based on dislocation d1-emission in silicon," in *Microelectronics Technology and Devices (SBMicro), Symposium on* IEEE, Curitiba.

Koshida, N., and Koyama, H. (1992). Visible electroluminescence from porous silicon. *Appl. Phys. Lett.* 60, 347–349. doi:10.1063/1.106652

Kurdi, M. E., Bertin, H., Martincic, E., de Kersauson, M., Fishman, G., Sauvage, S., et al. (2010). Control of direct band gap emission of bulk germanium by mechanical tensile strain. *Appl. Phys. Lett.* 96, 041909. doi:10.1063/1.3297883

Kurdi, M. E., Davida, S., Checourya, X., Fishmana, G., Boucauda, P., Kermarrecb, O., et al. (2008). Two-dimensional photonic crystals with pure germanium-on-insulator. *Opt. Commun.* 281, 846–850. doi:10.1016/j.optcom.2007.10.008

Laasonen, K., Pasquarello, A., Car, R., Lee, C., and Vanderbilt, D. (1993). Car-Parrinello molecular dynamics with Vanderbilt ultrasoft pseudopotentials. *Phys. Rev. B* 47, 10142. doi:10.1103/PhysRevB.47.10142

Liang, D., and Bowers, J. E. (2010). Recent progress in lasers on silicon recent progress in lasers on silicon recent progress in lasers on silicon recent progress in lasers on silicon. *Nat. Photonics* 4, 511–517. doi:10.1088/0034-4885/76/3/034501

Lim, P. H., Park, S., Ishikawa, Y., and Wada, K. (2009). Enhanced direct bandgap emission in germanium by micromechanical strain engineering. *Opt. Express* 17, 16358–16365. doi:10.1364/OE.17.016358

Linnros, J. (1998). Carrier lifetime measurements using free carrier absorption transients. I. Principle and injection dependence. *J. Appl. Phys.* 84, 275. doi:10.1063/1.368024

Liu, J. (2014). Monolithically integrated Ge-on-Si active photonics. *Photonics* 1, 162–197. doi:10.3390/photonics1030162

Liu, J., Camacho-Aguilera, R., Bessette, J. T., Sun, X., Wang, X., Cai, Y., et al. (2012). Ge-on-Si optoelectronics. *Thin Solid Films* 520, 3354–3360. doi:10.1016/j.tsf.2011.10.121

Liu, J., Cannon, D. D., Wada, K., Ishikawa, Y., Jongthammanurak, S., Danielson, D. T., et al. (2005). Tensile strained Ge *p-i-n* photodetectors on Si platform for C and L band telecommunications. *Appl. Phys. Lett.* 87, 011110. doi:10.1063/1.2037200

Liu, J., Sun, X., Camacho-Aguilera, R., Kimerling, L. C., and Michel, J. (2010). Ge-on-Si laser operating at room temperature. *Opt. Lett.* 35, 679–681. doi:10.1364/OL.35.000679

Liu, J., Sun, X., Pan, D., Wang, X., Kimerling, L. C., Koch, T. L., et al. (2007). Tensile-strained, n-type Ge as a gain medium for monolithic laser integration on Si. *Opt. Express* 15, 11272–11277. doi:10.1364/OE.15.011272

Liu, Y., Deal, M. D., and Plummer, J. D. (2004). High-quality single-crystal Ge on insulator by liquid-phase epitaxy on Si substrates. *Appl. Phys. Lett.* 84, 2563. doi:10.1063/1.1691175

Lo Savio, R., Portalupi, S. L., Gerace, D., Shakoor, A., Krauss, T. F., O'Faolain, L., et al. (2011). Room-temperature emission at telecom wavelengths from silicon photonic crystal nanocavities. *Appl. Phys. Lett.* 98, 201106. doi:10.1063/1.3591174

Luan, H.-C., Lim, D. R., Lee, K. K., Chen, K. M., Sandland, J. G., Wada, K., et al. (1999). High-quality Ge epilayers on Si with low threading-dislocation densities. *Appl. Phys. Lett.* 75, 2909. doi:10.1063/1.125187

Matsue, M., Yasutake, Y., Fukatsu, S., Hosoi, T., Shimura, T., and Watanabe, H. (2014). Strain-induced direct band gap shrinkage in local Ge-on-insulator structures fabricated by lateral liquid-phase epitaxy. *Appl. Phys. Lett.* 104, 031106. doi:10.1063/1.4862890

Menéndez, J., and Kouvetakis, J. (2004). Type-I Ge/Ge$_{1-x-y}$Si$_x$Sn$_y$ strained-layer heterostructure with a direct Ge bandgap. *Appl. Phys. Lett.* 85, 1175. doi:10.1063/1.1784032

Michel, J., Liu, J., and Kimerling, L. C. (2010). High-performance ge-on-si photodetectors. *Nat. Photonics* 4, 527–534. doi:10.1038/nphoton.2010.157

Michel, J., and Romagnoli, M. (2012). *Germanium as the Unifying Material for Silicon Photonics*. Bellingham: SPIE. doi:10.1117/2.1201206.004285

Miller, D. A. B. (2009). Device requirements for optical interconnections to silicon chips. *Proc. IEEE* 97, 1166–1185. doi:10.1109/JPROC.2009.2014298

Miyao, M., Tanaka, T., Toko, K., and Tanaka, M. (2009). Giant Ge-on-insulator formation by Si-Ge mixing-triggered liquid-phase epitaxy. *Appl. Phys. Express* 2, 045503. doi:10.1143/APEX.2.045503

Nakayama, S., Ishida, S., Iwamoto, S., and Arakawa, Y. (2012). Effect of cavity mode volume on photoluminescence from silicon photonic crystal nanocavities. *Appl. Phys. Lett.* 98, 171102.

Nakayama, S., Iwamoto, S., Ishida, S., and Arakawa, Y. (2010a). "Demonstration of a silicon photonic crystal slab led with efficient electroluminescence," in *International Conference on Solid State Devices and Materials (SSDM 2010)* D-4-3, Tokyo.

Nakayama, S., Iwamoto, S., Ishida, S., Bordel, D., Augedre, E., Calvelier, L., et al. (2010b). Enhancement pf photoluminescence from germanium by utilizing air-bridge-type photonic crytsal slab. *Physica E* 42, 2556–2559. doi:10.1016/j.physe.2010.05.026

Nakayama, S., Iwamoto, S., Kako, S., Ishida, S., and Arakawa, Y. (2011). "Demonstration of silicon nanocavity led with enhanced luminescence," in *International Conference on Solid State Devices and Materials*, Nagoya, I–8–2.

Nama, D., Sukhdeo, D. S., Kang, J.-H., Petykiewicz, J., Lee, J. H., Jung, W. S., et al. (2013). Strain-induced pseudoheterostructure nanowires confining carriers at room temperature with nanoscale-tunable band profiles. *Nano Lett.* 13, 3118–3123. doi:10.1021/nl401042n

Nassiopoulou, A. G. (2004). *Silicon Nanocrystals in SiO$_2$ Thin Layers*, Vol. 9. Valencia: American Scientific Publishers, 793–813.

Nelson, H. (1963). Epitaxial growth from the liquid state and its application to the fabrication of tunnel and laser diodes. *RCA Rev.* 24, 603–615.

Ng, W. L., Lourenço, M. A., Gwilliam, R. M., Ledain, S., Shao, G., and Homewood, K. P. (2001). An efficient room-temperature silicon-based light-emitting diode. *Nature* 410, 192–194. doi:10.1038/35069092

Ngo, T.-P., Kurdi, M. E., Checoury, X., Boucaud, P., Damlencourt, J. F., Kermarrec, O., et al. (2008). Two-dimensional photonic crystals with germanium on insulator obtained by a condensation method. *Appl. Phys. Lett.* 93, 241112. doi:10.1063/1.3054332

Noborisaka, J., Nishiguchi, K., Ono, Y., Kageshima, K., and Fujiwara, A. (2011). Strong stark effect in electroluminescence from phosphorous-doped silicon-on-insulator metal-oxide-semiconductor field-effect transistors. *Appl. Phys. Lett.* 98, 033503. doi:10.1063/1.3543849

Oda, K., Tani, K., Saito, S.-I., and Ido, T. (2014). Improvement of crystallinity by post-annealing and regrowth of Ge layers on Si substrates. *Thin Solid Films* 550, 509–514. doi:10.1016/j.tsf.2013.10.136

Ohta, Y., Tanaka, T., Toko, K., Sadoh, T., and Miyao, M. (2011). Growth-direction-dependent characteristics of Ge-on-insulator by Si-Ge mixing triggered melting growth. *Solid State Electron.* 60, 18–21. doi:10.1016/j.sse.2011.01.039

Okada, Y., and Tokumaru, Y. (1984). Precise determination of lattice parameter and thermal expansion coefficient of silicon between 300 and 1500 K. *J. Appl. Phys.* 56, 314. doi:10.1063/1.333965

Ortolland, C., Okuno, Y., Veheyen, P., Kerner, C., Stapelmann, C., Aoulaiche, M., et al. (2009). Stress memorization technique-fundamental understanding and low-cost integration for advanced CMOS technology using a nonselective process. *IEEE Trans. Electron Devices* 56, 1690–1697. doi:10.1109/TED.2009.2024021

Ossicini, S., Pavesi, L., and Priolo, F. (2006). *Light Emitting Silicon for Microphotonics*. Berlin: Springer.

Panish, M. B., Hayashi, I., and Sumski, S. (1970). Double heterostructure injection lasers with room temperature thresholds as low as 2300 a/cm$^2$. *Appl. Phys. Lett.* 16, 326. doi:10.1063/1.1653213

Pavesi, L., and Guillot, G. (eds) (2006). *Optical Interconnects The Silicon Approach*. Berlin: Springer.

Pavesi, L., and Lockwood, D. J. (eds) (2004). *Silicon Photonics*. Berlin: Springer.

Pavesi, L., Negro, L. D., Mazzoleni, C., Franzò, G., and Priolo, F. (2000). Optical gain in silicon nanocrystals. *Nature* 408, 440–444.

Perdew, J. P., Burke, K., and Wang, Y. (1996). Generalized gradient approximation for the exchange-correlation hole of a many-electron system. *Phys. Rev. B* 54, 16533. doi:10.1103/PhysRevB.54.16533

Purcell, E. M. (1946). Spontaneous emission probabilities at radio frequencies. *Phys. Rev* 69, 681.

Reed, G. (2008). *Silicon Photonics the State of the Art*. West Sussex: Wiley.

Reed, G. T., and Knights, A. P. (2004). *Silicon Photonics*. West Sussex: Wiley.

Rylyakov, A., Schow, C., Lee, B., Green, W., Campenhout, J. V., Yang, M., et al. (2011). "A 3.9ns 8.9mW 4×4 silicon photonic switch hybrid integrated with CMOS driver," in *Int. Solid-State Circuits Conference (ISSCC)* (San Francisco: IEEE), 222–224.

Saito, S. (2011). "Silicon and germanium quantum well light-emitting diode," in *IEEE 8th Int. Conf. Group IV Photonics*, London, 166–168.

Saito, S., Hisamoto, D., Shimizu, H., Hamamura, H., Tsuchiya, R., Matsui, Y., et al. (2006a). Electro-luminescence from ultra-thin silicon. *Jpn. J. Appl. Phys.* 45, L679–L682. doi:10.1143/JJAP.45.L679

Saito, S., Hisamoto, D., Shimizu, H., Hamamura, H., Tsuchiya, R., Matsui, Y., et al. (2006b). Silicon light-emitting transistor for on-chip optical interconnection. *Appl. Phys. Lett.* 89, 163504. doi:10.1063/1.2360783

Saito, S., Oda, K., Takahama, T., Tani, K., and Mine, T. (2011a). Germanium fin light-emitting diode. *Appl. Phys. Lett.* 99, 241105. doi:10.1364/OE.22.005927

Saito, S., Takahama, T., Tani, K., Takahashi, M., Mine, T., Suwa, Y., et al. (2011b). Stimulated emission of near-infrared radiation in silicon fin light-emitting diode. *Appl. Phys. Lett.* 98, 261104. doi:10.1063/1.3605255

Saito, S., Sakuma, N., Suwa, Y., Arimoto, H., Hisamoto, D., Uchiyama, H., et al. (2008). "Observation of optical gain in ultra-thin silicon resonant cavity light-emitting diode," in *IEEE Int. Conf. Electron Devices Meeting (IEDM)*, San Francisco, 19.5.

Saito, S., Suwa, Y., Arimoto, H., Sakuma, N., Hisamoto, D., Uchiyama, H., et al. (2009). Stimulated emission of near-infrared radiation by current injection into silicon (100) quantum well. *Appl. Phys. Lett.* 95, 241101. doi:10.1063/1.3273367

Saul, R. H., and Roccasecca, D. D. (1973). Vapor-doped multislice LPE for efficient GaP green LED's. *J. Electrochem. Soc.* 120, 1128–1131. doi:10.1149/1.2403644

Savio, R. L., Miritello, M., Shakoor, A., Cardile, P., Welna, K., Andreani, L. C., et al. (2013). Enhanced 1.54 $\mu$m emission in Y-Er disilicate thin films on silicon photonic crystal cavities. *Opt. Express* 21, 10278–10288. doi:10.1364/OE.21.010278

Shakoor, A., Lo Savio, R., Cardile, P., Portalupi, S. L., Gerace, D., Welna, K., et al. (2013). Room temperature all-silicon photonic crystal nanocavity light emitting diode at sub-bandgap wavelengths. *Laser Photon. Rev.* 7, 114–121. doi:10.1002/lpor.201200043

Sheng, J. J., Leonhardt, D., Han, S. M., Johnston, S. W., Cederberg, J. G., and Carroll, M. S. (2013). Empirical correlation for minority carrier lifetime to defect density profile in germanium on silicon grown by nanoscale interfacial engineering. *J. Vac. Sci. Technol. B Microelectron.* 31, 051201. doi:10.1116/1.4816488

Singh, H. P. (1968). Determination of thermal expansion of germanium, rhodium and iridium by x-rays. *Acta Crystallogr. A* 24, 469–471. doi:10.1107/S056773946800094X

Sproul, A. B. (1994). Dimensionless solution of the equation describing the effect of surface recombination on carrier decay in semiconductors. *J. Appl. Phys.* 76, 2851. doi:10.1063/1.357521

Steger, M., Yang, A., Sekiguchi, T., Saeedi, K., Thewalt, M. L. W., Henry, M. O., et al. (2011). Photoluminescence of deep defects involving transition metals i Si: new insights from highly enriched $^{28}$Si. *J. Appl. Phys.* 110, 081301. doi:10.1063/1.3651774

Süess, M. J., Geiger, R., Minamisawa, R. A., Schiefler, G., Frigerio, J., Chrastina, D., et al. (2013). Analysis of enhanced light emission from highly strained germanium microbridges. *Nat. Photonics* 7, 466–472. doi:10.1038/nphoton.2013.67

Sukhdeo, D. S., Nam, D., Kang, J. H., Brongersma, M. L., and Saraswat, K. C. (2014). Direct bandgap germanium-on-silicon inferred from 5.7% <100> uniaxial tensile strain. *Photon. Res.* 2, A8–A13. doi:10.1364/PRJ.2.0000A8

Sumikura, H., Kuramochi, E., Taniyama, H., and Notomi, M. (2014). Ultrafast spontaneous emission of copper-doped silicon enhanced by an optical nanocavity. *Sci. Rep.* 4, 5040. doi:10.1038/srep05040

Suwa, Y., and Saito, S. (2009). Intrinsic optical gain of ultrathin silicon quantum wells from first-principles calculations. *Phys. Rev. B* 79, 233308. doi:10.1103/PhysRevB.79.233308

Suwa, Y., and Saito, S. (2010). "First-principles study of light emisson from germanium guantum-well," in *IEEE 22nd Int. Semiconductor Laser Conference (ISLC)*, Kyoto, 131–132.

Suwa, Y., and Saito, S. (2011). "First-principles study of light emission from silicon and germanium due to direct transitions," in *IEEE 8th Int. Conf. Group IV Photonics*, London, 222–224.

Sze, S. M., and Lee, M. K. (2012). *Semiconductor Devices: Physics and Technology*, 3rd Edn. Singapore: John Wiley and Sons.

Tani, K., Oda, K., Kasai, J., Okumura, T., Mine, T., Saito, S., et al. (2013a). "Germanium waveguides on lateral silicon-on-insulator diodes for monolithic light emitters and photo detectors," in *Group IV Photonics (GFP), IEEE 10th International Conference* (Seoul: IEEE), 134–135.

Tani, K., Oda, K., Okumura, T., Takezaki, T., Kasai, J., Mine, T., et al. (2013b). "Enhanced electroluminescence from germanium waveguides by local tensile strain with silicon nitride stressors," in *International Conference on Solid State Devices and Materials (SSDM)* (Fukuoka: SSDM), K–6–3.

Tani, K., Saito, S., Oda, K., Miura, M., Mine, T., Sugawara, T., et al. (2011). "Ge(111)-fin light-emitting diodes," in *IEEE 8th Int. Conf. Group IV Photonics*, London, 217–219.

Tani, K., Saito, S., Oda, S., Okumura, T., Mine, T., and Ido, T. (2012). "Lateral carrier injection to germanium for monolithic light sources," in *Group IV Photonics (GFP), IEEE 9th Int. Conf*, San Diego, 328–330.

Taur, Y., and Ning, T. H. (2013). *Fundamentals of Modern VLSI Devices*, 2nd Edn. Cambridge: Cambridge University Press.

Tezuka, T., Toyoda, E., Irisawa, T., Hirashita, N., Moriyama, Y., Sugiyama, N., et al. (2009). Structural analyses of strained SiGe wires formed by hydrogen thermal etching and Ge-condensation processes. *Appl. Phys. Lett.* 94, 081910. doi:10.1063/1.3086884

Tseng, C.-K., Chen, W.-T., Chen, K.-H., Liu, H.-D., Kang, Y., Na, N., et al. (2013). A self-assembled microbonded germanium/silicon heterojunction photodiode for 25 Gb/s high-speed optical interconnects. *Sci. Rep.* 3, 3225. doi:10.1038/srep03225

Tweet, D. J., Lee, J. J., Maa, J.-S., and Hsu, S. T. (2005). Characterization and reduction of twist in Ge on insulator produced by localized liquid phase epitaxy. *Appl. Phys. Lett.* 87, 141908. doi:10.1063/1.2077860

Uchida, K., and Takagi, S. (2003). Carrier scattering induced by thickness fluctuation of silicon-on-insulator film in ultrathin-body metal–oxide–semiconductor field-effect transistors. *Appl. Phys. Lett.* 82, 2916. doi:10.1063/1.1571227

Ujihara, K. (1995). "Effects of atomic broadening on spontaneous emission in an optical microcavity," in *Spontaneous Emission and Laser Oscillation in Microcavities*, eds H. Yokoyama and K. Ujihara (Boca Raton, FL: CRC Press), 81–107.

Urino, Y., Horikawa, T., Nakamura, T., and Arakawa, Y. (2013). "High density optical interconnects integrated with lasers, optical modulators and photodetectors on a single silicon chip," in *Optical Fiber Communication Conference (OFC)*, (Anaheim: OSA), OM2J.

Vanderbilt, D. (1990). Soft self-consistent pseudopotentials in a generalized eigenvalue formalism. *Phys. Rev. B* 41, 7892(R). doi:10.1103/PhysRevB.41.7892

Virgilio, M., Manganelli, C. L., Grosso, G., Pizzi, G., and Capellini, G. (2013a). Radiative recombination and optical gain spectra in biaxially strained *n*-type germanium. *Phys. Rev. B* 87, 235313. doi:10.1103/PhysRevB.87.235313

Virgilio, M., Manganelli, C. L., Grosso, G., Schroeder, T., and Capellini, G. (2013b). Photoluminescence, recombination rate, and gain spectra in optically excited *n*-type and tensile strained germanium layers. *J. Appl. Phys.* 114, 243102. doi:10.1063/1.4849855

Vivien, L., and Pavesi, L. (eds) (2013). *Handbook of Silicon Photonics*. Taylor and Francis.

Wada, K., Liu, J., Jongthammanurak, S., Cannon, D. D., Danielson, D. T., Ahn, D., et al. (2006). *Si Microphotonics for Optical Interconnection*. Berlin: Springer Verlag.

Wale, M. (2008). "Photonic integration – an industrial perspective," in *European Photonic Integration Forum* (Brussels: European Photonics Integration Forum).

Wang, X., Li, H., Camacho-Aguilera, R., Cai, Y., Kimerling, L. C., Michel, J., et al. (2013). Infrared absorption of *n*-type tensile-strained Ge-on-Si. *Opt. Lett.* 38, 652–654. doi:10.1364/OL.38.000652

Wang, Y., Zhang, J., Wu, Y., An, J., Li, J., Wang, H., et al. (2012). Light emission enhancement from Er-doped silicon photonic crystal double-heterostructure microcavity. *IEEE Photon. Technol. Lett.* 24, 1041–1135. doi:10.1109/LPT.2011.2173183

Wieder, H. H., Clawson, A. R., and McWilliams, G. E. (1977). In$_x$Ga$_{1-x}$As$_y$P$_{1-y}$/InP heterojunction photodiodes. *Appl. Phys. Lett.* 31, 468. doi:10.1063/1.89718

Zelsmann, M., Picard, E., Charvolin, T., Hadji, E., Heitzmann, M., Dalzotto, B., et al. (2003). Seventy-fold enhancement of light extraction from a defectless photonic crystal made on silicon-on-insulator. *Appl. Phys. Lett.* 83, 2542–2544. doi:10.1063/1.1614832

Zimmermann, H. (2000). *Integrated Silicon Opto-Electronics*. Berlin: Springer.

**Conflict of Interest Statement:** The authors declare that the research was conducted in the absence of any commercial or financial relationships that could be construed as a potential conflict of interest.

# 3

# A brief review on syntheses, structures, and applications of nanoscrolls

*Eric Perim, Leonardo Dantas Machado and Douglas Soares Galvao**

*Applied Physics Department, State University of Campinas, Campinas, Brazil*

**Edited by:**
*Federico Bosia, University of Turin, Italy*

**Reviewed by:**
*Pratyush Tiwary, ETH Zürich, Switzerland*
*Shangchao Lin, Florida State University, USA*
*Simone Taioli, Bruno Kessler Foundation, Italy*

*\*Correspondence:*
*Douglas Soares Galvao, Applied Physics Department, State University of Campinas, Campinas, SP 13083-970, Brazil*
*e-mail: galvao@ifi.unicamp.br*

Nanoscrolls are papyrus-like nanostructures, which present unique properties due to their open ended morphology. These properties can be exploited in a plethora of technological applications, leading to the design of novel and interesting devices. During the past decade, significant advances in the synthesis and characterization of these structures have been made, but many challenges still remain. In this mini review, we provide an overview on their history, experimental synthesis methods, basic properties, and application perspectives.

**Keywords: nanoscrolls, review, carbon, boron nitride, applications, graphene**

## INTRODUCTION

Nanoscrolls consist of layered structures rolled into a papyrus-like form, as shown in the bottom frame of **Figure 1A**. They are morphologically similar to multiwalled nanotubes, except for the fact that they present open extremities and side, since the edges of the scrolled sheets are not fused. Similarly to nanotubes, nanoscrolls can be constructed from different materials, such as carbon, boron nitride, and others. These scrolled structures were first proposed by Bacon in 1960, as a consequence of his observations of the products from arc discharge experiments using graphite electrodes (Bacon, 1960). Decades later, the advances in imaging techniques made possible the confirmation of his structural model (Dravid et al., 1993). Iijima's synthesis of multiwalled carbon nanotubes (MWCNTs) brought renewed attention to nanoscrolls (Iijima, 1991), as they were later proposed to be part of MWCNTs synthesis (Zhang et al., 1993; Amelinckx et al., 1995). However, only more recently nanoscrolls have started attracting attention as individual and unique structures (Tománek, 2002), as a consequence of the development of efficient methods especially designed toward the production of carbon nanoscrolls (CNSs) (Shioyama and Akita, 2003; Viculis et al., 2003). This motivated a significant number of theoretical and experimental studies on the properties and possible CNS applications.

The characteristic which sets CNSs apart from MWNTs are their unique open ended morphology. These structures are kept from unrolling back into a planar morphology due to van der Waals (vdW) interactions between overlapping layers (Braga et al., 2004). As a consequence, CNSs can have their diameter easily tuned (Shi et al., 2010b), can be easily intercalated (Mpourmpakis et al., 2007). They offer wide solvent accessible surface area (Coluci et al., 2007) while sharing some of electronic and mechanical properties with MWCNTs (Zaeri and Ziaei-Rad, 2014) and preserving the high carrier mobility exhibited by graphene.

## SYNTHESIS

Carbon scrolls were first reported as byproducts of arc discharge experiments using graphite electrodes (Bacon, 1960). In this kind of experiment, the extremely high energies allow the formation of several different carbon structures besides nanoscrolls, such as, nanotubes and fullerenes (Krätschmer et al., 1990; Ugarte, 1992; Saito et al., 1993). However, the high cost, low yield, and non-selectivity of this method limits its wide use. The first method designed to produce CNSs at high yield, reaching over 80%, was developed only decades later (Mack et al., 2005). This process consists in three consecutive steps. Firstly, high quality graphite is intercalated with potassium metals, then it is exfoliated via a highly exothermal reaction with aqueous solvents. Lastly, the resulting dispersion of graphene sheets is sonicated resulting in CNSs (Viculis et al., 2003) – see **Figure 1A**. The strong deformations caused by the sonication process leads the solvated sheets to bend and, in case of overlapping layers, to scroll. As calculations pointed out (Braga et al., 2004), once significant layer overlap occurs the scrolling process is spontaneous and driven by vdW forces. The efficiency of this method has lead it to be adopted in other studies (Roy et al., 2008). A very similar method was shortly after developed (Shioyama and Akita, 2003), the most significant difference being the absence of sonication. In this case, longer times are necessary for graphene sheets in solution to spontaneously scroll.

Both these chemical methods use donor-type intercalation compounds, which are highly reactive, demanding the use of inert atmosphere during the process. In order to avoid this limitation, a variation of Viculis et al.'s method was devised utilizing acceptor-type intercalation compounds, namely graphite nitrate, which is much more stable and thus eliminates the need for an inert atmosphere (Savoskin et al., 2007). However, the most significant drawbacks of this chemical approach are the poor morphologies

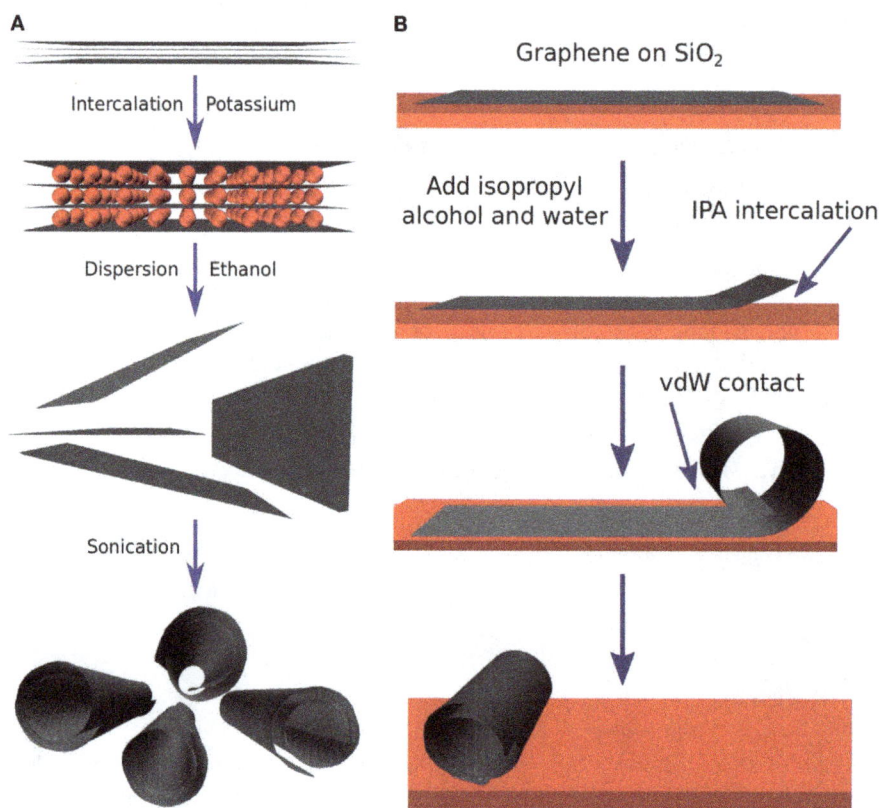

**FIGURE 1 | Scheme of some methods used to synthesize carbon nanoscrolls. (A)** Graphite is intercalated with potassium, then dispersed in ethanol and then the mixture is sonicated. **(B)** Graphene is first mechanically exfoliated from graphite and deposited on $SiO_2$. Then a droplet of isopropyl alcohol (IPA) and water is placed on the monolayer and evaporated. Both methods lead to well-formed nanoscrolls.

of the resulting nanoscrolls, the inability to control the number of scrolled graphene layers, and also the possibility of defects being introduced during the chemical process. In order to overcome these issues, a new method was later developed, offering higher control over the final product. In this new method, graphite is mechanically exfoliated using the scotch tape method and the extracted graphene layers are then deposited over $SiO_2$ substrates. Then a drop of a solution of water and isopropyl alcohol is applied over the structures and the system let to rest for a few minutes. After this, the system is dried out and spontaneously formed CNSs can be observed (Xie et al., 2009) – see **Figure 1B**. It is believed that surface strain is induced on the graphene layer as a consequence of one side being in contact with the solution and the other being in direct contact with the substrate. Once this strain causes the edges to lift, solvent molecules can occupy the space between layer and substrate, further bending the graphene sheets. As some deformation causes overlap on the layers, the scrolling process becomes spontaneous. While this method offers higher control over produced CNSs, on the other hand it is difficult to scale and more sensitive to defects in the graphene layers.

Synthesis of high quality CNSs from microwave irradiation has also been reported (Zheng et al., 2011). In this method, graphite flakes are immersed into liquid nitrogen and then heated under microwave radiation for a few seconds. As graphite presents very good microwave absorption, sparks are produced, which are believed to play a key role in the process, as their absence hinders high CNS yields. The resulting product is then sonicated and centrifugated, resulting in well-formed CNSs.

More recently, a purely physical route to CNSs synthesis method was proposed on theoretical grounds (Xia et al., 2010). In this method, a carbon nanotube (CNT) is used to trigger the scrolling of a graphene monolayer. Due to vdW interactions, the sheet rolls itself around the CNT in order to lower the surface energy in a spontaneous process. The advantage of this method would be being a dry, non-chemical, room-temperature method. However, it has been shown that the presence of a substrate can significantly affect the efficiency of this method (Zhang and Li, 2010). In order to circumvent these limitations, simple changes in substrate morphology have been proposed (Perim et al., 2013). The same principle has been used to propose a method for producing CNS-sheathed Si nanowires (Chu et al., 2011). However, an experimental realization of such process has yet to be reported. Even more recently, CNSs have been proposed to form from diamond nanowires upon heating (Sorkin and Su, 2014).

Nitrogen-doped graphene oxide nanoscrolls have also been successfully produced (Sharifi et al., 2013). Maghemite ($\gamma$-$Fe_2O_3$) particles were used to trigger the scrolling process. In this case, it is believed that the magnetic interaction between the nitrogen defects and maghemite particles is the governing effect in this process. This is supported by the fact that removal of $\gamma$-$Fe_2O_3$ particles causes the scroll to unroll in a reversible process, different from the observed for pure CNSs.

## STRUCTURE

From a topological point of view, scrolls can be considered as sheets rolled up into Archimedean spirals. Hence, the polar equation used to describe these spirals,

$$r = r_0 + \frac{h}{2\pi}\phi, \tag{1}$$

can be used to determine the points $r$ that belong to the scroll, for a given core radius $r_0$, interlayer spacing $h$, and number of turns $N$ ($\phi$ varies from 0 to $2\pi N$). See **Figure 2A**. In addition, in order to fully determine the geometry of the scroll, the axis around which the scroll was wrapped must be given (see **Figure 2B**). Therefore, armchair, zigzag, and chiral nanoscrolls exist, although scroll type is not fixed during synthesis and interconversion can occur in mild conditions, due to the open ended topology (Braga et al., 2004).

Particular values of $r_0$, $h$, and $N$ for the general scroll geometry described above will depend on the properties of the composing scroll material. As shown by Shi et al. (2011), the core radius $r_0$ can be determined from the interaction energy between layers ($\gamma$), the bending stiffness of the composing material ($D$), the interlayer

separation ($h$), the length of the composing sheet ($B$), and the difference between the inner ($p_i$) and outer pressure ($p_e$), as described by the following equation:

$$p_i - p_e = \frac{D}{2h}\left(\frac{2\gamma h}{D} - \frac{1}{r_0} + \frac{1}{R}\right)\left(\frac{1}{r_0} + \frac{1}{R}\right), \tag{2}$$

where $R = \sqrt{\frac{Bh}{\pi} + r_0^2}$.

Density Functional Theory calculations, carried out without pressure difference, predicted that the minimum stable core diameter is 23 Å (Chen et al., 2007). The interlayer spacing, however, depends mostly on the interaction energy between layers, although several factors can alter its value, like the presence of defects (Tojo et al., 2013). Given a core size and an interlayer distance, the number of turns can be obtained after fully wrapping a given sheet width.

In order to form a nanoscroll, there is an elastic energy cost associated with bending the sheet and a vdW energy gain associated with the creation of regions of sheet overlap. For graphene, in particular, if the core size and number of turns is appropriate, the vdW energy gain is large enough to make the final scrolled structure even more stable than its initial planar configuration (Braga et al., 2004). There is, however, an energy barrier associated with the formation of scrolls, since the initial bending cost is not followed by energy gains. Various theoretical and experimental methods have been devised to overcome this barrier, as discussed in the previous section.

## MECHANICAL PROPERTIES

In this and the next section, we will restrict ourselves to the discussion of carbon-based scrolls. In Section "Nanoscrolls from some other Materials", non-carbon scrolls will be also addressed.

So far, we have discussed mainly equilibrium geometry properties, but there are several studies addressing the mechanical response of scrolls to applied agents and/or forces. The first of these studies was carried out by Zhang et al. (2012) using molecular mechanics, and studied the response of CNSs to axial compression, twisting, and bending. With regard to compression, the authors found that the axial stiffness of CNTs and CNSs with similar diameters and number of layers is about the same, but that nanoscrolls buckle under a significantly smaller strain. The authors argued that the free ends of the CNSs tend to wrinkle and are vulnerable to further buckling, which then propagates inward from the ends. With relation to torsions, the paper reported that both the torsional rigidity and critical strain are much lower for CNSs when compared to CNTs, a result that was attributed again to the open topology. Finally, regarding bending, the authors found about the same response for size-similar CNTs and CNSs. To explain this result, the authors reasoned that bending buckling began in their simulations at localized kinks, and that therefore the global topology of the structure did not matter as much in this case. Also, they reported that increasing the number of turns increases the bending rigidity, but decreases the critical buckling strain.

Song et al. (2013) also used molecular mechanics to study the response of CNSs to compressive stresses. Similarly to the previously discussed work, they also found that compressive buckling

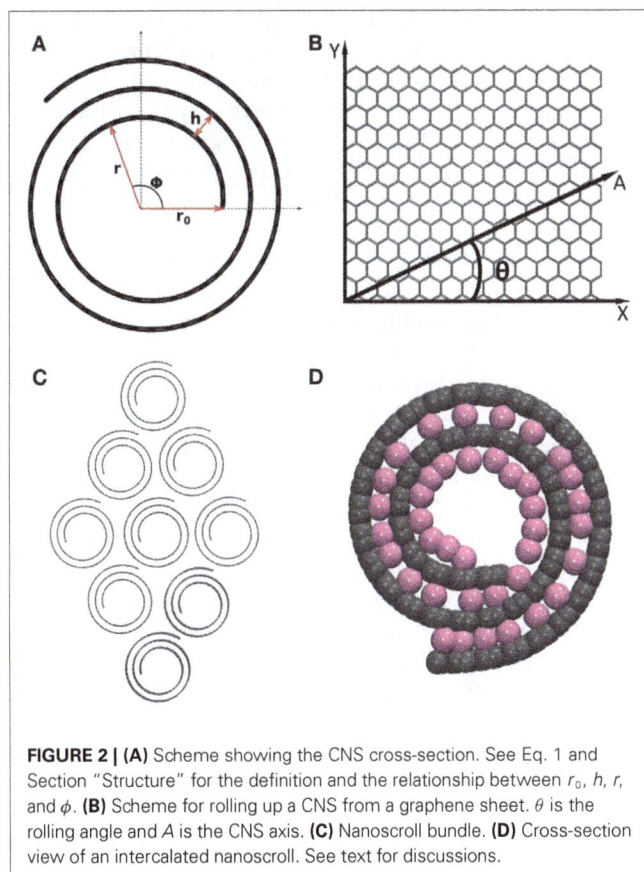

**FIGURE 2 | (A)** Scheme showing the CNS cross-section. See Eq. 1 and Section "Structure" for the definition and the relationship between $r_0$, $h$, $r$, and $\phi$. **(B)** Scheme for rolling up a CNS from a graphene sheet. $\theta$ is the rolling angle and $A$ is the CNS axis. **(C)** Nanoscroll bundle. **(D)** Cross-section view of an intercalated nanoscroll. See text for discussions.

started at the free ends. By adding nickel nanoparticles to the ends, they managed to stabilize the dangling bonds, preventing the wrinkling of the edges. This resulted in a slight increase in the elastic modulus (from 950–970 to 1000–1025 GPa) and in a decent increase in the compression strength of the nanoscrolls (from 40–47 to 45–51 GPa). The reason the values above are presented in a range is that the modulus and strength also were found to depend slightly on core radius and on chirality. Also, note that the compression strength corresponds to the point in the stress–strain curve in which increasing the strain leads to a decrease in the stress. The authors also studied the influence of adding a CNT to the inside of the scroll, and found that it did not significantly influence the critical strain value or the deformation morphology.

A third paper on mechanical properties of CNSs by Zaeri and Ziaei-Rad (2014), studied their response under tensile and torsional stresses. Unlike the previous studies, in this one a finite element based approach was used in the description of the elastic properties. Calculations were performed with and without vdW interactions. Regarding the tensile studies, the authors reported a Young's modulus of about 1100/1040 GPa with/without considering vdW interactions. They also studied the influence of changing chirality, core radius, number of turns, and scroll length on the Young's modulus, but found only a small dependence for each case. Regarding the application of the torsional stresses without explicitly taking into account vdW interactions, the authors found no shear modulus dependence on chirality and a moderate decrease of its value as the core radius size increased (from 48 to 36 GPa). More importantly, the shear modulus greatly increased as the number of layers increased (from 20 to 100 GPa) and greatly decreased as the length increased (from 95 to 10 GPa). For the first effect, the authors first argued that the inner and outer layers could not resist torsion well due to the open edges, and then explained that the shear modulus increased as the number of torsion resisting intermediate layers increased. To explain the second observation, the authors suggested that longer inner and outer edges increased the weakening effect, though no explanation was given as to why this should happen. Regarding the influence of vdW interactions, it was found that they increased the shear modulus 10 times, from about 50 to 500 GPa.

The vibrational properties of CNSs were studied by Shi et al. (2009). Using theoretical modeling and molecular dynamics simulations, they showed that the "breathing" (radial) oscillations can described by the interaction energy between layers ($\gamma$), the bending stiffness of the composing material ($D$), the interlayer separation ($h$), the length of the composing sheet ($B$), the density of the material ($r_o$), the internal radius ($r_0$), and the difference between the inner ($p_i$) and outer pressure ($p_e$). The vibrational frequency is described by the following equation:

$$\omega_0 = \frac{\pi D}{4 r_o B^3 h^2} \left( \frac{\alpha^3 (\alpha+3)}{(\alpha+1)^2} - 2 \frac{\gamma}{D} \sqrt{\frac{Bh^3}{\pi}} \frac{\alpha^2 \sqrt{\alpha}}{(\alpha+1)^{3/2}} - 2 \frac{Bh^2 p}{\pi D} \alpha \right),$$

(3)

where $\alpha = \frac{Bh}{\pi r_0^2}$. For a CNS of 10 nm length the aforementioned equation leads to a frequency of almost 60 GHz.

Much has yet to be done regarding the mechanical properties of CNSs. For instance, the ultimate tensile strength and strain and the

fracture pattern of scrolls has yet to be reported. Moreover, both studies regarding the application of compressive stress remark that the results might depend on the scroll length – it is possible that the CNS might bend under compression if they are long enough. One last example is that it remains to be tested whether elements other than nickel could improve the mechanical properties of CNSs.

## APPLICATIONS

Xie et al. (2009) have built a CNS based electronic device in which a nanoscroll was placed between two metallic contacts over a $SiO_2/Si$ substrate. The advantage of using a CNS instead of a CNT lies in the ability of nanoscrolls to carry current through all of its layers, while MWCNTs only carry current through the outermost layer, since the inner ones do not make direct contact. It was shown that a CNS was able to withstand a current density up to $5.10^7$ A/cm$^2$, indicating its suitability for circuit interconnects. A detailed theoretical study on the quantum electron transport in CNSs was carried by Li et al. (2012), showing a strong dependence of the conductance on the nanoscroll radius as well as on the temperature.

Another possible application of CNSs is as electroactuators (structures which present mechanical response to adding/removing charges) (Rurali et al., 2006). The authors carried their study by adding/removing charges (up to ±0.055 $e$/atom) and then performing geometrical optimization using density functional theory (DFT) methods. In the axial direction, the reported length variation for CNSs (~0.4%) was comparable to those reported for CNTs (Verissimo-Alves et al., 2003). In the radial direction, however, the authors reported diameter variations of up to 2.5%, a result that is an order of magnitude larger than what has been reported for CNTs. These results can be understood by considering the open geometry of CNSs, which enable large increases of interlayer distances without breaking bonds. Although for low values of charge injection the calculated response depended on whether electrons or holes had been injected, for large enough values the mechanical response was found to be dominated by the electrostatic layer repulsions. The calculations also showed that the extra charge accumulated at the edges and at the central part of the scroll. Since the authors considered little more than one turn of scroll, only charges at the edge contributed to the interlayer expansion, suggesting that even larger expansion could be possible for CNSs with more layers, in which the central charges would also play a role. This large radial actuation has been proposed as a method to control the flow rate in nanoscopic water channels, nanofilters, and ion channels (Shi et al., 2010b). Other possible method to produce mechanical response from scrolls is by applying electric fields. Shi et al. (2010a) reported that the application of an electric field causes a decrease in the interaction between scroll layers, which in turn could be used to controllable roll/unroll CNSs.

Another application that has received considerable attention is the use of CNSs in gas storage, particularly hydrogen. Coluci et al. (2007) were the first to investigate the use of CNSs as a medium to store hydrogen, using classical grand-canonical Monte Carlo simulations. The authors reported that the interlayer galleries of the scrolls were only available for $H_2$ storage for interlayer spacings larger than 4.4 Å. For instance, at 150 K, the gravimetric storage of CNSs was predicted to increase from 0.9% to about 2.8% for crystal packed scrolls and from 1.5 to 5.5% for scroll bundles (see **Figure 2C**) when the interlayer spacing increased from 3.4 to 6.4 Å. The authors used a fixed pressure of 1 MPa in all their calculations. One possible way to experimentally realize this increase in interlayer spacing is by intercalating scroll layers with alkali atoms (**Figure 2D**), and Mpourmpakis et al. (2007) reported that this method indeed works well. Coluci et al. (2007) also reported that the gravimetric storage decreased greatly by increasing the temperature to 300 K, and Braga et al. (2007) used molecular dynamics simulations to show that it is possible to cyclically absorb and reabsorb hydrogen from scrolls by decreasing and then increasing the temperature. Huang and Li (2013) also studied possible mechanisms for delivering hydrogen from the scrolls, and reported that other effective methods include twisting the scrolls and decreasing their interlayer distance. Finally, note that calculations performed by Peng et al. (2010) reported that CNSs with expanded interlayer distances could also be used to store methane or trap carbon dioxide.

It should be noted that recently the use of CNSs in a variety of experimental devices has been gaining momentum. For instance, scrolls have already been used as supercapacitors (Zeng et al., 2012; Yan et al., 2013), in batteries (Tojo et al., 2013; Yan et al., 2013), in catalysis (Zhao et al., 2014), and in sensors (Li et al., 2013a). When compared to similar planar graphene based devices, the ones based on CNS were found to present superior performance. The difference was attributed to the open ended CNS topology.

## NANOSCROLLS FROM SOME OTHER MATERIALS

The successful isolation of single layer graphene (Novoselov et al., 2004) has created a revolution in carbon-based materials. In part because of this, there is renewed interest in other two-dimensional materials, which has led to some very significant synthesis advances. Hexagonal boron nitride (hBN) (Jin et al., 2009; Meyer et al., 2009), graphene-like carbon nitride (Li et al., 2007), graphyne (Kehoe et al., 2000; Haley, 2008), silicene (Vogt et al., 2012) and, very recently, germanene (Dávila et al., 2014) monolayers have been already produced. Each one of these structures present their own unique properties, therefore it is a natural question whether it is possible to use them to form novel nanoscrolls.

Hexagonal boron nitride nanoscrolls (BNNSs) were theoretically predicted some years ago (Perim and Galvao, 2009). These structures were predicted to be even more stable than their carbon counterparts, due to stronger interlayer vdW interactions, and should present analogous morphology. Thus, BNNSs should present many of the CNSs properties, like easily tunable diameter and very large solvent accessible surface area, as well as, new ones due to the electric insulating and chemically inert hBN nature. Quite recently, these predictions were confirmed as three independent groups reported the successful BNNS synthesis (Chen et al., 2013; Li et al., 2013b; Hwang da and Suh, 2014). The first reported method consists in exposing hBN crystals to an intense solvent flow inside a spinning disk processor. The shear forces are believed to exfoliate hBN layers, which are then scrolled, forming BNNSs (Chen et al., 2013). However, the yield of this method is considerably low (~5%). A more simple method has been proposed, in which molten hydroxides are used to exfoliate hBN

crystals (Li et al., 2013b). A NaOH/KOH mixture was added to hBN and then thoroughly ground, subsequently being heated at 180°C. The analysis of the product revealed presence of BNNSs, however, at even lower yields than the previous method. BNNSs have also been produced by the interaction between exfoliated hBN and lithocholic acid (Hwang da and Suh, 2014), in a self-assembly process. The considerably lower yields of these methods when compared to the ones utilized for producing CNSs indicate the higher difficulty of exfoliating hBN crystals due to stronger interlayer interactions.

Carbon nitride nanoscrolls (CNNSs) have similarly been predicted to be stable (Perim and Galvao, 2014). Three different graphene-like carbon nitride structures have been successfully synthesized (Li et al., 2007) with varying pore sizes and simulations predict that all three of them should be able to form stable nanoscroll structures. The existence of pores in these structures reduces the contact area between overlapping layers, leading to weaker vdW interactions, and thus less stable nanoscrolls. On the other hand, it also means lower mass density and also easier intercalations, which means these scrolls could be even more suited to hydrogen storage and similar applications. Up to now, no CNNS synthesis has been achieved.

It should be stressed that, in principle, under favorable circumstances, any layered material should be able to form scrolls. Therefore, we should expect new forms of nanoscrolls to be reported in the next years, opening the possibility for exciting novel technological applications.

## SUMMARY

In summary, nanoscrolls are very unique nanostructures due to their open ended morphology. Such morphology creates the possibility of many different technological applications. However, synthesis difficulties had precluded these structures from being more widely investigated. Recent advances in synthesis techniques had led to a change in this scenario, as interest in nanoscrolls is re-emerging and practical applications are becoming a reality.

The recent experimental realization of BNNSs and other scroll materials open new perspectives in the study of these nanostructures. Also, there are many other potential candidates for the formation of novel scrolled structures. In light of this, we can expect in the near future not only significant advances in the production and applications of CNSs, but also the emergence of nanoscrolls from different materials with their own unique properties that can be exploited to be the basis of new applications. We hope the present work can stimulate further studies along these lines.

## ACKNOWLEDGMENTS

Douglas Soares Galvao would like to thank and to acknowledge former group members who worked on the scroll projects, Dr. S. Braga, R. Giro, Profs. V. R. Coluci, S. O. Dantas, and S. B. Legoas. The authors would also like to thank Profs. R. H. Baughman, N. M. Pugno, R. Ruralli, and D. Tomanek for many helpful discussions. *Funding*: Work partially funded by Brazilian agencies CNPq, CAPES, and FAPESP. The authors acknowledge the Center for Computational Engineering and Sciences at Unicamp for financial support through the FAPESP/CEPID Grant #2013/08293-7.

# REFERENCES

Amelinckx, S., Bernaerts, D., Zhang, X., Van Tendeloo, G., and Van Landuyt, J. (1995). A structure model and growth mechanism for multishell carbon nanotubes. *Science* 267, 1334–1338. doi:10.1126/science.267.5202.1334

Bacon, R. (1960). Growth, structure, and properties of graphite whiskers. *J. Appl. Phys.* 31, 283–290. doi:10.1063/1.1735559

Braga, S., Coluci, V., Baughman, R., and Galvao, D. (2007). Hydrogen storage in carbon nanoscrolls: an atomistic molecular dynamics study. *Chem. Phys. Lett.* 441, 78–82. doi:10.1016/j.cplett.2007.04.060

Braga, S. F., Coluci, V. R., Legoas, S. B., Giro, R., Galvão, D. S., and Baughman, R. H. (2004). Structure and dynamics of carbon nanoscrolls. *Nano Lett.* 4, 881–884. doi:10.1021/nl0497272

Chen, X., Boulos, R. A., Dobson, J. F., and Raston, C. L. (2013). Shear induced formation of carbon and boron nitride nano-scrolls. *Nanoscale* 5, 498–502. doi:10.1039/c2nr33071g

Chen, Y., Lu, J., and Gao, Z. (2007). Structural and electronic study of nanoscrolls rolled up by a single graphene sheet. *J. Phys. Chem. C* 111, 1625–1630. doi:10.1021/jp066030r

Chu, L., Xue, Q., Zhang, T., and Ling, C. (2011). Fabrication of carbon nanoscrolls from monolayer graphene controlled by p-doped silicon nanowires: a md simulation study. *J. Phys. Chem. C* 115, 15217–15224. doi:10.1021/jp2030768

Coluci, V., Braga, S., Baughman, R., and Galvao, D. (2007). Prediction of the hydrogen storage capacity of carbon nanoscrolls. *Phys. Rev. B* 75, 125404. doi:10.1103/PhysRevB.75.125404

Dávila, M., Xian, L., Cahangirov, S., Rubio, A., and Lay, G. L. (2014). Germanene: a novel two-dimensional germanium allotrope akin to graphene and silicene. *New J. Phys.* 16, 095002.

Dravid, V., Lin, X., Wang, Y., Wang, X., Yee, A., Ketterson, J., et al. (1993). Buckytubes and derivatives: their growth and implications for buckyball formation. *Science* 259, 1601–1604. doi:10.1126/science.259.5101.1601

Haley, M. M. (2008). Synthesis and properties of annulenic subunits of graphyne and graphdiyne nanoarchitectures. *Pure Appl. Chem.* 80, 519–532. doi:10.1351/pac200880030519

Huang, Y., and Li, T. (2013). Molecular mass transportation via carbon nanoscrolls. *J. Appl. Mech.* 80, 040903. doi:10.1115/1.4024167

Hwang da, Y., and Suh, D. H. (2014). Formation of hexagonal boron nitride nanoscrolls induced by inclusion and exclusion of self-assembling molecules in solution process. *Nanoscale* 6, 5686–5690. doi:10.1039/c4nr00897a

Iijima, S. (1991). Helical microtubules of graphitic carbon. *Nature* 354, 56–58. doi:10.1038/354056a0

Jin, C., Lin, F., Suenaga, K., and Iijima, S. (2009). Fabrication of a freestanding boron nitride single layer and its defect assignments. *Phys. Rev. Lett.* 102, 195505. doi:10.1103/PhysRevLett.102.195505

Kehoe, J. M., Kiley, J. H., English, J. J., Johnson, C. A., Petersen, R. C., and Haley, M. M. (2000). Carbon networks based on dehydrobenzoannulenes. 3. Synthesis of graphyne substructures 1. *Org. Lett.* 2, 969–972. doi:10.1021/ol005623w

Krätschmer, W., Lamb, L. D., Fostiropoulos, K., and Huffman, D. R. (1990). Solid C60: a new form of carbon. *Nature* 347, 354–358. doi:10.1038/347354a0

Li, H., Wu, J., Qi, X., He, Q., Liusman, C., Lu, G., et al. (2013a). Graphene oxide scrolls on hydrophobic substrates fabricated by molecular combing and their application in gas sensing. *Small* 9, 382–386. doi:10.1002/smll.201202358

Li, X., Hao, X., Zhao, M., Wu, Y., Yang, J., Tian, Y., et al. (2013b). Exfoliation of hexagonal boron nitride by molten hydroxides. *Adv. Mater. Weinheim* 25, 2200–2204. doi:10.1002/adma.201204031

Li, J., Cao, C., and Zhu, H. (2007). Synthesis and characterization of graphite-like carbon nitride nanobelts and nanotubes. *Nanotechnology* 18, 115605. doi:10.1088/0957-4484/18/11/115605

Li, T., Lin, M., Huang, Y., and Lin, T. (2012). Quantum transport in carbon nanoscrolls. *Phys. Lett. A* 376, 515–520. doi:10.1016/j.physleta.2011.10.049

Mack, J. J., Viculis, L. M., and Kaner, R. B. (2005). Chemical manufacture of nanostructured materials. U.S. Patent No. 6,872,330.

Meyer, J. C., Chuvilin, A., Algara-Siller, G., Biskupek, J., and Kaiser, U. (2009). Selective sputtering and atomic resolution imaging of atomically thin boron nitride membranes. *Nano Lett.* 9, 2683–2689. doi:10.1021/nl9011497

Mpourmpakis, G., Tylianakis, E., and Froudakis, G. E. (2007). Carbon nanoscrolls: a promising material for hydrogen storage. *Nano Lett.* 7, 1893–1897. doi:10.1021/nl070530u

Novoselov, K. S., Geim, A. K., Morozov, S., Jiang, D., Zhang, Y., Dubonos, S., et al. (2004). Electric field effect in atomically thin carbon films. *Science* 306, 666–669. doi:10.1126/science.1102896

Peng, X., Zhou, J., Wang, W., and Cao, D. (2010). Computer simulation for storage of methane and capture of carbon dioxide in carbon nanoscrolls by expansion of interlayer spacing. *Carbon N. Y.* 48, 3760–3768. doi:10.1016/j.carbon.2010.06.038

Perim, E., and Galvao, D. S. (2009). The structure and dynamics of boron nitride nanoscrolls. *Nanotechnology* 20, 335702. doi:10.1088/0957-4484/20/33/335702

Perim, E., and Galvao, D. S. (2014). Novel nanoscroll structures from carbon nitride layers. *Chemphyschem* 15, 2367–2371. doi:10.1002/cphc.201402059

Perim, E., Paupitz, R., and Galvao, D. S. (2013). Controlled route to the fabrication of carbon and boron nitride nanoscrolls: a molecular dynamics investigation. *J. Appl. Phys.* 113, 054306. doi:10.1063/1.4790304

Roy, D., Angeles-Tactay, E., Brown, R., Spencer, S., Fry, T., Dunton, T., et al. (2008). Synthesis and Raman spectroscopic characterisation of carbon nanoscrolls. *Chem. Phys. Lett.* 465, 254–257. doi:10.1016/j.cplett.2008.09.044

Rurali, R., Coluci, V., and Galvao, D. (2006). Prediction of giant electroactuation for papyruslike carbon nanoscroll structures: first-principles calculations. *Phys. Rev. B* 74, 085414. doi:10.1103/PhysRevB.74.085414

Saito, Y., Yoshikawa, T., Inagaki, M., Tomita, M., and Hayashi, T. (1993). Growth and structure of graphitic tubules and polyhedral particles in arc-discharge. *Chem. Phys. Lett.* 204, 277–282. doi:10.1016/0009-2614(93)90009-P

Savoskin, M. V., Mochalin, V. N., Yaroshenko, A. P., Lazareva, N. I., Konstantinova, T. E., Barsukov, I. V., et al. (2007). Carbon nanoscrolls produced from acceptor-type graphite intercalation compounds. *Carbon N. Y.* 45, 2797–2800. doi:10.1016/j.carbon.2007.09.031

Sharifi, T., Gracia-Espino, E., Barzegar, H. R., Jia, X., Nitze, F., Hu, G., et al. (2013). Formation of nitrogen-doped graphene nanoscrolls by adsorption of magnetic γ-Fe2O3 nanoparticles. *Nat. Commun.* 4, 2319. doi:10.1038/ncomms3319

Shi, X., Cheng, Y., Pugno, N. M., and Gao, H. (2010a). A translational nanoactuator based on carbon nanoscrolls on substrates. *Appl. Phys. Lett.* 96, 053115. doi:10.1063/1.3302284

Shi, X., Pugno, N. M., and Gao, H. (2010b). Tunable core size of carbon nanoscrolls. *J. Comput. Theor. Nanosci.* 7, 517–521. doi:10.1166/jctn.2010.1387

Shi, X., Pugno, N. M., Cheng, Y., and Gao, H. (2009). Gigahertz breathing oscillators based on carbon nanoscrolls. *Appl. Phys. Lett.* 95, 163113. doi:10.1063/1.3253423

Shi, X., Pugno, N. M., and Gao, H. (2011). Constitutive behavior of pressurized carbon nanoscrolls. *Int. J. Fract.* 171, 163–168. doi:10.1007/s10704-010-9545-y

Shioyama, H., and Akita, T. (2003). A new route to carbon nanotubes. *Carbon N. Y.* 41, 179–181. doi:10.1016/S0008-6223(02)00278-6

Song, H., Geng, S., An, M., and Zha, X. (2013). Atomic simulation of the formation and mechanical behavior of carbon nanoscrolls. *J. Appl. Phys.* 113, 164305. doi:10.1063/1.4803034

Sorkin, A., and Su, H. (2014). The mechanism of transforming diamond nanowires to carbon nanostructures. *Nanotechnology* 25, 035601. doi:10.1088/0957-4484/25/3/035601

Tojo, T., Fujisawa, K., Muramatsu, H., Hayashi, T., Kim, Y. A., Endo, M., et al. (2013). Controlled interlayer spacing of scrolled reduced graphene nanotubes by thermal annealing. *RSC Adv.* 3, 4161–4166. doi:10.1039/c3ra22976a

Tománek, D. (2002). Mesoscopic origami with graphite: scrolls, nanotubes, peapods. *Physica B Condens. Matter* 323, 86–89. doi:10.1016/S0921-4526(02)00989-4

Ugarte, D. (1992). Morphology and structure of graphitic soot particles generated in arc-discharge C60 production. *Chem. Phys. Lett.* 198, 596–602. doi:10.1016/0009-2614(92)85035-9

Verissimo-Alves, M., Koiller, B., Chacham, H., and Capaz, R. (2003). Electromechanical effects in carbon nanotubes: Ab initio and analytical tight-binding calculations. *Phys. Rev. B* 67, 161401. doi:10.1103/PhysRevB.67.161401

Viculis, L. M., Mack, J. J., and Kaner, R. B. (2003). A chemical route to carbon nanoscrolls. *Science* 299, 1361–1361. doi:10.1126/science.1078842

Vogt, P., De Padova, P., Quaresima, C., Avila, J., Frantzeskakis, E., Asensio, M. C., et al. (2012). Silicene: compelling experimental evidence for graphenelike two-dimensional silicon. *Phys. Rev. Lett.* 108, 155501. doi:10.1103/PhysRevLett.108.155501

Xia, D., Xue, Q., Xie, J., Chen, H., Lv, C., Besenbacher, F., et al. (2010). Fabrication of carbon nanoscrolls from monolayer graphene. *Small* 6, 2010–2019. doi:10.1002/smll.201000646

Xie, X., Ju, L., Feng, X., Sun, Y., Zhou, R., Liu, K., et al. (2009). Controlled fabrication of high-quality carbon nanoscrolls from monolayer graphene. *Nano Lett.* 9, 2565–2570. doi:10.1021/nl900677y

Yan, M., Wang, F., Han, C., Ma, X., Xu, X., An, Q., et al. (2013). Nanowire templated semihollow bicontinuous graphene scrolls: designed construction, mechanism, and enhanced energy storage performance. *J. Am. Chem. Soc.* 135, 18176–18182. doi:10.1021/ja409027s

Zaeri, M. M., and Ziaei-Rad, S. (2014). Elastic properties of carbon nanoscrolls. *RSC Adv.* 4, 22995–23001. doi:10.1039/c4ra01931h

Zeng, F., Kuang, Y., Liu, G., Liu, R., Huang, Z., Fu, C., et al. (2012). Supercapacitors based on high-quality graphene scrolls. *Nanoscale* 4, 3997–4001. doi:10.1039/c2nr30779k

Zhang, X., Zhang, X., Van Tendeloo, G., Amelinckx, S., Op de Beeck, M., and Van Landuyt, J. (1993). Carbon nano-tubes; their formation process and observation by electron microscopy. *J. Cryst. Growth* 130, 368–382. doi:10.1016/0022-0248(93)90522-X

Zhang, Z., Huang, Y., and Li, T. (2012). Buckling instability of carbon nanoscrolls. *J. Appl. Phys.* 112, 063515. doi:10.1063/1.4754312

Zhang, Z., and Li, T. (2010). Carbon nanotube initiated formation of carbon nanoscrolls. *Appl. Phys. Lett.* 97, 081909. doi:10.1063/1.3479050

Zhao, J., Yang, B., Zheng, Z., Yang, J., Yang, Z., Zhang, P., et al. (2014). Facile preparation of one-dimensional wrapping structure: graphene nanoscroll-wrapped of $Fe_3O_4$ nanoparticles and its application for lithium ion battery. *ACS Appl. Mater Interfaces* doi:10.1021/am502574j

Zheng, J., Liu, H., Wu, B., Guo, Y., Wu, T., Yu, G., et al. (2011). Production of high-quality carbon nanoscrolls with microwave spark assistance in liquid nitrogen. *Adv. Mater. Weinheim* 23, 2460–2463. doi:10.1002/adma.201004759

**Conflict of Interest Statement:** The authors declare that the research was conducted in the absence of any commercial or financial relationships that could be construed as a potential conflict of interest.

**4**

# Transition wave in the collapse of the San Saba Bridge

*Michele Brun[1,2], Gian Felice Giaccu[3], Alexander B. Movchan[2]\* and Leonid I. Slepyan[4,5]*

[1] Dipartimento di Ingegneria Meccanica, Chimica e dei Materiali, Università di Cagliari, Cagliari, Italy
[2] Department of Mathematical Sciences, University of Liverpool, Liverpool, UK
[3] Dipartimento di Architettura, Design e Urbanistica, Facoltà di Architettura, Università di Sassari, Alghero, Italy
[4] School of Mechanical Engineering, Tel Aviv University, Tel Aviv, Israel
[5] Department of Mathematics and Physics, Aberystwyth University, Aberystwyth, UK

**Edited by:**
*Davide Bigoni, University of Trento, Italy*

**Reviewed by:**
*Francesco Dal Corso, University of Trento, Italy*
*Andrea Piccolroaz, University of Trento, Italy*

**\*Correspondence:**
*Alexander B. Movchan, Department of Mathematical Sciences, University of Liverpool, Peach Street, L69 7ZL Liverpool, UK*
*e-mail: abm@liverpool.ac.uk*

A domino wave is a well-known illustration of a transition wave, which appears to reach a stable regime of propagation. Nature also provides spectacular cases of gravity-driven transition waves at large scale observed in snow avalanches and landslides. On a different scale, the micro-structure level interaction between different constituents of the macro-system may influence critical regimes leading to instabilities in avalanche-like flow systems. Most transition waves observed in systems, such as bulletproof vests, racing helmets under impact, shock-wave-driven fracture in solids, are transient. For some structured waveguides, a transition wave may stabilize to achieve a steady regime. Here, we show that the failure of a long bridge is also driven by a transition wave that may allow for steady-state regimes. The recent observation of a failure of the San Saba Bridge in Texas provides experimental evidence supporting an elegant theory based on the notion of transition failure wave. No one would think of an analogy between a snow avalanche and a collapsing bridge. Despite an apparent controversy of such a comparison, both these phenomena can be described in the framework of a model of the dynamic gravity driven transition fault.

Keywords: transition waves, dynamics, structural mechanics, Wiener–Hopf functional equation, failure analysis

## INTRODUCTION

The long San Saba railway bridge (**Figure 1C**) in the Central Texas (also known as *Harmony Ridge Bridge*, 31°14′07″ North, 98°33′52″ West) collapsed in May 2013 as a result of initial damage caused by fire. A 300-yard bridge fell apart after catching fire in a dramatic collapse captured on video (https://www.youtube.com/watch?v= LLVKb1HxhAY). This dramatic event was the subject of attention worldwide when it was featured on BBC News and other News programs across the globe. The video footage provides the data for measuring the speed of propagation of the failure, and it is apparent that this failure reaches a steady-state regime.

Although the phenomenon of collapse of a long bridge is extraordinarily complicated, we show that it can be analyzed in the framework of an analytical model, which refers to gravity driven transition waves. Furthermore, an explicit simple formula has been derived for the speed $\overline{V}$ of the steady-state propagating fault:

$$\overline{V} = \left( \frac{D\kappa}{\rho^2} \right)^{\frac{1}{4}} = 24.3 \, \text{m/sec} \tag{1}$$

where $D$ is the flexural rigidity, $\rho$ is the linear mass density of the bridge, and $\kappa$ is the stiffness of the supporting pillars.

Several examples of transition faults, included in **Figure 1**, incorporate an avalanche flow (**Figure 1A**), gravity-friction-driven domino effect (**Figure 1B**), the Tay Bridge in Scotland (**Figure 1D**), and San Saba burning bridge in Texas (**Figure 1C**). The failures of long bridges were recorded on a number of occasions in the last 200 years. Perhaps, one of the most dramatic events was the Tay

Bridge Disaster of 1879, featured in the poetry by McGonnagall, as caused by an impact of a derailed train:

> *"But when the train came near to Wormit Bay,*
> *Boreas he did loud and angry bray,*
> *And shook the central girders of the Bridge of Tay*
> *On the last Sabbath day of 1879,*
> *Which will be remember'd for a very long time."*

The substantial damage shown in **Figure 1D** includes multiple sections destroyed, apparently as a result of a failure wave propagating along the bridge. Due to a lack of recording technology, no video footage was available at the time, and no experimental observations were made.

Similar to a slab snow avalanche (Bartelt et al., 2006; Heierli et al., 2008), where the fault (moving snow powder versus solid base) reaches a steady regime, the collapse of the Harmony Ridge Bridge also fits into the framework of transition waves, and moreover, an elegant analytical model enables one to predict the steady regime of propagation of this fault. Although the physical background in these two cases is different, we note that the balance of energy is required, which includes potential, kinetic, and internal energy. The dissipation mechanism in the avalanche flow is explained via heat transfer and friction (Bartelt et al., 2006), whereas in the collapsing bridge the energy is taken away from failure region by radiating waves (Brun et al., 2013a). Also, elegant crack (or anti-crack) propagation models are applicable to both cases (Slepyan, 1981, 2002, 2010; Heierli et al., 2008). A simple gravity-friction-driven failure wave is also known for a domino

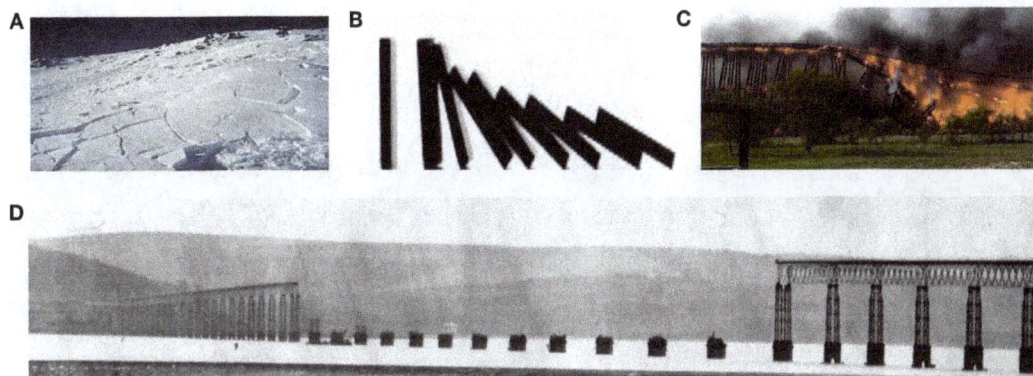

**FIGURE 1 | Examples of gravitationally driven failure waves occurring in nature and in structural systems are shown.** **(A)** A slab snow avalanche. **(B)** A falling domino. **(C)** The failure of the San Saba Bridge (Texas, May 2013). **(D)** The Tay Bridge disaster. Dundee, Scotland, 1879 (from http://en.wikipedia.org/wiki/The_Tay_Bridge_Disaster).

row (**Figure 1B**), and the steady-state regime is independent on the initial conditions (Maddox, 1987; Stronge, 1987).

The theoretical background developed by Brun et al. (Brun et al., 2012, 2013b,c) refers to long bridge structures as waveguides rather than finite size elastic bodies. That approach enables one to bring the notion of so-called Floquet–Bloch waves from Physics (Brillouin, 1953; Kittel, 1996) in the areas such as Metamaterials pioneered by Veselago (1968) and Pendry (2000). Important related areas in Applied Mathematics involve averaging and high-frequency homogenization (Movchan and Slepyan, 2007; Craster et al., 2010). Dispersion of waves and the pass band structure are of paramount importance in understanding of fundamental mechanism of vibration of long bridges. For a model example, a recent analytical work (Brun et al., 2013a) presented the analysis of a class of functional equations of the Wiener–Hopf type that describes transition waves in a periodic flexural system. The notion of configurational forces enables us to develop the model further to take into account the non-linear features of the physical problem. The new mathematical approach has delivered an accurate estimate of the failure wave speed in the collapse of the Harmony Ridge Bridge as compared with the rare footage, which was taken during propagation of the fault.

In this paper, we show an unusual, and unexpected to a certain degree, phenomenon of a transition wave in a long collapsing bridge. We also demonstrate the link with a certain class of solutions, known as Bloch waves, in infinite periodic systems, which provide the best description of the influence of individual constituents and their interaction within the macro-system of the bridge. Transition waves and failure phenomena are also considered in relation to damage and impact of structured solids, such as honeycombs and bistable lattices. The approach to failure as a transition wave was advocated through lattice models (Slepyan and Troyankina, 1984; Fineberg and Marder, 1999; Balk et al., 2001; Cherkaev et al., 2005; Slepyan et al., 2005) including the advanced molecular dynamics simulation (Abraham and Gao, 2000).

The importance of such problems is also apparent to elucidate ways to prevent such destructions in earthquake protection systems.

## MATERIALS AND METHODS
### UNWANTED VIBRATIONS OF LONG BRIDGES
An unusual example of a failure wave occurred in the unfortunate collapse of the San Saba railway bridge. One would assume that the structure was optimally designed and capable of withstanding both quasi-static and dynamic loads. Nevertheless, a 300-yard bridge fell apart after catching fire in a dramatic collapse.

Even advanced engineering analysis and optimal design were not sufficient to prevent a collapse of this relatively modern system. This was a hard and dramatic lesson to learn and a mathematical model offered here shows an unexpected link to a notion of transition waves, which would not be commonly used by structural engineers and architects. An accurate finite element model (FEM) of San Saba Bridge has been developed for the eigenvalue analysis presented here. The bridge in its actual dimensions is displayed in **Figure 2A**. The standard procedures of engineering analysis would require identification of low-frequency resonance vibrations. As in **Figure 2**, the "dangerous" vibrations would normally be associated with horizontal motion of the main deck of the bridge. One would not expect the vertical vibrations of the main deck to be of any concern, as the frequencies involved are relatively high. The detailed discussion of this data is given below, and the surprising outcome is that the vertical vibrations play a significant role in formation of the transition wave.

### ANALYSIS OF THE SAN SABA BRIDGE FAILURE
In elastic waveguides, the notion of Floquet–Bloch waves is commonly used (Mead, 1970; Graff, 1991; Brun et al., 2013b) to describe the rate of transmission of energy and to visualize the vibrating structure. Such waves are also proved to be essential in understanding the failure of systems with embedded structural elements like lattices (Slepyan and Troyankina, 1984; Marder and Gross, 1995; Fineberg and Marder, 1999; Slepyan, 2010) or supporting pillars (Brun et al., 2013a) as common for long bridges. Namely, the dispersion relations for Floquet–Bloch waves are embedded into the structure of the Wiener–Hopf equation that describes propagation of a failure wave, which may occur in the form of fracture or a transition wave. In the particular case of

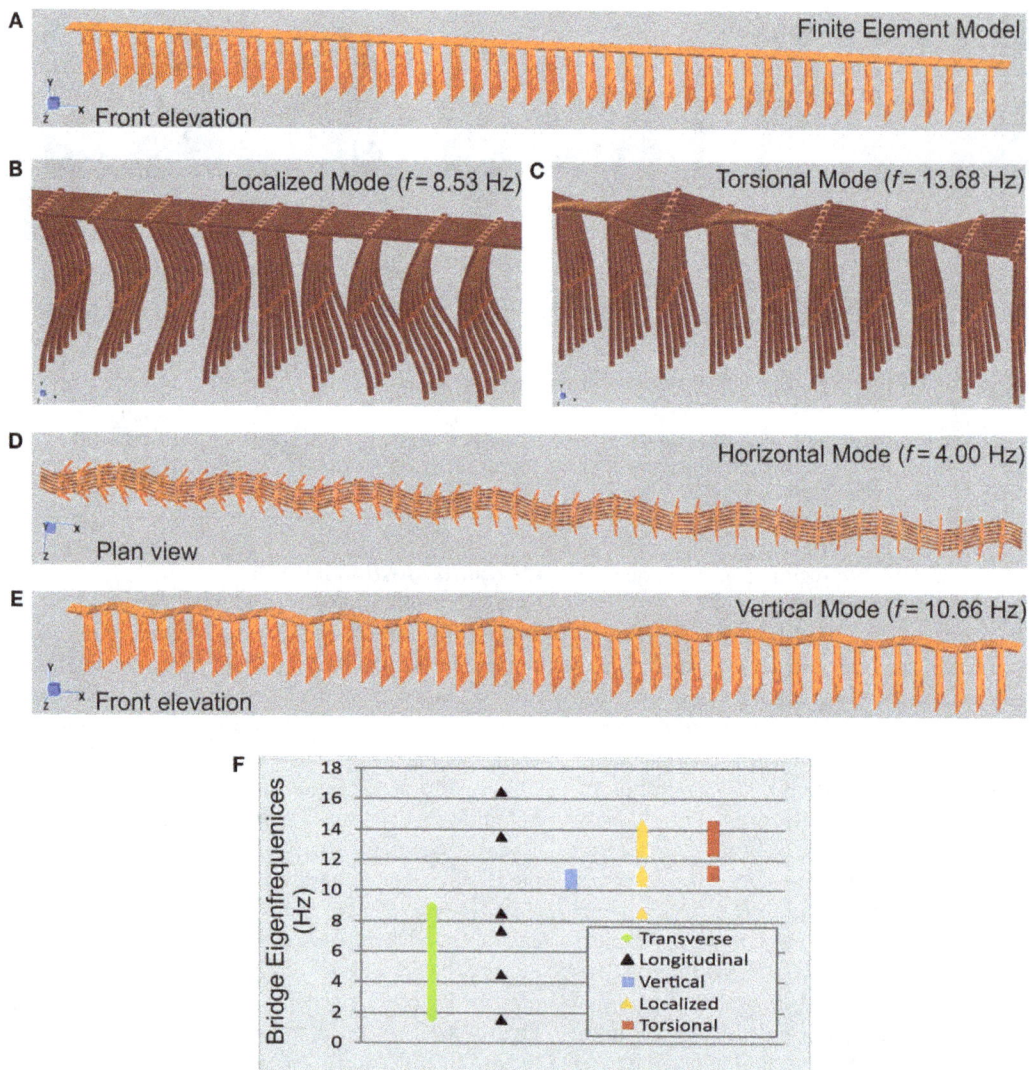

**FIGURE 2 | Analyses of the first 100 eigenmodes of the San Saba Bridge.**
**(A)** Finite element model of the "millipede-like" bridge structure implemented in Strand7. **(B)** Localized mode at frequency $f = 8.53$ Hz. The upper deck of the bridge acts as a waveguide, while relative large amplitude vibrations are localized in the pillars. **(C)** Torsional mode, $f = 13.68$ Hz.

**(D)** Horizontal flexural mode involving transverse displacement of the upper deck; $f = 4.00$ Hz. **(E)** Vertical flexural mode involved in the bridge failure; $f = 10.66$ Hz. **(F)**, Eigenfrequencies of the five types of vibration modes: transverse flexural and longitudinal horizontal modes, vertical, localized in the supporting pillars and torsional.

the San Saba Bridge, when we refer to two different phases, these are the intact bridges in front of the moving failure region and the bridge devoid of its support behind the failure front. The transition can excite elastic waves propagating in both directions. In this case, the bridge can be compared to a flexural elastic beam rested on an elastic foundation (continuous or discrete as in **Figures 3D,E**).

In the framework of fracture mechanics, we consider the failure wave as a "negative exfoliation" of the bridge from its support, where the bridge is represented as a heavy string on a rigid foundation (**Figure 5**). In this case, not only the critical force but also the energy release rate can play the role of the transition criterion.

The observation of the failure of the San Saba Bridge suggests that two main stages can be identified: (1) the transient accelerating propagation of the failure (**Figure 3A**) and (2) the steady-state regime (**Figure 3B**), where the speed of the failure

wave is approximately constant. The velocity of propagation of the failure wave in **Figure 3C** shows that the steady-state regime is reached after the failure of the first 31 pillars.

Whereas the first transient stage (shown in **Figure 3A**) is a highly complex non-linear phenomenon, the steady-state regime can be analyzed in relatively simple analytical terms. It also appears that the structure would be substantially damaged by the time the steady regime is reached and any attempt to stop the propagation of the failure wave should be made at the initial transient stage.

Theoretically, we consider the steady-state regime. The propagation of a failure wave is modeled for an infinite beam supported by a (non-homogeneous) piecewise continuous elastic foundation (**Figure 3D**). As a result of the damage, the stiffness of the left and right parts of the foundation is different. The prediction of this simple continuous model has a good applicability to the

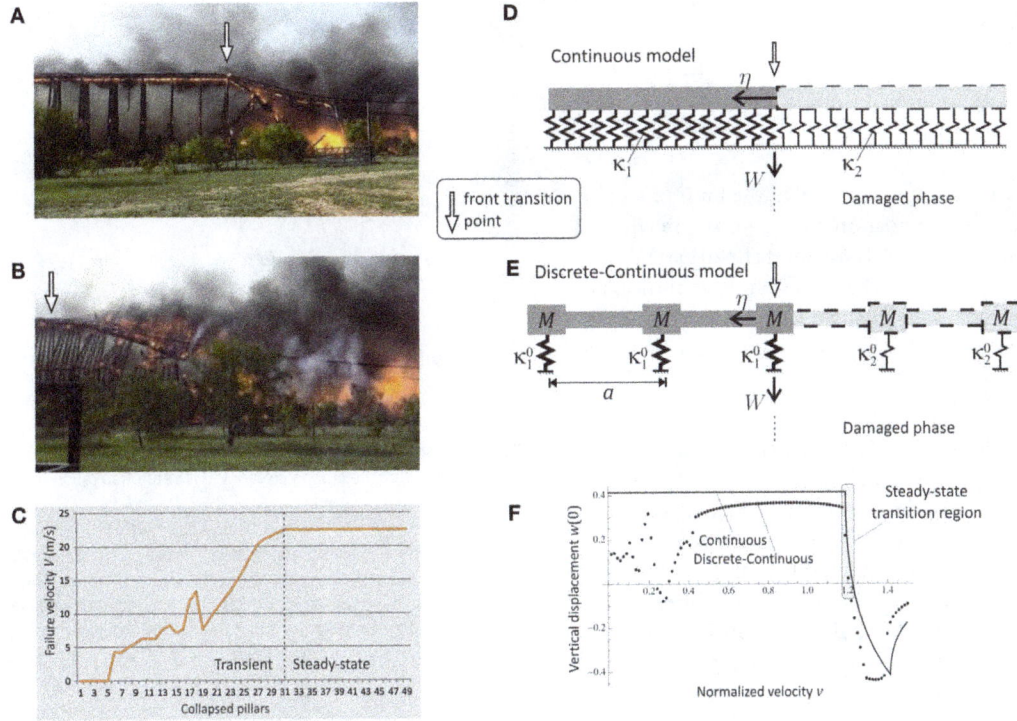

**FIGURE 3 | The collapse of San Saba Bridge**. Snapshots of the bridge failure: **(A)** The initial transient regime, **(B)** the consequent steady-state regime. **(C)** Failure velocity $V$ shown as a function of the front failure position (number of collapsed pillars). Steady-state propagation is reached when the pillar no. 31 collapses. **(D,E)** Structural models adopted for the analysis of the steady-state propagation of the failure wave [linear density $\rho$ (kg/m), bending stiffness $D$ (N·m²)]. **(D)** Continuous beam on continuous elastic foundation [stiffness per unit length $\kappa_{1,2}$ (N/m²), before and after failure, respectively]. **(E)** Discrete–continuous model on discrete elastic foundation [participating mass $M$ (kg), and stiffness $\kappa_{1,2}^0$ (N/m)]. **(F)** Normalized vertical displacement $w$ at the transition point $\eta = 0$ as a function of the normalized velocity $v$ for the continuous **(D)** and discrete–continuous **(E)** structural models (see the Section "Three Dynamic Regimes of Interfacial Waves" in Supplementary Material for the normalization). Steady-state propagation of the failure is possible only in the intersonic velocity interval $\sqrt[4]{4\kappa_2/\kappa_1} < V/(\xi/\tau) < \sqrt{(1 + \kappa_2/\kappa_1)}$.

cases when the bridge structure is of a semi-discrete nature and the supporting elements are distributed periodically (**Figure 3E**). The speed of a steady-state propagation of the interphase damage appears to be well identified by the continuous model as shown in **Figure 3F**. On the other hand, transient stage requires more insight, which is gained via the semi-discrete model involving point masses at the junction points as well as distributed inertia along the flexural beam.

## RESULTS

### EVALUATION OF THE SPEED OF THE FAILURE WAVE

The solution leads to the expression of the vertical flexural displacement $W(0)$ (see **Figures 3D,E**) measured in the moving system of coordinates centered at the front of the transition wave:

where $\kappa_1$ and $\kappa_2 < \kappa_1$ are the stiffness of the supporting pillars on the left (before collapse) and on the right (collapsed) parts with respect to the moving front transition point $\eta = 0$, $g$ is the gravitational force, and $\rho$ the linear density of the beam structure (see **Figure 3D**).

The generic approach applies to a wide range of parameters of elastic systems and hence the results are linked to characteristic length $\xi = (D/\kappa_1)^{1/4}$ and time $\tau = \sqrt{\rho/\kappa_1}$ with $D$, the bending stiffness of the beam (see **Figures 3D,E**). Three velocity regimes are present: subsonic $\left(V \leq \sqrt[4]{\frac{4\kappa_2}{\kappa_1}}\left(\frac{\xi}{\tau}\right)\right)$, intersonic $\left(\sqrt[4]{4\frac{\kappa_2}{\kappa_1}}\left(\frac{\xi}{\tau}\right) \leq V \leq \sqrt{2}\left(\frac{\xi}{\tau}\right)\right)$, and supersonic $\left(V \geq \sqrt{2}\left(\frac{\xi}{\tau}\right)\right)$. In particular, the important regime is identified for the intersonic propagation as the steady-state corresponding to $w(0) \geq 0$, which

$$
W(0) = \begin{cases}
\left(\sqrt{\frac{\kappa_1}{\kappa_2}} - 1\right)\frac{\rho g}{\kappa_1} & \text{for} \quad V \leq \sqrt[4]{4\frac{\kappa_2}{\kappa_1}}\left(\frac{\xi}{\tau}\right), \\[3mm]
\frac{V^2\tau^2 - 2(\kappa_2/\kappa_1)\xi^2 - \sqrt{V^4\tau^4 - 4(\kappa_2/\kappa_1)\xi^4}}{2(\kappa_2/\kappa_1)\xi^2}\frac{\rho g}{\kappa_1} & \text{for} \quad \sqrt[4]{4\frac{\kappa_2}{\kappa_1}}\left(\frac{\xi}{\tau}\right) \leq V \leq \sqrt{2}\left(\frac{\xi}{\tau}\right), \\[3mm]
-\frac{V^2\tau^2 - \sqrt{V^4\tau^4 - 4(\kappa_2/\kappa_1)\xi^4}}{\sqrt{V^4\tau^4 - 4\xi^4} + \sqrt{V^4\tau^4 - 4(\kappa_2/\kappa_1)\xi^4}}\frac{\kappa_1 - \kappa_2}{\kappa_2}\frac{\rho g}{\kappa_1} & \text{for} \quad V \geq \sqrt{2}\left(\frac{\xi}{\tau}\right),
\end{cases}
\tag{2}
$$

is a monotonically decreasing function of the crack speed. In particular, when $w(0) = 0$ we have

$$V = \overline{V} = \left(\frac{\xi}{\tau}\right)\sqrt{1 + \frac{\kappa_2}{\kappa_1}}, \qquad (3)$$

which becomes $\overline{V} = (D\kappa_1/\rho^2)^{1/4} = \xi/\tau$ in the limit of a complete failure of the damaged foundation (i.e., $\kappa_2 \to 0$) as in **Figure 4** and in Eq. (1). The analytical model of Brun et al. (2013a) predicts the speed of steady propagation to be in the left neighborhood of the upper limit $\overline{V}$. In the following, we will show that the predicted speed matches amazingly with the observation recorded during the failure of the bridge.

### RESONANCE MODES AND "INSIGNIFICANT" FLEXURAL VIBRATIONS

A direct transient analysis for a failing bridge would involve a large-scale computational model and is not considered to be feasible in the engineering practice. Of course, the choice of initial conditions and evolution of the structure becomes an important and challenging part of the computational procedure. A conventional engineering approach would allow an extensive and detailed analysis of eigenfrequencies and resonant modes. How useful would this information be in the circumstances related to San Saba Bridge?

To answer this question, we have done a complete eigenfrequency analysis in the framework of an FEM based on the industrial grade tool Strand7. The computational model has not been simplified in any way, and every technical detail has been embedded in a full three-dimensional FEM computational domain as shown in **Figure 2A**. The details of computational parameters are supplied in the supplementary material. The computed eigenfrequencies accurately represent the resonant vibrations of the actual San Saba Bridge.

For an undamaged structure, the vertical flexural modes would not attract much significant attention of a Structural Engineer, since they correspond to relatively high-frequency range ($f = 10.4$–$11.1$ Hz) compared to the modes involving a horizontal motion of the upper deck of the bridge ($f = 1.6$–$8.9$ Hz). In **Figure 2**, the first types of eigenmodes are shown: an example of a transverse flexural mode is represented in **Figure 2D**, a mode where vibrations are localized within the supporting beam is shown in **Figure 2B**, a typical torsional mode is reported in **Figure 2C** and, finally, a vertical flexural mode is given in **Figure 2E**. The overall diagram with eigenfrequencies (**Figure 2F**) suggests that the vertical flexural vibrations would be in the highest frequency range among the identified vibrations. The three-dimensional computation has revealed that the low-frequency vibrations of the San Saba Bridge correspond to horizontal modes, transverse (**Figure 2D**) and longitudinal. On the contrary, the vertical flexural vibrations (**Figure 2E**) occur in a narrow band at much higher frequencies and also take into account the effect of the longitudinal stiffness of the supporting pillars. The corresponding resonance frequencies are in the same range of the frequencies associated with localized vibration of the pillars ($f = 8.5$–$10.6$ Hz) and torsional vibrations ($f = 10.9$–$14.4$ Hz), making difficult to distinguish different eigenmodes.

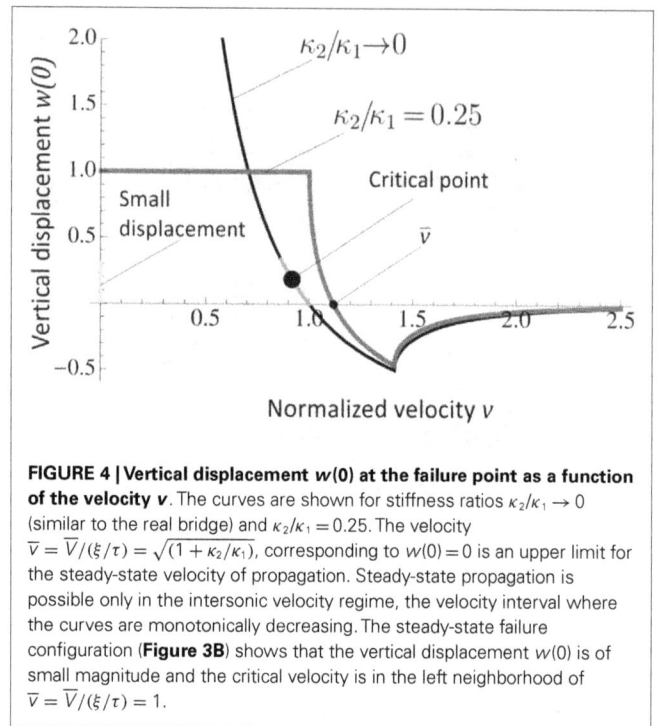

**FIGURE 4 | Vertical displacement $w(0)$ at the failure point as a function of the velocity $v$.** The curves are shown for stiffness ratios $\kappa_2/\kappa_1 \to 0$ (similar to the real bridge) and $\kappa_2/\kappa_1 = 0.25$. The velocity $\overline{v} = \overline{V}/(\xi/\tau) = \sqrt{(1 + \kappa_2/\kappa_1)}$, corresponding to $w(0) = 0$ is an upper limit for the steady-state velocity of propagation. Steady-state propagation is possible only in the intersonic velocity regime, the velocity interval where the curves are monotonically decreasing. The steady-state failure configuration (**Figure 3B**) shows that the vertical displacement $w(0)$ is of small magnitude and the critical velocity is in the left neighborhood of $\overline{v} = \overline{V}/(\xi/\tau) = 1$.

### DISCUSSION
#### PREDICTION OF THE SPEED OF THE TRANSITION WAVE

The first impression gained from the computational model is that the vertical flexural motion is less relevant to the identification of dangerous vibrations within the dynamic design process, and the main attention should be given to the low-frequency transverse modes. As follows from the physical evidence, the vertical flexural mode that is driven by gravitational forces, is the one, which leads to a failure wave. This also suggests that the standard, although advanced, engineering techniques would not lead to the right conclusion in the explanation of the failure wave in the San Saba Bridge. However, the information provided by the finite element computations, combined with the knowledge of flexural Bloch waves, leads to the correct answer, and prediction of the steady regime also includes an accurate estimate of the speed of the failure wave. The comparison between the FEM and the simplified waveguide model for an elastic beam structure, supported by the elastic foundation (**Figures 3D,E**), shows that the parameters of the system are chosen so that the frequencies generated by FEM (**Figure 2F**) match well with the pass band interval identified for Floquet–Bloch waves in the periodic waveguide model (Brillouin, 1953; Brun et al., 2013a).

The movie taken for the wooden section of the San Saba Bridge, together with the measurements, show that the length of the failed section is around 209 m, and the speed of the steady-state propagation approaches $V = 22.4$ m/s (see **Figure 3C**). To compare with the analytical model, we require the evaluation of the internal unit length $\xi = 0.382$ m and unit time $\tau = 0.157 \times 10^{-1}$ s, which has been estimated form the FEM implementation as detailed in the supplementary material (in **Figures 3D,E** the parameters are: $a = 4.26$ m, $D = 0.8 \times 10^6$ Nm$^2$, $\kappa_1 = 37.5$ MPa,

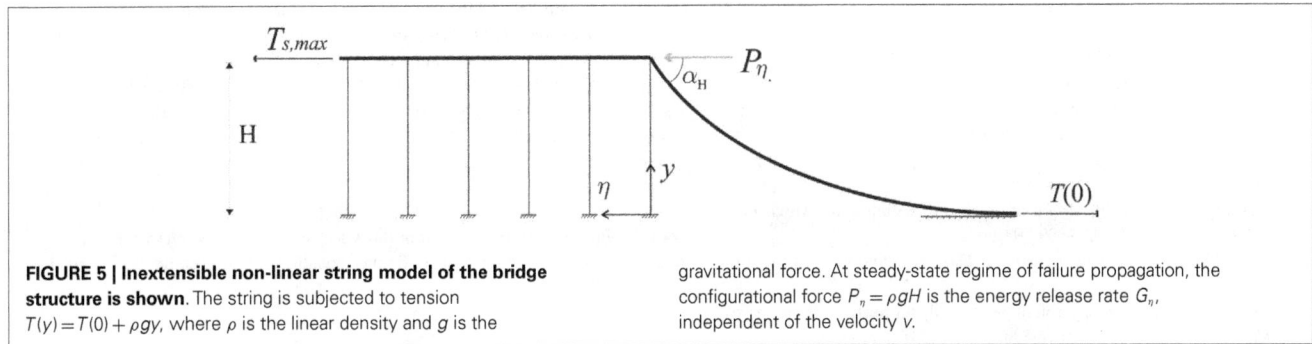

**FIGURE 5 | Inextensible non-linear string model of the bridge structure is shown**. The string is subjected to tension $T(y) = T(0) + \rho g y$, where $\rho$ is the linear density and $g$ is the gravitational force. At steady-state regime of failure propagation, the configurational force $P_\eta = \rho g H$ is the energy release rate $G_\eta$, independent of the velocity $v$.

$\kappa_2 \to 0$, $\rho = 9.23 \times 10^3$ kg/m). The interphase wave speed for the failing bridge appears to be $v = V/(\xi/\tau) = V/\overline{V} = (22.4 \, \text{m/sec})/(24.3 \, \text{m/sec}) = 0.922$, which is exactly within the predicted range, as shown in **Figures 3F** and **4**.

It is noted that an extremely complex phenomenon, which is transient and highly non-linear, has been explained in the framework of the propagating failure wave. On a practical note, it is also worth mentioning that the initial transient stage would allow an intervention by "removing certain sections of the bridge" and hence stopping the propagation of the fault.

### CONFIGURATIONAL FORCE ACCOMPANYING THE BRIDGE FAILURE

We have shown that the failure wave speed is bounded by speeds of the flexural waves, which can propagate in front and behind the transition point.

A simplified model based on the analysis of an inextensible string on a rigid foundation (**Figure 5**) shows that the energy release rate can be considered as the transition criterion similarly to that used in fracture mechanics (Slepyan, 1981, 2002). This approach gives an elegant and explicit approximation of the force acting on failing bridge at the front of the transition wave. Such a force is the Eshelby configurational force (Eshelby, 1951; Maugin, 1993; Bigoni et al., 2014).

The horizontal configurational force (as in **Figure 5**) acting at the transition front is equal to

$$P_\eta = \rho g H. \tag{4}$$

It is independent of the speed of the transition wave, and is simply the potential energy with respect to the ground level. Furthermore, by adding the condition $\rho g H = G_c$, where $G_c$ stands for the critical energy release rate required to fail a single section of the bridge, we observe that if $\rho H$ is maintained below $G_c/g$ the failure will not propagate, and, as expected, the lighter and lower structure appears to be more resistant to failure compared to heavier and taller bridge systems.

Both linear and non-linear regimes of the transition wave of failure are fully covered by the model presented here. The formulae (1) and (3) for the velocity of failure propagation are simple and accurate, but their derivation is highly non-trivial and cannot be achieved by an intuitive *ad hoc* effort. Availability of the experimental data was a unique occurrence for a large-scale failure such as the widely featured in press San Saba Bridge. The theory, which is in full agreement with the observational data, paves the way to the design of highly robust and dynamically resistant structures, which includes long bridges and skyscrapers.

### ACKNOWLEDGMENTS

Alexander B. Movchan, Michele Brun, and Leonid I. Slepyan acknowledge the financial support of the European Community's Seven Framework Programme under contract numbers PIAP-GA-2011-286110-INTERCER2, PIEF-GA-2011-302357-DYNAMETA, and IAPP-2011-284544-PARM-2, respectively. Gian Felice Giaccu and Michele Brun acknowledge the financial support of the Regione Autonoma della Sardegna (LR 7 2010, grant "M4" CRP-27585).

### REFERENCES

Abraham, F. D., and Gao, H. (2000). How fast cracks propagate? *Phys. Rev. Lett.* 84, 3113. doi:10.1103/PhysRevLett.84.3113

Balk, A., Cherkaev, A., and Slepyan, L. (2001). Dynamics of chains with non-monotone stress-strain relations. I. Model and numerical experiments. *J. Mech. Phys. Solids* 49, 131–148. doi:10.1016/S0022-5096(00)00026-0

Bartelt, P., Buser, O., and Platzer, K. (2006). Fluctuation-dissipation relations for granular snow avalanches. *J. Glaciol.* 52, 631–643. doi:10.3189/172756506781828476

Bigoni, D., Bosi, F., Dal Corso, F., and Misseroni, D. (2014). Instability of a penetrating blade. *J. Mech. Phys. Solids* 64, 411–425. doi:10.1016/j.jmps.2013.12.008

Brillouin, L. (1953). *Wave Propagation in Periodic Structures*, 2nd Edn. New York: Dover.

Brun, M., Giaccu, G. F., Movchan, A. B., and Movchan, N. V. (2012). Asymptotics of eigenfrequencies in the dynamic response of elongated multi-structures. *Proc. R. Soc. Lond.* 468, 378–394. doi:10.1098/rspa.2011.0415

Brun, M., Movchan, A. B., and Slepyan, L. I. (2013a). Transition wave in a supported heavy beam. *J. Mech. Phys. Solids* 61, 2067–2085. doi:10.1016/j.jmps.2013.05.004

Brun, M., Movchan, A. B., Jones, I. S., and McPhedran, R. C. (2013b). Bypassing shake, rattle and roll. *Phys. World* 26, 32–36.

Brun, M., Movchan, A. B., and Jones, I. S. (2013c). Phononic band gap systems in structural mechanics: slender elastic structures versus periodic waveguides. *J. Vib. Acoust.* 135, 041013. doi:10.1115/1.4023819

Cherkaev, A., Cherkaev, E., and Slepyan, L. (2005). Transition waves in bistable structures. I. Delocalization of damage. *J. Mech. Phys. Solids* 53, 383–405. doi:10.1016/j.jmps.2004.08.002

Craster, R. V., Kaplunov, J., and Postnova, J. (2010). High-frequency asymptotics, homogenisation and localisation for lattices. *Q. J. Mech. Appl. Math.* 63, 497–519. doi:10.1093/qjmam/hbq015

Eshelby, J. D. (1951). The force on an elastic singularity. *Philos. Trans. R. Soc. Lond. A* 244, 87–112. doi:10.1098/rsta.1951.0016

Fineberg, J., and Marder, M. (1999). Instability in dynamic fracture. *Phys. Rep.* 313, 1–108. doi:10.1016/S0370-1573(98)00085-4

Graff, K. F. (1991). *Wave Motion in Elastic Solids.* New York: Dover.

Heierli, J., Gumbsch, P., and Zaiser, M. (2008). Anticrack nucleation as triggering mechanism for snow slab. *Science* 321, 240–243. doi:10.1126/science.1153948

Kittel, C. (1996). *Introduction to Solid State Physics.* New York: Wiley.

Maddox, R. (1987). The domino effect explained. *Nature* 325, 191. doi:10.1038/325191a0

Marder, M., and Gross, S. (1995). Origin of crack tip instabilities. *J. Mech. Phys. Solids* 43, 1–48. doi:10.1016/0022-5096(94)00060-I

Maugin, G. A. (1993). *Material Inhomogeneities in Elasticity.* London: Chapman and Hall.

Mead, D. J. (1970). Free wave propagation in periodically supported infinite beams. *J. Sound Vib.* 11, 181–197. doi:10.1016/S0022-460X(70)80062-1

Movchan, A. B., and Slepyan, L. I. (2007). Band gap Green's functions and localized oscillations. *Proc. R. Soc. Lond. A* 463, 2709–2727. doi:10.1098/rspa.2007.0007

Pendry, J. B. (2000). Negative refraction makes a perfect lens. *Phys. Rev. Lett.* 85, 3966. doi:10.1103/PhysRevLett.85.3966

Slepyan, L., Cherkaev, A., and Cherkaev, E. (2005). Transition waves in bistable structures. II. Analytical solution: wave speed and energy dissipation. *J. Mech. Phys. Solids* 53, 407–436. doi:10.1016/j.jmps.2004.08.001

Slepyan, L. I. (1981). Dynamics of a crack in a lattice. *Sov. Phys.* 26, 538–540.

Slepyan, L. I. (2002). *Models and Phenomena in Fracture Mechanics.* Berlin: Springer.

Slepyan, L. I. (2010). Wave radiation in lattice fracture. *Acoust. Phys.* 56, 962–971. doi:10.1134/S1063771010060217

Slepyan, L. I., and Troyankina, L. V. (1984). Fracture wave in a chain structure. *J. Appl. Mech. Tech. Phys.* 25, 921–927. doi:10.1007/BF00911671

Stronge, W. J. (1987). The domino effect: a wave of destabilizing collisions in a periodic array. *Proc. R. Soc. Lond.* 409, 199–208. doi:10.1098/rspa.1987.0013

Veselago, G. V. (1968). The electrodynamics of substances with simultaneously negative values of $\varepsilon$ and $\mu$. *Sov. Phys. Uspekhi* 10, 509. doi:10.1070/PU1968v010n04ABEH003699

**Conflict of Interest Statement:** The authors declare that the research was conducted in the absence of any commercial or financial relationships that could be construed as a potential conflict of interest.

# 5

# Electrically conductive polyaniline-coated electrospun poly(vinylidene fluoride) mats

*Claudia Merlini[1]\*, Guilherme Mariz de Oliveira Barra[1]\*, Sílvia Daniela Araújo da Silva Ramôa[1], Giseli Contri[1], Rosemeire dos Santos Almeida[2], Marcos Akira d'Ávila[2] and Bluma G. Soares[3]*

[1] Department of Mechanical Engineering, Universidade Federal de Santa Catarina, Florianópolis, Brazil
[2] School of Mechanical Engineering, Universidade Estadual de Campinas (UNICAMP), Campinas, Brazil
[3] Instituto de Macromoléculas, Universidade Federal do Rio de Janeiro, Rio de Janeiro, Brazil

**Edited by:**
Kyriaki Kalaitzidou, Georgia Institute of Technology, USA

**Reviewed by:**
Peng-Cheng Ma, Chinese Academy of Sciences, China
Chung Hae Park, Ecole Nationale Supérieure des Mines de Douai, France

**\*Correspondence:**
Claudia Merlini and Guilherme Mariz de Oliveira Barra, Department of Mechanical Engineering, Universidade Federal de Santa Catarina, Campus Universitário Trindade, Caixa Postal 476, Florianópolis, Santa Catarina 88040-900, Brazil
e-mail: dra.claudiamerlini@yahoo.com.br; g.barra@ufsc.br

Electrically conductive polyaniline (PANI)-coated electrospun poly(vinylidene fluoride) (PVDF) mats were fabricated through aniline (ANI) oxidative polymerization on electrospun PVDF mats. The effect of polymerization condition on structure and property of PVDF/PANI mats was investigated. The electrical conductivity and PANI content enhanced significantly with increasing ANI concentration due to the formation of a conducting polymer layer that completely coated the PVDF fibers surface. The PANI deposition on the PVDF fibers surface increased the Young modulus and the elongation at break reduced significantly. Attenuated total reflectance-Fourier transform Infrared spectroscopy revealed that the electrospun PVDF and PVDF/PANI mats display a polymorph crystalline structure, with absorption bands associated to the $\beta$, $\alpha$, and $\gamma$ phases.

**Keywords: polyaniline, poly(vinylidene fluoride), electrospinnig, conductive membranes, *in situ* polymerization**

## INTRODUCTION

The development of electrospun mats based on intrinsically conducting polymer (ICP), such as polypyrrole (PPy) and polyaniline (PANI) is interesting due to the possibility of combining the optical, electrical, and magnetic properties of ICP with the three-dimensional fiber network of the electrospun mats (Merlini et al., 2014a). Additionally, the electrospun mats display very large surface area (Merlini et al., 2014b) small pore size and very high porosity (Chronakis et al., 2006), superior mechanical performances (e.g., stiffness and tensile strength), and surface functionalities when compared with any other known form of the material (Huang et al., 2003).

Conducting polymer electrospun mats have been produced by different approaches. The first one is the direct electrospinning, in which fibers are electrospun from a conducting polymer suspension (MacDiarmid et al., 2001; Yu et al., 2008b; Srinivasan et al., 2010) or a mixture comprising insulating polymer and ICP suspension in a common solvent (Norris et al., 2000; Bagheri and Aghakhani, 2012; Lin et al., 2012; Merlini et al., 2014b). It is well know that ICP are insoluble in the most organic solvents, hence it is difficult to obtain an electrospun mat of neat ICP. On the other hand, the electrospinning of an insulating polymer and ICP suspension is an easy alternative to produce conducting electrospun mats with good mechanical properties, but lower electrical conductivities are obtained when compared with those found for electropun ICP (Yu et al., 2008a; Merlini et al., 2014b; Sarvi et al., 2014). The second one is an indirect technique, in which the

insulating polymer is electrospun and after that, the polymer fiber mat is coated with the ICP through *in situ* oxidative polymerization (Ji et al., 2010; Yu et al., 2011; Aznar-Cervantes et al., 2012; Chen et al., 2013). The *in situ* oxidative polymerization consists to swell the insulating polymer mat in a monomer or oxidant aqueous solution. The polymerization is carried out by adding an oxidant or monomer solution in the vessel containing monomer or oxidant and the polymer mat (Malinauskas, 2001). By choosing the appropriated reaction conditions, such as the monomer concentration, oxidant-to-monomer molar ratio, reaction temperature and time, the polymerization takes place preferentially on the polymer fiber mat (Merlini et al., 2014a). The deposition of ICP on electrospun mats through *in situ* oxidative polymerization provides the possibility of obtaining a new hybrid material that displays functional properties not available in any single material. ICP-coated electrospun insulating polymer mats can display electrical conductivity similar to neat ICP without a significant reduction on the mechanical properties of insulating matrix (Huang et al., 2004; Merlini et al., 2014a).

Among the ICP, PANI has been widely studied and used in many technological fields because of relatively high conductivity ($\sigma$) (emeraldine salt form $\sigma > 1\,\mathrm{S\,cm^{-1}}$) (Huang et al., 2004), redox reversibility, and environmental stability. Furthermore, its easy polymerization and low monomer cost are also attractive (Chen et al., 2013). According to the desired properties, a wide range of insulating polymers can be electrospun and used as template for PANI coating. Among insulating polymers, the poly(vinylidene fluoride) (PVDF) is a suitable polymer for development of electrospun conductive mats due to its unique pyroelectric/piezoelectric features, mechanical properties, and easy processability (Huang et al., 2010; Merlini et al., 2014a). In our previous study (Merlini et al., 2014a), conductive PPy-coated PVDF mats with electrosensitive properties were prepared through pyrrole (Py) oxidative polymerization on electrospun PVDF mats. However, to our best knowledge, there are no studies concerning on the PANI-coated PVDF. Based on this context, the focus of this study is to develop PANI-coated electrospun PVDF mats through aniline (ANI) oxidative polymerization on the electrospun PVDF fibers surface, by using Iron (III) chloride hexahydrate ($FeCl_3 \cdot H_2O$) as oxidant. The influence of synthesis condition, such as monomer concentration and reaction time, on the structure and physical properties was investigated.

## EXPERIMENTAL
### MATERIALS
Aniline (analytical grade, Merck) was distilled under vacuum and stored in a refrigerator. Iron (III) chloride hexahydrate ($FeCl_3 \cdot 6H_2O$) (analytical grade, Vetec) was used without further purification. PVDF commercially designated Solef 11010/1001 was kindly supplied by *Solvay do Brasil Ltda*. Dymethylformamide (DMF) (99.8%) and acetone (99.5%), both P.A.–ACS reagents, were purchased from Synth. Hydrochloric acid, ACS reagent grade, 37%, was purchased from Nuclear.

### PREPARATION OF PVDF MATS BY ELECTROSPINNING
Poly(vinylidene fluoride) was dissolved in DMF under stirring for 2 h at 70°C, resulting in a solution of 20 wt%. After cooling to room temperature, acetone was added to the solution under stirring in a DMF/acetone proportion of 75/25 by weight. The solution was electrospun through a 5 mL syringe (needle with internal diameter of 0.5 mm) using a syringe pump (KD-100, KD Scientific) at a flow rate 2.5 mL h$^{-1}$. The electric field was generated using a high voltage supply (Testtech), which generates DC fields from 0 to 30 kV. The positive pole was connected to the syringe needle and the collector plate was grounded. PVDF fibers mats were obtained using an electric potential of 15 kV and needle-collector distance of 30 cm, at 25°C and humidity of 54%.

### POLYANILINE COATING
Poly(vinylidene fluoride) mats obtained by electrospinning were coated with PANI through the *in situ* oxidative polymerization, using Iron (III) chloride hexahydrate ($FeCl_3 \cdot 6H_2O$), as an oxidant. First, dried electrospun PVDF mats with width of 20 mm and length of 35 mm were immersed in 0.062 L of hydrochloric acid solution (HCl 0.1 mol L$^{-1}$) under stirring at room temperature and then an appropriate amount of ANI was added. After 10 min, the $FeCl_3 \cdot 6H_2O$ dissolved in 0.05 L of HCl acid solution was slowly added. In order to achieve the optimal reaction condition, the polymerization was performed using different ANI concentrations (0.05–0.5 mol L$^{-1}$) and reaction times (from 3 to 24 h). The oxidant-to-monomer molar ratio was 3/1. After the reaction, PANI-coated PVDF mats were washed thoroughly with HCl solution in order to extract the byproducts and wastes of the reaction and vacuum dried at room temperature. The samples

were denoted as PVDF/PANI [$x$], where $x$ represents the ANI concentration in the polymerization medium. For comparison, neat PANI was also synthesized using similar methodology, by using 0.05 mol L$^{-1}$ of ANI and 6 h of reaction.

### CHARACTERIZATION
Electrical conductivity measurements of PANI and PANI-coated electrospun PVDF mats were performed using the four probe standard method with a Keithley 6220 current source to apply the current and a Keithley Model 6517A electrometer to measure the potential difference. For neat PVDF, the measurements were performed using a Keithley 6517A electrometer connected to a Keithley 8009 test fixture. Sample measurements were performed at least five times at room temperature.

The morphology of the samples was analyzed using a scanning electron microscope (SEM), Zeiss–Evoma-15. The samples were coated with gold and observed using an applied tension of 10 kV.

An elemental analysis (carbon, hydrogen, and nitrogen) was performed with a Perkin–Elmer CHN 2400 analyzer. The combustion process was held at 925°C using pure oxygen (99.995%).

The tensile properties of electrospun PVDF and PANI-coated electrospun PVDF fibers mats with thicknesses in the range 500 $\mu$m were performed in a Dynamic mechanical analyzer (Q-800, TA Scientific) equipped with clamp for films. The analysis was performed six times for each sample on different specimens, with speed of 3 mm min$^{-1}$ at room temperature. The tensile strength at break, Young's modulus, and elongation at break were calculated from the stress–strain curves.

Attenuated total reflectance-Fourier transform infrared spectroscopy (ATR-FTIR) was performed in a Bruker spectrometer, model TENSOR 27, in the range of 4000–600 cm$^{-1}$ by accumulating 32 scans at a resolution of 4 cm$^{-1}$.

Thermogravimetric analysis (TGA) were carried out on a STA 449 F1 Jupiter® (Netzsch) instrument at a heating rate of 10°C min$^{-1}$, from 35 to 700°C, under nitrogen flow of 50 cm$^3$ min$^{-1}$.

## RESULTS AND DISCUSSION
The effect of the ANI content on the electrical conductivity was investigated in order to produce PANI-coated electrospun PVDF mats with the highest electrical conductivity using the lowest ANI content. **Figure 1** shows the electrical conductivity and digital photographs of PVDF/PANI mats with different ANI concentrations after 6 h of reaction. From the photographs, it is possible to observe that the color of mats changes from white (neat PVDF) to green, and dark green with increasing the monomer concentration. The PVDF/PANI mat synthesized by using ANI concentration of 0.05 mol L$^{-1}$ displays an irregular coating on the PVDF surface and consequently lower electrical conductivity of $(2.6 \pm 0.3) \times 10^{-5}$ S cm$^{-1}$ is obtained. However, with increasing the monomer concentration the electrical conductivity of the mats increases significantly. When ANI content used in the polymerization is higher than 0.2 mol L$^{-1}$, the electrical conductivity of the PVDF/PANI mats is 16 orders of magnitude higher than that found for the neat PVDF [$(3.2 \pm 0.4) \times 10^{-16}$ S cm$^{-1}$], which is quite similar to that found for the neat PANI ($3.3 \pm 0.2$ S cm$^{-1}$).

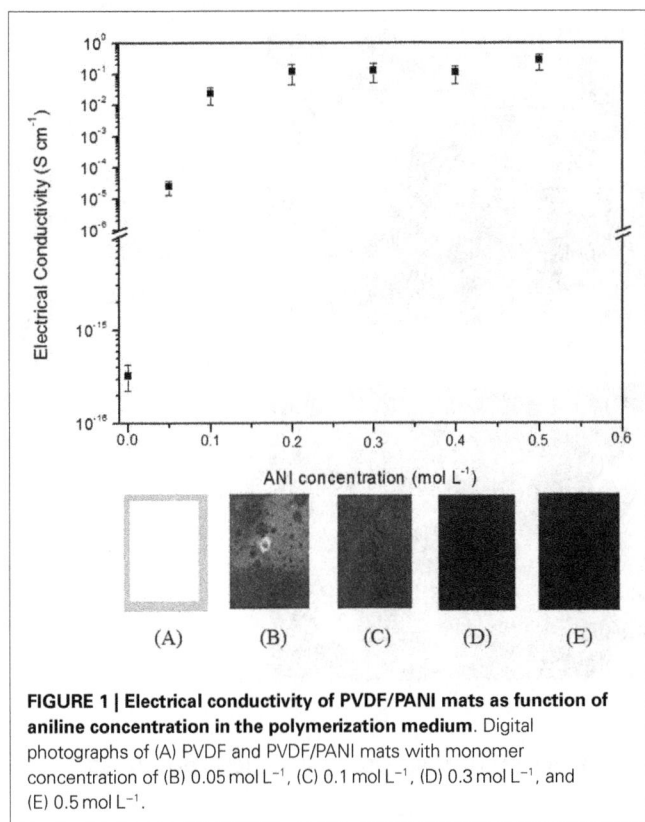

FIGURE 1 | Electrical conductivity of PVDF/PANI mats as function of
aniline concentration in the polymerization medium. Digital
photographs of (A) PVDF and PVDF/PANI mats with monomer
concentration of (B) 0.05 mol L$^{-1}$, (C) 0.1 mol L$^{-1}$, (D) 0.3 mol L$^{-1}$, and
(E) 0.5 mol L$^{-1}$.

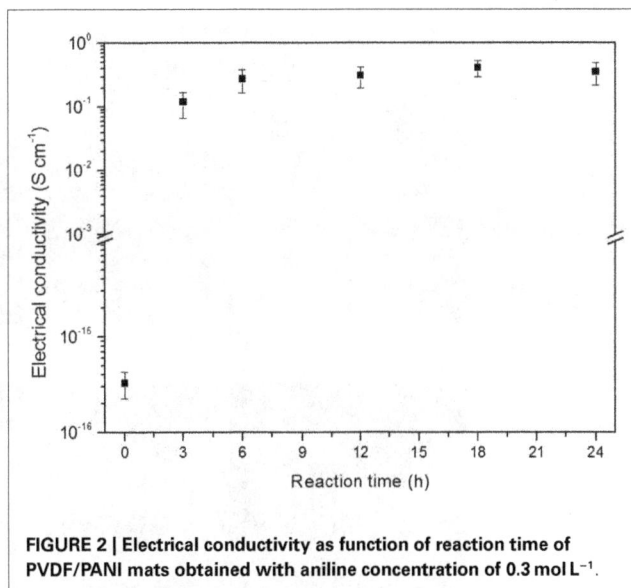

FIGURE 2 | Electrical conductivity as function of reaction time of
PVDF/PANI mats obtained with aniline concentration of 0.3 mol L$^{-1}$.

On the other hand, for ANI content up to 0.5 mol L$^{-1}$, the
polymerization also occurs outside of the PVDF fibers mats, due
to the large amount of monomer in the reaction medium. This
result indicates that a monomer concentration corresponding to
0.3 mol L$^{-1}$ is enough to producing efficient coating of the membranes with electrical conductivity quite similar to that found for
the neat PANI.

The influence of polymerization time on the electrical conductivity of PVDF/PANI was also evaluated (**Figure 2**) by using a
monomer concentration of 0.3 mol L$^{-1}$. During polymerization,
the PVDF mats turn from white to green within 3 h of reaction,
indicating a PANI coating on the PVDF fibers. However, this membrane displays a slightly low electrical conductivity. After 6 h, the
electrical conductivity does not depend on the reaction time and
remaining practically constant.

The SEM micrographs of PVDF/PANI mats obtained by using
different ANI concentration in the polymerization and electrospun PVDF are shown in **Figure 3**. The electrospun PVDF mat
(**Figure 3A**) is constituted of three-dimensional network structure
with randomly oriented fibers, with diameter of 0.8 ± 0.2 μm. The
PVDF/PANI mats synthesized through different ANI concentration (**Figures 3B–D**) comprising PANI continuous layer on the
PVDF fibers surfaces. With increasing the ANI content, the conducting layer thickness increases and some PANI in the form of
agglomerates can be observed on the fiber surfaces (**Figure 3D**).
This morphology is responsible for the electrical conductivities
of the membranes, which are quite similar to that found for
neat PANI.

Table 1 shows the elemental analysis obtained by CHN, the
PANI content estimated by elemental analysis and the tensile properties of PVDF before and after coating with PANI. Considering
that nitrogen is absent in the PVDF structure, the PANI amount
incorporated on the PVDF surface during the *in situ* oxidative
polymerization was determined by the nitrogen difference in the
neat PANI and PANI-coated PVDF fibers mats, according to the
procedure described in the literature (Merlini et al., 2014b). The
PANI amount deposited on the PVDF fibers increases significantly with increasing monomer concentration. PVDF/PANI mats
exhibit PANI content of 11.6, 28.4, and 39.8 wt% for ANI concentration of 0.1, 0.3, and 0.5 mol L$^{-1}$, respectively. PVDF/PANI
[0.1] membrane displays small amount of PANI and hence, it
presents lower electrical conductivity [(2.4 ± 0.4) × 10$^{-2}$ S cm$^{-1}$]
than that found for membranes with higher PANI content. On the
other hand, PVDF/PANI mats with approximately 28 and 40 wt%
of PANI display electrical conductivity quite similar to that found
for neat PANI. This behavior indicates that for these ANI concentrations, a uniform layer is formed on the fiber mats reaching
maximum electrical conductivity of 0.1 S cm$^{-1}$.

Figure 4 shows the representative stress–strain curves of electrospun PVDF and PANI-coated PVDF mats. PVDF mat shows
typical ductile behavior with 21.2 ± 0.6% of elongation at break,
while the tensile strength and Young modulus are 0.6 ± 0.1 and
6.5 ± 0.4 MPa, respectively. The tensile strength of the PANI-
coated PVDF composites remains almost the same, regardless the
amount of ANI used in the synthesis, as shown in the **Table 1**.
On the other hand, a significant increase in the Young modulus
and a decrease in elongation at break are observed with increasing the PANI concentration. These results may be attributed to
the rigidity of PANI and to the presence of adhered PANI layers
that improve the interfiber bonding, reducing the mobility of the
PVDF fibers.

The thermogravimetric curves of PANI, electrospun PVDF, and
PVDF/PANI mats are reported in **Figure 5**. The electrospun PVDF
mat exhibits only one weight loss at 469°C (midpoint from DTG),

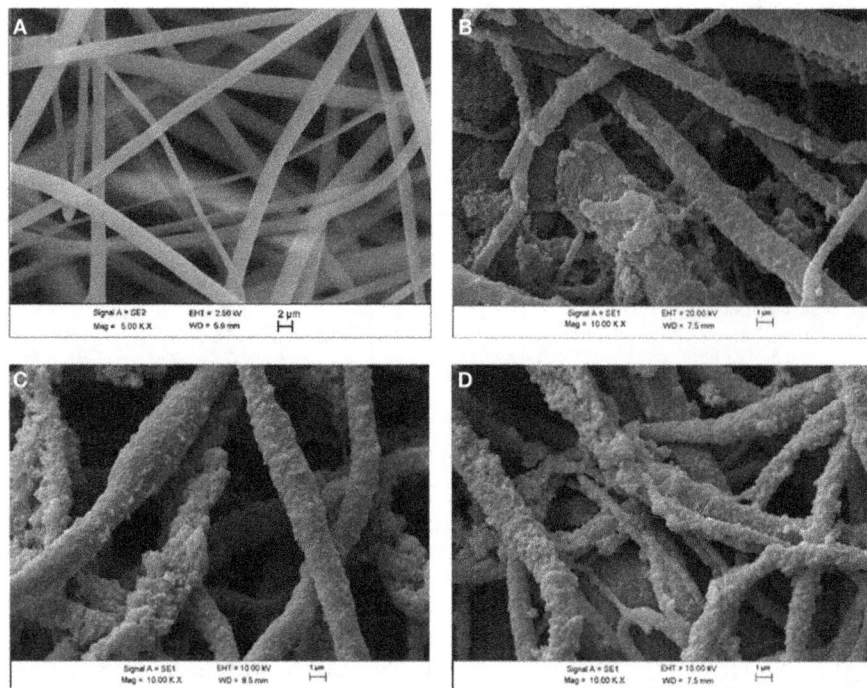

FIGURE 3 | Scanning electron microscope micrographs of (A) electrospun PVDF mats, (B) PVDF/PANI [0.1], (C) PVDF/PANI [0.3], (D) PVDF/PANI [0.5] after 6 h of reaction.

Table 1 | Elemental analysis, PANI content, and tensile properties of electrospun PVDF, PVDF/PANI mats.

| Sample | Proportion of | | | Pani content (wt%)[a] | σ (MPa) | ε (%) | E (MPa) |
|---|---|---|---|---|---|---|---|
| | C | H | N | | | | |
| PVDF | – | – | – | 0.0 | 0.6 ± 0.1 | 21.2 ± 0.6 | 6.5 ± 0.4 |
| PANI | 46.2 ± 0.1 | 4.3 ± 0.1 | 9.5 ± 0.1 | 100.0 | – | – | – |
| PVDF/PANI [0.1] | 34.8 ± 0.2 | 3.0 ± 0.1 | 1.1 ± 0.2 | 11.6 ± 0.2 | 0.5 ± 0.1 | 14.5 ± 0.4 | 6.9 ± 0.3 |
| PVDF/PANI [0.3] | 37.2 ± 0.3 | 3.8 ± 0.1 | 2.7 ± 0.1 | 28.4 ± 0.1 | 0.4 ± 0.1 | 6.5 ± 0.5 | 7.9 ± 0.4 |
| PVDF/PANI [0.5] | 39.1 ± 0.7 | 3.5 ± 0.2 | 3.8 ± 0.2 | 39.8 ± 0.2 | 0.4 ± 0.1 | 3.6 ± 0.3 | 17.1 ± 0.5 |

[a] From CHN elemental analysis.

which is attributed to the decomposition of polymer chains with a residual weight of 30.5% at 700°C (Zhong et al., 2012; Ramôa et al., 2014). The neat PANI presents two main weight loss stages. The first stage with maximum at 105°C is related to the presence of moisture. The second one with weight loss from 277°C up to 390°C is attributed to degradation polymer backbone. The PVDF/PANI mats start to decompose at lower temperatures than neat PVDF. The degradation profile of the PVDF/PANI [0.1] is quite similar to the PVDF. However, increasing the PANI content on the fibers surface, the onset temperature of the mats decreases. However, it is possible to observe that, the PANI deposited on the fibers surface starts to decompose at higher temperature when compared with neat PANI. This behavior can be attributed to the dipole–dipole interaction between the $-F^-$ ($-C-F$) groups of PVDF and the $H^+$ ($=N-H$) groups of PANI (Merlini et al., 2014b).

The FTIR spectra of neat PANI, electrospun PVDF, and PVDF/PANI mats are shown in **Figure 6**. In the FTIR spectrum of PVDF, absorption bands associated to the β, α, and γ phases are identified. The absorption bands at 1402 and 876 $cm^{-1}$ are attributed to the C–F stretching vibration of amorphous phase, while the bands at 1172 $cm^{-1}$ assigned to the C–C bond (Gregorio and Borges, 2008; Yu and Cebe, 2009). The electrospun mats spectrum shows the presence of β phase (bands at 1274 and 837 $cm^{-1}$), which offers the highest piezo-, pyro-, and ferro-electric properties and it is mainly interesting for sensing applications (Merlini et al., 2014a).The PANI spectrum exhibits absorption bands at 1540 and 1400 $cm^{-1}$ associated to the quinone and benzene ring stretching deformations. The band at 1274 $cm^{-1}$ corresponds to C–N+ stretching vibration (Trchová et al., 2004). The PVDF/PANI [0.1] exhibited overlapped absorption bands of both components.

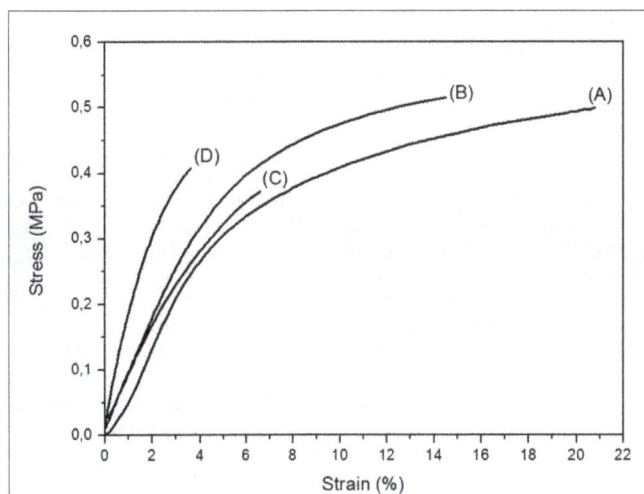

FIGURE 4 | Representative stress–strain curves of (A) electrospun PVDF mats, (B) PVDF/PANI [0.1], (C) PVDF/PANI [0.3], and (D) PVDF/PANI [0.5].

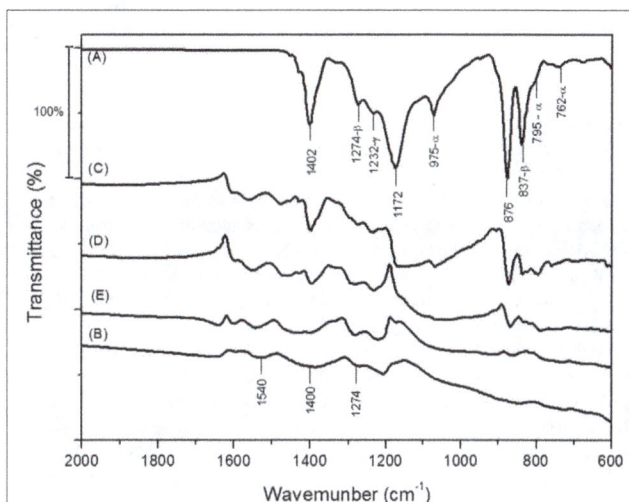

FIGURE 6 | FTIR spectra of (A) electrospun PVDF mats, (B) neat PANI, (C) PVDF/PANI [0.1], (D) PVDF/PANI [0.3], and (E) PVDF/PANI [0.5].

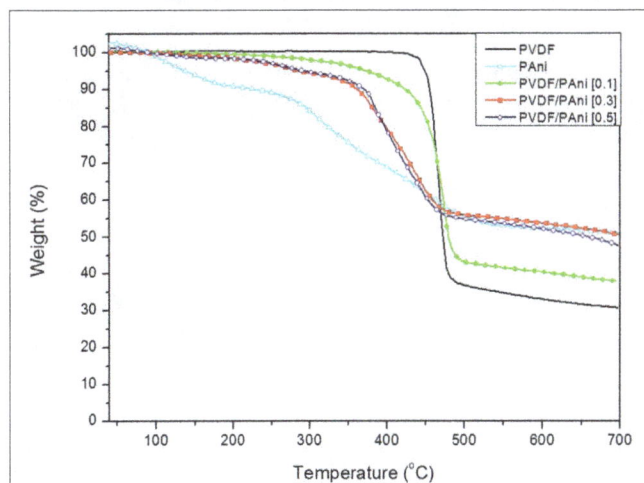

FIGURE 5 | TG curves of neat PANI, electrospun PVDF, and PVDF/PANI mats prepared with different ANI concentrations.

Furthermore, with increasing PANI content, the bands assigned to the PVDF groups practically disappear suggesting that PVDF fibers were completed coated with an external PANI layer.

## CONCLUSION

In this study, an electrically conductive mat based on PVDF and PANI was successfully obtained through electrospinning technique and *in situ* oxidative polymerization. An efficient coating of the PVDF mats was obtained with ANI concentration higher than $0.1 \, mol \, L^{-1}$, wherein the fibers surface was coated with a PANI layer adhered on PVDF fibers. The electrical conductivity and the amount of PANI on the PVDF fibers surface increased with the increasing of monomer concentration. After $0.2 \, mol \, L^{-1}$ of ANI, the electrical conductivity of the PVDF/PANI mats reached a constant value, which was quite similar to that found for the neat

PANI. After coating with PANI, the Young modulus enhanced significantly due to the presence of adhered PANI on the PVDF fibers, which increases the interfiber bonding. On the other hand, the elongation at break was reduced indicating that the membranes become more fragile probably because of the brittle nature of the PANI component. PVDF/PANI mats display a polymorph crystalline structure, with absorption bands associated to the β, α, and γ phases. PANI-coated electrospun PVDF mats developed in this work display, within others interesting features, three-dimensional fibers network structure, and good electrical conductivity, which demonstrates the potential to use as a chemicals sensor.

The two-step procedure described in this work, allow the development of electrospun mats with electrical conductivity similar to the neat ICP. This process is encouraged to be used when higher electrical conductivity is required, since studies in the literature (Merlini et al., 2014b) have reported that electrospun mats obtained from direct electrospinning show electrical conductivities as low as $10^{-15} \, S \, cm^{-1}$. Furthermore, the spinnability of the neat insulating polymer can be carry out more easily than its solution containing the ICP.

## ACKNOWLEDGMENTS

The authors gratefully acknowledge the financial support by Conselho Nacional de Desenvolvimento Científico e Tecnológico – CNPq, Coordenação de Aperfeiçoamento de Pessoal de Ensino Superior – CAPES, and Fundação de Amparo à Pesquisa e Inovação do Estado de Santa Catarina – FAPESC. We are sincerely thankful to Central Electronic Microscopy Laboratory, Santa Catarina Federal University (LCME-UFSC).

## REFERENCES

Aznar-Cervantes, S., Roca, M. I., Martinez, J. G., Meseguer-Olmo, L., Cenis, J. L., Moraleda, J. M., et al. (2012). Fabrication of conductive electrospun silk fibroin scaffolds by coating with polypyrrole for biomedical applications. *Bioelectrochemistry* 85, 36–43. doi:10.1016/j.bioelechem.2011.11.008

Bagheri, H., and Aghakhani, A. (2012). Polyaniline-nylon-6 electrospun nanofibers for headspace adsorptive microextraction. *Anal. Chim. Acta* 713, 63–69. doi:10.1016/j.aca.2011.11.027

Chen, D., Miao, Y. E., and Liu, T. (2013). Electrically conductive polyaniline/polyimide nanofiber membranes prepared via a combination of electrospinning and subsequent in situ polymerization growth. *ACS Appl. Mater. Interfaces* 5, 1206–1212. doi:10.1021/am303292y

Chronakis, I. S., Grapenson, S., and Jakob, A. (2006). Conductive polypyrrole nanofibers via electrospinning: electrical and morphological properties. *Polymer* 47, 1597–1603. doi:10.1016/j.polymer.2006.01.032

Gregorio, R., and Borges, D. S. (2008). Effect of crystallization rate on the formation of the polymorphs of solution cast poly(vinylidene fluoride). *Polymer* 49, 4009–4016. doi:10.1016/j.polymer.2008.07.010

Huang, J., Virji, S., Weiller, B. H., and Kaner, R. B. (2004). Nanostructured polyaniline sensors. *Chemistry* 10, 1314–1319. doi:10.1002/chem.200305211

Huang, W., Edenzon, K., Fernandez, L., Razmpour, S., Woodburn, J., and Cebe, P. (2010). Nanocomposites of poly(vinylidene fluoride) with multiwalled carbon nanotubes. *J. Appl. Polym. Sci.* 115, 3238–3248. doi:10.1002/app.31393

Huang, Z.-M., Zhang, Y. Z., Kotaki, M., and Ramakrishna, S. (2003). A review on polymer nanofibers by electrospinning and their applications in nanocomposites. *Compos. Sci. Technol.* 63, 2223–2253. doi:10.1016/s0266-3538(03)00178-7

Ji, L., Lin, Z., Li, Y., Li, S., Liang, Y., Toprakci, O., et al. (2010). Formation and characterization of core-sheath nanofibers through electrospinning and surface-initiated polymerization. *Polymer* 51, 4368–4374. doi:10.1016/j.polymer.2010.07.042

Lin, Q. Q., Li, Y., and Yang, M. J. (2012). Polyaniline nanofiber humidity sensor prepared by electrospinning. *Sens. Actuators B Chem.* 161, 967–972. doi:10.1016/j.snb.2011.11.074

MacDiarmid, A. G., Jones, W. E., Norris, I. D., Gao, J., Johnson, A. T., Pinto, N. J., et al. (2001). Electrostatically-generated nanofibers of electronic polymers. *Synth. Met.* 119, 27–30. doi:10.1016/s0379-6779(00)00597-x

Malinauskas, A. (2001). Chemical deposition of conducting polymers. *Polymer* 42, 3957–3972. doi:10.1016/S0032-3861(00)00800-4

Merlini, C., Almeida, R. S., d'Ávila, M. A., Schreiner, W. H., and Barra, G. M. O. (2014a). Development of a novel pressure sensing material based on polypyrrole-coated electrospun poly(vinylidene fluoride) fibers. *Mater. Sci. Eng. B* 179, 52–59. doi:10.1016/j.mseb.2013.10.003

Merlini, C., Barra, G. M. O., Medeiros Araujo, T., and Pegoretti, A. (2014b). Electrically pressure sensitive poly(vinylidene fluoride)/polypyrrole electrospun mats. *RSC Adv.* 4, 15749–15758. doi:10.1039/C4RA01058B

Norris, I. D., Shaker, M. M., Ko, F. K., and Macdiarmid, A. G. (2000). Electrostatic fabrication of ultrafine conducting fibers: polyaniline/polyethylene oxide blends. *Synth. Met.* 114, 109–114. doi:10.1016/s0379-6779(00)00217-4

Ramôa, S. D. A. S., Merlini, C., Barra, G. M. O., and Soares, B. G. (2014). Obtenção de nanocompósitos condutores de montmorilonita/polipirrol: efeito

da incorporação do surfactante na estrutura e propriedades. *Polímeros* 24, 57–62. doi:10.4322/polimeros.2014.051

Sarvi, A., Chimello, V., Da Silva, A. B., Bretas, R. E. S., and Sundararaj, U. (2014). Novel semiconductive coaxial electrospun nanofibers. *Plast. Res. Online.* doi:10.2417/spepro.005274

Srinivasan, S. S., Ratnadurai, R., Niemann, M. U., Phani, A. R., Goswami, D. Y., and Stefanakos, E. K. (2010). Reversible hydrogen storage in electrospun polyaniline fibers. *Int. J. Hydrogen Energy* 35, 225–230. doi:10.1016/j.ijhydene.2009.10.049

Trchová, M., Šedenková, I., Tobolková, E., and Stejskal, J. (2004). FTIR spectroscopic and conductivity study of the thermal degradation of polyaniline films. *Polym. Degrad. Stab.* 86, 179–185. doi:10.1016/j.polymdegradstab.2004.04.011

Yu, L., and Cebe, P. (2009). Crystal polymorphism in electrospun composite nanofibers of poly(vinylidene fluoride) with nanoclay. *Polymer* 50, 2133–2141. doi:10.1016/j.polymer.2009.03.003

Yu, Q.-Z., Dai, Z.-W., and Lan, P. (2011). Fabrication of high conductivity dual multi-porous poly (l-lactic acid)/polypyrrole composite micro/nanofiber film. *Mater. Sci. Eng. B* 176, 913–920. doi:10.1016/j.mseb.2011.05.017

Yu, Q.-Z., Shi, M.-M., Deng, M., Wang, M., and Chen, H.-Z. (2008a). Morphology and conductivity of polyaniline sub-micron fibers prepared by electrospinning. *Mater. Sci. Eng. B* 150, 70–76. doi:10.1016/j.mseb.2008.02.008

Yu, Q. Z., Li, Y., Wang, M., and Chen, H. Z. (2008b). Polyaniline nanobelts, flower-like and rhizoid-like nanostructures by electrospinning. *Chin. Chem. Lett.* 19, 223–226. doi:10.1016/j.cclet.2007.12.005

Zhong, Z., Cao, Q., Jing, B., Wang, X., Li, X., and Deng, H. (2012). Electrospun PVdF-PVC nanofibrous polymer electrolytes for polymer lithium-ion batteries. *Mater. Sci. Eng. B* 177, 86–91. doi:10.1016/j.mseb.2011.09.008

**Conflict of Interest Statement:** The authors declare that the research was conducted in the absence of any commercial or financial relationships that could be construed as a potential conflict of interest.

# 6

# Mechanical behavior of osteoporotic bone at sub-lamellar length scales

*Ines Jimenez-Palomar[1], Anna Shipov[2], Ron Shahar[2] and Asa H. Barber[1,3] \**

[1] School of Engineering and Materials Science, Queen Mary University of London, London, UK
[2] Koret School of Veterinary Medicine, The Hebrew University of Jerusalem, Jerusalem, Israel
[3] School of Engineering, University of Portsmouth, Portsmouth, UK

**Edited by:**
*Federico Bosia, University of Torino, Italy*

**Reviewed by:**
*Guy M. Genin, Washington University in St. Louis, USA*
*Fei Hang, South China University of Technology, China*

**\*Correspondence:**
*Asa H. Barber, School of Engineering, University of Portsmouth, Portsmouth PO1 2UP, UK*
*e-mail: asa.barber@port.ac.uk*

Osteoporosis is a disease known to promote bone fragility but the effect on the mechanical properties of bone material, which is independent of geometric effects, is particularly unclear. To address this problem, micro-beams of osteoporotic bone were prepared using focused ion beam microscopy and mechanically tested in compression using an atomic force microscope while observing them using *in situ* electron microscopy. This experimental approach was shown to be effective for measuring the subtle changes in the mechanical properties of bone material required to evaluate the effects of osteoporosis. Osteoporotic bone material was found to have lower elastic modulus and increased strain to failure when compared to healthy bone material, while the strength of osteoporotic and healthy bone was similar. Surprisingly, the increased strain to failure for osteoporotic bone material provided enhanced toughness relative to the control samples, suggesting that lowering of bone fragility due to osteoporosis is not defined by material performance. A mechanism is suggested based on these results and previous literature that indicates degradation of the organic material in osteoporosis bone is responsible for resultant mechanical properties.

**Keywords: bone, osteoporosis, micromechanics, AFM, FIB**

## INTRODUCTION

Osteoporosis is one of the most significant types of bone disease that causes degradation of bone's mechanical function. Osteoporosis is characterized by significant changes in bone structure causing increases in bone fragility and therefore an increase in fracture risk (Kilbanski et al., 2001). The clinical importance of osteoporosis has been vigorously investigated in recent years due to the amount of people affected. In the United States alone, the costs of fractures resulting from osteoporosis have been estimated to be from 10 to 18 billion dollars per year and are expected to increase to 60 billion by the year 2020 (Iacono, 2007). The effects of osteoporosis on bone are characterized by two distinct forms; Type I, which refers to the loss of trabecular bone mass after menopause due to lack of estrogen and Type II, which refers to loss of cortical and trabecular bone in both men and women as a result of aging (Marcus and Bouxsein, 2010). The disturbances in osteoporotic bone structure are known to be due to changes in metabolic conditions such as hormonal changes (decrease in estrogen levels, growth hormone deficiency, increase in parathyroid hormone), steroids (glucocorticoid deficiency), diet, and lifestyle (reduction in calcium intake, lack of vitamin D, sedentary lifestyles) (Hauge et al., 2003; Iacono, 2007). Both Type I and Type II osteoporosis share the common effect of increased susceptibility to catastrophic fracture in bone.

Bone fragility due to osteoporosis has been examined in terms of changes in bone structure and resultant influence on mechanical properties. The ability of bone to resist catastrophic fracture depends on structure including bone mass, spatial distribution such as shape and micro-architecture, as well as the intrinsic properties of the bone material (Bouxsein, 2001). Bone fragility is therefore determined by the complex interaction between these parameters and, ultimately, the failure of mechanical function due to the diseased osteoporotic bone state (Turner, 2002). However, a range of bone mechanical properties are known to control catastrophic failure, particularly the strength (ultimate stress), stiffness (elastic modulus), and energy absorption quality (work-to-fracture) of bone structures (Turner, 2002). Structural changes in osteoporotic bone at the macrostructural, architectural, and microstructural levels are typically diagnosed as a reduction of bone density. This density loss is consistent across the variety of organizations found in bone such that cortical bone displays a reduction in bone mass and trabecular bone exhibits thinning and loss of the number of trabecular struts across the body (Carter and Hayes, 1976; Wu et al., 2008; Kennedy et al., 2009; Zebaze et al., 2010) from the onset of osteoporosis. The stiffness and strength of trabecular bone are typically related to bone density in a non-linear fashion with either a squared (Rice et al., 1988), cubic (Carter and Hayes, 1976) or more complex (Marcus and Bouxsein, 2010) relationship to the change in density. However, this non-linear relationship between bone density and resultant mechanical properties for trabecular bone has been explained by considering variations in bone volume fraction, trabecular orientation, trabecular interconnectivity, and structural anisotropy, which result in a linear relationship between bone density and mechanical properties for loading along the main trabecular orientation (Silva and Gibson, 1997; Keaveny et al., 2001). The geometry and organizational structure of osteoporotic bone therefore clearly define mechanical properties. Changes in material composition have been additionally identified as lowering strength in osteoporotic bone, particularly due to decreases in the degree of mineralization as porosity increases (Currey, 1988; Schaffler and Burr, 1988) or increases in mineralization due to the continuous aggregations of mineral without resorption (Grynpas, 1993). Additional compositional changes in cortical bone induced by osteoporosis include collagen content and orientation of collagen fibrils, the extent and nature of collagen cross-linking (Burr, 2002), as well as the number and composition of cement lines (Burr et al., 1988) that cause fatigue-induced micro-damage (Burr et al., 1997; Burr, 2003). Many of these latter factors affecting osteoporotic bone mechanical properties are more closely related to compositional changes at the smaller length scales existing at lower hierarchical levels. The structural changes in osteoporotic bone at higher hierarchical levels provide mechanical performance that is thus dependent on both the constituent material properties and the changes in bone geometry due to reduction in bone mass. However, the effect of osteoporosis on bone mechanics remains uncertain such that density alone cannot, for example, account for the decrease in stiffness and strength of trabecular bone, with strain to failure almost independent of density (Keaveny et al., 2001). The material properties of bone dominate at the lower hierarchical levels of bone where geometric and structural factors can be ignored. Thus, evaluating the quality of the bone material requires suitable mechanical tests at relatively small length scales.

The effect of the quality of osteoporotic bone on mechanical properties is important as current diagnosis methods purely based on bone density scales from x-ray scans are not optimal. For example, the National Osteoporosis Guideline Group (NOGG) has placed guidelines for the diagnosis of osteoporosis, which take into account the patient's medical history along with the x-ray measured bone mineral density (BMD) index. BMD alone has been a poor indicator for potential increases in bone fragility and is only able to predict 60% of the variations in bone strength (Ammann and Rizzoli, 2003). The quality of the bone material brought on by a patient's lifestyles and other factors affecting the quality of bone material has been suggested as being an important consideration in determining bone fragility due to osteoporosis (WHO, 2012). Further techniques have been developed in order to quantify the mineral content of bone and assess the quality of bone material. These techniques include microradiography (Boivin and Baud, 1984; Boivin and Meunier, 2002), quantitative backscattered electron imaging (qBEI) (Roschger et al., 2003) and synchrotron radiation micro computed tomography (SRμCT) (Borah et al., 2005). All of these methods perform measurements in what is referred to as bone mineralization density distribution (BMDD). BMDD is a measure of the mineral content in small areas defined as image pixels or voxels and can distinguish local variations in mineral content. BMD is a potentially poorer description of osteoporosis as an estimate of the total amount of mineral in a scanned area of whole bone, but is the current method used clinically (Roschger et al., 2008). Imaging techniques used to quantify bone mineral distribution have been previously combined with addition structural or mechanical testing, notably nanoindentation (Guo and Goldstein, 2000), scanning acoustic microscopy (SAM) (Katz and Meunier, 1993), Raman spectroscopy (McCreadie et al., 2006), and Fourier transform infrared imaging (FTIR) (Paschalis

et al., 2004) in order to correlate mineral content to structure and function relationships (Roschger et al., 2008).

Compositional changes in bone material due to osteoporosis have been shown to decrease the degree of mineralization and collagen cross-linking, resulting in bone fragility (Paschalis et al., 2004; Marcus and Bouxsein, 2010). Reductions in the degree of mineralization have been further emphasized as detrimental to the material properties of bone (Ciarelli et al., 2003). The stiffness versus toughness of bone is determined in part by the mineral content (Currey, 1988; McCreadie et al., 2006) and exhibits significant degradation in mechanical properties with relatively small mineral content changes, which increase bone fragility (Roschger et al., 2008). In the case of osteoporosis, a decrease or an increase in mineralization may therefore be detrimental to the mechanical properties of bone (Ciarelli et al., 2003; Roschger et al., 2008). Low mineralization levels, or hypomineralization, cause reductions in stiffness and strength while high mineralization levels, or hypermineralization, reduce fracture toughness (Ciarelli et al., 2003). Hypomineralization occurs either due to lack of time for secondary mineralization to occur after bone remodeling or due to pathological conditions affecting mineralization. Conversely, hypermineralization only occurs when changes in crystal size or shape provide increased packing for a higher mineral density (Roschger et al., 2008). The significance of changes in the properties of bone material has led to works that attempt to measure mechanical properties of bone at small length scales, thus ignoring geometric effects at higher hierarchical levels. Notable experiments at the microstructural level were performed applying nanoindentation on trabecule from the lumbar region of 17-month-old control and ovariectomized (OVH) Sprague Dawley rats. These results showed no change in elastic modulus or hardness at the microscopic level between control and diseased specimens (Guo and Goldstein, 2000), suggesting that osteoporosis does not change the material properties of bone but instead only induces changes in bone density. A similar study by Maïmoun et al. (2012) showed a reduction in bone density due to a depletion of oestrogen in Sprague Dawley OVH rats, but a reduction in the elastic modulus in trabecular bone, which contradicted previous results. Additional studies attempting to assess the effect of osteoporosis on the mechanical properties of bone have also shown significant mechanical variations. Nanoindentation applied to cross-sections of osteoporotic and healthy bone of female human femurs was shown to give no change in elastic modulus even though the results of the qBEI analysis showed a lower mineralization level for the osteoporotic samples (Fratzl-Zelman et al., 2009). This lack of a decrease in elastic modulus with lower mineralization in osteoporotic bone was attributed to changes in the organic matrix determining mechanical performance. Specifically, increasing the stiffness of the collagen fibrils of the organic matrix can occur with an increase in the cross-linking between protein chains, which may compensate for the low mineral content or a change in the mineral–organic interface during osteoporosis. Such an observation is important as consideration of changes in the softer organic phase in addition to variations in the volume fraction of the harder mineral must be considered in osteoporotic bone. The lack of clarity in osteoporotic bone mechanics is emphasized when considering aged bone, with nanoindentation showing an increase in the elastic modulus of osteoporotic bone (Silva et al., 2004) or decreasing strength, stated as due to decreasing mineral content and size distribution but increasing average crystal size (Boskey, 2003). The resultant variability in nanoindentation data makes correlation with structural and biochemical observations difficult. The testing environment may also contribute to listed variation in mechanical performance for osteoporotic, and indeed healthy, bone with a number of studies evaluating bone in a dehydrated state (Guo and Goldstein, 2000; Silva et al., 2004; Fratzl-Zelman et al., 2009). The diversity of literature evaluations of osteoporotic bone is summarized in **Table 1** and includes comparative data on healthy bone. This table highlights the range of loading conditions used to determine the mechanical properties of bone. General comparisons between healthy and osteoporotic bone suggest small losses in strength and elastic modulus due to osteoporosis, with smaller length scale measurements providing higher absolute values than larger length scale measurements. The current paper therefore attempts to address the conflict in defining the effects of osteoporosis on bone by mechanically testing the material properties of bone. Techniques to isolate specific constituents and discrete volumes of bone have been previously shown to be effective in characterizing the material behavior (Hang and Barber, 2011; Hang et al., 2011, 2014; Jimenez-Palomar et al., 2012). Of these studies, the ability to mechanically test discrete volumes of bone is particularly beneficial for understanding the synergy between constituents while removing the effects of sample geometry (Jimenez-Palomar et al., 2012). Micro-beams selected using focused ion beam (FIB) microscopy have been previously employed to understand the mechanical properties of biological materials including teeth (Chan et al., 2009) and bone (Jimenez-Palomar et al., 2012). This work therefore exploits micro-beams from cortical bone but expands on the technique

to evaluate the effects of osteoporosis on resultant bone material mechanics.

## MATERIALS AND METHODS

The bone of rats is commonly used as a model for osteoporosis, which can be induced through estrogen depravation by performing an ovariectomy (Frost and Jee, 1992; Guo and Goldstein, 2000). OVH and control rat femurs were obtained from the Hebrew University of Jerusalem with ethical approval in order to compare the mechanical properties of osteoporotic and healthy bone. The diaphysis from the extracted rat femur was first isolated using a water-cooled diamond blade slow speed circular saw (Buehler, USA) to produce a bone sample with approximate dimensions of 12 mm × 1 mm × 1 mm. The bone sample was stored in 70% ethanol: 30% water solution overnight followed by submerging within progressively increasing ethanol solutions of 85, 95, and 100% for 60, 30, and 120 min, respectively, to provide sample dehydration. The dehydrated sample was then transferred to the chamber of a small dual beam system (SDB, Quanta 3D, FEI Company, EU/USA) for subsequent FIB milling. Micro-beams of bone were created using FIB as detailed in Jimenez-Palomar et al. (2012) using conditions to remove FIB damage in soft materials as described previously (Bailey et al., 2013). The FIB preparation method can be summarized in a series of steps where bone material was first removed rapidly, followed by more precision FIB removal at smaller length scales to produce micro-beams with regular geometries. The corner of the macroscopic bone sample was first cleaned to produce orthogonal surfaces at the sample edge using a high current ion beam of 65 nA and accelerating voltage of 30 kV. Flattening of the bone surfaces was achieved by further FIB removal of smaller bone volumes using smaller ion beam currents down to 0.1 nA. These smaller ion beam currents

**Table 1 | Mechanical properties of osteoporotic and comparative healthy rat bone taken from the literature.**

|  | State | Testing method | Strength (MPa) | Elastic modulus (GPa) | Reference |
|---|---|---|---|---|---|
| **MACROSTRUCTURAL LEVEL** | | | | | |
| Whole bone | Healthy | 3-Point bending | 180 ± 6 | 6.9 ± 0.3 | Jorgensen et al. (1991) |
| Whole bone | Healthy | 3-Point bending | 134 ± 4 | 8 ± 0.4 | Barengolts et al. (1993) |
| Whole bone | Healthy | 3-Point bending | 153 ± 45 | 4.9 ± 4 | Ejersted et al. (1993) |
| **ARCHITECTURAL LEVEL** | | | | | |
| Cross section (1 mm thick) | Healthy (cortical) | Compression | 139.5 ± 19.14 | 8.8 ± 2.5 | Cory et al. (2010) |
| Cross section (1 mm thick) | Osteoporotic (cortical) | Compression | 127.24 ± 35.04 | 7.3 ± 2.7 | Cory et al. (2010) |
| Cross section (1 mm thick) | Healthy (trabecular) | Compression | 35.95 ± 15.62 | 2.2 ± 0.92 | Cory et al. (2010) |
| Cross section (1 mm thick) | Osteoporotic (trabecular) | Compression | 26.89 ± 22.35 | 1.02 ± 0.79 | Cory et al. (2010) |
| Beams (1 mm thick) | Healthy | 3-Point bending | – | 5.12 ± 0.77 | Kasra et al. (1997) |
| Beams (1 mm thick) | Osteoporotic | 3-Point bending | – | 4.70 ± 0.98 | Kasra et al. (1997) |
| **SUB-MICROSTRUCTURAL LEVEL (SUB-LAMELLAR/MATERIAL LEVEL)** | | | | | |
| Lamellar (25 μm² indent) | Healthy (trabecular) | Nanoindentation | – | 16.1 ± 3.9 | Guo and Goldstein (2000) |
| Lamellar (25 μm² indent) | Osteoporotic (trabecular) | Nanoindentation | – | 15.8 ± 3.9 | Guo and Goldstein (2000) |
| Lamellar | Healthy (cortical) | Nanoindentation | – | 18.98 ± 4.78 | Cory et al. (2010) |
| Lamellar | Healthy (trabecular) | Nanoindentation | – | 18.27 ± 4.26 | Cory et al. (2010) |
| Lamellar | Healthy (trabecular) | Nanoindentation | – | 18.73 ± 0.71 | Maïmoun et al. (2012) |
| Lamellar | Osteoporotic (trabecular) | Nanoindentation | – | 16 ± 0.85 | Maïmoun et al. (2012) |
| Lamellar | Healthy (cortical) | Nanoindentation | – | 21.27 ± 1.2 | Maïmoun et al. (2012) |
| Lamellar | Osteoporotic (cortical) | Nanoindentation | – | 21.12 ± 1.12 | Maïmoun et al. (2012) |

avoid observable ion beam damage. FIB milling was additionally performed parallel to the produced sample faces in all preparation steps to reduce embedding the impinging gallium ions from the FIB within the discrete beam volumes produced. A short column between each of the micro-beams was retained in order to prevent the re-deposition of milled material and gallium ions on to neighboring beams. Thus, bone material sputtered from the FIB is more likely to redeposit on the short columns instead of the sample micro-beams. The average micro-beam dimensions produced at the end of the bone material sample were $8\,\mu m \times 2\,\mu m \times 2\,\mu m$, with the long axis of the micro-beam aligned in the direction of the long axis of the femur. These dimensions at micron length scales are comparable to bone lamellae and therefore remove structural features present at larger length scales.

Resultant bone micro-beams were removed from the SDB setup and placed in a closed vessel containing Hank's buffer solution for 2 h to allow bone rehydration. Samples were then returned to the SDB system for subsequent mechanical testing, with the prior sample rehydration shown to preserve the mechanical properties of the wet bone for up to 2 h in such an environment (Jimenez-Palomar et al., 2012). Mechanical testing of the bone micro-beams was carried out using an atomic force microscope (AFM) integrated within the SDB (Hang et al., 2011). A physiologically relevant compressive loading configuration was used, with the load applied in the direction of the long axis of the micro-beams. The AFM allowed the application of load to the micro-beams while scanning electron microscopy (SEM) within the SDB system was used to observe the deformation and resultant failure of the samples. Mechanical testing was achieved by first translating the AFM tip toward individual bone micro-beams until compressive load was applied parallel to the micro-beam long axis as shown in **Figure 1**. *In situ* SEM was used to observe the movement of the AFM tip toward the end of the micro-beam and ensure that the AFM tip fully contacted the top of the beam as shown in **Figure 2**. Compression of four osteoporotic and six healthy bone micro-beams was carried out under quasi-static loading rates.

## RESULTS

Compression of bone micro-beams using the AFM produced corresponding data for the force applied to the sample and resultant deformation. **Figure 2** shows SEM imaging highlighting the loading of the micro-beam sample in compression until failure of the sample occurred. The stress and strain induced in the bone micro-beams were calculated using the force-deflection curves generated from the AFM system. Stress is calculated by dividing the force applied to the sample from the AFM tip by the cross-sectional area of the micro-beam sample, measured from SEM, whereas strain is calculated by dividing the change in length by the total micro-beam sample length, as shown in Eqs 1 and 2 below.

$$\sigma = \frac{f}{A} \qquad (1)$$

$$\varepsilon = \frac{\Delta L}{L_0} \qquad (2)$$

Where $\sigma$ is the stress in the compressed micro-beam sample, $f$ is the force applied by the AFM, $A$ is the micro-beam cross-sectional

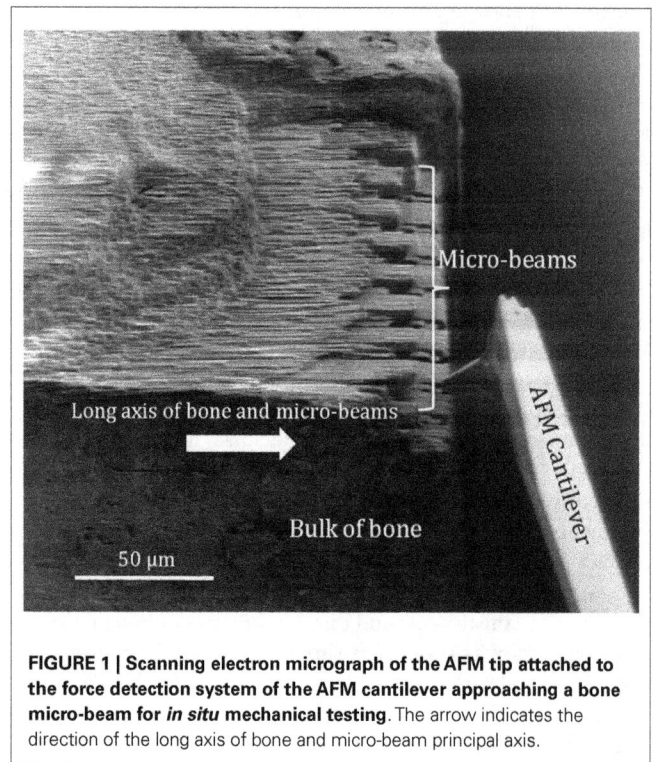

**FIGURE 1 | Scanning electron micrograph of the AFM tip attached to the force detection system of the AFM cantilever approaching a bone micro-beam for *in situ* mechanical testing.** The arrow indicates the direction of the long axis of bone and micro-beam principal axis.

area, $\varepsilon$ is the strain in the bone micro-beam, $\Delta L$ is change in length of the beam, and $L_0$ is the original length of the bone micro-beam prior to mechanical deformation. The resultant stress–strain behavior for the osteoporotic and control bone micro-beams is shown in **Figure 3**. The stress generally increases in a relatively linear manner with increasing strain until failure of the micro-beam. We note that local non-linearity is due to the interferometer measurement system of the AFM as described previously (Hang et al., 2011). The elastic modulus, strength, and the strain to failure values calculated from the bone micro-beam compression tests in **Figure 3** are shown in **Table 2** for the healthy bone control and the osteoporotic OVH model samples. The OVH bone has an average elastic modulus of $1.59 \pm 1.26$ GPa, almost half the value of the elastic modulus of $2.9 \pm 1.45$ GPa for the control sample. The OVH and control samples exhibit similar strength values of $169.23 \pm 21.35$ and $169.51 \pm 66.19$ MPa, respectively. A larger average strain to failure of 10% is recorded for the OVH samples when compared to ~6% for control samples. The toughness of the bone defined by the area under the stress–strain curves is ~8 and 5 J·m$^{-3}$ for OVH and control samples, respectively. The increased toughness displayed for the OVH samples is surprising as osteoporotic bone is commonly associated with brittle failure. However, we reiterate that the work presented here examines the material properties of bone and the enhanced toughness is due to increased strain to failure of the material. The fragility of osteoporotic bone associated with larger or whole bone samples is therefore absent when evaluating the small-scale material performance.

The mechanical property values recorded from the micro-beam compression of this work generally lie within the architectural range of previous literature as listed in **Table 1**. The

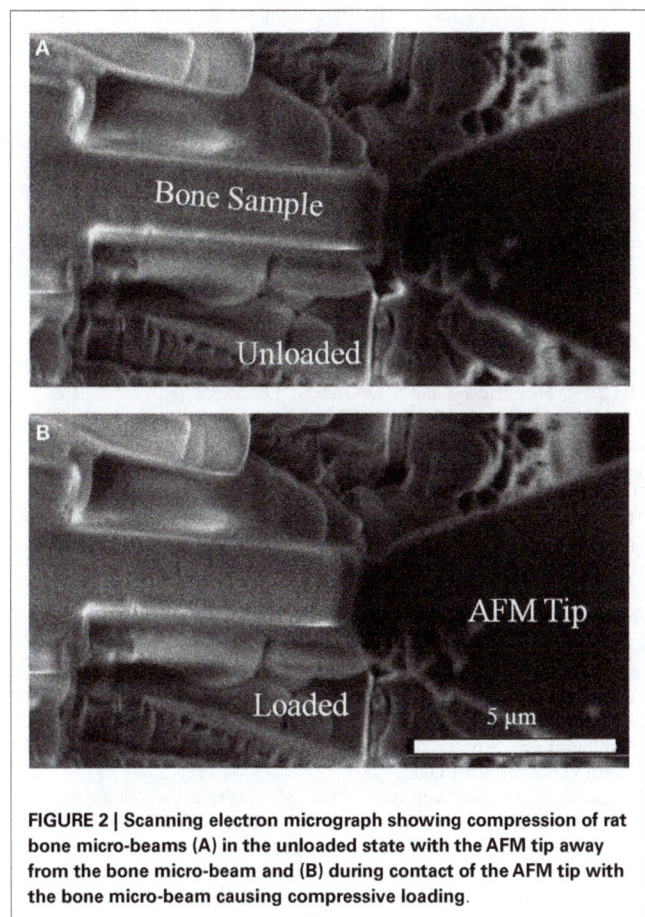

FIGURE 2 | Scanning electron micrograph showing compression of rat bone micro-beams (A) in the unloaded state with the AFM tip away from the bone micro-beam and (B) during contact of the AFM tip with the bone micro-beam causing compressive loading.

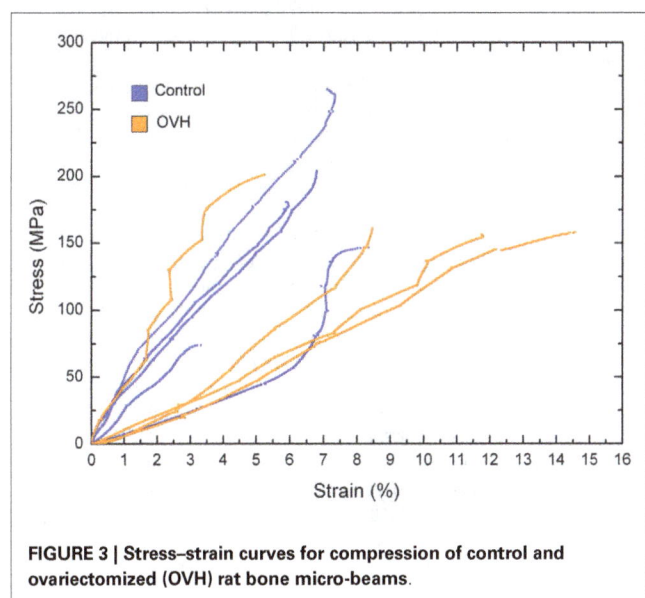

FIGURE 3 | Stress–strain curves for compression of control and ovariectomized (OVH) rat bone micro-beams.

Table 2 | Elastic modulus, strength, and strain to failure values of both control and ovariectomized (OVH) rat femur bone micro-beams tested in compression.

| Beam no. | Elastic modulus (GPa) | Strength (MPa) | Strain (%) |
|---|---|---|---|
| Control | | | |
| Average | $2.9 \pm 1.45$ | $169.51 \pm 66.19$ | $6.3 \pm 1.89$ |
| 1 | 3.06 | 204.22 | 6.8 |
| 4 | 3.62 | 180.53 | 5.88 |
| 5 | 2.37 | 73.91 | 3.32 |
| 6 | 4.65 | 248.35 | 7.16 |
| 8 | 0.78 | 140.52 | 8.35 |
| OVH | | | |
| Average | $1.59 \pm 1.26$ | $169.23 \pm 21.35$ | $10 \pm 4.04$ |
| 1 | 3.46 | 201.09 | 5.24 |
| 2 | 1.08 | 156.17 | 8.46 |
| 3 | 1.07 | 161.38 | 11.74 |
| 4 | 0.74 | 158.26 | 14.57 |

material due to osteoporosis. Indeed, error associated with the average elastic modulus values is expected to be due to variability in the orientation of collagen fibrils within the micro-beam that is measurable using the experimental setup. We suggest that this enhanced mechanical sensitivity of the technique is due to the removal of sample geometry effects, such as shape of the bone or porosity that is found at larger length scales, using FIB. Many bone structures also have strain to failure values considerably lower than our values and this may be attributed to the lack of geometric effects, such as porosity, which provide strain concentrations locally whereas the bulk of the material remains at lower strain. Indeed, glassy polymers such as polystyrene are analogous to this potential situation where mechanical testing of smaller volumes of material removes the effect of defects, causing increases in strain to failure (van der Sanden et al., 1993). Other mechanical testing techniques such as nanoindentation typically probe significantly smaller volumes than the micro-beams of this work and are potentially sensitive to more variability from the location of the indenting probe at the sample surface, with considerable issues related to the uncertainty in the composition of the material and resultant contact area with the indenting probe previously reviewed (Lewis and Nyman, 2008). The enhanced sensitivity of the micro-beam compression in our work indicates a clear decrease of bone elastic modulus properties with osteoporosis as shown in **Figure 3**, which potentially contradicts some works in **Table 1** that indicate little variability in the elastic modulus of osteoporotic bone compared to healthy bone. This lowering of the elastic modulus of osteoporotic bone has been suggested as being due to mechanical degradation of the collagen in osteoporotic bone from reductions in the level of immature collagen crosslinks and decreases in collagen fibril diameters (Currey, 2003). Compositional changes in collagen, such as the ratio of α1 to α2 chains in different phenotypes of *COLIA1* found in Type 1 collagen, appear to influence the fracture risk of bone that is independent to the changes in bone mass (McGuigan et al., 2001). A corresponding decrease in bone strength is not observed in our

average elastic modulus values in our micro-beam compression show larger variations between osteoporotic and healthy bone samples than the results in **Table 1**, highlighting the sensitivity of the technique in elucidating mechanical changes in bone work and suggests that the failure of the material does not change with osteoporosis although the interactions between constituents are affected. Indeed, molecular modeling has indicated considerable variation in mechanical behavior of collagen fibrils as a function of cross-linking density, including significant changes in elastic modulus as well as regime of cross-linking densities that provide minimal changes in strength (Buehler, 2008). We would therefore expect a decrease in elastic modulus of OVH samples as the stress transfer between protein molecules becomes inefficient from the previously reported changes in cross-linking density. Such a mechanism can additionally describe the lack of a loss in strength as the same collagen protein molecules are failing in both the healthy and OVH samples. The proposed mechanism here therefore describes deformation and failure of osteoporotic bone in terms of protein from the collagen in bone. Non-collagenous proteins (NCPs) present between collagen fibrils have also been shown to control fracture behavior of bone (Hang et al., 2014) and are known to chemically change in osteoporotic bone (Sroga and Vashishth, 2012). While the performance of osteoporotic bone material has been defined in this paper, the origin of the mechanical changes is still contentious, with collagen, mineral, NCPs, and their interactions all potentially contributing to mechanics. Such a complex synergy has been highlighted previously when considering compensation mechanisms where increased stiffness for the organic phase is balanced by a decrease in mineral content that results in similar nanoindentation hardness for diseased and healthy bone (Fratzl-Zelman et al., 2009). Future development of mechanistic explanations for osteoporotic bone therefore requires a comprehensive understanding of all constituent materials together.

## CONCLUSION

The compressive elastic modulus, strength, and strain to failure of bone micro-beams were measured in order to assess the effect of osteoporosis on the mechanical properties of bone as a material at the sub-lamellar level. Although compression testing herein cannot be directly compared to previous studies in the literature, results showed a decrease in the elastic modulus of osteoporotic bone compared to a control. This decrease in the elastic modulus with osteoporosis was additionally associated with relatively constant micro-beam strength and a small increase in failure strain, with associated changes in material toughness. The origin of osteoporotic induced decreases in bone elastic modulus was suggested as being due to mechanical degradation of the collagen within the bone material.

## REFERENCES

Ammann, P., and Rizzoli, R. (2003). Bone strength and its determinants. *Osteoporos. Int.* 14, 13–18. doi:10.1007/s00198-002-1345-4

Bailey, R. J., Geurts, R., Stokes, D. J., Jong, F. D., and Barber, A. H. (2013). Evaluating focused ion beam induced damage in soft materials. *Micron* 50, 51–56. doi:10.1016/j.micron.2013.04.005

Barengolts, E. I., Curry, D. J., Bapna, M. S., and Kukreja, S. C. (1993). Effects of endurance exercise on bone mass and mechanical properties in intact and ovariectomized rats. *J. Bone Miner. Res.* 8, 937–942. doi:10.1002/jbmr.5650080806

Boivin, G., and Baud, C. A. (1984). "Microradiographic methods for calcified tissues," in *Methods for Calcified Tissue Preparation*, ed. G. R. Dickson (Amsterdam: Elsevier), 391–411.

Boivin, G., and Meunier, P. J. (2002). The degree of mineralization of bone tissue measured by computerized quantitative contact microradiography. *Calcif. Tissue Int.* 70, 503–511. doi:10.1007/s00223-001-2048-0

Borah, B., Ritman, E. L., Dufresne, T. E., Jorgensen, S. M., Liu, S., Sacha, J., et al. (2005). The effect of risedronate on bone mineralization as measured by micro-computed tomography with synchrotron radiation: correlation to histomorphometric indices of turnover. *Bone* 37, 1–9. doi:10.1016/j.bone.2005.03.017

Boskey, A. (2003). Bone mineral crystal size. *Osteoporos. Int.* 14, 16–21. doi:10.1007/s00198-003-1468-2

Bouxsein, M. (2001). "Biomechanics of age-related fractures," in *Osteoporosis*, 2nd Edn, eds R. Marcus, D. Feldman, and J. Kelsey (San Diego, CA: Academic Press), 509–534.

Buehler, M. J. (2008). Nanomechanics of collagen fibrils under varying cross-link densities: atomistic and continuum studies. *J. Mech. Behav. Biomed. Mater.* 1, 59–67. doi:10.1016/j.jmbbm.2007.04.001.

Burr, D. (2003). Microdamage and bone strength. *Osteoporos. Int.* 14, S67–S72. doi:10.1007/s00198-003-1476-2

Burr, D. B. (2002). The contribution of the organic matrix to bone's material properties. *Bone* 31, 8–11. doi:10.1016/S8756-3282(02)00815-3

Burr, D. B., Forwood, M. R., Fyhrie, D. P., Martin, R. B., Schaffler, M. B., and Turner, C. H. (1997). Bone microdamage and skeletal fragility in osteoporotic and stress fractures. *J. Bone Miner. Res.* 12, 6–15. doi:10.1359/jbmr.1997.12.1.6

Burr, D. B., Schaffler, M. B., and Frederickson, R. G. (1988). Composition of the cement line and its possible mechanical role as a local interface in human compact bone. *J. Biomech.* 21, 939–945. doi:10.1016/0021-9290(88)90132-7

Carter, D. R., and Hayes, W. C. (1976). Bone compressive strength: the influence of density and strain rate. *Science* 149, 1174–1176. doi:10.1126/science.996549

Chan, Y. L., Ngan, A. H. W., and King, N. M. (2009). Use of focused ion beam milling for investigating the mechanical properties of biological tissues: a study of human primary molars. *J. Mech. Behav. Biomed. Mater.* 2, 375–383. doi:10.1016/j.jmbbm.2009.01.006

Ciarelli, T. E., Fyhrie, D. P., and Parfitt, A. M. (2003). Effects of vertebral bone fragility and bone formation rate on the mineralization levels of cancellous bone from white females. *Bone* 32, 311–315. doi:10.1016/S8756-3282(02)00975-4

Cory, E., Nazarian, A., Entezari, V., Vartanians, V., Müller, R., and Snyder, B. D. (2010). Compressive axial mechanical properties of rat bone as functions of bone volume fraction, apparent density and micro-ct based mineral density. *J. Biomech.* 43, 953–960. doi:10.1016/j.jbiomech.2009.10.047

Currey, J. D. (1988). The effect of porosity and mineral content on the young's modulus of elasticity of compact bone. *J. Biomech.* 21, 131–139. doi:10.1016/0021-9290(88)90006-1

Currey, J. D. (2003). Role of collagen and other organics in the mechanical properties of bone. *Osteoporos. Int.* 14, S29–S36. doi:10.1007/s00198-003-1470-8

Ejersted, C., Andreassen, T. T., Oxlund, H., Jorgensen, P. H., Bak, B., Haggblad, J., et al. (1993). Human parathyroid hormone (1-34) and (1-84) increase the mechanical strength and thickness of cortical bone in rats. *J. Bone Miner. Res.* 8, 1097–1101. doi:10.1002/jbmr.5650080910

Fratzl-Zelman, N., Roschger, P., Gourrier, A., Weber, M., Misof, B. M., Loveridge, N., et al. (2009). Combination of nanoindentation and quantitative backscattered electron imaging revealed altered bone material properties associated with femoral neck fragility. *Calcif. Tissue Int.* 85, 335–343. doi:10.1007/s00223-009-9289-8

Frost, H. M., and Jee, W. S. S. (1992). On the rat model of human osteopenias and osteoporoses. *Bone Miner.* 18, 227–236. doi:10.1016/0169-6009(92)90809-R

Grynpas, M. (1993). Age and disease-related changes in the mineral of bone. *Calcif. Tissue Int.* 53, S57–S64. doi:10.1007/BF01673403

Guo, X. E., and Goldstein, S. A. (2000). Vertebral trabecular bone microscopic tissue elastic modulus and hardness do not change in ovariectomized rats. *J. Orthop. Res.* 18, 333–336. doi:10.1002/jor.1100180224

Hang, F., and Barber, A. H. (2011). Nano-mechanical properties of individual mineralized collagen fibrils from bone tissue. *J. R. Soc. Interface* 8, 500–505. doi:10.1098/rsif.2010.0413

Hang, F., Gupta, H. S., and Barber, A. H. (2014). Nanointerfacial strength between non-collagenous protein and collagen fibrils in antler bone. *J. R. Soc. Interface* 11, 20130993. doi:10.1098/rsif.2013.0993

Hang, F., Lu, D., Bailey, R. J., Jimenez-Palomar, I., Stachewicz, U., Cortes-Ballesteros, B., et al. (2011). In situ tensile testing of nanofibers by combining atomic force microscopy and scanning electron microscopy. *Nanotechnology* 22, 365708. doi:10.1088/0957-4484/22/36/365708

Hauge, E. M., Steiniche, T., and Andreassen, T. T. (2003). "Histomorphometry of metabolic bone conditions," in *Handbook of Histology Methods for Bone and Cartilage*, eds Y. H. An and K. L. Martin (Totowa, NJ: Human Press Inc), 391–410.

Iacono, M. V. (2007). Osteoporosis: a national public health priority. *J. Perianesth. Nurs.* 22, 175–183. doi:10.1016/j.jopan.2007.03.009

Jimenez-Palomar, I., Shipov, A., Shahar, R., and Barber, A. H. (2012). Influence of SEM vacuum on bone micromechanics using in situ AFM. *J. Mech. Behav. Biomed. Mater.* 5, 149–155. doi:10.1016/j.jmbbm.2011.08.018

Jorgensen, P. H., Bak, B., and Andreassen, T. T. (1991). Mechanical properties and biochemical composition of rat cortical femur and tibia after long-term treatment with biosynthetic human growth hormone. *Bone* 12, 353–359. doi:10.1016/8756-3282(91)90022-B

Kasra, M., Vanin, C. M., Maclusky, N. J., Casper, R. F., and Grynpas, M. D. (1997). Effects of different estrogen and progestin regimens on the mechanical properties of rat femur. *J. Orthop. Res.* 15, 118–123. doi:10.1002/jor.1100150117

Katz, J. L., and Meunier, A. (1993). Scanning acoustic microscope studies of the elastic properties of osteons and osteon lamellae. *J. Biomech. Eng.* 115, 543–548. doi:10.1115/1.2895537

Keaveny, T. M., Morgan, E. F., Niebur, G. L., and Yeh, O. C. (2001). Biomechanics of trabecular bone. *Annu. Rev. Biomed. Eng.* 3, 307–333. doi:10.1146/annurev.bioeng.3.1.307

Kennedy, O. D., Brennan, O., Rackard, S. M., Staines, A., O'Brien, F. J., Taylor, D., et al. (2009). Effects of ovariectomy on bone turnover, porosity, and biomechanical properties in ovine compact bone 12 months postsurgery. *J. Orthop. Res.* 27, 303–309. doi:10.1002/jor.20750

Kilbanski, A., Adams-Campbell, L., Bassford, T., Blair, S. N., Boden, S. D., Dickersin, K., et al. (2001). Osteoporosis prevention, diagnosis, and therapy. *JAMA* 285, 785–795. doi:10.1001/jama.285.6.785

Lewis, G., and Nyman, J. S. (2008). The use of nanoindentation for characterizing the properties of mineralized hard tissues: state-of-the art review. *J. Biomed. Mater. Res. B Appl. Biomater.* 87, 286–301. doi:10.1002/jbm.b.31092

Maïmoun, L., Brennan-Speranza, T. C., Rizzoli, R., and Ammann, P. (2012). Effects of ovariectomy on the changes in microarchitecture and material level properties in response to hind leg disuse in female rats. *Bone* 51, 586–591. doi:10.1016/j.bone.2012.05.001

Marcus, R., and Bouxsein, M. L. (2010). "The nature of osteoporosis," in *Fundamentals of Osteoporosis*, eds R. Marcus, D. Feldman, D. A. Nelson, and C. J. Rosen (San Diego, CA: Academic Press), 25–34.

McCreadie, B. R., Morris, M. D., Chen, T., Sudhaker, R. D., Finney, W. F., Widjaja, E., et al. (2006). Bone tissue compositional differences in women with and without osteoporotic fracture. *Bone* 39, 1190–1195. doi:10.1016/j.bone.2006.06.008

McGuigan, F. E. A., Armbrecht, G., Smith, R., Felsenberg, D., Reid, D. M., and Ralston, S. H. (2001). Prediction of osteoporotic fractures by bone densitometry and COLIA1 genotyping: a prospective, population-based study in men and women. *Osteoporos. Int.* 12, 91–96. doi:10.1007/s001980170139

Paschalis, E. P., Shane, E., Lyritis, G., Skarantavos, G., Mendelsohn, R., and Boskey, A. L. (2004). Bone fragility and collagen cross-links. *J. Bone Miner. Res.* 19, 2000–2004. doi:10.1359/jbmr.040820

Rice, J. C., Cowin, S. C., and Bowman, J. A. (1988). On the dependence of the elasticity and strength of cancellous bone on apparent density. *J. Biomech.* 21, 155–168. doi:10.1016/0021-9290(88)90008-5

Roschger, P., Gupta, H. S., Berzlanovich, A., Ittner, G., Dempster, D. W., Fratzl, P., et al. (2003). Constant mineralization density distribution in cancellous human bone. *Bone* 32, 316–323. doi:10.1016/S8756-3282(02)00973-0

Roschger, P., Paschalis, E. P., Fratzl, P., and Klaushofer, K. (2008). Bone mineralization density distribution in health and disease. *Bone* 42, 456–466. doi:10.1016/j.bone.2007.10.021

Schaffler, M. B., and Burr, D. B. (1988). Stiffness of compact bone: effects of porosity and density. *J. Biomech.* 21, 13–16. doi:10.1016/0021-9290(88)90186-8

Silva, M. J., Brodt, M. D., Fan, Z., and Rho, J. Y. (2004). Nanoindentation and whole-bone bending estimates of material properties in bones from the senescence accelerated mouse SAMP6. *J. Biomech.* 37, 1639–1646. doi:10.1016/j.jbiomech.2004.02.018

Silva, M. J., and Gibson, L. J. (1997). Modeling the mechanical behavior of vertebral trabecular bone: effects of age-related changes in microstructure. *Bone* 21, 191–199. doi:10.1016/S8756-3282(97)00100-2

Sroga, G. E., and Vashishth, D. (2012). Effects of bone matrix proteins on fracture and fragility in osteoporosis. *Curr. Osteoporos. Rep.* 10, 141–150. doi:10.1007/s11914-012-0103-6

Turner, C. H. (2002). Biomechanics of bone: determinants of skeletal fragility and bone quality. *Osteoporos. Int.* 13, 97–104. doi:10.1007/s001980200000

van der Sanden, M. C. M., Meijer, H. E. H., and Lemstra, P. J. (1993). Deformation and toughness of polymeric systems: 1. The concept of critical thickness. *Polymer* 34, 2148–2154. doi:10.1016/0032-3861(93)90249-A

WHO. (2012). *National Osteoporosis Guideline Group (NOGG) [Online]*. Sheffield: University of Sheffield. Available at: http://www.shef.ac.uk/NOGG/

Wu, Z.-X., Lei, W., Hu, Y.-Y., Wang, H.-Q., Wan, S.-Y., Ma, Z.-S., et al. (2008). Effect of ovariectomy on BMD, micro-architecture and biomechanics of cortical and cancellous bones in a sheep model. *Med. Eng. Phys.* 30, 1112–1118. doi:10.1016/j.medengphy.2008.01.007

Zebaze, R. M. D., Ghasem-Zadeh, A., Bohte, A., Iuliano-Burns, S., Mirams, M., Price, R. I., et al. (2010). Intracortical remodelling and porosity in the distal radius and post-mortem femurs of women: a cross-sectional study. *Lancet* 375, 1729–1736. doi:10.1016/S0140-6736(10)60320-0

**Conflict of Interest Statement:** The authors declare that the research was conducted in the absence of any commercial or financial relationships that could be construed as a potential conflict of interest.

# 7

# Hydrogels with micellar hydrophobic (nano)domains

*Miloslav Pekař\**

Faculty of Chemistry, Brno University of Technology, Brno, Czech Republic

**Edited by:**
P. Davide Cozzoli, University of
Salento, Italy

**Reviewed by:**
Gerardino D'Errico, Università degli
Studi di Napoli Federico II, Italy
Yatender Kumar Bhardwaj,
Government of India, India

**\*Correspondence:**
Miloslav Pekař, Faculty of Chemistry,
Brno University of Technology,
Purkyňova 118, Brno 612 00, Czech
Republic
e-mail: pekar@fch.vutbr.cz

Hydrogels containing hydrophobic domains or nanodomains, especially of the micellar type, are reviewed. Examples of the reasons for introducing hydrophobic domains into hydrophilic gels are given; typology of these materials is introduced. Synthesis routes are exemplified and properties of a variety of such hydrogels in relation with their intended applications are described. Future research needs are identified briefly.

**Keywords: amphiphilic materials, hydrogel, hydrophobic domains, micelle, nanostructuring, surfactant**

## GENERAL INTRODUCTION

Gels are well-known materials of colloidal type in which the dispersion medium (phase) is in liquid state and the dispersed phase (colloidal "particles") is in solid state. The dispersed phase forms a network throughout the whole gel sample and the liquid is, in fact, entrapped in this mesh. In other words, the liquid swells the solid matter. The entrapment of a liquid within a solid network is the base of specific and unique properties of gels. Gels combine the behavior of liquids and solids in a soft matter, which usually retains its shape, but is relatively easily deformable. Specific behavior of a particular gel sample is determined by the density of its network or the density of crosslinks connecting the network. The crosslinks provide the structure and physical integrity of the gel network. Typically, gels can be characterized as viscoelastic solids (and liquids in a broader sense). If the liquid part is water, the gel is called hydrogel.

Gels in general are soft materials of significant interest for applications in various areas of industry. Hydrogels, composed usually of crosslinked chains of hydrophilic polymers, exhibit distinct property of good swelling capacity in aqueous media. If there are no problems of biocompatibility, they can have widespread applications in different fields related to human health, appearance, or nutrition including pharmaceutical, bioengineering, food industry, medicine, cosmetics, agriculture and horticulture, and so on. Hydrogels, prepared both from synthetic polymers and biopolymers, which are similar to biological tissues and compatible with them have been important materials for drug delivery and tissue engineering. Hydrogels are applied as soft contact lenses, artificial implants, actuators, wound healing dressings, etc. (Shukla et al., 2009). Specifically designed hydrogels capable of a sensitive response to external stimuli such as pH, temperature, ionic strength, magnetic field, electric field, ultraviolet light, and internal stimuli-like chemical architecture, initiators conditions enable them to deserve as potential candidates for biomaterial applications (Shukla et al., 2009).

Hydrogels can be thus be viewed as elastic polymer networks that absorb large amounts of water. Typical hydrogels are prepared by chemical or physical crosslinking of hydrophilic polymers, such as polyacrylamide, poly(ethylene oxide), or poly(2-hydroxyethyl methyl acrylate) (Hao and Weiss, 2011), or biopolymers such as native or modified polysaccharides. Chemically, crosslinked hydrogels are formed from covalent bonds, and thus, they have a permanent shape, which resists to changes in temperature within the range of hydrogel thermal stability. The shape can be deformed by the application of stress and recovered when the stress is removed or if the stress is low enough. Because of the irreversible feature of a covalent bond, such hydrogels cannot be manufactured into controlled shapes by melt processing methods such as injection molding or compression molding, except for cases where the covalent network is formed during that process.

Physical hydrogels have transient crosslinks formed by associations among polymeric chains bearing complementary functional groups that exhibit, for example, electrostatic interactions, chain entanglements, hydrophobic interactions, and hydrogen bonding or produce crystallizing segments (Hao and Weiss, 2011). Because of the reversible nature of non-covalent intermolecular interactions upon temperature changes, the shape of physical hydrogels can be reversibly changed (destroyed and renewed) by changes in temperature and these gels can be melt processed into a desired shape. Although physical hydrogels do not possess a permanent (covalent) network structure, they do exhibit elasticity if the relaxation time of the network is much longer than the application time of stress. That is, physical elastic networks can be achieved when the stresses are either small, the temperature is sufficiently low, or the stress application is of short duration (i.e., short time or high frequency). Under such conditions, a physical gel may be mechanically indistinguishable from a covalent gel, and the network appears to be permanent. The reversible aspect of the physical network structure provides unique properties, such as viscous flow above a critical stress or time that allows the gel to be injectable and improved mechanical toughness compared with a covalent gel.

In general, three-dimensional network structure of (hydro)gels is responsible for their mechanical properties and porous microstructure formed in this network provide permeability and filtration effect.

With the advances of nanotechnology besides the traditional macroscopic gels, macrogels, also micro-, and recently nanogels (Yallapu et al., 2007) have been designed, synthetized, and studied. Microgel is a dispersion of discrete polymeric gel particles, which are in the size range typically from 1 mm to 1 μm. Similarly, nanogels can be described as gel macromolecules in the size range of tens to hundreds nanometers though various definitions of nanogels have been given.

After this Section "General Introduction," let us move to the main topic of this tractate.

## WHY HYDROPHOBIC DOMAINS IN HYDROGELS

This review is focused on hydrogels containing hydrophobic domains or nanodomains, especially of the micellar type. Why to combine hydrogels with something of opposite character and properties? The answer can be simple – to modify, upgrade, or tailor properties of hydrogels. The improved properties are usually dictated by hydrogel applications, many of them coming from the fields of medicine and health-care. Let us review the main application requirements in general and with selected reference examples. The purpose is not to give an exhaustive overview but to point to leading approaches and "hot topics" in this area including their variability and to describe them rather thoroughly. Hydrogels with micellar domains are hybrid colloidal materials of specific properties, which deserve more attention and study, are seen. For example, the research topic phrase "hydrogels incorporating micelles" returned in SciFinder database only about 20 hits (in July 2014); similarly, the phrases "hydrogels and micelles" or "hydrogels and micelles." It is hoped that this review will encourage more intensive research in this interesting material field.

Nature is a great inspiration and in the same time puts the principal requirements. Biocompatible hydrogels are in dramatically increasing demand that exhibit complex behavior, but are nonetheless conceptually easy to prepare and to process, while having accessible handles with which their mechanical and physiological properties can be tailored (Guo et al., 2014). In general, gels that are highly swollen by a liquid are very weak because they do not have many intrinsic mechanisms to dissipate energy during their deformation. While the frictional properties of hydrogels may be desirable, many hydrogels suffer from low shear strength. Many hydrogels lack the required mechanical properties to be useful as articulating and weight bearing materials. To obtain a hydrogel with a high degree of mechanical toughness, additional dissipative mechanisms at the molecular level is needed. Proper molecular design can impart hydrogels with a suite of attractive properties: improvements in tensile toughness, in strength and resilience while saving the high water uptake, introduction of dynamic physical (i.e., usually non-covalent) crosslinks to promote processability and self-healing, stimuli responsiveness, or shape memory behavior.

Manipulating and controlling hydrogel structures at nanometer level are very effective ways in the design of precisely defined three-dimensional structures, which enable tailoring their mechanical properties. The nanometric scaled structures with different morphologies incorporated into the hydrogel contribute to the control over the hydrogel mechanical and physical properties. The development of hydrogel nanostructuring methods is still challenging because the high water content excludes the use of lithographic techniques. On the other hand, hydrogels nanostructuring can be approached generally and relatively easily by the self-assembly capability of hydrophobic moieties in aqueous environment. Domains formed by the self-assembly impart hydrogels' new properties not only from the mechanical but also from the functional point of view as will be demonstrated in this review.

The behavior of hydrogels with incorporated hydrophobic (nano)domains resemble the behavior of semi-crystalline polymers, which contain both crystalline regions and amorphous regions (Thomas et al., 2009). Crystallinity is known to have desired effect on the mechanical properties and solubility of polymers. It can be proposed that the hydrophilic segments (resembling amorphous regions) in such a hydrogel could provide the water absorption, fluid flow, and lubricious properties needed while the hydrophobic segments (playing the role of crystalline regions) provide the strength, tear, and shear resistance. Properties may be varied based on the ratio of hydrophobic to hydrophilic composition. The polymers must overcome the heat and entropy factors associated with the crystalline or hydrophobic regions during dissolution, the hydrophobized hydrogels are thus expected to have excellent dissolution resistance in aqueous environment (e.g., in the body). On the other hand, it must be considered that the creation of hydrophobic domains can normally result in a change in the maximum hydration (swelling) as well as in the organization of water molecules within hydrogel structure.

Micelles formed by surfactants at and above the critical micellar concentration are well-known examples of the self-assembly due to hydrophobic interactions in aqueous environment. Moreover, not only low molecular surfactants but also properly structured polymers (or modified biopolymers) may form micelles, called polymeric micelles in this case. Nowadays generally, surfactant and polymer systems play a more and more important role in modern drug delivery, where they may be used to control drug release rate, enhance effective drug solubility, minimize drug degradation, reduce drug toxicity, or facilitate control of drug uptake (Liu and Li, 2005). Moreover, the hydrophobic domains of surfactant micelles allow the solubilization of lipophilic substances. The amphiphilic surfactant structure also provides a potential in increasing the permeability of the drug through biological membranes, resulting in an augmented intracellular drug concentration. In fact, micellar solutions have been proposed as efficient strategies for drug delivery a couple of decades ago. They are able to modulate both the pharmacokinetics and the bioavailability of the drug to result in an overall increase in the drug therapeutic index. Similarly, the incorporation of micelles into hydrogel is expected to result in changes not only of material properties but also in molecular transport (diffusion) within a hydrogel.

With the reference to the medical applications, stimuli-sensitive (sometimes called "intelligent") and injectable (*in situ* gelling) hydrogels should be mentioned. The latters can be administered into the body via a minimally invasive route, which is one of their principal advantages. The injectable hydrogels are mostly based on self-assembled nanosized polymeric micelles. When injected into the body using a syringe, sudden changes of surrounding environments such as temperature, pH, and other biochemical signals alter their conformation and/or their non-covalent interactions, resulting in a sol-to-gel phase transition at critical conditions.

These general reasons for the incorporation of hydrophobic (nano)domains into hydrogels will be now illustrated by a selection of published specific examples. Although the main focus is on the micellar domains, related inspiring systems could also be included.

Liu and Li (2005) used micelles in hydrogel for the purpose of drug solubilization, immobilization, and protection. The drug was camptothecin, which exhibits high anti-cancer activity against a wide spectrum of human malignancies. Because it is practically water-insoluble, its full therapeutic potential had not been achieved. Moreover, *in vivo* and in aqueous solutions buffered at pH above seven it rapidly converts to the respective water-soluble, pharmacologically less active carboxylate form. Camptothecin was solubilized by means of an ionic surfactant, sodium dodecyl sulfate (SDS), and the loaded micelles were mixed into agarose hydrogel. This method not only helped to increase the solubility of the drug but also promoted a good and stable distribution of the drug molecules in the hydrogel, which participated in the controlled released function as well.

Ju et al. (2013) solubilized paclitaxel, a well-known anti-cancer agent with a poor water-solubility, in hydrophobically modified chitosan capable of forming (polymeric) micelles. The loaded micelles were dispersed in an interpenetrated hydrogel formed by a hydrophilic linear triblock polymer and carboxymethyl chitosan, specifically designed for the *in situ* cancer treatment. The triblock polymer provided gelation at elevated temperature when applied intratumorally. The carboxymethylated chitosan was chemically crosslinked to generate a network preventing the fast erosion of the triblock polymer gel. The gelation and structure are schematically shown in **Figure 1**. The micellar encapsulation reduced

the side-toxicity of the drug and the whole system demonstrated prolonging the *in vitro* drug release to a large extent.

Wang et al. (2014) tried to develop a system for oral delivery of docetaxel, another anti-cancer agent. This system should provide solution to two main tasks – to overcome the low water-solubility of docetaxel and to target the drug to the intestinal track. Docetaxel was encapsulated in polymeric micelles based on amphiphilic block copolymer, which improved its solubility and permeability. The loaded micelles were incorporated into pH-sensitive hydrogel. The ideal oral delivery hydrogel should shrink with drug entrapped in at low pH environment of stomach, and swell to release drug at high pH environment of intestine. Docetaxel micellar encapsulation should also prevent the identification of docetaxel by proteins, which would cause its efflux and change the transport channel of this drug in the gastrointestinal walls.

Yom-Tov et al. (2014) utilized the self-assembly capability of block and graft copolymers driven by hydrophobic interactions between the blocks for nanostructuring of hydrogels. One of the main observed problems was the limited stability of micelles formed in the hydrogels, mostly attributed to their diffusion out of the hydrogels. A novel method for nanostructuring was therefore developed based on embedding micelles in a hydrogel while anchoring some of their molecules to the surrounding network through their endgroups. The anchored molecules should provide a means to further crosslink the hydrogel and stabilize the network.

Shukla et al. (2009) grafted hydrophobic moieties onto hydrophilic polymer, which forms physical hydrogels. Grafted nanohydrogel was produced, which structurally contained hydrophobic nanodomains. The nanodomains enabled to control water sorption properties and alter physical properties of the hydrogel. Similarly, Inoue et al. (1997) grafted a hydrophobic oligomer on a lightly crosslinked polyelectrolytic hydrogel. The resulting hydrophobic domain structure was intended to be applied as a bioadhesive hydrogel for controlled release of either cationic drugs or hydrophobic drugs. The hydrophobic domains were also supposed to strengthen and control the pore structure

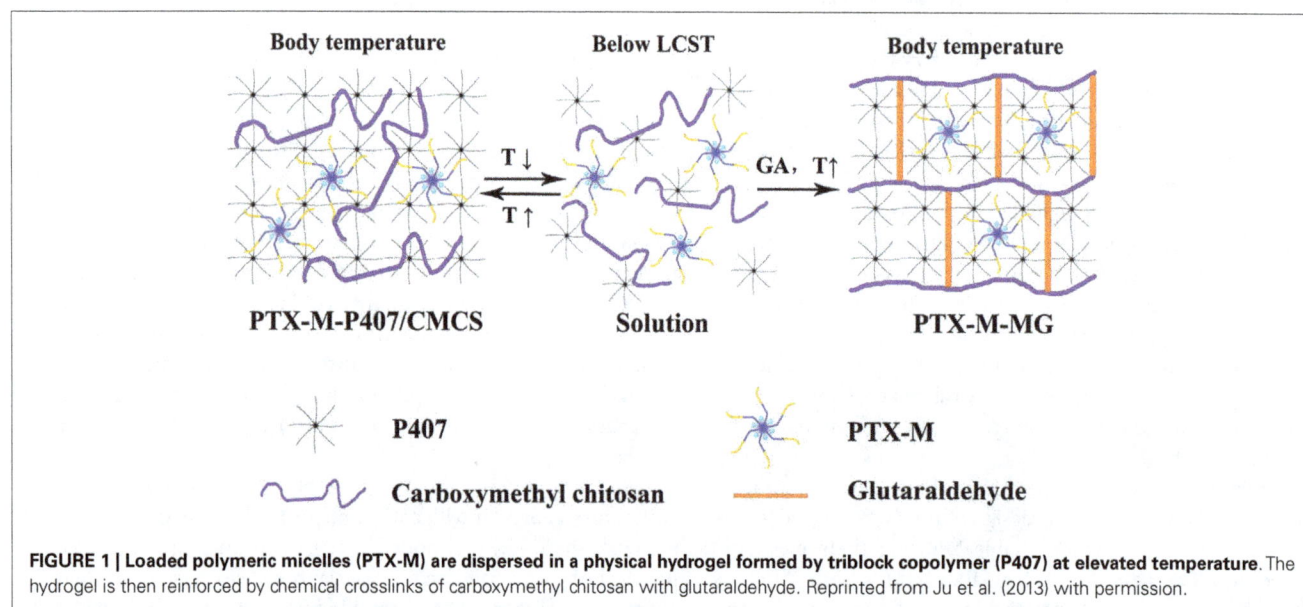

**FIGURE 1 | Loaded polymeric micelles (PTX-M) are dispersed in a physical hydrogel formed by triblock copolymer (P407) at elevated temperature.** The hydrogel is then reinforced by chemical crosslinks of carboxymethyl chitosan with glutaraldehyde. Reprinted from Ju et al. (2013) with permission.

within the hydrogel. Also Kim et al. (2014) grafted hydrophobic chains on the hydrophilic backbone of physically crosslinked hydrogel. Micelles were formed from the hydrophobic grafts in the resulting material, which were capable to solubilize a water-insoluble macrophage recruitment agent. The whole system was designed for biomaterial scaffolds, which could provide a suitable environment for cell-based bone regeneration.

Tedeschi et al. (2006) incorporated cationic surfactant micelles (decyltrimethylammonium bromide) into physical poly(vinyl alcohol) hydrogels prepared by the well-known freeze/thaw method. These hydrogels are known for their ability to include inside their porous network variety of water-soluble molecules. This study was aimed at the analysis of the behavior of surfactant molecules in the hydrogel pore matrix and at the study of diffusion of micelles within the hydrogel. The latter is important to understand the dynamics of guest molecules within hydrogels containing micelles especially in relation to their potential use as molecular carriers.

Bromberg (2005) gives an overview of intelligent hydrogels for the oral delivery of (hydrophobic) anti-cancer drugs. Micellar domains within such hydrogels serve to solubilize the hydrophobic agents and participate in their protection during transport within the gastrointestinal tract and in their controlled release. The micelles are of polymeric types and formed by Pluronics, which are a part of gelling copolymer.

Guo et al. (2014) aimed at the improvement of mechanical properties of hydrogels for biomaterial applications, particularly their weakness and brittleness, while retaining their processability. This was achieved by including strong, yet reversible interactions within segregated nanoscopic domains. A multiblock copolymer based on hydrophilic polymer was complemented by ureidopyrimidinone (UPy) moieties contained within the polymeric backbone. These self-complementary units were encapsulated within hydrophobic domains and assembled into strong dimers by fourfold hydrogen bonding, acting to reinforce networks, while the non-covalent nature promoted processability, in contrast to chemical crosslinks in conventional thermosets.

The work by Thomas et al. (2009) was motivated by the design of synthetic replacement for cartilage lost due to osteoarthritis. While hydrogels had shown promise for this application, many of them lacked the required mechanical properties. In this work, hydrogel-forming polymer was blended with a hydrophobic polymeric partner by a mechanical mixing. The hydrophobic regions formed by the hydrophobic component were believed to act as crystalline domains in semi-crystalline polymers with a positive effect on the mechanical properties but without negative effects on the coefficient of friction.

Injectable hydrogels are mostly based on self-assembled micelles composed of amphiphilic block copolymers. Upon injection of the initial sol, sudden changes of surrounding environment induce its gelation. Pluronic copolymers are example of such materials. Their sol–gel transition is induced by the changes in temperature leading to the self-association of these copolymers into spherical micelles. Above the transition temperature, which is around the body temperature, the spherical micelles are closely packed together, resulting in the physically crosslinked hydrogels. Several shortcomings including low mechanical strength, fast

dissolution, and rapid drug release under physiological conditions limited their biomedical applications. Chemical crosslinking is one approach to overcome them. To avoid side-effects due to the use of toxic crosslinking agents, Lee et al. (2011) used a bio-inspired, enzyme-mediated crosslinking of functionalized Pluronic micelles for fabricating high-strength injectable hydrogels.

Missirlis et al. (2005) primarily used Pluronic micelles as a solubilization medium for hydrophobic drugs with a prolonged circulation time. They tried to overcome the drawbacks of using these micelles directly (e.g., premature renal excretion or low stability upon dilution in the blood stream) by bridging between micelles to form macromolecular nanogels. This goal was achieved by inverse emulsion photochemical copolymerization with functionalized poly(ethylene glycol) (PEG). Hydrophobic nanodomains within the particles remained after copolymerization and were capable of absorbing large amounts of hydrophobic drugs.

In the work by Hao and Weiss (2011), strong hydrophobic interactions were used to physically crosslink acryl amide-based hydrogels in the swelled state. An acrylic monomer was copolymerized with a fluoroacrylate as the hydrophobic component. The hydrophobic associations of fluorocarbon hydrophobes in water medium created nanodomains that served as multifunctional physical crosslinks (see **Figure 2**). High stress or slow deformation could be used to achieve viscous flow, but gelation occurred when the stress was removed. These properties, resulting from the dynamic equilibrium nature of the hydrophobic associations, are attractive for injectable hydrogel materials. Moreover, fluorocarbon groups possess excellent chemical and biological inertness, good thermal stability, high fluidity, and low surface energy, which are useful for biomedical applications.

In the area of tissue engineering, an ideal scaffold should be able to provide a well-defined microenvironment to promote cell adhesion, proliferation, and differentiation with biodegradability and good biocompatibility. Fabrication of three-dimensional scaffold that can physically support cell infiltration and biologically direct

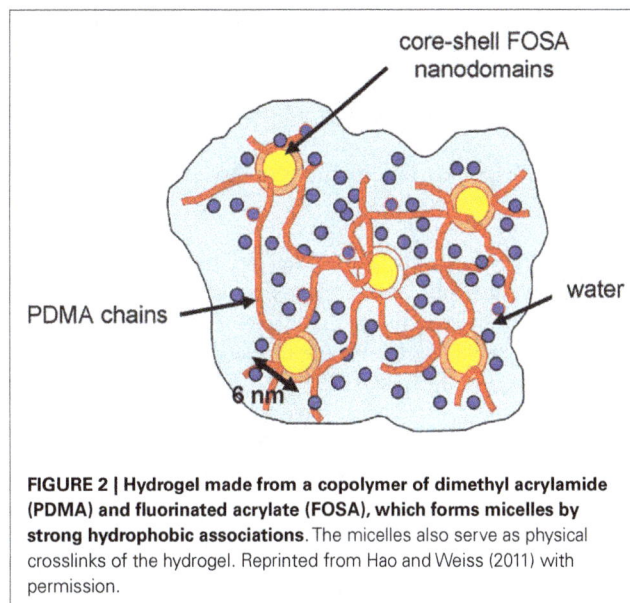

FIGURE 2 | Hydrogel made from a copolymer of dimethyl acrylamide (PDMA) and fluorinated acrylate (FOSA), which forms micelles by strong hydrophobic associations. The micelles also serve as physical crosslinks of the hydrogel. Reprinted from Hao and Weiss (2011) with permission.

cell behavior remains a challenge. Physical incorporation of polymeric micelles into hydrogels was used to tune the storage modulus, thereby influencing cell behavior in the hydrogel. Li et al. (2012) improved this approach by covalently incorporating nano-sized polymeric micelles self-assembled from an amphiphilic block copolymer with crosslinkable functional groups into hydrogel networks.

Frisman et al. (2012) introduced micellar nanostructuring into hydrogels intended as scaffolds for the use in tissue engineering. The nanostructuring should promote cell–matrix interactions and thus provide the ability to control cellular fate and tissue morphogenesis. Scaffold nanostructures guiding cell–matrix interactions mimic environment that is familiar to cells from their native extracellular matrix.

Autonomous damage repair and resulting self-healing in hydrogels require reversible breakable bonds, which prevent the fracture of the molecular backbone. Consequently, self-healing of permanently crosslinked hydrogels is a challenging task because of the irreversible nature of chemical crosslinks. Tuncaboylu et al. (2013) developed self-healing hydrogels by the creation of strong hydrophobic interactions between hydrophilic polymers. Hydrophobe monomer was copolymerized with hydrophilic monomer via micellar polymerization technique in aqueous surfactant solution. Incorporation of hydrophobic sequences within the hydrophilic chains generated strong hydrophobic interactions, which prevented dissolution of the physical gels in water, while the dynamic nature of the junction zones provided homogeneity and self-healing properties.

At the end of this section, let us summarize the reasons for incorporating the hydrophobic (micellar) domains into hydrogels. These domains can

- physically crosslink hydrogels and, when supplied with suitable functional groups, can further serve as structured chemical crosslinks;
- be used for the structuring of hydrogels, which is closely related to;
- controlling (improving, tailoring) of mechanical properties of hydrogels;
- solubilize hydrophobic, usually biologically active, substances, participate on their stabilization, protection, and controlled release;
- impart the self-healing and stimuli-responsive properties to hydrogels.

Moreover, when traditional surfactants are used in hydrogel preparation technology, the general benefits of applying surfactants can be explored like reducing the surface tension, increasing the conductivity, stabilizing the bubble formation, or improving the hydrogel or fiber uniformity.

## CLASSIFICATION OF HYDROGELS WITH MICELLAR HYDROPHOBIC (NANO)DOMAINS

As follows from the published examples reviewed in the preceding section, three basic types of hydrogels containing micelles, which form hydrophobic domains or nanodomains can be identified. Of course, materials represented by a combination of the basic types cannot be excluded.

1. Hydrogels with dispersed micelles. They are formed by a simple mixture of micelles and hydrogel matrix. As examples, the systems reported by Liu and Li (2005), Tedeschi et al. (2006), Mangiapia et al. (2007), Stoppel et al. (2011), Frisman et al. (2012), Ju et al. (2013), or Wang et al. (2014) can be given.
2. Hydrogels with integrated micelles. These may be further divided in two subtypes:
   a) at least some of dispersed micelles are anchored to the hydrogel network as, for example, in the work by Yom-Tov et al. (2014);
   b) micelles (domains) are formed by hydrophobic segments (usually blocks or grafts) located directly on the gel-forming precursor(s), for example, see in references Shukla et al. (2009), Inoue et al. (1997), Kim et al. (2014), Bromberg (2005), Guo et al. (2014), or Thomas et al. (2009).
3. Hydrogels with micellar crosslinks. Micelles (hydrophobic domains in general) form directly either physical or chemical crosslinks of the hydrogel network. Such materials are described, e.g., by Lee et al. (2011), Missirlis et al. (2005), Hao and Weiss (2011), Li et al. (2012), or Tuncaboylu et al. (2013). Here belong also physical hydrogels formed by mixing polyelectrolytes with oppositely charged surfactants under proper ratio and conditions. Surprisingly, these materials almost have not been studied and explored; currently, they are subject of intensive investigation in our laboratory.

## PREPARATION

Various procedures enabling to prepare hydrogels with hydrophobic domains are illustrated in this section by selected examples. In order to give a reader a good idea on their practical realizations, they will be reviewed with adequate details. The order follows the typology introduced in the preceding section.

The preparation used by Liu and Li (2005) was motivated by the solubilization and protection of camptothecin or its derivatives, anti-cancer agents against a wide spectrum of human malignancies. These substances are practically insoluble in water and in vivo or in aqueous solutions buffered at pH above 7, they rapidly convert to the respective water-soluble, pharmacologically less active carboxylate form. First, a homogeneous aqueous solution of camptothecin in micellar surfactant solution of an ionic surfactant, SDS, was prepared. Then, a weighed amount of agarose was added to the solution of camptothecin and surfactant and the mixture was slowly heated to around 90°C until an optically clear solution was obtained. The resultant solution was poured into a flat mold of about 1 mm in thickness, which was kept in an oven at 90°C before use. The warm polymer-surfactant-drug solution in the mold was allowed to cool down naturally at room temperature to form a sheet hydrogel. The experimental procedure and the expected structures are illustrated in **Figure 3**. For comparison, saturated camptothecin solutions in surfactant solutions below the critical micellar concentration were prepared and incorporated into hydrogels.

Poly(vinyl alcohol) hydrogel was prepared by Tedeschi et al. (2006) through the standard freeze/thaw procedure from an aqueous polymer solution (averaged $M_w$ of about 115,000, degree of hydrolysis of 98–99%, and concentration of 11% w/w). The polymer was dissolved at elevated temperature, slowly cooled

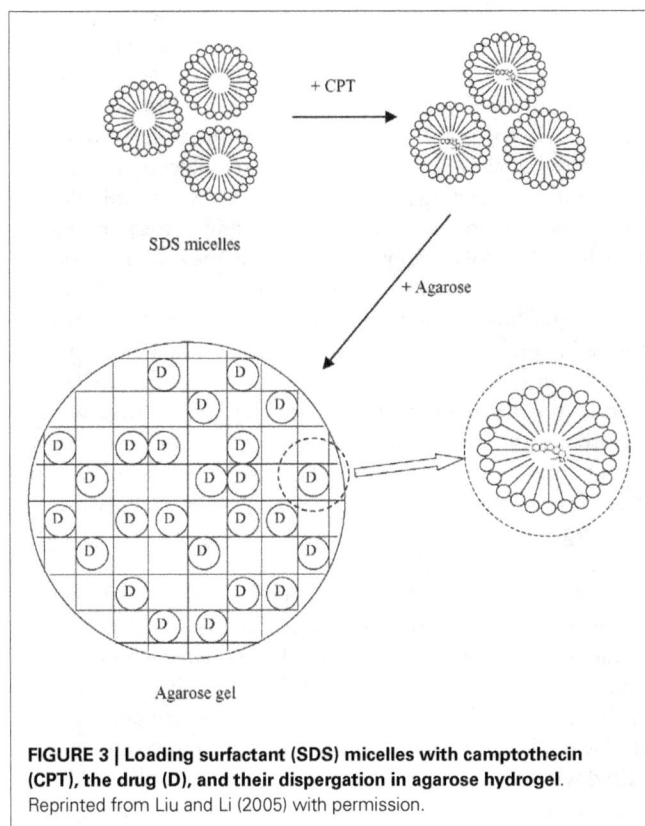

**FIGURE 3 | Loading surfactant (SDS) micelles with camptothecin (CPT), the drug (D), and their dispergation in agarose hydrogel.** Reprinted from Liu and Li (2005) with permission.

to room temperature, and left overnight at this temperature to enable the escape of air bubbles. The solution was then poured between glass slides with 1 mm spacers and subjected to three repeated freeze/thaw cycles −22°C/25°C, respectively. Dried gels were rehydrated for 24 h at 25°C by a large excess of surfactant aqueous solution at concentrations ranging between 0.006 and 0.276 mol kg$^{-1}$ (the critical micellar concentration in water of the used surfactant was 0.060 mol kg$^{-1}$). Very similar system was described also in the study by Mangiapia et al. (2007).

Stoppel et al. (2011) incorporated Pluronic F68 micelles in an alginate hydrogel. Sodium alginate solutions were prepared at 1% (w/v) with the addition of 0.1% glucose and HEPES buffer and stirred for 24 h at room temperature. Pluronic F68 was added to the alginate solution and stirred for an additional 24 h at room temperature. Two common methods were used to crosslink alginate: (1) an inhomogeneous or external method, using CaCl$_2$ or BaCl$_2$, (2) a homogeneous or internal method, which utilizes *in situ* release of calcium ions from pH-sensitive CaEDTA decomposition. In both methods, the Pluronic-containing alginate solutions were used. The external method was realized by placing the alginate solution into 10 mm dialysis tubing with a molecular weight cut-off of 3–5 kDa (the viscosity-averaged molecular weight of used alginate was declared as about 240 kmol/g) and submerged in either calcium (1 mol/l) or barium (0.5 mol/l) chloride with a specified Pluronic F68 concentration for 48 h. The obtained cylindrical hydrogels were freeze-dried. In the internal gelation technique, the alginate solution was mixed with CaEDTA (50 mmol/l) and glucono-δ-lactone (50 mmol/l) solution and after

12 h of initial crosslinking, hydrogels were immersed in calcium chloride (100 mmol/l) solution containing the same concentration of surfactant (to prevent surfactant leaching) in a humidified chamber (humidity 90% at least) at 25°C for an additional 48 h.

Frisman et al. (2012) prepared a biosynthetic hydrogel scaffold from crosslinked PEG–fibrinogen conjugates modified with the block copolymer Pluronic F127. Pluronic self-assembly into micelles created the desired nanostructures within the hydrogel. PEG-modified fibrinogen with acrylate end groups was used as hydrogel precursor and its solution was thoroughly mixed with Pluronic F127 at 4°C. After equilibration (to ensure micelle formation), the mixture was combined with a photoinitiator solution, equilibrated in a water bath at 37°C, and finally irradiated by UV light (365 nm) to induce crosslinking of the acrylate end groups on the pendant PEG chains.

Ju et al. (2013) upgraded the technique for the solubilization of hydrophobic drugs (paclitaxel in their case) in polymeric micelles formed by Poloxamer 407, which itself also forms hydrogel. Paclitaxel was dissolved in micelles of N-octyl-O-sulfate chitosan, the loaded micelles were dispersed in Poloxamer 407 gels, interpenetrated by a network formed by crosslinking carboxymethyl chitosan with glutaraldehyde. Paclitaxel was solubilized in the micelles by dropping its solution in dehydrated ethanol into the aqueous solution of N-octyl-O-sulfate chitosan (no characteristics given in the original reference) under constant stirring. The resulting mixture was dialyzed against deionized water overnight, filtered through a 450 nm membrane, and freeze-dried. To prepare micelles-containing hydrogel, first solutions of carboxymethyl chitosan (concentration 1.5%, w/v; no characteristics given in the original reference except the carboxylation degree of 45%) and paclitaxel-loaded micelles (4 mg/ml) were mixed and the mixture used to dissolve Poloxamer 407 ($M_w = 12600$ g/mol, concentration probably 19%, w/v). Finally, glutaraldehyde was added (0.05% in final concentration, w/v) under intense agitation. Gelation was achieved after an incubation at 37°C for 30 min.

Wang et al. (2014) used polymeric micelles formed by triblock copolymer poly(3-caprolactone)–PEG-poly(ε-caprolactone) synthesized by ring-opening polymerization of ε-caprolactone initiated by PEG of molecular weight ($M_n$) about 2000 g/mol. The final copolymer molar weight was around 3700 g/mol. This amphiphilic, micelles-forming copolymer was used to solubilize docetaxel by thin-film hydration method. Docetaxel and copolymer were co-dissolved in dehydrated ethanol, the solvent was evaporated under reduced pressure and elevated temperature (60°C). Resulting thin layer of homogenous film was hydrated with water under mild stirring. During the process, copolymer could self-assemble into micelles in which docetaxel was encapsulated. Finally, the suspension was filtered through a syringe filter with pore size of 220 nm. As a hydrogel, methoxyl PEG-poly(ε-caprolactone)-acryloyl chloride copolymer was used because it possesses the desired and tunable pH-responsive properties. The hydrogel copolymer was prepared by ring-opening polymerization initiated by PEG methyl ether ($M_n = 2000$ g/mol) and substitution reaction of acryloyl chloride. The molecular weight of resulting copolymer was about 3000 g/mol. The docetaxel-loaded micelles were incorporated in the hydrogel by water absorption. In fact, the hydrogel matrix was swelled with the micellar solution. A certain volume of docetaxel-loaded micelles solution was completely absorbed in hydrogel owing to the swelling capability of hydrogel. Then the mixture was freeze-dried to obtain the stable docetaxel-micelle-hydrogel system.

Yom-Tov et al. (2014) combined simple embedding of micelles with anchoring some of them to the surrounding hydrogel network. To achieve this goal, part of surfactant monomer molecules were functionalized to create endgroups, which would react with the network. Pluronic F127 was used as a surfactant in pure and diacrylated form. Acrylation was carried by reacting the surfactant with excess (relative to the hydroxyl groups) acryloyl chloride and triethylamine. Mixture of pure and acrylated surfactants at desired ratio was mixed with hydrogel precursor solution at low temperature (4°C) until complete dissolution and then a photoinitiator was added. The resulting solution was heated to induce the formation of micelles (37°C) and then irradiated with UV light to achieve chemically crosslinked material. The total concentration of Pluronics was kept constant at 10% (w/v). The hydrogel precursor solution used in this work was a diacrylated polyethylene glycol-protein (fibrinogen) mixture.

Shukla et al. (2009) prepared hydrogel containing nanosized domains (20–35 nm) of hydrophobic moieties of poly(methyl methacrylate) by grafting crosslinked poly(acrylic acid-co-methyl methacrylate) chains onto polyvinyl alcohol backbone using an efficient redox system. Typically, polyvinyl alcohol, acrylic acid, methyl methacrylate, and water were mixed in a round-bottom flask and homogenized by mechanical mixing. After nitrogen purge at elevated temperature (35°C), $N,N'$-methylenebis(acrylamide), potassium persulfate, and potassium metabisulfite were added and the mixture was poured into a Petri dish, covered, and kept at 35°C for 72 h to complete gelation.

Inoue et al. (1997) described preparation of hydrophobically modified pH-sensitive hydrogels based on hydrophobic oligomers grafted to a polyelectrolyte hydrogel matrix. Poly(acrylic acid) was used as the bioadhesive polyelectrolyte, oligo(methyl methacrylate) as the hydrophobic component. The grafted hydrogel was prepared by coupling the amino-terminated oligo(methyl methacrylate) onto the poly(acrylic acid) hydrogel backbone through the reaction of the amino group with an activated carboxyl group in the poly(acrylic acid) hydrogel using dicyclohexyl carbodiimide as an activation reagent. The amino-terminated methyl methacrylate oligomer was prepared by free-radical polymerization of methyl methacrylate using 2,2'-butyronitrile as an initiator and 2-amino ethanethiol hydrochloride as a chain transfer agent, respectively. The lightly crosslinked poly(acrylic acid) hydrogel was pre-synthesized by the copolymerization (in aqueous solution) of acrylic acid with added ethylene glycol dimethacrylate as a crosslinking agent (0.5 w/w% to the monomer) and ammonium persulfate as an initiator. The dried poly(acrylic acid) hydrogel was added with the DMF solution of the amino-terminated oligo(methyl methacrylate) and shaken for 24 h. Then, excess solution was removed and a DMF solution of dicyclohexyl carbodiimide was added. The reaction was completed after 24 h at room temperature giving the final graft level from 10 to 50% (w/w).

In the work by Kim et al. (2014), the non-water-soluble macrophage recruiting agent was solubilized in water through micelles formation with L-lactic acid-oligomer grafted gelatin,

and the resulting micelles with platelet-rich plasma were incorporated into gelatin hydrogels. First, the L-lactic acid oligomer with a number–average molecular weight of 1000 g/mol was synthesized by ring-opening reaction. The oligomer was grafted on gelatin (of isoelectric point equal to 5) through its amino groups in the average final ratio of about 3 moles/1 mole of gelatin. The macrophage recruiting agent was solubilized in the micelles formed by the grafted gelatin by simple mixing corresponding solutions in DMSO followed by 3 h of stirring; the solubilization was completed by dialysis, centrifugation of undissolved agent, and freeze-drying of final product. The micelles with the solubilized agent were incorporated in gelatin hydrogels by simply mixing a gelatin solution (5% by weight) with the pre-prepared micelles. The mixture was cast in molds, freeze-dried, and the hydrogel was then crosslinked by dehydrothermal treatment at 140°C for 48 h in a vacuum oven. The gelatin hydrogel incorporating the micelles was then impregnated with a solution of the platelet-rich plasma.

Bromberg et al. (2002) prepared Pluronic-poly(acrylic acid) copolymers and crosslinked them by ethylene glycol dimethacrylate to the form of microgel particles. Acrylic acid was partially neutralized in sodium hydroxide solution, Pluronic was dissolved in the resulting solution under nitrogen, and a desired amount of ethylene glycol dimethacrylate was added. Lauroyl peroxide and 4,4'-azobis(4-cyanovaleric acid) dissolved in acrylic acid were then added. The resulting solution was deaerated by nitrogen bubbling and mixed with dodecane containing a dispersion stabilizer. The reaction was run under vigorous stirring at 70°C under nitrogen purge for several hours.

The multisegmented amphiphilic copolymers were prepared by Guo et al. (2014) in a step-growth manner by reacting amino-telechelic PEG with diisocyanate functionalized UPy unit. The aminated PEG was synthesized from imidazole-telechelic precursor, which had been prepared by reaction of telechelic hydroxy-terminated PEG ($M_n = 10$ kg/mol) with 1,1-carbonyldiimidazole. The amine-terminated PEG contained $C_{12}H_{24}$ spacer and was prepared by reacting the imidazole precursor with 1,12-dodecyldiamine. The Upy-containing diisocyanate was made from the hydroxyl-functionalized uracil and 1,6-diisocyanatohexane in the presence of pyridine. The basic unit of the final chain-extended copolymers was of the structure [-PEG$_n$-C$_{12}$H$_{24}$-urea-C$_6$H$_{12}$-UPy-C$_6$H$_{12}$-urea-C$_{12}$H$_{24}$-], i.e., had a total of 36 methylene groups per macromolecular repeating unit).

Thomas et al. (2009) produced miscible blends of polymers, which contained hydrophilic and hydrophobic groups. Poly(vinyl alcohol) with an average molecular weight of 124–186 kDa (99% hydrolyzed) or poly(vinyl pyrrolidone) (no other specification except the trade name Plasdone K-90) were used as hydrophilic components, while ethylene units in poly(ethylene-co-vinyl alcohol) based its use as the hydrophobic component. A melt processable miscible blend was created of the polymers utilizing DMSO/water mixtures (80/20 or 90/10) between 80 and 160°C. Mechanical mixing was required at concentrations >30 wt% polymer. Mixing was performed utilizing a twin-screw mixing apparatus followed by extrusion through a small die to form a lyogel (a lyogel is a polymeric gel in which the continuous phase is a solvent other than water.) Poly(vinyl alcohol) (non-blended) was processed using the same method and tested as a control material. After processing, all materials were immersed in reagent grade alcohol for a minimum of 20 min followed by immersion in water. During this solvent-exchange process, the hydrophobic/hydrophilic blended materials turned from a transparent to an opaque, flexible hydrogel as the hydrophobic domains formed. All pure poly(vinyl alcohol) samples remained transparent after this treatment. Samples remained in deionized water for a minimum of 72 h to equilibrate prior to testing.

Lee et al. (2011) induced the enzyme-mediated crosslinking of Pluronic micelles for fabricating high-strength injectable hydrogels. To this end, Pluronic copolymers were end-terminated with tyramine. The tyramine units should be exposed on the surface of Pluronic micelles and can be enzymatically oxidized by tyrosinase giving a crosslinked material. First, Pluronic F127 was transformed to amine-reactive form by a reaction with $p$-nitrophenyl chloroformate with resulting substitution degree (of end hydroxyl groups) approaching 100%. The modified Pluronic was then reacted with tyramine giving the final substitution also close to 100%. The Pluronic–tyramine conjugate was crosslinked at room temperature by the action of tyrosinase for 48 h.

Pluronic-based chemically cross-linked nanoparticles were developed in Missirlis et al. (2005), which should provide stability toward dilution and drying. They employed photoinitiated polymerization of acrylates as a mean to cross-link aqueous solutions of Pluronic derivatives. Pluronic F127 diacrylate was synthesized by a reaction with acryloyl chloride as described by Cellesi et al. (2002) with 100% conversion of alcohols to acrylates. Hydrogels were prepared either in the form of nanoparticles or as disks. The nanoparticles were synthesized by inverse emulsion photopolymerization. The oil phase for the emulsion preparation was formed by $n$-hexane with Span 65 surfactant (2% by weight). The aqueous phase contained diacrylated Pluronic F127, PEG diacrylate (average $M_n = 575$), both in concentration of 6.75% by weight, triethanolamine (2% by weight), and eosin Y (0.02% by weight). Oil-to-water weight ratio was 65/35 and the emulsion was prepared

by short sonication. Polymerization occurred under the illumination by laser light of 480–520 nm wavelengths for 1 h. Similar procedure, but without the oil phase, was used to prepare hydrogel disks – the aqueous solution of precursors was placed between two glass slides, illuminated for 30 min, and the disks were finally exposed to water until equilibrium swelling was reached (24 h, at least).

Materials described by Hao and Weiss (2011) are physically crosslinked copolymer hydrogels synthesized from $N,N$-dimethylacrylamide and 2-($N$-ethylperfluorooctane sulfonamido) ethyl acrylate (FOSA). Aggregation of the fluorocarbon hydrophobe moieties produces a physically crosslinked microstructure of core-shell nanodomains, which are composed of an FOSA core surrounded by a water-depleted shell of dimethylacrylamide. The water-poor shell of the dimethylacrylamide presumably comprises the chain segments attached to the FOSA repeat units that have restricted mobility due to the covalent attachment to the FOSA aggregates. These core-shell nanodomains constitute the crosslink junctions in the physical hydrogel that is essentially water-swollen poly(dimethylacrylamide). The copolymer of dimethylacrylamide and FOSA was synthesized by free-radical polymerization of the monomers in dioxane. The resulting molecular weight ($M_w$) ranged between 50 and 80 kmol/g (with polydispersity between 1.5 and 1.9). The hydrogel samples were prepared by compression-molding dry copolymer films at 150–180°C under vacuum and then immersing the films in deionized water for 7 days to ensure equilibrium hydration.

Li et al. (2012) reported on biodegradable micelles-containing PEG hydrogels synthesized via Michael addition chemistry. For schematic representation of hydrogel synthesis and structure, see **Figure 4**. The basic polymeric unit of the hydrogel was formed by the four-arm acrylated pol(yethylene glycol) (tetraacrylated PEG), whereas micelles were formed from PEG-polycarbonate diblock copolymer end-capped with vinyl sulfone group (on the PEG block). Due to the scaffold application, also the RGD peptide

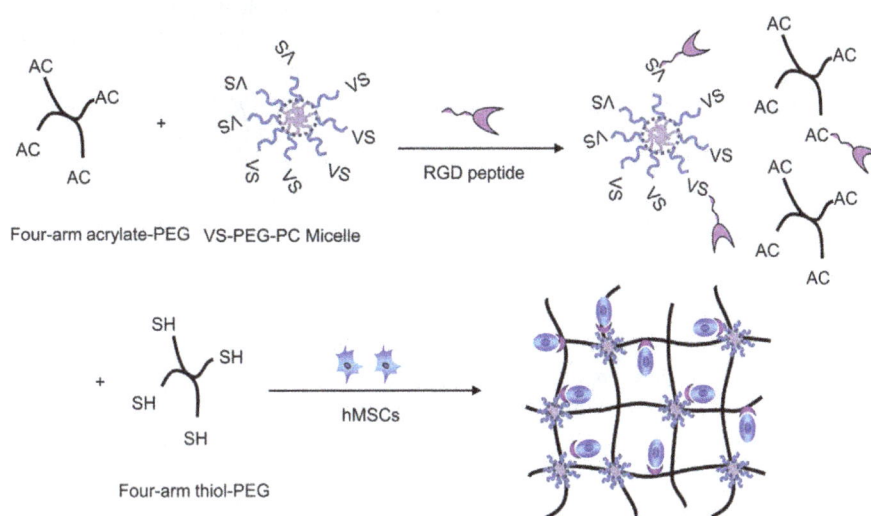

**FIGURE 4 | Synthesis of hydrogel with micellar crosslinks and a peptidic structure to bind human mesenchymal stem cells (hMSCs); VS stands for vinyl sulfone**. Reprinted from Li et al. (2012) with permission.

was chemically built into the hydrogel network as a cell adhesion enhancer. The tetra acrylate-terminated PEG was dissolved in buffer and to this solution, the pre-prepared micellar solution of the vinyl sulfone terminated PEG-polycarbonate in the same buffer was added followed by the addition of peptide solution. The mixture was incubated at 37°C and then added with the buffered solution of tetra-thiolated PEG crosslinker (four-arm tetra sulfhydryl PEG). The reaction mixture was kept at this temperature and the hydrogel formed quickly. Cells in the growth medium (human mesenchymal stem cells) could be added to the gel precursors before the addition of the thiolated crosslinker and thus encapsulated in the hydrogel. Alternatively, a bolaamphiphile/DNA complex as a gene transfection vector could be mixed with the cell suspension and then added to the gel precursors before the addition of the crosslinker. In this way, the effects on gene expression efficiency could be studied directly in the hydrogel matrix.

Tuncaboylu et al. (2013) copolymerized large hydrophobes such as stearyl methacrylate with the hydrophilic monomer acrylamide in aqueous SDS surfactant solutions. The reaction medium was prepared as a solution of NaCl at the concentration of 0.5 mol/l. The salt led to micellar growth, and hence, solubilization of large hydrophobes within the grown worm-like SDS micelles. A series of hybrid hydrogels was prepared using stearyl methacrylate (2% molar) as a physical crosslinker and various amounts of $N,N'$-methylenebis(acrylamide) as a chemical crosslinker. The reaction mixture contained SDS at the concentration of 0.24 mol/l, i.e., more than 30 times higher than its critical micellar concentration in water (in the used salt solution, the critical micellar concentration would be about one order of magnitude yet lower). The total concentration of stearyl methacrylate and acrylamide monomers was fixed at 10% (w/v). Specifically, surfactant was dissolved in NaCl solution at elevated temperature (35°C) and then stearyl methacrylate was solubilized in this solution. Acrylamide and bis(acrylamide) were added and after their dissolution, the redox initiator system was added. The micellar copolymerization ran at 25°C for 24 h.

## PROPERTIES

In this section, concrete examples of the benefits coming from the incorporation of hydrophobic domains into hydrogels are given. The examples correspond to works referred in preceding sections and thus complete this review.

Liu and Li (2005) confirmed that solubilization of camptothecin in surfactant micelles dispersed in agarose hydrogel greatly affected the release behavior of drug molecules. The release of camptothecin was slowed down significantly with increasing surfactant concentration as a result of lowered drug diffusion coefficient. The drug diffusion coefficient could be approximately expressed as an exponential function of surfactant concentration. The authors separated the effects of drug in the aqueous and micellar phases on its diffusion coefficient. At lower surfactant concentrations (up to about twice its critical micellar concentration in pure water), the free drug molecules showed a great influence on diffusion coefficient. With the concentration of surfactant increasing, this effect decreased sharply due to the solubilization of drug molecules into micelles. The influence of the micellarly solubilized

drug on its diffusion became prominent particularly at surfactant concentration about three times higher than its critical micellar concentration in pure water. The diffusion coefficient of micelles might be lowered as the amounts of micelles in the system increase, due to the increased friction in crowded environment or due to other possible interaction between micelles. The notable influence of temperature on the release was also observed and was even stronger than that in the case when no surfactant was used. The release was significantly accelerated increasing the temperature from 23 to 37°C; at higher temperature, its effect became negligible.

Results published by Tedeschi et al. (2006) on physical poly(vinyl alcohol) hydrogels with incorporated micelles showed that surfactant micellization was almost unperturbed regardless the confinement of its molecules within the hydrogel network. No direct surfactant–polymer interactions were detected. Thus, the hydrogel maximum swelling was not affected by addition of surfactant and was also independent of surfactant concentration. Surfactant concentrations measured in the hydrogel samples were equal to that in the surfactant aqueous solution used to rehydrate the gels – no surfactant preference for the hydrogel matrix was found. The critical micellar concentration was found to slightly decrease in the polymer solution and it decreased even more in hydrogel. This finding was attributed to the increased surfactant effective concentration in the limited space created by the excluded volume of the polymer. EPR spin probe and small-angle neutron scattering techniques showed that the micelles formed within the pores of the hydrogel matrix possess the same properties as micelles formed in bulk water. NMR diffusion study revealed that diffusion of micelles in hydrogels is much slower in comparison to bulk water. Because no surfactant–hydrogel interactions were found, this was explained by the effect of the intrinsic structural features of hydrogels in which the pores available for the diffusion are filled by the polymer-rich phase.

From the point of view of possible applications, this work confirmed that micelle-containing hydrogels have a complex multidomain structure, which is able to host both hydrophilic and hydrophobic molecules. Diffusion of the guest molecules from such carrier matrix is quite slow. Such systems can be thus promising candidates for the design of new systems for controlled release and/or adsorption of various guest molecules. Very similar results on the surfactant role in poly(vinyl alcohol) hydrogels were reported in the study by Mangiapia et al. (2007). The main aim of this work was detailed structural study of pure hydrogels by small (SANS) and ultrasmall (USANS) angle neutron scattering techniques. Some hydrogel samples were rehydrated using surfactant solution as in the work by Tedeschi et al. (2006) but in this case anionic sodium decylsulfate was used. The authors reported on the lack of any detectable interaction between the polymer chains and surfactant micelles. Also, the characteristics of micelles like the structural parameters, size, aggregation number, and actual charge were not affected by the presence of the poly(vinyl alcohol). Thus, the behavior of surfactant micelles within the poly(vinyl alcohol) hydrogels prepared by the freeze/thaw technique seems to be general and independent on the surfactant nature.

The study by Stoppel et al. (2011) was devoted to the effect of non-ionic surfactants on protein and small molecule transport within hydrogel matrices. Understanding this effect is crucial for the use of these systems in the design and formulation of drug delivery systems, biomaterials, and tissue engineering scaffolds. In their study, Pluronic F68 micelles were incorporated in an alginate hydrogel. The experimental results showed that small molecule transport within hydrogels improved slightly by the addition of Pluronic F68 (evidenced by the increase in their effective diffusion coefficient). The authors expressed expectation that use of a higher surfactant concentration or larger surfactant molecular weight, e.g., Pluronic F108 or F127, would continue to enhance diffusion coefficient and improve transport characteristics, as long as the material properties of the hydrogel would not be strongly affected by the change in formulation. They also note that in the case of a more hydrophobic small molecule, changes in the diffusion behavior would be highly influenced by the size of the complex micelle-solubilized hydrophobic small molecule. From the point of view of mechanical properties, it was found that addition of Pluronic F68 above 2% (w/v) may lead to undesirable material properties. Thus, the incorporation of 5% Pluronic F68 (w/v) reduced the storage modulus by 57% and increased the loss angle tangent by 24% (at a frequency of 1 Hz). Interestingly, while significant changes in solution viscosity were observed for Pluronic F68 addition below 2% (w/v), these interactions were not dominant in the hydrogel form. However, the solution interactions did translate to the gel state through the increased loss angle tangent and increased shrinkage ratio, which ultimately had an effect on the loading capacity through changes in water retention properties. Pluronic addition had minimal effects on riboflavin transport and loading. Conversely, significant effects were observed for bovine serum albumin where loading capacity decreased and release either increased or decreased depending on the gelation method.

Frisman et al. (2012) nanostructured hydrogels based on fibrinogen with Pluronic F127 micelles. Resulting materials were comprehensively analyzed using small-angle X-ray scattering (SAXS) and transmission electron cryo-microscopy and leaching test of the Pluronic micelles from the crosslinked hydrogel network under tissue culture conditions. Cryo-microscopy and SAXS experiments revealed formation of partially ordered micellar structures, surrounded by a PEG–fibrinogen network. Micelles diameter was determined to be approximately 14 nm, which was close to the dimension of micelles (10 nm) formed in the solution of modified fibrinogen (hydrogel precursor). The structure and the ordering of the micelles in the hydrogels were dependent on the concentration of Pluronic F127. At a relatively low concentration of Pluronic, micelles were distributed homogenously and randomly throughout the hydrogel matrix. Organized arrangement of the micelles was observed in samples containing higher concentrations of Pluronic. The arrangement was proposed in the form of clusters of micelles surrounded by areas containing mainly the modified fibrinogen backbone. The shape and the size of the micelles and the distances between them were found not to be affected by the hydrogel crosslinking density. Pluronic F127 also decreased the size distribution of the fibrinogen precursor in the hydrogel. One of the radii determined for the fibrinogen structure was reduced from 3.8 nm in the neat hydrogel to 2.3 nm at the

highest Pluronic F127 concentration. The other (smaller) radii of 1.5 nm remained unchanged.

Swelling increased upon the addition of Pluronic and with its increasing concentration – probably due to the hydrophilic nature of this additive. Leaching tests showed that more than 50% of the added Pluronic was released from the hydrogel during the first few hours. The remaining fraction was leached over the next 2–3 days and after 4 days only a very small amount of Pluronic remained entrapped inside the hydrogel matrix.

A fibroblast cell culture was inoculated in the prepared hydrogels. A non-linear dependence of the cell shape on the amount of the added Pluronic was found; hydrogels containing 7% (by weight) of Pluronic provided an optimal environment for cell growth. Increased amount of Pluronic resulted in the decrease of the cell shape factor as a consequence of the above mentioned effect of the surfactant on the size of the fibrinogen precursor. Smaller radius of the fibrinogen proteins backbone reflects a lesser tendency of the hydrogel precursor toward aggregation. This means that more bioactive fibrinogen polypeptidic structures can be exposed to resident cells.

Ju et al. (2013) presented an upgraded system based on Poloxamer 407 for localized delivery of paclitaxel. Although paclitaxel can be solubilized in Poloxamer 407 solution, the solubility is too low to achieve the desired treatment threshold. Therefore, paclitaxel was solubilized in $N$-octyl-$O$-sulfate chitosan micelles and these were dispersed in Poloxamer hydrogels interpenetrated by crosslinked network of carboxymethyl chitosan. The introduction of paclitaxel-loaded micelles had no influence on the lower critical solution temperature and gelation time of Poloxame hydrogel, whereas the loading capability for paclitaxel was greatly enhanced. The incorporation of chitosan-based network resulted in higher swelling ratio, stronger mechanical properties, and longer term drug release both *in vitro* and *in vivo*. Prolonged retention was revealed at tumor sites, lasting for 20 days, as well as superior tumor inhibition rate with reduced toxicity compared to Taxol (a commercial paclitaxel preparation) or Poloxamer hydrogels not reinforced by chitosan-based networks and containing the same paclitaxel-loaded micelles.

Wang et al. (2014) designed a system for oral delivery of anticancer drugs, docetaxel in their case. The pH-responsive hydrogel carrying micelles with solubilized drug was selected as an ideal candidate with following hypothesis on the delivery:

1. Primarily, the loaded micelles pass through the stomach under the protection of pH-shrinked hydrogel and diffuse from the hydrogel in intestinal tract.
2. The hydrogel swells at high pH environment of intestine. The micelles are released and absorbed by Peyer's patches in small intestine owing to absorption enhancer and subsequently they circulate longer in blood.
3. The micelles are passively targeted to tumor tissue by enhanced permeation and retention effect.
4. At last, docetaxel is released from micelles and suppresses the growth of tumor.

Hydrogels containing loaded micelles showed similar mesh size compared to the blank hydrogels and their pH-responsibility was not impacted. With the pH rising from 1 to 8, more micelles were exposed on the surface of the network, which was primarily ascribed to the expansion of the hydrogel. The diffusion rate of micelles in pH equal to 6.8 was much faster than that in acidic pH (1.2) and also the diffusion in simulated intestinal fluid was much faster than that in simulated gastric fluid. The docetaxel-loaded micelles were confirmed to be absorbed in the intestine, which was attributed to their small size (about 20 nm). The oral bioavailability of docetaxel reached up to 75.6% after oral administration of docetaxel-micelles-hydrogel, about 10 times higher than that of free docetaxel-loaded micelles. The pharmacokinetic area under curve of docetaxel improved accordingly as the administration does increased from 10 to 30 mg/kg in mice. In the subcutaneous breast cancer model, oral docetaxel-micelles-hydrogel inhibited tumor growth dose-dependently; in high dose group, the oral hydrogel system was comparable with intravenous injection of Taxotere (a marketed formulation of docetaxel).

Yom-Tov et al. (2014) covalently linked part of the Pluronic micelles into the hydrogel matrix by replacing part of the Pluronic F127 molecules with a diacryl derivative. They found that the influence of the diacryl derivative on the hydrogel nanostructure is pronounced only after covalently integrating its micelles in the network. By fine tuning the ratio of Pluronic-diacrylate to unmodified Pluronic molecules, different degrees of crosslinking were achieved, which influenced the hydrogels weight gain ability and Young's modulus. Larger diacryl derivative percentages led to smaller swelling degree, which is in line with the integration of acrylated micelles within the hydrogel network, thus enhancing its crosslinking density and diminishing its weight gain ability. It was also observed that after about 2 h of swelling, the unbound (unmodified) Pluronic molecules leached out of the hydrogel creating cavities and network imperfections. An increase in Pluronic diacryl derivative percentage gave higher values of Young's modulus as expected due to the increased network density. The authors concluded that the higher porosity and mesh size resulting from the release of unbound unmodified Pluronic molecules from the fully hydrated hydrogels had the most pronounced impact on the hydrogels' characteristics.

An innovative hydrogel synthesized by Shukla et al. (2009) contained nanosized domains of the size 20–35 nm of hydrophobic moieties of poly(methyl methacrylate) grafted onto hydrophilic poly(vinyl alcohol) backbone and investigated its water sorption properties. Increasing concentration of hydrophobic monomer gave materials with decreasing swelling degree while the diffusion coefficient of water increased. The effects of various parameters on water sorption were then studied on hydrogels prepared with a fixed ("medium") amount of hydrophobic monomer. The water imbibition capacity increased when pH of the swelling media varied in the range 2–11, but a significant rise in swelling was seen only beyond pH equal to 9. The swelling was favored by increasing temperature or by the presence of chloride or carbonate ions; on contrary, a fall in water sorption was observed during the presence of trivalent phosphate ions.

Inoue et al. (1997) grafted oligo(methyl methacrylate) as a hydrophobic domains forming unit onto hydrophilic poly(acrylic acid) as a hydrogel-forming polymer and tested swelling and drug release properties of final materials. The graft level and molecular weight of this oligomer influenced swelling of resulted hydrogels. Higher graft levels and lower molecular weights of the grafted chains yielded lower swelling ratios. This was explained by the higher concentration of hydrophobic grafts and their aggregation into domains within the hydrogel. Drugs of various polarities and a model protein (positively charged), lysozyme, were solubilized in the hydrogels. Hydrophobic drugs and lysozyme were more slowly released from the hydrogels when compared to the ungrafted poly(acrylic acid) hydrogel. This was probably a consequence of favored absorption of the hydrophobes into the oligomer-formed domains and the lack of interconnections between these domains. Positively charged lysozyme might associate both with the negative charges on acrylic acid moieties and the hydrophobic domains. Consequently, also its release was decelerated. On the other hand, uncharged hydrophilic drugs released faster because of their lower hydrophobic or ionic interactions within the hydrogel. The formation of hydrophobic domains might also enlarge the aqueous pore sizes of the hydrogel, thus permitting the hydrophilic drug to diffuse out more rapidly.

Kim et al. (2014) solubilized a non-water-soluble macrophage recruitment agent in micelles formed within hydrogels of gelatin grafted with hydrophobic chains. Platelet-rich plasma was also incorporated into these hydrogels, which were designed for the bone tissue engineering. Dual release of the agent and plasma from hydrogel was expected to stimulate macrophage recruitment and modulate inflammation, resulting in enhanced bone regeneration. The results of that study experimentally confirmed this hypothesis and dual release in a controlled fashion. The higher number of macrophages recruited was observed around gelatin hydrogels incorporating mixture of micelles-loaded agent and plasma compared with those incorporating either the loaded micelles or the plasma. The hydrogels also decreased the production of proinflammatory cytokine, but increased that of anti-inflammatory cytokines. The controlled dual release thus recruited macrophages into the bone defect and modulated inflammatory responses thereat, resulting in promoted bone regeneration.

The development of novel systems for the oral chemotherapy was overviewed by Bromberg (2005). Solubilization in micelles was combined with large macromolecular structures preventing such carriers from being transported into the systemic circulation. The systems were based on the copolymers of Pluronics and acrylic acid prepared in the form of spherical microgel particles with controllable diameter from tens to hundreds of microns. Dangling chains of Pluronics created the hydrophobic domains in microgels. At elevated temperatures, the dangling Pluronic chains rearranged creating tighter micellar aggregates that acted as physical crosslinks, lowering the equilibrium swelling of the microgels. The microgels were then crosslinked by both chemical and physical crosslinks. The size of micellar crosslinks filled the gap between macromolecules (Stokes radii below 10 nm) and microparticles (size 100 nm and higher). These micelles were shown to enhance solubility of hydrophobic drugs such as steroid hormones or camptothecins. The solubilizing micelles also provided an effective barrier against drug decomposition. The gelled material enabled a prolonged contact with mucous tissue due to enhanced bioadhesion and the decelerated diffusion enhanced the bioavailability of the drug.

Formation of hydrophobic domains within a multiblock PEG-based hydrogel was combined with strong hydrogen-bonding associations of UPy units contained in the same copolymer (Guo et al., 2014). The strength of hydrogels significantly improved and could additionally be adjusted by implementing different lengths and composition of PEG precursor. The hydrophobic segments as observed by transmission electron microscopy appeared to be dispersed as small spherical compartments with diameter about 2–5 nm. Although the weight fraction of PEG in these materials was around 0.9, they did not dissolve (at ambient temperature) in water in contrast to pure PEG; only swelling was observed with equilibrium water weight fraction somewhat above 0.8. A short-chain hydrophobic oligo-methylene spacers were used to shield the UPy units and to facilitate their aggregation such that hydrogen-bonding disruption caused by the hydrated matrix could be prevented. The integrity of crosslinking was thus maintained through this molecular shielding. When combined with long PEG segments, this feature translated to low crosslink densities, allowing relatively large elongation-recovery to be achieved with minimal diminishment of modulus compared with pristine PEG. Tensile tests revealed a remarkably strong hydrogel that exhibited nearly perfect strength recovery even at large deformation (exceeding 300%). The cumulative mechanical properties were driven in part by the incompatibility between hydrophilic and hydrophobic chain segments in concert with the strong tendency for hydrogen bonding of the UPy motifs. For example, these hydrogels were substantially stronger and tougher than similar hydrogels formed by chemically crosslinked copolymer micelles with very similar hydrophobic versus hydrophilic content. The developed materials

also showed a shape memory behavior. The shape recovery from temporary deformed states was demonstrated to be stimulated either by heating above the melting temperature of PEG or by immersing in water.

Thomas et al. (2009) prepared hydrogels combining both hydrophilic and hydrophobic structures by injection molding with the intention of their application as cartilage replacement materials. Unfortunately, many of their investigated hydrogels lacked the required shear, tear, and creep properties. However, the introduction of hydrophobic groups in the materials was found to have a positive effect on mechanical properties with a minimal effect on coefficient of friction and contact angle. Particularly, the strength and creep resistance were greatly improved. Thus, hydrophobic domains-containing hydrogels were proved to exhibit signs of hydrodynamic lubrication, which are stronger than the reference poly(vinyl alcohol) hydrogels used in this study.

Lee et al. (2011) reinforced poor mechanical properties of Pluronic hydrogels by using Pluronic copolymers terminally conjugated with tyramine, which can be enzymatically crosslinked between neighboring micelles containing the modified copolymers (**Figure 5**). The enzyme crosslinked hydrogels showed far lower critical gelation concentration, concomitantly showing enhanced gel strength compared to unmodified Pluronic copolymer hydrogels. Rheological studies demonstrated that the enzyme crosslinked hydrogels exhibited a fast and reversible sol–gel transition in response to temperature while maintaining sufficient mechanical strength at the gel state. These materials were thus suitable for injectable, *in situ* gelling applications. *In situ* formed hydrogels were eroded gradually, releasing fluorescently

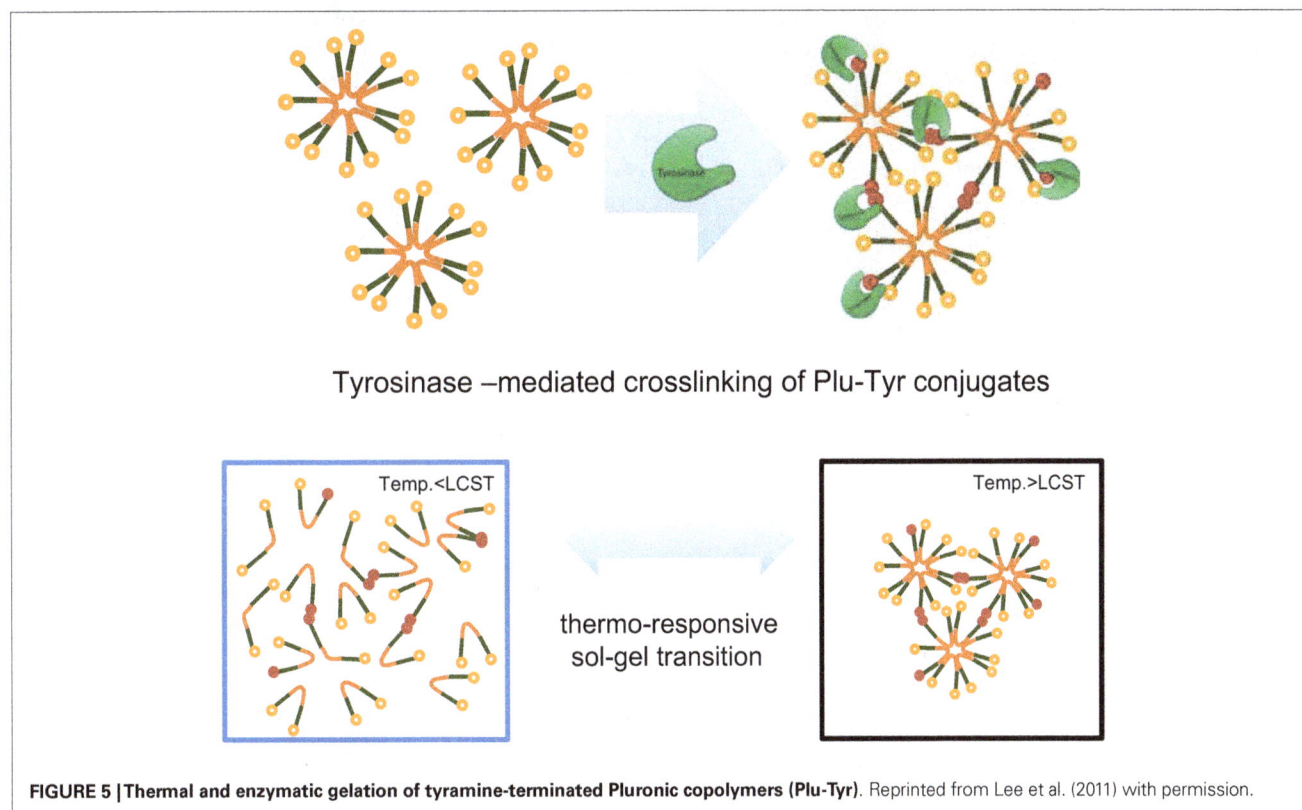

**FIGURE 5 | Thermal and enzymatic gelation of tyramine-terminated Pluronic copolymers (Plu-Tyr)**. Reprinted from Lee et al. (2011) with permission.

Tyrosinase –mediated crosslinking of Plu-Tyr conjugates

Temp.<LCST

Temp.>LCST

thermo-responsive
sol-gel transition

labeled dextran in an erosion-controlled manner. Moreover, they showed tissue-adhesive properties due to the presence of unreacted catechol groups in the gel structure.

Pluronic-based nanoparticles were obtained in Missirlis et al. (2005) by inverse emulsion photocopolymerization with a PEG diacrylate. The size of nanoparticles could be varied between 50 and 500 nm. The hydrophobic domains within these nanoparticles solubilized doxorubicin, an anti-cancer drug, up to almost 10% (w/w). The nanoparticles also appeared stable to freeze-drying and re-suspending in water.

Hao and Weiss (2011) used a fluoroacrylate, capable of strong hydrophobic interactions, as the hydrophobic modifier in poly(alkylacrylamide) hydrogels. Microphase separation of a hydrophobic nanophase due to incompatibility of the fluoroacrylate groups with the hydrogel network served as multifunctional physical crosslinks. This resulted in excellent mechanical properties of final materials, providing the water concentration in the gel was not too low (in this case the hydrogel became more brittle). The strength of these physical hydrogels increased with increasing concentration of the fluoroacrylate moieties. The unusually high stiffness and toughness was attributed also to the ability of these gels to dissipate energy when deformed because of a reversible disengagement of the hydrophobic groups from the nanodomains. That provided a viscous mechanism for stress relaxation and effective energy dissipation. The physical nature of the hydrogels also allowed for their viscous flow if the applied stress was sufficiently high or the deformation time was sufficiently long (i.e., low deformation frequency). The physical gels were more viscous than comparable chemical gels and were much more efficient at dissipating stress, which resulted in higher tensile toughness. At 25°C, the hydrogels were highly elastic (the storage modulus was higher than the loss modulus), but as the temperature increased, viscous behavior increased and a moduli crossover occurred at 55°C and the rheological characteristics of the material changed from a viscoelastic solid to a viscoelastic liquid.

The temperature behavior reflected the glass transition of these hydrogels temperature of which was determined at 45°C. At room temperature, the hydrophobic nanodomains were glassy and the relaxation time of interactions holding them together could be expected to be relatively long. As temperature increased above the glass transition, the hydrophobic domains started to be more flexible and their relaxation times decreased. Observed mechanical behavior is useful for designing an injectable hydrogel where a high stress or slow deformation can be used to achieve viscous flow, but gelation occurs when the stress is removed or sufficiently decreased.

Li et al. (2012) covalently incorporated nanosized polymeric micelles into PEG hydrogel network to tune its properties relevant for cell-related applications in tissue engineering. Increasing the content of the micelles from 0 to 80% led to increased porosity and tunable mechanical property of the hydrogels. It was observed that the size of pores in hydrogel increased with increasing micelle content due to decreased crosslinking degree in this system. At the content of the micelles about 80%, the pores were highly interconnected. The hydrogel with 20% micelles provided the best balance among hydrogel stiffness and porosity for cell survival, leading to the highest viability of human mesenchymal stem cells. When the micelle content was increased up to 80%, storage modulus and gel yield decreased significantly, whereas swelling ratio increased dramatically. The cationic bolaamphiphile/DNA complexes induced higher gene expression efficiency in the hydrogels than the traditional polyethylenimine/DNA complexes, yet showed no cytotoxicity. The gene expression level in the cells in the hydrogel with 20% micelles was the highest as compared to that in the other hydrogels. Overall, the hydrogel with 20% micelles offered an optimal scaffold with ideal physical properties for cell growth and transfection. Nanoparticles incorporated into the hydrogels thus represent a useful strategy to control cellular behavior in a three-dimensional environment. Developed biodegradable nanostructured hydrogels can be an excellent platform for the delivery of human mesenchymal stem cells.

A series of hybrid hydrogels was prepared by the micellar copolymerization of acrylamide stearyl methacrylate as a physical crosslinker forming micelle-like domains and a bis(acrylamide)-based chemical crosslinker, in the presence of a surfactant (Tuncaboylu et al., 2013). Rheological measurements showed that the dynamic reversible physical crosslinks consisting of hydrophobic associations surrounded by surfactant micelles are also effective within the covalent network of the hybrid hydrogels. A significant enhancement in the compressive mechanical properties of the hybrid gels was observed with increasing chemical crosslinker content as expected. Cyclic tensile and compression tests showed that the fraction of hydrophobic associations reversibly broken under an external force decreased with increasing the ratio of chemical crosslinker to the monomer. The results showed that self-healing in hybrid gels can be observed at sufficiently low chemical crosslink densities where the network chains are sufficiently flexible to allow the re-formation of broken hydrophobic associations on the cut surfaces. The hydrophobic associations surrounded by surfactant micelles acted as reversible breakable crosslinks responsible for rapid self-healing of the hydrogels at room temperature without the need for any stimulus or healing agent. No self-healing observed in hybrids formed at the chemical crosslinker ratio equal and greater than 0.01 was explained by the decreased mobility of hydrophobic blocks in the inhomogeneous network structure. The authors also proposed that the use of hydrophobic monomers with longer alkyl side chains would produce enhanced self-healing properties due to the increased mobility of the hydrophobes.

## CONCLUSION

This review focused on rather unexplored colloidal materials. Hydrogels with hydrophobic, particularly micellar, (nano)domains are hybrid materials from the point of view of polarity of their constituents. This hybrid character is the base of their specific material, structural, and colloidal properties. Manipulating the two basic constituents – their contents, chemistry, size, incorporation into hydrogel matrix – enables tailoring of characteristics of these hybrid gels. They can be then favorably employed in variety of applications, especially in the fields of (bio)medicine, biomaterials, drug delivery, cosmetics, but even in agriculture or environment protection. The number of published works is still low and most of them are focused on synthetic and compositional aspects. More information on their physico-chemical or colloidal properties, like (long-term) stability or transport of both hydrophobic and hydrophilic molecules within and from (into) these systems, is needed badly. Further, there is a lack of detailed biology- and cell-related studies in the case of materials intended for biomedical applications.

## ACKNOWLEDGMENTS

The support from the project no. LO1211 (National Programme for Sustainability I, Ministry of Education, Youth and Sports) is acknowledged.

## REFERENCES

Bromberg, L. (2005). Intelligent hydrogels for the oral delivery of chemotherapeutics. *Expert. Opin. Drug. Deliv.* 2, 1003–1013. doi:10.1517/17425247.2.6.1003

Bromberg, L., Temchenko, M., and Hatton, T. A. (2002). Dually responsive microgels from polyether-modified poly(acrylic acid): swelling and drug loading. *Langmuir* 18, 4944–4952. doi:10.1021/la011868l

Cellesi, F., Tirelli, N., and Hubbell, J. A. (2002). Materials for cell encapsulation via a new tandem approach combining reverse thermal gelation and covalent crosslinking. *Macromol. Chem. Phys.* 203, 1466–1472. doi:10.1002/1521-3935(200207)203:10/11<1466::AID-MACP1466>3.0.CO;2-P

Frisman, I., Seliktar, D., and Havazelet, B.-P. (2012). Nanostructuring biosynthetic hydrogels for tissue engineering: a cellular and structural analysis. *Acta Biomater.* 8, 51–60. doi:10.1016/j.actbio.2011.07.030

Guo, M., Pitet, L. M., Wyss, H. M., Vos, M., Dankers, P. Y. W., and Meijer, E. W. (2014). Tough stimuli-responsive supramolecular hydrogels with hydrogen-bonding network junctions. *J. Am. Chem. Soc.* 136, 6969–6977. doi:10.1021/ja500205v

Hao, J., and Weiss, R. A. (2011). Viscoelastic and mechanical behavior of hydrophobically modified hydrogels. *Macromolecules* 44, 9390–9398. doi:10.1021/ma202130u

Inoue, T., Chen, G., Nakamae, K., and Hoffman, A. S. (1997). A hydrophobically-modified bioadhesive polyelectrolyte hydrogel for drug delivery. *J. Control. Release* 49, 167–176. doi:10.1016/S0168-3659(97)00072-2

Ju, C., Sun, J., Zi, P., Jin, X., and Zhang, C. (2013). Thermosensitive micelles–hydrogel hybrid system based on poloxamer 407 for localized delivery of paclitaxel. *J. Pharm. Sci.* 102, 2707–2717. doi:10.1002/jps.23649

Kim, Y.-H., Furuya, H., and Tabata, Y. (2014). Enhancement of bone regeneration by dual release of a macrophage recruitment agent and platelet-rich plasma from gelatin hydrogels. *Biomaterials* 35, 214–224. doi:10.1016/j.biomaterials.2013.09.103

Lee, S. H., Lee, Y., Lee, S.-W., Ji, H.-Y., Lee, D. S., and Park, T. G. (2011). Enzyme-mediated cross-linking of pluronic copolymer micelles for injectable and in situ forming hydrogels. *Acta Biomater.* 7, 1468–1476. doi:10.1016/j.actbio.2010.11.029

Li, Y., Yang, C., Khan, M., Liu, S., Hedrick, J. L., Yang, Y.-Y., et al. (2012). Nanostructured PEG-based hydrogels with tunable physical properties for gene delivery to human mesenchymal stem cells. *Biomaterials* 33, 6533–6541. doi:10.1016/j.biomaterials.2012.05.043

Liu, J., and Li, L. (2005). SDS-aided immobilization and controlled release of camptothecin from agarose hydrogel. *Eur. J. Pharm. Sci.* 25, 237–244. doi:10.1016/j.ejps.2005.02.013

Mangiapia, G., Ricciardi, R., Auriemma, F., De Rosa, C., Lo Celso, F., Triolo, R., et al. (2007). Mesoscopic and microscopic investigation on poly(vinyl alcohol) hydrogels in the presence of sodium decylsulfate. *J. Phys. Chem. B* 111, doi:10.1021/jp0663107

Missirlis, D., Tirelli, N., and Hubbell, J. A. (2005). Amphiphilic hydrogel nanoparticles. preparation, characterization, and preliminary assessment as new colloidal drug carriers. *Langmuir* 21, 2605–2613. doi:10.1021/la047367s

Shukla, S. K., Shaikh, W., Gunari, N., Bajpai, A. K., and Kulkarni, R. A. (2009). Self assembled hydrophobic nanoclusters of poly(methylmethacrylate) embedded into polyvinyl alcohol based hydrophilic matrix: preparation and water sorption study. *J. Appl. Polym. Sci.* 111, 1300–1310. doi:10.1002/app.29155

Stoppel, W. L., White, J. C., Horava, S. D., Bhatia, S. R., and Roberts, S. A. (2011). Transport of biological molecules in surfactant–alginate composite hydrogels. *Acta Biomater.* 7, 3988–3998. doi:10.1016/j.actbio.2011.07.009

Tedeschi, A., Auriemma, F., Ricciardi, R., Mangiapia, G., Trifuoggi, M., Franco, L., et al. (2006). A Study of the microstructural and diffusion properties of poly(vinyl alcohol) cryogels containing surfactant supramolecular aggregates. *J. Phys. Chem. B* 110, 23031–23040. doi:10.1021/jp061941m

Thomas, B. H., Fryman, J. C., Liu, K., and Mason, J. (2009). Hydrophilic–hydrophobic hydrogels for cartilage replacement. *J. Mech. Behav. Biomed. Mater.* 2, 588–595. doi:10.1016/j.jmbbm.2008.08.001

Tuncaboylu, D. C., Argun, A., Algi, M. P., and Okay, O. (2013). Autonomic self-healing in covalently crosslinked hydrogels containing hydrophobic domains. *Polymer* 54, 6381–6388. doi:10.1016/j.polymer.2013.09.051

Wang, Y. J., Chen, L. J., Tan, L. W., Zhao, Q., Luo, F., Wei, Y. Q., et al. (2014). PEG-PCL based micelle hydrogels as oral docetaxel delivery systems for breast cancer therapy. *Biomaterials* 35, 6972–6985. doi:10.1016/j.biomaterials.2014.04.099

Yallapu, M. M., Reddy, M. K., and Labhasetwar, V. (2007). "Nanogels: chemistry to drug delivery," in *Biomedical Applications of Nanotechnology*, eds V. Labhasetwar and D. L. Leslie-Pelecky (Hoboken: Wiley), 131–171.

Yom-Tov, O., Frisman, I., Seliktar, D., and Bianco-Peled, H. (2014). A novel method for hydrogel nanostructuring. *Eur. Polym. J.* 52, 137–145. doi:10.1016/j.eurpolymj.2014.01.004

**Conflict of Interest Statement:** The author declares that the research was conducted in the absence of any commercial or financial relationships that could be construed as a potential conflict of interest.

# 8

# Highly selective mercury detection at partially oxidized graphene/poly(3,4-ethylenedioxythiophene):poly (styrenesulfonate) nanocomposite film-modified electrode

*Nael G. Yasri[1,2,3†], Ashok K. Sundramoorthy[2*†], Woo-Jin Chang[1,4]\* and Sundaram Gunasekaran[2]\**

[1] Department of Mechanical Engineering, University of Wisconsin-Milwaukee, Milwaukee, WI, USA
[2] Department of Biological Systems Engineering, University of Wisconsin-Madison, Madison, WI, USA
[3] Department of Chemistry, Faculty of Science, University of Aleppo, Aleppo, Syria
[4] Great Lakes WATER Institute, School of Freshwater Sciences, University of Wisconsin-Milwaukee, Milwaukee, WI, USA

**Edited by:**
Emilia Morallon, Universidad de
Alicante, Spain

**Reviewed by:**
Santosh Kumar Yadav, Drexel
University, USA
Cesar Alfredo Barbero, Universidad
Nacional de Rio Cuarto, Argentina

**\*Correspondence:**
Ashok K. Sundramoorthy and
Sundaram Gunasekaran, Department
of Biological Systems Engineering,
University of Wisconsin-Madison, 460
Henry Mall, Madison, WI 53706, USA
e-mail: sundramoorth@wisc.edu,
ashok.research@outlook.com;
guna@wisc.edu;
Woo-Jin Chang, Department of
Mechanical Engineering, University of
Wisconsin-Milwaukee, 3200 N.
Cramer Street, Milwaukee, WI
53211, USA
e-mail: wjchang@uwm.edu

[†] Nael G. Yasri and Ashok K.
Sundramoorthy have contributed
equally to this work.

Partially oxidized graphene flakes (po-Gr) were obtained from graphite electrode by an electrochemical exfoliation method. As-produced po-Gr flakes were dispersed in water with the assistance of poly(3,4-ethylenedioxythiophene)/poly(styrenesulfonate) (PEDOT:PSS). The po-Gr flakes and the po-Gr/PEDOT:PSS nanocomposite (po-Gr/PEDOT:PSS) were characterized by Raman spectroscopy, Fourier transform-infrared spectroscopy (FT-IR), UV–Vis spectroscopy, X-ray diffraction (XRD), and scanning electron microscopy (SEM). In addition, we demonstrated the potential use of po-Gr/PEDOT:PSS electrode in electrochemical detection of mercury ions ($Hg^{2+}$) in water samples. The presence of po-Gr sheets in PEDOT:PSS film greatly enhanced the electrochemical response for $Hg^{2+}$. Cyclic voltammetry measurements showed a well-defined $Hg^{2+}$ redox peaks with a cathodic peak at 0.23 V, and an anodic peak at 0.42 V. Using differential pulse stripping voltammetry, detection of $Hg^{2+}$ was achieved in the range of 0.2–14 $\mu$M ($R^2 = 0.991$), with a limit of detection of 0.19 $\mu$M for $Hg^{2+}$. The electrode performed satisfactorily for sensitive and selective detection of $Hg^{2+}$ in real samples, and the po-Gr/PEDOT:PSS film remains stable on the electrode surface for repeated use. Therefore, our method is potentially suitable for routine $Hg^{2+}$ sensing in environmental water samples.

**Keywords: graphene flakes, mercury determination, PEDOT:PSS-modified electrode, heavy metal analysis, electrochemical exfoliation**

## INTRODUCTION

Mercury (Hg) is an essential element in the industry; however, the metal and its compounds are extremely dangerous to human health and to the environment. Although the use of Hg is regulated in many countries, it is still used in several domestic and industrial applications, which has led to the accumulation of Hg residues in landfills, soils, and streams (Seco-Reigosa et al., 2014). As a result, the focus of numerous investigations has been on closely monitoring Hg present in the environment (Pesavento et al., 2009). Standard methods for Hg analysis include: cold vapor atomic absorption spectrometry (CVAAS) (EPA, 2007a) and inductively coupled plasma-mass spectrometry (ICP-MS) (EPA, 2007b). Some Hg-monitoring applications, without requiring sample pretreatment, depend upon point-of-use sensors that are simple, rapid, stable, reliable, and inexpensive. Accordingly, various methods have been developed such as colorimetric (Liu et al., 2010), fluorometric (Wang et al., 2014a), magnetic (Najafi et al., 2013), electrochemical (Martin-Yerga et al., 2013), etc. The electroanalytical techniques have played a major role in simplified testing for Hg

and have been approved by many regulatory bodies (EPA, 1996). In general, many electroanalytical or colorimetric methods rely upon a change in electrical signal following a reaction (amalgamation) of gold (Au) or other precious metals with Hg, either at the electrode or within the sample solution to selectively bind with the target mercury ions ($Hg^{2+}$) (Welch et al., 2004; Martin-Yerga et al., 2013). The major drawbacks in using Au for sensing Hg are the significant effect of the sample matrix (Botasini et al., 2013) and the structural changes on the sensor material caused by the amalgam formation (Welch et al., 2004; Martin-Yerga et al., 2013), which require some additional chemical, electrochemical, and mechanical pretreatment of the sample (Anandhakumar et al., 2012).

Some electrochemical methods for sensing Hg use the conjugation of functional groups, such as the donor ligands of nitrogen (N) or sulfur (S) present in amino acids or conducting polymers, such as poly (3,4-ethylenedioxythiophene) (PEDOT), which have a strong binding preference for $Hg^{2+}$ (Chow and Gooding, 2006; Giannetto et al., 2011; Anandhakumar et al., 2012). In the case of

PEDOT, although it contains S, which can endow two unpaired electrons, it may not be possible to use PEDOT by itself due to the low signal sensitivity, high insolubility, and intractability (Martin-Yerga et al., 2013). Therefore, to produce a stable and flexible polymer, a conventional poly(styrenesulfonate) (PSS) is incorporated into PEDOT to form PEDOT:PSS (Vacca et al., 2008), which is an excellent copolymer because of its high conductivity, environmental stability (Wang et al., 2014b). To date, the potentials of PEDOT:PSS for electrode modification and electrochemical sensing of toxic metal ions have been seldom reported (Anandhakumar et al., 2011; Yasri et al., 2011; Rattan et al., 2013).

Due to high electrical conductivity, biocompatibility, and the exceptional surface-to-volume ratio, graphene (Gr) received more attention in various applications, including chemical and biosensors (Hill et al., 2011; Sundramoorthy and Gunasekaran, 2014). Recently, incorporating Gr on the working electrode surface to enhance the electrochemical signal for analytical applications has become fairly common (Mikolaj and Zbigniew, 2012; Sundramoorthy and Gunasekaran, 2014). Generally, Gr or reduced graphene oxide (rGO) are used for sensor applications (Yang and Gunasekaran, 2013; Yang et al., 2013). Gr can be obtained from graphite by mechanical cleavage (Jayasena and Subbiah, 2011), chemical exfoliation (Zhang et al., 2010), thermal decomposition (Wang et al., 2012), or electrochemical exfoliation (Low et al., 2013). Among other methods, electrochemical exfoliation of graphite electrode is considered a simple, rapid, and "green" method, as the use of toxic or corrosive reducing reagents or stabilizers are avoided in this method (Su et al., 2011; Singh et al., 2012; Chang et al., 2013; Gee et al., 2013; Mao et al., 2013).

The electrochemical exfoliation of graphite can be achieved by a one or two-step process. In a two-step process, in the first step, graphite electrode is activated in an electrolyte solution at a relatively low bias voltage. During this part of the process, the electrode expands due to the intercalation of electrolyte (usually $Li^+$ or $Na^+$) ions into the graphite lattice (Zhong and Swager, 2012). The second step involves applying a higher bias voltage to ensure exfoliation of the expanded graphite electrode and separating Gr flakes into the solution (Qi et al., 2011; Su et al., 2011; Zhang et al., 2012; Gee et al., 2013; Li et al., 2013). For example, when lithium ion ($Li^+$) is used as an electrolyte, $-3.0\,V$ of static bias voltage is applied to intercalate $Li^+$ into graphite electrode (Wang et al., 2011). In this work, we have electrochemically synthesized partially oxidized Gr (po-Gr) flakes to study its potential application in detecting mercury ($Hg^{2+}$) ions with PEDOT:PSS. The po-Gr/PEDOT:PSS conducting film readily conjugates with $Hg^{2+}$ in water and allowed us to do selective detection of $Hg^{2+}$ in real samples.

## MATERIALS AND METHODS
### REAGENTS
We used analytical-grade chemicals from Fisher, Acros Organics, and Sigma-Aldrich (USA). Supporting electrolytes were prepared using $HNO_3$ or NaCl. Stock solution of $10 \times 10^{-4}\,M$ $Hg^{2+}$ was prepared using mercury nitrate [$Hg(NO_3)_2$] and used after further dilution. All aqueous solutions were prepared using deionized water with $18.2\,M\Omega\,cm$ (EMD, Millipore). The

PEDOT:PSS sample was received from CIDETEC research group (San Sebastian, Spain) (Istamboulie et al., 2010).

### ELECTROCHEMICAL SYNTHESIS OF po-Gr
The po-Gr flakes were obtained by electrochemical exfoliation of graphite sheet. Briefly, a two-electrode cell was used with a piece of flexible graphite sheet (Graphitestore, Inc., USA) as a working electrode and a platinum (Pt) wire were placed parallel to and about 10 mm away from the graphite electrode which served as a counter electrode. The electrodes were connected to a DC power supply (Tektronix PS 280). About 10 mm of the working and counter electrodes were immersed into 25 mL containing $0.1\,M\,HClO_4$ and $0.1\,M\,NaCl$ which served as an electrolyte. The exfoliation was performed by applying DC bias on the working electrode. Initially, a potential of $-2.5\,V$ was applied for 60 min to facilitate the electrochemical expansion of graphite electrode by intercalation of $Na^+$ ions into graphite layers; after which, a potential of $+10\,V$ was applied for another 60 min to achieve exfoliation. The electrolyte solution containing the exfoliated po-Gr flakes was vacuum filtrated (using membrane with a pore size $0.4\,\mu m$) and washed with deionized water many times to remove the residual electrolyte.

### PREPARATION OF po-Gr/PEDOT:PSS DISPERSION
The obtained po-Gr flakes were dried at 60°C for 30 min, and then dispersed in 10 mL PEDOT:PSS solution by a probe sonicator (Sonics, VibraCell VCX130) for 15 min. Later, the po-Gr/PEDOT:PSS dispersion was centrifuged for 30 min at 2,000 rpm (Sorvall Super T21) and the supernatant was used for further characterization. The po-Gr flakes were also dispersed in water (without PEDOT:PSS) for 15 min by probe sonicator and centrifuged at 2,000 rpm for 30 min to collect supernatant for control studies. All experiments were performed at room temperature ($25 \pm 3$°C).

### PREPARATION OF po-Gr/PEDOT:PSS-MODIFIED ELECTRODE
A $10\,\mu L$ sample of po-Gr/PEDOT:PSS or po-Gr dispersion was placed on a well cleaned (after mirror-like polishing with alumina powder) glassy carbon electrode (GCE) surface and dried in an air-oven for 30 min to evaporate solvents. By this method, po-Gr/PEDOT:PSS or po-Gr film-coated electrode was obtained and gently washed by immersing in water for about 5 min to remove unbounded materials from the electrode surface. For Raman measurements, the po-Gr film prepared on glass substrate was reduced with hydrazine in pH 9.0 water solution at 80°C for 2 h (Park et al., 2011).

### ELECTROCHEMICAL MEASUREMENTS AND CHARACTERIZATION
Electrochemical measurements were performed by using an electrochemical workstation (660D, CH Instruments). A 10-mL volume, three-electrode system was used with GCE, Ag/AgCl (3 M KCl), and Pt wire as working, reference, and counter electrode, respectively. The GCE was used either bare or after modification with po-Gr film, PEDOT:PSS film, or po-Gr/PEDOT:PSS film. The electrochemical responses of the bare GCE and modified GCE's toward $Hg^{2+}$ were examined with cyclic voltammetry (CV) in $0.05\,M\,HNO_3$ as an electrolyte.

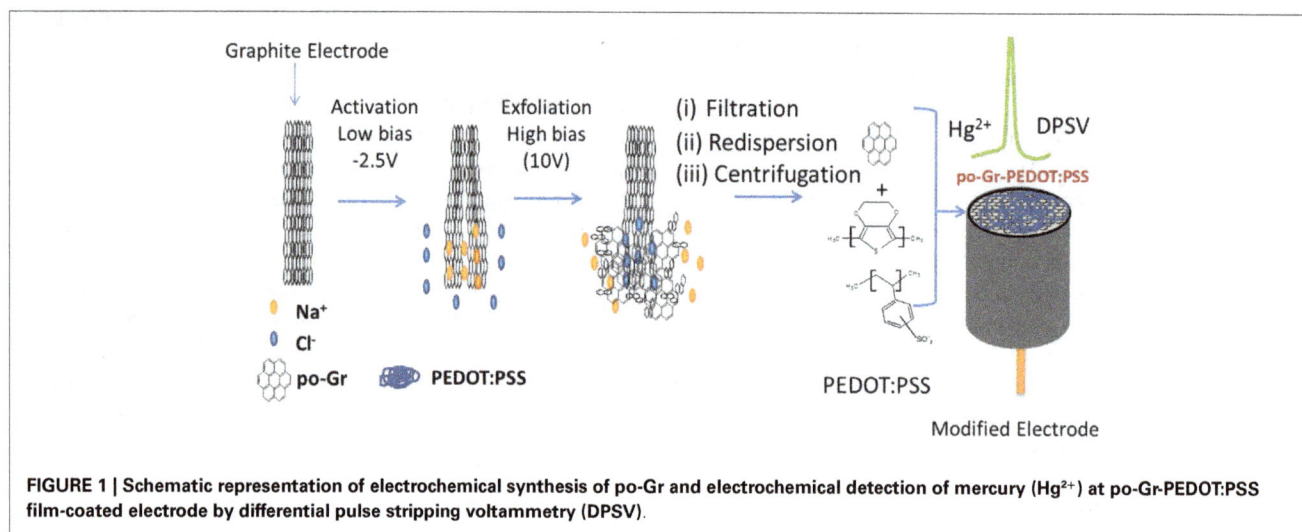

**FIGURE 1 | Schematic representation of electrochemical synthesis of po-Gr and electrochemical detection of mercury (Hg²⁺) at po-Gr-PEDOT:PSS film-coated electrode by differential pulse stripping voltammetry (DPSV).**

Electrochemical impedance spectroscopy (EIS) measurements were performed in a solution containing 2.5 mM $[Fe(CN)_6]^{4-/3-}$ and 0.1 M KCl supporting electrolyte in the frequency range of 1–106 Hz. Differential pulse stripping voltammetry (DPSV) was performed by applying deposition and then stripping steps with the following parameters: initial potential, −0.2 V; final potential, 0.8 V; amplitude, 50 mV; pulse width, 0.2 s; pulse period, 0.5 s; sample width, 0.0169 s; and deposition time, 2 min. During the deposition period, the solution was stirred at 800 rpm, and the potential was held at −0.30 V. The DPSV voltammograms were recorded upon injection of $Hg^{2+}$ ions in the range of 0.2–14.0 μM in 0.05 M $HNO_3$. The interferences of some metal ions ($Ca^{2+}$, $Fe^{2+}$, $Co^{2+}$, $Ni^{2+}$, $Cd^{2+}$, $Zn^{2+}$, $Cr^{6+}$, and $Pb^{2+}$) on the determination of 6.0 μM $Hg^{2+}$ in 0.05 M $HNO_3$ were also investigated by adding their respective nitrate salts.

The surface morphology of po-Gr flakes was studied by scanning electron microscopy (SEM) (LEO1530, Gemini FESEM, Carl Zeiss). Further characterizations were performed on dry samples using Raman spectroscopy (LabRAM Aramis Horiba Jobin Yvon Confocal Raman Microscope, wavelength: 532 nm) and attenuated total reflectance-Fourier transform-infrared spectroscopy (ATR-FT-IR) (Spectrum 100, PerkinElmer). UV–Vis spectra of po-Gr and po-Gr/PEDOT:PSS dispersed in water were obtained using a spectrophotometer (Lambda 25, PerkinElmer). X-ray diffraction (XRD) pattern of samples were measured using Bruker D8 Discover diffractometer.

## RESULTS AND DISCUSSION
### CHARACTERIZATION OF po-Gr AND po-Gr/PEDOT:PSS FILM
The scheme for electrochemical exfoliation of po-Gr flakes and modification of GCE surface with po-Gr/PEDOT:PSS dispersion for $Hg^{2+}$ detection using DPSV is shown in **Figure 1**. After successful exfoliation, po-Gr flakes were dispersed separately in water, and PEDOT:PSS solution (**Figure 2A**). The po-Gr solution (**Figure 2A**, image a) was light yellowish brown in color and the po-Gr dispersed in PEDOT:PSS was light bluish (**Figure 2A**, image b). UV–Vis spectra of po-Gr solution showed a strong absorption band at 261 nm, which can be assigned to the partially oxidized

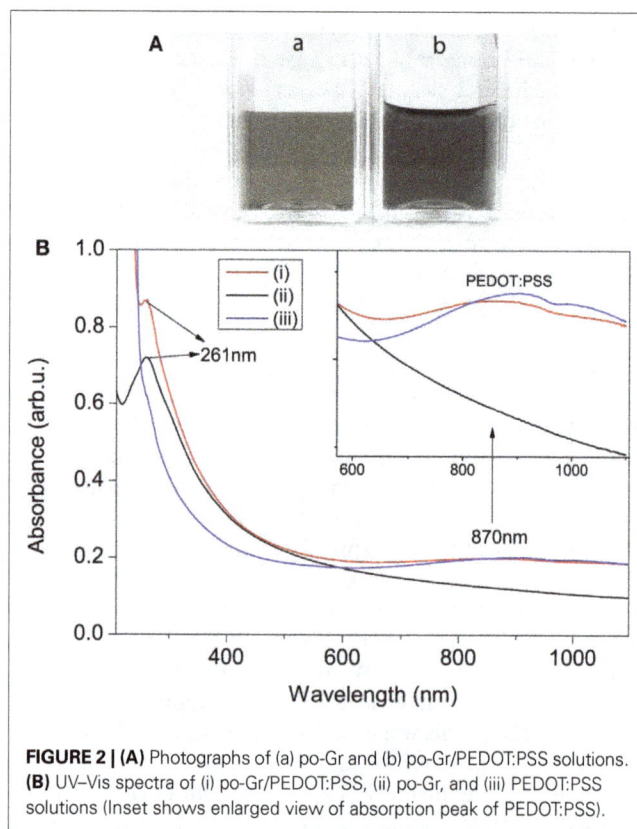

**FIGURE 2 | (A)** Photographs of (a) po-Gr and (b) po-Gr/PEDOT:PSS solutions. **(B)** UV–Vis spectra of (i) po-Gr/PEDOT:PSS, (ii) po-Gr, and (iii) PEDOT:PSS solutions (Inset shows enlarged view of absorption peak of PEDOT:PSS).

graphene sheets. It has been shown that graphene oxide (GO) and rGO have absorption peaks at 230 nm, and 270 nm (π–π* transition of aromatic C–C bonds), respectively (**Figure 2B**) (Li et al., 2008; Choi et al., 2010).

The optical absorption peak of our synthesized po-Gr solution (after centrifugation) is in the range between that for GO and rGO; thus, we describe it as po-Gr. However, UV–Vis spectra of po-Gr/PEDOT:PSS solution showed two major absorption bands, first band observed at 261 nm and second broad band centered at 870 nm. The first peak corresponds to the optical absorption

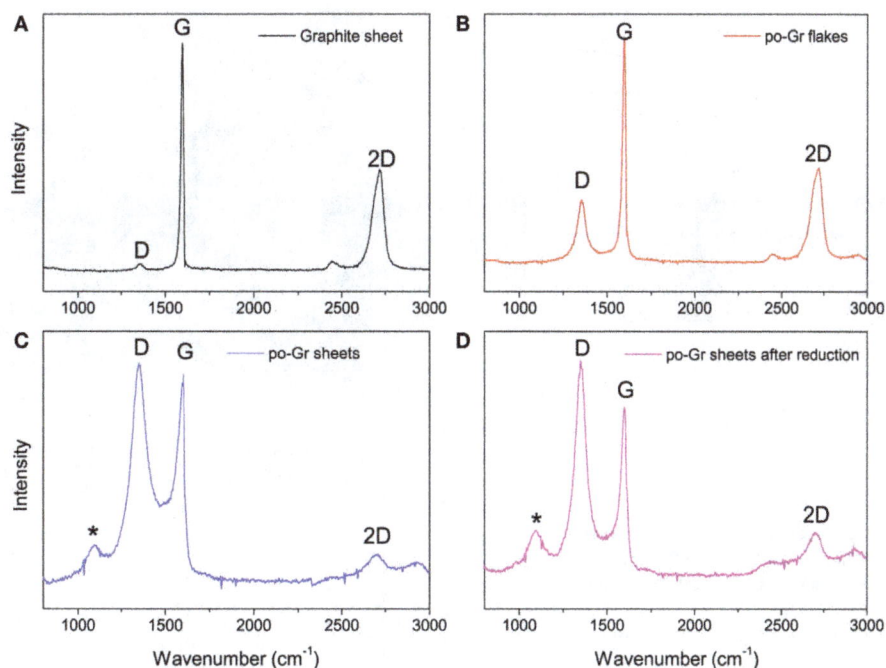

**FIGURE 3 |** Raman spectra of **(A)** pristine graphite sheet, **(B)** po-Gr flakes (as-synthesized from electrochemical exfoliation), **(C)** po-Gr film (after centrifugation), and **(D)** po-Gr film after reduction with hydrazine (532 nm laser was used) (*peak is assigned to the glass substrate).

**Table 1 |** The D, G, 2D bands and $I_D/I_G$ ratio were estimated from the Raman spectrum of pristine graphite sheet, po-Gr flakes, po-Gr film (prepared after centrifugation), and po-Gr film after chemical reduction using hydrazine.

| Samples | D band intensity at ~1353 cm$^{-1}$ | G band intensity at ~1594 cm$^{-1}$ | $I_D/I_G$ | 2D band intensity at ~2718 cm$^{-1}$ |
|---|---|---|---|---|
| Graphite sheets | 38.58 | 752.02 | 0.051 | 345.24 |
| po-Gr flakes (without centrifuging) | 196.08 | 631.87 | 0.310 | 282.18 |
| po-Gr film (prepared after centrifuging) | 574.58 | 535.62 | 1.073 | 165.68 |
| po-Gr film after reduction with hydrazine | 400.60 | 324.64 | 1.234 | 122.50 |

of the po-Gr and the second peak corresponds to the oxidized PEDOT:PSS (light blue) (Gustafsson-Carlberg et al., 1995; Pettersson et al., 1999; Tarabella et al., 2012). For comparison, we also measured UV–Vis spectra of PEDOT:PSS solution, which shows a broad optical absorption band centered at 870 nm (**Figure 2B**). This study provides evidence that po-Gr sheets are stabilized in PEDOT:PSS solution.

Raman spectra of the (a) pristine graphite sheet, (b) as-synthesized po-Gr flakes, (c) po-Gr film prepared from water dispersion after centrifugation, and (d) po-Gr film after reduction with hydrazine are presented in **Figure 3**. The spectra show D, G, and 2D bands for all samples (**Table 1**). For pristine graphite sheet, almost insignificant D band was observed, because of highly crystalline structure of graphite without defects (**Figure 3A**). However, Raman spectrum of the other three samples (**Figures 3B–D**) showed significant changes upon exfoliation, following redispersion in water and after reduction with hydrazine (**Table 1**). The G band is characteristic of sp2-hybridized C $=$ C bonds in graphene sheets (Childres et al., 2013); whereas, the D band is associated with

structural defects and partially disordered structures of the sp2 domain (Childres et al., 2013). The 2D band located at 2718 cm$^{-1}$ originates from a double-resonance process (Krauss et al., 2009; Yan et al., 2011). The calculated values of $I_D/I_G$ listed in **Table 1**. Su et al. (2011)) show that it changed significantly from 1.073 to 1.234 after reduction with hydrazine, indicating that restoration of C–C bonds in the graphene lattice, and a decrease in the average size of graphene domains (Lee et al., 2014). This study supports our understanding that as synthesized graphene flakes contain significant defects, due to functional groups generated upon electrochemical exfoliation at high voltage (Morales et al., 2011). $I_D/I_G$ ratio (0.310) of as-synthesized graphene flakes is relatively small compared to po-Gr film (after centrifugation), which may be due to the presence of large graphene flakes without complete exfoliation (**Figure 3B**). Therefore, it was necessary to disperse po-Gr flakes in a suitable dispersant with a probe sonicator to achieve complete exfoliation of graphene sheets. In this work, we used PEDOT:PSS to disperse po-Gr flakes in water.

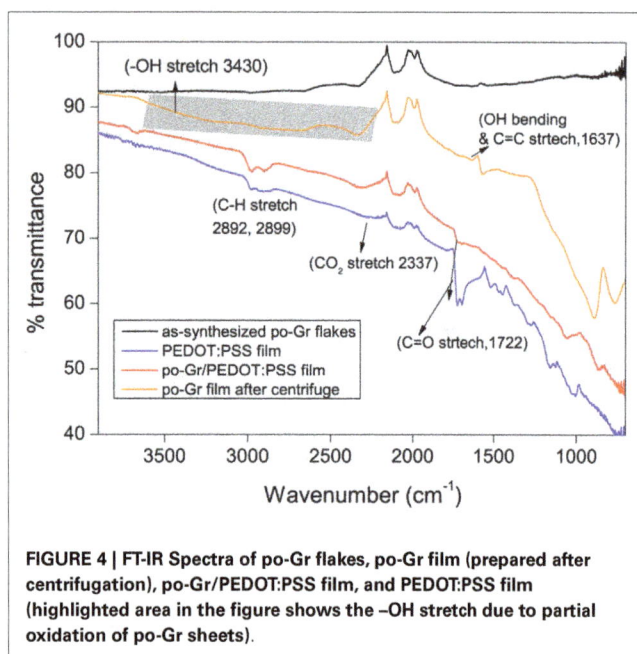

**FIGURE 4 | FT-IR Spectra of po-Gr flakes, po-Gr film (prepared after centrifugation), po-Gr/PEDOT:PSS film, and PEDOT:PSS film (highlighted area in the figure shows the –OH stretch due to partial oxidation of po-Gr sheets).**

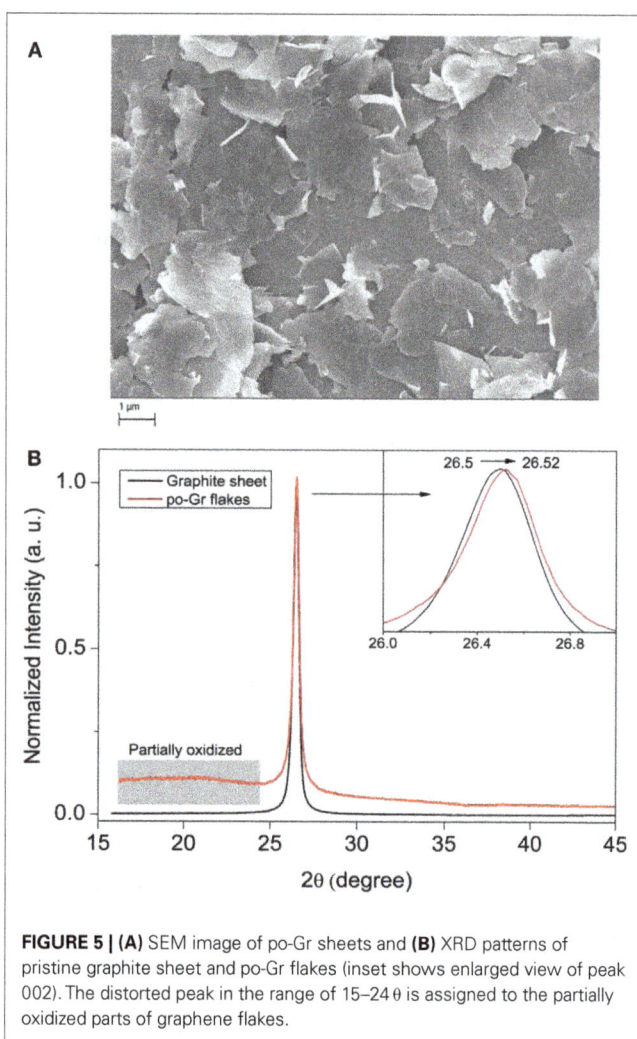

**FIGURE 5 | (A)** SEM image of po-Gr sheets and **(B)** XRD patterns of pristine graphite sheet and po-Gr flakes (inset shows enlarged view of peak 002). The distorted peak in the range of 15–24 θ is assigned to the partially oxidized parts of graphene flakes.

The as-synthesized po-Gr flakes, po-Gr film (after centrifugation), PEDOT:PSS film, and po-Gr/PEDOT:PSS film were characterized by FT-IR (**Figure 4**). For as-synthesized po-Gr flakes, no significant bands were observed, which may be due to the incomplete exfoliation process. Indeed, after successful redispersion in water and following centrifugation, po-Gr film on glass showed significant bands at ~3430 (OH stretch), 2337 ($CO_2$ stretch), 1722 (C $=$ O stretch), and 1637 cm$^{-1}$(OH bending and C $=$ C stretch)(Pham et al., 2011). The FT-IR spectrum of PEDOT:PSS showed peaks at 1372, 1289, 1124, 1023, 1002 cm$^{-1}$, which are derived from PEDOT:PSS (Alemu Mengistie et al., 2013; Yoo et al., 2014). The IR bands at 1160 and 1023 cm$^{-1}$ are assigned to $SO_3^-$ of the PSS. Further, the FT-IR spectrum of po-Gr/PEDOT:PSS consists bands of PEDOT:PSS, which proves that po-Gr sheets wrapped with the polymer structure (**Figure 4**).

Surface morphology of po-Gr sheets was also studied by SEM. SEM images of the exfoliated po-Gr film show wrinkled or folded thin sheets with the lateral dimension of 1–3 μm (**Figure 5A**). The XRD spectrum of graphite sheet exhibits an intense peak at 26.5°, corresponding to *d*-spacing of 0.34 nm (**Figure 5B**). However, after electrochemical exfoliation, po-Gr flakes exhibit a broad peak at ~16 to 23° and another intense peak at 26.52°. The broad peak at ~16 to 23° indicates the presence of functional groups containing oxygen, which is formed during electrochemical exfoliation (**Figure 5B**) (Fang et al., 2009). There is also slight shift in the peak position of po-Gr flakes from graphite after electrochemical exfoliation from 26.50 to 26.52 (inset of **Figure 5B**), which may due to intercalation of ions into the graphene layers.

## ELECTROCHEMICAL IMPEDANCE STUDIES

**Figure 6** is the Nyquist plot of the modified GCE's in 2.5 mM [Fe(CN)$_6$]$^{4-/3-}$ in 0.1 M KCl. The semicircular part in the high-frequency region represents electron-transfer-limiting process with its effective diameter equal to Faradaic charge transfer resistance ($R_{ct}$), which is responsible for the electron transfer kinetics of redox reactions at the electrode-electrolyte interface (Kumar et al., 2010; Yang and Gunasekaran, 2013). The $R_{ct}$ values of po-Gr/GCE ($R_{ct} = 144 \, \Omega$), PEDOT:PSS/GCE ($R_{ct} = 65 \, \Omega$), and po-Gr/PEDOT:PSS/GCE ($R_{ct} = 54 \, \Omega$), modified electrodes were lower than that for bare electrode ($R_{ct} = 228 \, \Omega$), indicating higher conductivity as a result of modification processes (**Figure 6**). Generally poor conductivity of GO-modified electrode is due to the presence of excessive oxygenated species, which accentuates the insulating characteristics (Yang and Gunasekaran, 2013). The $R_{ct}$ for po-Gr film is also higher than for PEDOT:PSS/GCE and po-Gr/PEDOT:PSS/GCE's perhaps due to the presence of oxygenated species associated with po-Gr, which may affect conductivity of the electrode. However, when po-Gr sheets present in PEDOT:PSS, it improves the conductivity of the electrode (**Figure 6**).

## DETECTING Hg$^{2+}$
### Linear sweep voltammetry
The linear sweep voltammograms (LSVs) of Hg$^{2+}$ recorded on bare, po-Gr/PEDOT:PSS-, PEDOT:PSS-, and po-Gr-modified GCE's in 0.05 M HNO$_3$ show two electrochemical oxidation peaks

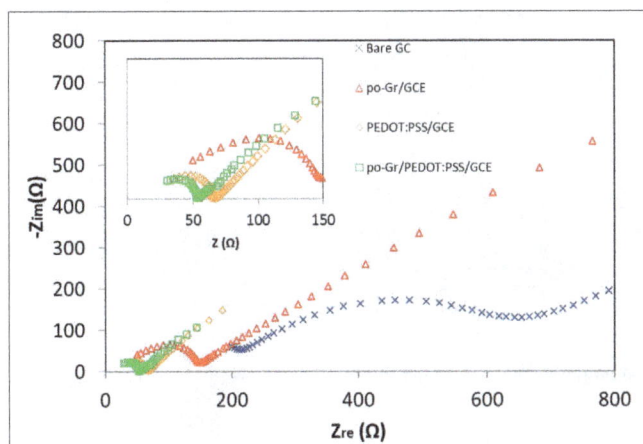

**FIGURE 6 | Nyquist ($Z_{re}$ vs. $Z_{img}$) impedance spectra collected using bare GCE, po-Gr/GCE, PEDOT:PSS/GCE, and po-Gr/PEDOT:PSS/GCE electrodes in 2.5 mM [Fe(CN)$_6$]$^{4-/3-}$ and 0.1 M KCl. Inset: enlarged high-frequency region**.

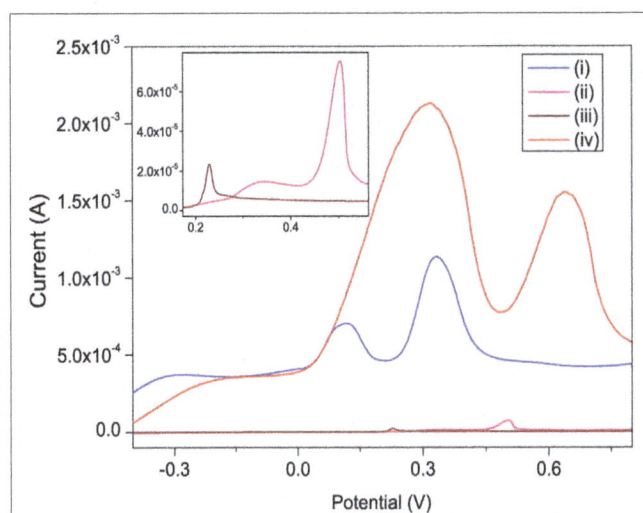

**FIGURE 7 | Linear sweep voltammograms recorded for 0.8 mM Hg$^{2+}$ in 0.05 M HNO$_3$ solution on (i) PEDOT:PSS-, (ii) po-Gr-, (iii) bare-, and (iv) po-Gr/PEDOT:PSS/modified GCE's. Inset: enlarged bare and po-Gr-modified GCE's**. Scan rate = 100 mV/s.

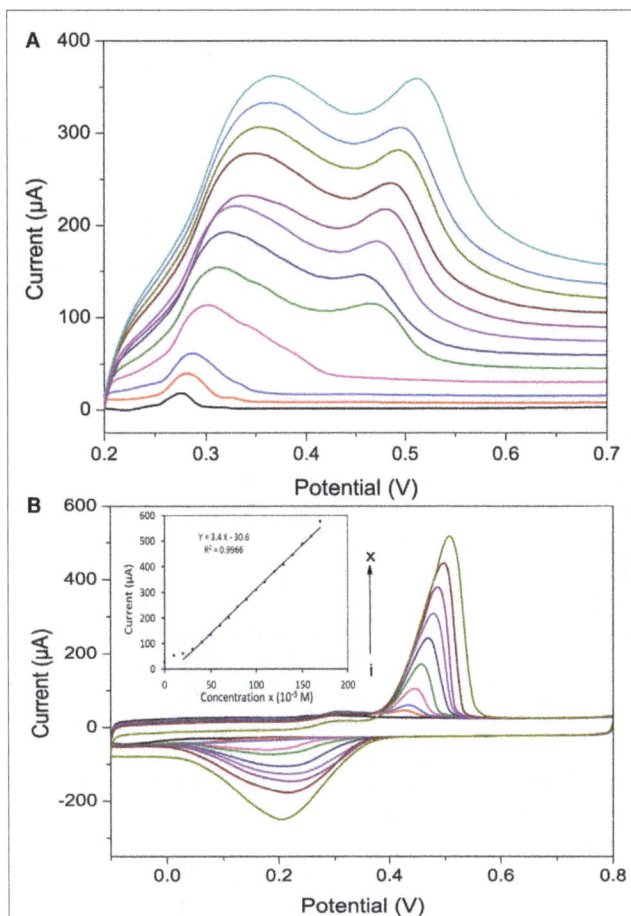

**FIGURE 8 | (A)** A linear sweep voltammograms recorded using po-Gr/PEDOT:PSS-modified GCE in 0.4 mM Hg$^{2+}$ + 0.05 M HNO$_3$ at various scan rates (1–100 mV/s) and **(B)** cyclic voltammograms of different concentration of Hg$^{2+}$. (i) 0.0, (ii) 0.1, (iii) 0.2, (iv) 0.4, (v) 0.6, (vi) 0.8, (vii) 1.0, (viii) 1.2, (ix) 1.4, and (x) 1.6 mM (inset shows the corresponding calibration curve) in 0.05 M HNO$_3$ medium using po-Gr/PEDOT:PSS-modified GCE at a scan rate of 20 mV/s.

for Hg$^{2+}$ (**Figure 7**). These peaks are attributed to two well-defined one-electron steps, according to the two-step equation: Hg$^{2+}$ + e$^-$ = Hg$^+$ + e$^-$ = Hg$^0$ (Orlik and Galus, 2007). The presence of po-Gr on both bare GCE and PEDOT:PSS-modified GCE improved the peak current (I$_{pa}$) compared to that in the absence of po-Gr. For example, 3-fold, 23-fold, and 100-fold increases in oxidation currents were obtained with po-Gr-, PEDOT:PSS-, and po-Gr-PEDOT:PSS-modified GCE's, respectively compared to 23.4 μA obtained with bare GCE (**Figure 7**). These I$_{pa}$ increases are attributed to the combined effects of sulfonic (Pillay et al., 2013) and thiol (Chandrasekhar et al., 2007; Kadarkaraisamy et al., 2011; Mandal et al., 2013) functional groups on PEDOT:PSS matrix/po-Gr nanocomposite (Shao et al., 2010; Anandhakumar et al., 2012).

The effect of scan rate on the voltammograms of Hg$^{2+}$ at 0.4 mM concentration is shown in **Figure 8A**. Significant increases in I$_{pa}$ with increasing scan rate from 1 to 100 mV/s were obtained using po-Gr/PEDOT:PSS-modified GCE (**Figure 8A**). The relation between I$_{pa}$ vs. square root of scan rate was linear, which indicated that the electrode reaction was diffusion-controlled (Bard and Faulkner, 2001). However, at lower scan rates of up to 20 mV/s, the I$_{pa}$ showed only single oxidation peak, which may be due to a slower process, as the reaction occurs in one step: Hg$^{2+}$ + 2e$-$ = Hg$^0$ (Orlik and Galus, 2007). The po-Gr/PEDOT:PSS-modified GCE showed a linear response for various Hg$^{2+}$ concentrations from 0.3 to 1.6 mM ($R^2$ = 0.997) (**Figure 8B**).

### Differential pulse stripping voltammetry measurements

The DPSV is a highly sensitive technique for electroanalysis of trace metals in different samples. It involves two steps for the detection of Hg$^{2+}$ (Somerset et al., 2010) such as (i) deposition of Hg$^0$ at an

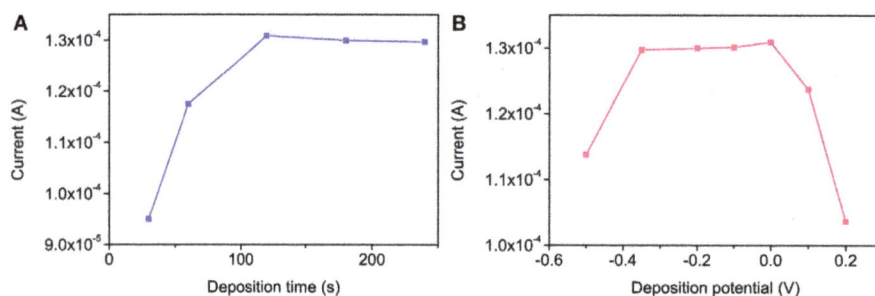

**FIGURE 9 |** The effect of (A) deposition time at −0.3 V vs. Ag/AgCl and (B) deposition potential for 120 s on the DPSV stripping responses of 6 μM of Hg²⁺ at po-Gr/PEDOT:PSS-modified GCE in 0.05 M HNO₃.

optimized potential for a certain duration and (ii) anodic stripping of deposited Hg⁰. As shown in **Figure 8B**, the reduction of Hg²⁺ occurs at 0.21 V vs. Ag/AgCl. The effects of the deposition potential and time on the Hg stripping responses were investigated accordingly. The stripping peak currents at different deposition potentials (from −0.5 to +0.2 V) showed that the best stripping signal was obtained in the range of −0.30 to 0.0 V vs. Ag/AgCl (**Figure 9B**). Based on these results, a deposition potential of −0.30 V was selected for further investigation. **Figure 9A** shows the relationship between the Hg stripping signal against the deposition time at −0.3 V vs. Ag/AgCl using po-Gr/PEDOT:PSS-modified electrode. The peak current becomes fairly stable after 120 s of deposition.

The anodic DPSV response of the po-Gr/PEDOT:PSS-modified electrode with successive Hg²⁺ concentrations in 0.05 M HNO₃ solution are presented in **Figure 10**. A linear variation of the $I_{pa}$ was observed for concentrations ranging from 0.2 to 14.0 μM ($R^2 = 0.991$), with a sensitivity of 8.72 μA/μM. The limit of detection (LOD) and limit of quantification (LOQ) were calculated as LOD = 3.3 SD/$b$ and LOQ = 10 SD/$b$, where SD is the standard deviation of five reagent blank determinations and $b$ is the slope of the calibration curve (Shrivastava and Gupta, 2011). Using po-Gr/PEDOT:PSS-modified GCE, the LOD and LOQ were found to be 0.19 and 0.58 μM for Hg²⁺, respectively. This LOD is lower than dithiodianiline-derivative-modified electrode (~2.1 μM) (Somerset et al., 2010) and silver ink screen-printed electrode (~0.5 μM) (Chiu et al., 2008). The po-Gr-PEDOT:PSS-modified GCE also exhibited excellent stability and reproducibility with relative standard deviation (RSD) of 0.93% for 10 successive measurements of 6.0 μM Hg²⁺ in 0.05 M HNO₃ solution. The RSD for six similarly prepared electrodes tested under the same conditions was 1.6%. LOD of Hg²⁺ at Au-PEDOT carbon composite film was ~5 μM (Anandhakumar et al., 2012). In addition, Au-PEDOT film-modified electrode may not be suitable for repeated use because of amalgam formation with AuNPs. Amalgam formation on the electrode surface affects reproducibility of analytical response. It is necessary to regenerate or activate the electrode surface each time with ethylenediaminetetraacetic acid (Giannetto et al., 2011). Electrochemical response of Hg²⁺ at po-Gr/PEDOT:PSS-modified GCE is not dependent on amalgam formation, so it offers reproducible measurements without the need for regeneration or electrode activation.

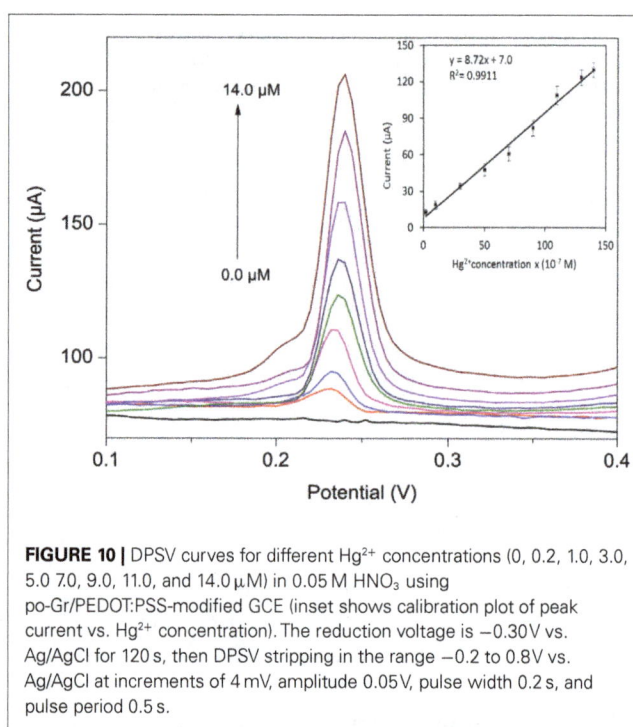

**FIGURE 10 |** DPSV curves for different Hg²⁺ concentrations (0, 0.2, 1.0, 3.0, 5.0 7.0, 9.0, 11.0, and 14.0 μM) in 0.05 M HNO₃ using po-Gr/PEDOT:PSS-modified GCE (inset shows calibration plot of peak current vs. Hg²⁺ concentration). The reduction voltage is −0.30 V vs. Ag/AgCl for 120 s, then DPSV stripping in the range −0.2 to 0.8 V vs. Ag/AgCl at increments of 4 mV, amplitude 0.05 V, pulse width 0.2 s, and pulse period 0.5 s.

## EFFECT OF INTERFERENCE

The influence of various common interfering metal cations (Ca²⁺, Zn²⁺, Ni²⁺, Cr⁶⁺, As³⁺, Cd²⁺, Co²⁺, Fe²⁺, and Cu²⁺) in the presence of 6.0 μM Hg²⁺ in 0.05 M HNO₃ were tested using the po-Gr/PEDOT:PSS-modified GCE. The interference effect was calculated as:

$$Interference \ (\%) = \frac{I_{Hg} - I_{Hg+interferent}}{I_{Hg}} \times 100$$

where, $I_{Hg}$ = peak current for Hg²⁺, $I_{Hg+ interferent}$ = peak current for Hg²⁺ plus added interferent. The data in **Table 2** indicate that the performance of the po-Gr-PEDOT:PSS-modified GCE was unaffected by the tested interferents. The stripping peak current of Hg²⁺ exhibited no change in the presence of Zn²⁺, Cd²⁺, Ca²⁺, As³⁺, or Ni²⁺ ions even at concentrations each at more than 300 times that of Hg²⁺. Assuming an acceptable interference

**Table 2 | Interference study of other metal ions on the DPSV measurement of 6.0 μM $Hg^{2+}$ using po-Gr/PEDOT:PSS-modified GCE.**

| Metal ion | Ratio of interfering cation/$Hg^{2+}$ | $Hg^{2+}$ measured (μM) (Mean ± SD) | RSD (%) | Interference (%) |
|---|---|---|---|---|
| $Hg^{2+}$ | 0 | 6.09 ± 0.112 | 1.83 | 0.0 |
| $Ca^{2+}$ | 330 | 5.94 ± 0.014 | 0.24 | −2.5 |
| $Zn^{2+}$ | 330 | 5.93 ± 0.022 | 0.38 | −2.6 |
| $Ni^{2+}$ | 330 | 5.97 ± 0.021 | 0.36 | −2.0 |
| $Cr^{6+}$ | 33 | 5.88 ± 0.010 | 0.18 | −3.4 |
| $As^{3+}$ | 330 | 5.98 ± 0.020 | 0.41 | −1.8 |
| $Cd^{2+}$ | 330 | 5.90 ± 0.023 | 0.39 | −3.1 |
| $Co^{2+}$ | 330 | 5.78 ± 0.024 | 0.42 | −5.1 |
| $Fe^{2+}$ | 10 | 6.49 ± 0.036 | 0.97 | +6.7 |
| $Cu^{2+}$ | 16 | 6.38 ± 0.052 | 0.82 | +4.8 |

**Table 3 | Comparison of $Hg^{2+}$ content determined by using po-Gr/PEDOT:PSS/GCE and by standard cold vapor atomic absorption spectroscopy (CVAAs) method in unknown samples.**

| Sample | Our sensor (ppm), (mean ± SD; $n = 3$) | CVAAS method (ppm), (mean ± SD; $n = 3$) |
|---|---|---|
| 1 | 15.08 ± 0.042 | 15.00 ± 0.05 |
| 2 | 18.94 ± 0.084 | 19.00 ± 0.05 |
| 3 | 4.17 ± 0.112 | 420 ± 0.05 |

of ±5% (Fifield and Kealey, 2000), only $Cu^{2+}$ at 16 times that of $Hg^{2+}$ and $Fe^{2+}$, at 10 times that of $Hg^{2+}$, can be considered as interferents. DPSV of a solution containing $Fe^{2+}$, $Cu^{2+}$, and $Hg^{2+}$, exhibited oxidation peaks for $Fe^{2+}$ at 0.44 V and for $Cu^{2+}$ at −0.024 V vs. Ag/AgCl (Figure S1 in Supplementary Material).

## VALIDATION WITH REAL SAMPLE TESTS

The performance of the po-Gr/PEDOT:PSS-modified GCE was evaluated by comparing test results with those determined according to US Environmental Protection Agency (EPA) Method 245.1 (EPA, 2007a) using CVAAS performed at the Wisconsin State Laboratory of Hygiene. The data obtained by measuring $Hg^{2+}$ content in three unknown laboratory waste samples (**Table 3**) provide an excellent validation of proposed method.

## CONCLUSION

We synthesized po-Gr from graphite sheets by electrochemical exfoliation using $HClO_4$/NaCl solution as an electrolyte. The po-Gr and po-Gr-PEDOT:PSS film were characterized using SEM, Raman, FT-IR, XRD, and UV–Vis spectroscopy. Both EIS and CV measurements proved that the presence of po-Gr enhanced the electrochemical catalytic properties of PEDOT:PSS material. The po-Gr/PEDOT:PSS-modified GCE exhibited higher catalytic peak current for $Hg^{2+}$ compared to bare and PEDOT:PSS-modified GCE. The Gr-PEDOT:PSS/GCE was stable and reproducible for determining $Hg^{2+}$ at micromolar levels. The LOD determined by DPSV was 0.19 μM, and the detection was linear in the range of 0.2–14.0 μM ($R^2 = 0.991$). The sensor response was not affected

by other metal ions. Accurate selective detection of $Hg^{2+}$ in laboratory water samples showed that our method is suitable for routine $Hg^{2+}$ sensing in environmental samples.

## ACKNOWLEDGMENTS

The authors are thankful to funding support to Nael G. Yasri through Scholar Rescue Fund Fellowship. Authors would also like to thank Prof. Thierry Noguer (Université de Perpignan Via Domitia) for cooperation and supplying the sensitized PEDOT:PSS sample.

## REFERENCES

Alemu Mengistie, D., Wang, P.-C., and Chu, C.-W. (2013). Effect of molecular weight of additives on the conductivity of PEDOT:PSS and efficiency for ITO-free organic solar cells. *J. Mater. Chem. A* 1, 9907–9915. doi:10.1039/c3ta11726j

Anandhakumar, S., Mathiyarasu, J., Lakshmi, K., Phani, N., and Yegnaraman, V. (2011). Simultaneous determination of cadmium and lead using PEDOT:PSS modified glassy carbon electrode. *Am. J. Anal. Chem.* 2, 470–474. doi:10.4236/ajac.2011.24056

Anandhakumar, S., Mathiyarasu, J., and Phani, K. L. N. (2012). Anodic stripping voltammetric detection of mercury(ii) using Au-PEDOT modified carbon paste electrode. *Anal. Methods* 4, 2486–2489. doi:10.1039/c2ay25170a

Bard, A. J., and Faulkner, L. R. (2001). *Electrochemical Methode; Fundamentals and Applications.* New York: John Wiley & Sons, INC.

Botasini, S., Heijo, G., and Mendez, E. (2013). Toward decentralized analysis of mercury (II) in real samples. A critical review on nanotechnology-based methodologies. *Anal. Chim. Acta* 800, 1–11. doi:10.1016/j.aca.2013.07.067

Chandrasekhar, S., Chopra, D., Gopalaiah, K., and Guru Row, T. N. (2007). The generalized anomeric effect in the 1,3-thiazolidines: evidence for both sulphur and nitrogen as electron donors. Crystal structures of various N-acylthiazolidines including mercury(II) complexes. Possible relevance to penicillin action. *J. Mol. Struct.* 837, 118–131. doi:10.1016/j.molstruc.2006.10.034

Chang, C. F., Truong, Q. D., and Chen, J. R. (2013). Graphene sheets synthesized by ionic-liquid-assisted electrolysis for application in water purification. *Appl. Surf. Sci.* 264, 329–334. doi:10.1016/j.apsusc.2012.10.022

Childres, I., Jauregui, L. A., Park, W., Cao, H., and Chen, Y. (2013). "Raman spectroscopy of graphene and related materials," in *Developments in Photon and Materials Research*, ed. J. I. Jang (New York: Nova Science Publishers).

Chiu, M.-H., Zen, J.-M., Kumar, A. S., Vasu, D., and Shih, Y. (2008). Selective cosmetic mercury analysis using a silver ink screen-printed electrode with potassium iodide solution. *Electroanalysis* 20, 2265–2270. doi:10.1002/elan.200804307

Choi, E.-Y., Han, T. H., Hong, J., Kim, J. E., Lee, S. H., Kim, H. W., et al. (2010). Noncovalent functionalization of graphene with end-functional polymers. *J. Mater. Chem.* 20, 1907–1912. doi:10.1039/b919074k

Chow, E., and Gooding, J. J. (2006). Peptide modified electrodes as electrochemical metal ion sensors. *Electroanalysis* 18, 1437–1448. doi:10.1002/elan.200603558

EPA. (1996). *Mercury in Aqueous Samples and Extracts by Anodic Stripping Voltammetry (ASV) (Method 7472).* Boston: EPA. Available from: http://www.epa.gov/osw/hazard/testmethods/sw846/pdfs/7472.pdf

EPA. (2007a). *Determination of Mercury in Water by Cold Vapor Atomic Absorption Spectrometry (Method 245.1).* Cincinnati, OH: U.S. EPA. Available at: http://water.epa.gov/scitech/methods/cwa/bioindicators/upload/2007_07_10_methods_method_245_1.pdf

EPA. (2007b). *Determination of Trace Elements In Waters And Wastes by Inductively Coupled Plasma – Mass Spectrometry (Method 200.8).* Cicinnati, OH: US EPA. Available at: http://water.epa.gov/scitech/methods/cwa/bioindicators/upload/2007_07_10_methods_method_200_8.pdf

Fang, M., Wang, K., Lu, H., Yang, Y., and Nutt, S. (2009). Covalent polymer functionalization of graphene nanosheets and mechanical properties of composites. *J. Mater. Chem.* 19, 7098–7105. doi:10.1039/b908220d

Fifield, F. W., and Kealey, D. (2000). *Principles and Practice of Analytical Chemistry.* Oxford: Blackwell Science Ltd.

Gee, C. M., Tseng, C. C., Wu, F. Y., Chang, H. P., Li, L. J., Hsieh, Y. P., et al. (2013). Flexible transparent electrodes made of electrochemically exfoliated graphene sheets from low-cost graphite pieces. *Displays* 34, 315–319. doi:10.1016/j.displa.2012.11.002

Giannetto, M., Mori, G., Terzi, F., Zanardi, C., and Seeber, R. (2011). Composite PEDOT/Au nanoparticles modified electrodes for determination of mercury at trace levels by anodic stripping voltammetry. *Electroanalysis* 23, 456–462. doi:10.1002/elan.201000469

Gustafsson-Carlberg, J. C., Inganäs, O., Andersson, M. R., Booth, C., Azens, A., and Granqvist, C. G. (1995). Tuning the bandgap for polymeric smart windows and displays. *Electrochim. Acta* 40, 2233–2235. doi:10.1016/0013-4686(95)00169-F

Hill, E. W., Vijayaragahvan, A., and Novoselov, K. (2011). Graphene sensors. *Sens. J. IEEE* 11, 3161–3170. doi:10.1109/JSEN.2011.2167608

Istamboulie, G., Sikora, T., Jubete, E., Ochoteco, E., Marty, J.-L., and Noguer, T. (2010). Screen-printed poly(3,4-ethylenedioxythiophene) (PEDOT): a new electrochemical mediator for acetylcholinesterase-based biosensors. *Talanta* 82, 957–961. doi:10.1016/j.talanta.2010.05.070

Jayasena, B., and Subbiah, S. (2011). A novel mechanical cleavage method for synthesizing few-layer graphenes. *Nanoscale Res. Lett.* 6, 95. doi:10.1186/1556-276X-6-95

Kadarkaraisamy, M., Thammavongkeo, S., Basa, P. N., Caple, G., and Sykes, A. G. (2011). Substitution of thiophene oligomers with macrocyclic end caps and the colorimetric detection of Hg(II). *Org. Lett.* 13, 2364–2367. doi:10.1021/ol200442k

Krauss, B., Lohmann, T., Chae, D. H., Haluska, M., Von Klitzing, K., and Smet, J. H. (2009). Laser-induced disassembly of a graphene single crystal into a nanocrystalline network. *Phys. Rev. B* 79, 165428. doi:10.1103/Physrevb.79.165428

Kumar, S. A., Cheng, H.-W., Chen, S.-M., and Wang, S.-F. (2010). Preparation and characterization of copper nanoparticles/zinc oxide composite modified electrode and its application to glucose sensing. *Mater. Sci. Eng. C* 30, 86–91. doi:10.1016/j.msec.2009.09.001

Lee, K. H., Lee, B., Hwang, S.-J., Lee, J.-U., Cheong, H., Kwon, O.-S., et al. (2014). Large scale production of highly conductive reduced graphene oxide sheets by a solvent-free low temperature reduction. *Carbon N. Y.* 69, 327–335. doi:10.1016/j.carbon.2013.12.031

Li, D., Muller, M. B., Gilje, S., Kaner, R. B., and Wallace, G. G. (2008). Processable aqueous dispersions of graphene nanosheets. *Nat. Nanotechnol.* 3, 101–105. doi:10.1038/nnano.2007.451

Li, Y.-F., Chen, S.-M., Lai, W.-H., Sheng, Y.-J., and Tsao, H.-K. (2013). Superhydrophilic graphite surfaces and water-dispersible graphite colloids by electrochemical exfoliation. *J. Chem. Phys.* 139, 64703–64714. doi:10.1063/1.4817680

Liu, D., Qu, W., Chen, W., Zhang, W., Wang, Z., and Jiang, X. (2010). Highly sensitive, colorimetric detection of mercury(II) in aqueous media by quaternary ammonium group-capped gold nanoparticles at room temperature. *Anal. Chem.* 82, 9606–9610. doi:10.1021/ac1021503

Low, C. T. J., Walsh, F. C., Chakrabarti, M. H., Hashim, M. A., and Hussain, M. A. (2013). Electrochemical approaches to the production of graphene flakes and their potential applications. *Carbon N. Y.* 54, 1–21. doi:10.1016/j.carbon.2012.11.030

Mandal, S., Banerjee, A., Lohar, S., Chattopadhyay, A., Sarkar, B., Mukhopadhyay, S. K., et al. (2013). Selective sensing of Hg2+ using rhodamine–thiophene conjugate: red light emission and visual detection of intracellular Hg2+ at nanomolar level. *J. Hazard. Mater.* 261, 198–205. doi:10.1016/j.jhazmat.2013.07.026

Mao, M., Wang, M., Hu, J., Lei, G., Chen, S., and Liu, H. (2013). Simultaneous electrochemical synthesis of few-layer graphene flakes on both electrodes in protic ionic liquids. *Chem. Commun. (Camb)* 49, 5301–5303. doi:10.1039/c3cc41909f

Martin-Yerga, D., Gonzalez-Garcia, M. B., and Costa-Garcia, A. (2013). Electrochemical determination of mercury: a review. *Talanta* 116, 1091–1104. doi:10.1016/j.talanta.2013.07.056

Mikolaj, D., and Zbigniew, S. (2012). "*Nanoparticles and Nanostructured Materials Used in Modification of Electrode Surfaces*," in *Functional Nanoparticles for Bioanalysis, Nanomedicine, and Bioelectronic Devices*, Vol. 1. Washington: American Chemical Society, 313–325. doi:10.1021/bk-2012-1112.ch012

Morales, G. M., Schifani, P., Ellis, G., Ballesteros, C., Martínez, G., Barbero, C., et al. (2011). High-quality few layer graphene produced by electrochemical intercalation and microwave-assisted expansion of graphite. *Carbon N. Y.* 49, 2809–2816. doi:10.1016/j.carbon.2011.03.008

Najafi, E., Aboufazeli, F., Zhad, H. R., Sadeghi, O., and Amani, V. (2013). A novel magnetic ion imprinted nano-polymer for selective separation and determination of low levels of mercury(II) ions in fish samples. *Food Chem.* 141, 4040–4045. doi:10.1016/j.foodchem.2013.06.118

Orlik, M., and Galus, Z. (2007). "Electrochemistry of mercury," in *Encyclopedia of Electrochemistry*, Vol. 7 (New York: Wiley-VCH Verlag GmbH & Co. KGaA), 958–991. doi:10.1002/9783527610426.bard072406

Park, S., An, J., Potts, J. R., Velamakanni, A., Murali, S., and Ruoff, R. S. (2011). Hydrazine-reduction of graphite- and graphene oxide. *Carbon N. Y.* 49, 3019–3023. doi:10.1016/j.carbon.2011.02.071

Pesavento, M., Alberti, G., and Biesuz, R. (2009). Analytical methods for determination of free metal ion concentration, labile species fraction and metal complexation capacity of environmental waters: a review. *Anal. Chim. Acta* 631, 129–141. doi:10.1016/j.aca.2008.10.046

Pettersson, L. A. A., Johansson, T., Carlsson, F., Arwin, H., and Inganäs, O. (1999). Anisotropic optical properties of doped poly(3,4-ethylenedioxythiophene). *Synth. Met.* 101, 198–199. doi:10.1016/S0379-6779(98)01215-6

Pham, V. H., Cuong, T. V., Hur, S. H., Oh, E., Kim, E. J., Shin, E. W., et al. (2011). Chemical functionalization of graphene sheets by solvothermal reduction of a graphene oxide suspension in N-methyl-2-pyrrolidone. *J. Mater. Chem.* 21, 3371–3377. doi:10.1039/c0jm02790a

Pillay, K., Cukrowska, E. M., and Coville, N. J. (2013). Improved uptake of mercury by sulphur-containing carbon nanotubes. *Microchem. J.* 108, 124–130. doi:10.1016/j.microc.2012.10.014

Qi, B., He, L., Bo, X. J., Yang, H. J., and Guo, L. P. (2011). Electrochemical preparation of free-standing few-layer graphene through oxidation-reduction cycling. *Chem. Eng. J.* 171, 340–344. doi:10.1016/j.cej.2011.03.078

Rattan, S., Singhal, P., and Verma, A. L. (2013). Synthesis of PEDOT:PSS (poly(3,4-ethylenedioxythiophene))/poly(4-styrene sulfonate))/ngps (nanographitic platelets) nanocomposites as chemiresistive sensors for detection of nitroaromatics. *Polym. Eng. Sci.* 53, 2045–2052. doi:10.1002/pen.23466

Seco-Reigosa, N., Cutillas-Barreiro, L., Nóvoa-Muñoz, J. C., Arias-Estévez, M., Fernández-Sanjurjo, M., Álvarez-Rodríguez, E., et al. (2014). Mixtures including wastes from the mussel shell processing industry: retention of arsenic, chromium and mercury. *J. Clean. Prod.* 84, 680–690. doi:10.1016/j.jclepro.2014.01.050

Shao, Y., Wang, J., Wu, H., Liu, J., Aksay, I. A., and Lin, Y. (2010). Graphene based electrochemical sensors and biosensors: a review. *Electroanalysis* 22, 1027–1036. doi:10.1002/elan.200900571

Shrivastava, A., and Gupta, V. (2011). Methods for the determination of limit of detection and limit of quantitation of the analytical methods. *Chron. Young Sci.* 2, 21–25. doi:10.4103/2229-5186.79345

Singh, V. V., Gupta, G., Batra, A., Nigam, A. K., Boopathi, M., Gutch, P. K., et al. (2012). Greener electrochemical synthesis of high quality graphene nanosheets directly from pencil and its spr sensing application. *Adv. Funct. Mater.* 22, 2352–2362. doi:10.1002/adfm.201102525

Somerset, V., Leaner, J., Mason, R., Iwuoha, E., and Morrin, A. (2010). Development and application of a poly(2,2'-dithiodianiline) (PDTDA)-coated screen-printed carbon electrode in inorganic mercury determination. *Electrochim. Acta* 55, 4240–4246. doi:10.1016/j.electacta.2009.01.029

Su, C. Y., Lu, A. Y., Xu, Y., Chen, F. R., Khlobystov, A. N., and Li, L. J. (2011). High-quality thin graphene films from fast electrochemical exfoliation. *ACS Nano* 5, 2332–2339. doi:10.1021/nn200025p

Sundramoorthy, A. K., and Gunasekaran, S. (2014). Applications of graphene in quality assurance and safety of food. *Trends Anal Chem* 60, 36–53. doi:10.1016/j.trac.2014.04.015

Tarabella, G., Nanda, G., Villani, M., Coppede, N., Mosca, R., Malliaras, G. G., et al. (2012). Organic electrochemical transistors monitoring micelle formation. *Chem Sci* 3, 3432–3435. doi:10.1039/c2sc21020g

Vacca, P., Petrosino, M., Miscioscia, R., Nenna, G., Minarini, C., Della Sala, D., et al. (2008). Poly(3,4-ethylenedioxythiophene):poly(4-styrenesulfonate) ratio: structural, physical and hole injection properties in organic light emitting diodes. *Thin Solid Films* 516, 4232–4237. doi:10.1016/j.tsf.2007.12.143

Wang, J. Z., Manga, K. K., Bao, Q. L., and Loh, K. P. (2011). High-yield synthesis of few-layer graphene flakes through electrochemical expansion of graphite in propylene carbonate electrolyte. *J. Am. Chem. Soc.* 133, 8888–8891. doi:10.1021/ja203725d

Wang, L., Fang, G., Ye, D., and Cao, D. (2014a). Carbazole and triazole-containing conjugated polymer as a visual and fluorometric probe for iodide and mercury. *Sens Actuators B Chem* 195, 572–580. doi:10.1016/j.snb.2014.01.081

Wang, Z., Xu, J., Yao, Y., Zhang, L., Wen, Y., Song, H., et al. (2014b). Facile preparation of highly water-stable and flexible PEDOT:PSS organic/inorganic composite materials and their application in electrochemical sensors. *Sens Actuators B Chem* 196, 357–369. doi:10.1016/j.snb.2014.02.035

Wang, Z.-L., Xu, D., Huang, Y., Wu, Z., Wang, L.-M., and Zhang, X.-B. (2012). Facile, mild and fast thermal-decomposition reduction of graphene oxide in air and its application in high-performance lithium batteries. *Chem. Commun.* 48, 976–978. doi:10.1039/c2cc16239c

Welch, C. M., Nekrassova, O., Dai, X., Hyde, M. E., and Compton, R. G. (2004). Fabrication, characterisation and voltammetric studies of gold amalgam nanoparticle modified electrodes. *Chemphyschem* 5, 1405–1410. doi:10.1002/cphc.200400263

Yan, K., Peng, H., Zhou, Y., Li, H., and Liu, Z. (2011). Formation of bilayer Bernal graphene: layer-by-layer epitaxy via chemical vapor deposition. *Nano Lett.* 11, 1106–1110. doi:10.1021/nl104000b

Yang, J., and Gunasekaran, S. (2013). Electrochemically reduced graphene oxide sheets for use in high performance supercapacitors. *Carbon N. Y.* 51, 36–44. doi:10.1016/j.carbon.2012.08.003

Yang, J., Yu, J.-H., Rudi Strickler, J., Chang, W.-J., and Gunasekaran, S. (2013). Nickel nanoparticle–chitosan-reduced graphene oxide-modified screen-printed electrodes for enzyme-free glucose sensing in portable microfluidic devices. *Biosens Bioelectron* 47, 530–538. doi:10.1016/j.bios.2013.03.051

Yasri, N. G., Halabi, A. J., Istamboulie, G., and Noguer, T. (2011). Chronoamperometric determination of lead ions using PEDOT:PSS modified carbon electrodes. *Talanta* 85, 2528–2533. doi:10.1016/j.talanta.2011.08.013

Yoo, D., Kim, J., and Kim, J. (2014). Direct synthesis of highly conductive poly(3,4-ethylenedioxythiophene):poly(4-styrenesulfonate) (PEDOT:PSS)/graphene composites and their applications in energy harvesting systems. *Nano Res.* 7, 717–730. doi:10.1007/s12274-014-0433-z

Zhang, L., Li, X., Huang, Y., Ma, Y., Wan, X., and Chen, Y. (2010). Controlled synthesis of few-layered graphene sheets on a large scale using chemical exfoliation. *Carbon N. Y.* 48, 2367–2371. doi:10.1016/j.carbon.2010.02.035

Zhang, W. Y., Zeng, Y., Xiao, N., Hng, H. H., and Yan, Q. Y. (2012). One-step electrochemical preparation of graphene-based heterostructures for Li storage. *J. Mater. Chem.* 22, 8455–8461. doi:10.1039/c2jm16315b

Zhong, Y. L., and Swager, T. M. (2012). Enhanced electrochemical expansion of graphite for in situ electrochemical functionalization. *J. Am. Chem. Soc.* 134, 17896–17899. doi:10.1021/ja309023f

**Conflict of Interest Statement:** The authors declare that the research was conducted in the absence of any commercial or financial relationships that could be construed as a potential conflict of interest.

# 9

# Structural, electronic, and mechanical properties of inner surface modified imogolite nanotubes

**Maurício Chagas da Silva, Egon Campos dos Santos, Maicon Pierre Lourenço, Mateus Pereira Gouvea[†] and Hélio Anderson Duarte\***

Grupo de Pesquisa em Química Inorgânica Teórica, Departamento de Química, Instituto de Ciências Exatas (ICEx), Universidade Federal de Minas Gerais, Belo Horizonte, Brazil

**Edited by:**
Demircan Canadinc, Koç University, Turkey

**Reviewed by:**
Seda Keskin, Koç University, Turkey
Gotthard Seifert, Dresden University of Technology, Germany

**\*Correspondence:**
Hélio Anderson Duarte, Grupo de Pesquisa em Química Inorgânica Teórica, Departamento de Química, Instituto de Ciências Exatas (ICEx), Universidade Federal de Minas Gerais, Av. Antonio Carlos, 6627, Bairro Pampulha, Belo Horizonte, Minas Gerais 31.270-901, Brazil
e-mail: duarteh@ufmg.br

[†] In memorian.

The electronic, structural, and mechanical properties of the modified imogolites have been investigated using self consistent charge-density functional-tight binding method with "a posteriori" treatment of the dispersion interactions (SCC-DFTB-D). The zigzag (12,0) imogolite has been used as the initial structure for the calculations. The functionalization of the inner surface of (12,0) imogolite nanotubes (NTs) by organosilanes and by heat treatment leading to the dehydroxylation of the silanols were investigated. The reaction of the silanols with the trimethylmethoxysilanes is favored and the arrangement of the different substitutions leads to the most symmetrical structures. The Young moduli and band gaps (BGs) are slightly decreased. However, the dehydroxylation of the silanol groups in the inner surface of the imogolite leads to the increase of the Young moduli and a drastic decrease of the BG to about 4.4 eV. It has been shown that the degree of the dehydroxylation can be controlled by heat treatment and tune the BG, eventually, leading to a semiconductor material with well-defined NT structure.

**Keywords: SCC-DFTB, imogolite, functionalization, dehydroxylation, band gap, Young modulus**

## INTRODUCTION

Imogolite, halloysite, and chrysotile are an important class of nanostructured clay mineral. They have been envisaged for several applications such as gas storage (Azzouz, 2012; Assima et al., 2013; Cavallaro et al., 2014), separation (Li et al., 2014; Murali et al., 2014; Zhong et al., 2014), controlled delivery systems (Tu et al., 2013; Wang et al., 2013; Lun et al., 2014; Lvov et al., 2014; Rao et al., 2014), composites (Tham et al., 2014; Peng et al., 2015; Xu et al., 2015), and catalysis (Machado et al., 2013; Gomez et al., 2014). They are natural nanotubes (NTs) and can be easily synthesized and modified (Price et al., 2001; Shchukin et al., 2005, 2006; Shchukin and Mohwald, 2007; Veerabadran et al., 2007; Lvov et al., 2008; Alhuthali and Low, 2013; Wang and Huang, 2013). Halloysite (Bates et al., 1950a) and chrysotile (Bates et al., 1950b; Whittaker, 1956) are polydisperse and multi-walled with diameters in the range of 15–100 nm and length of 500 nm up to mm range. Halloysite, $Al_2Si_2O_5(OH)_4$, is formed by a sheet of gibbsite $[Al(OH)_3]$ in the inner part and a layer of silicate outside and chrysotile, $Mg_3Si_2O_5(OH)_4$, is formed by a sheet of tridymite $(SiO_2)$ in the inner part and a sheet of brucite $[Mg(OH)_2]$ in the outer part of the NT. Imogolite (Cradwick et al., 1972) is very distinct from the other natural NTs. It is monodisperse, single-walled, with a well-defined diameter of 2.3 nm and its length is typically in the range around 100 nm. The strain energy curve defined as the energy to bend the ideal lamellar structure to form the NT has been used to explain the monodispersity of imogolite (Guimaraes et al., 2007). Actually, this curve presents a minimum explaining its well-defined diameter. Its formula $Al_2Si(OH)_4O_3$ is formed by a gibbsite sheet in the inner part of the NT and a silanol group in the center of the hexagons of the gibbsite. The mismatch of the bond distances explains the bending and NT formation of the

imogolite, which results the minimum in the strain energy curve. Carbon NTs and all other inorganic NTs present strain energy curves that decrease asymptotically to 0 with the increase of the diameter (Tenne et al., 1992).

Furthermore, it has been shown that the silanol group can be replaced by $Ge(OH)_4$ to form aluminogermanate imogolite-like (img-Ge) NTs with very similar structure but with larger diameter (Wada and Wada, 1982; Levard et al., 2008, 2009, 2010, 2011). It has been reported that img-Ge is easily synthesized and it is possible to control its diameter and the formation of single-walled or double-walled NTs during the syntheses (Thill et al., 2012a,b; Yucelen et al., 2012). It is expected that the electronic and mechanical properties are also affected by modifying its structure. Actually, it has been shown by SCC-DFTB calculations that different imogolite-like NTs based on phosphate, phosphite, arsenate, and arsenite substituents of the silanol can be synthesized if the pH is adequately controlled (Guimaraes et al., 2013). However, the syntheses of these NTs remain to be achieved.

Actually, single-walled NTs (SWNTs) have been considered important building blocks in the development of materials with enhanced properties. Therefore, imogolite and img-Ge are becoming an important target in the development of nanotechnology, since it has been shown to be easily modified externally and internally. As it has been suggested elsewhere (Qi et al., 2008; Kang et al., 2011, 2014; Ma et al., 2011, 2012), the outer surface modification of the SWNT can improve its compatibility with other materials and liquid phase and the inner surface modification can be used for molecular recognition, nanoreactors, and size selective catalysis. The main strategy is normally covalently immobilizing organic functional groups in the inner surface as it was made by Kang et al. (2011). They have modified the inner surface by the reaction of organosilane with the silanol groups of imogolite. The modified NT was detailed characterized by powder X-ray diffraction (XRD), thermogravimetric analysis (TGA), transmission electron microscopy (TEM), solid state NMR, nitrogen physisorption, and water adsorption. It was estimated that 35% of the silanols were substituted in the inner surface of the imogolite. Another modification is based on the dehydroxylation of the inner surface of imogolite by heat treatment up to 450°C (Kang et al., 2010). The process leads to the modification of the inner surface, which has been characterized in detail by FTIR, TEM and XRD, NMR, and TGA techniques.

In spite of the importance of these achievements, the structural and mechanical properties of modified NTs are very difficult to be evaluated. In the present work, SCC-DFTB-D calculations have been performed for investigating the effect of the inner surface modification of imogolite NTs in their electronic, structural, and mechanical properties.

## MATERIALS AND METHODS
### ELECTRONIC STRUCTURE CALCULATIONS

The electronic structure calculations of the modified imogolites were carried out using the self consistent charge-density functional-tight binding (SCC-DFTB) method (Porezag et al., 1995; Seifert et al., 1996; Elstner et al., 1998) as implemented in the DFTB+ software package (Aradi et al., 2007). This is an approximate density functional theory method that uses a minimal, localized, and confined atomic basis set and tight-binding-like approximations to the Hamiltonian (Oliveira et al., 2009). The linear combination of atomic orbitals (LCAO) *ansatz* is used:

$$\psi_i(\mathbf{r}_1) = \sum_\mu C_{\mu i}\phi_\mu(\mathbf{r}_1). \tag{1}$$

The overlap matrix and the Fock-like matrix are described by the Eqs 2 and 3:

$$S_{\mu\nu} = \langle \mu | \nu \rangle. \tag{2}$$

$$F_{\mu\nu}^{\text{DFTB}} = \begin{cases} \langle \mu \left| -\frac{1}{2} \nabla^2 + V_{\text{ef}}^A \right| \nu \rangle \text{ for } \mu = \nu \text{ and } \mu, \nu \in \{A\} \\ \langle \mu \left| -\frac{1}{2} \nabla^2 + V_{\text{ef}}^A + V_{\text{ef}}^B \right| \nu \rangle \text{ for } \mu \in \{A\} \text{ and } \nu \in \{B\} \\ = 0, \text{ otherwise} \end{cases} \tag{3}$$

where $\mu$ and $\nu$ refers to the minimal basis sets centered on the $A$ and $B$ nuclei. The effective potential is defined according to Eq. 4.

$$V_{\text{ef}}^A(\mathbf{r}_1) = v_{\text{ext}}^A(\mathbf{r}_1) + \frac{1}{2} \int \frac{\rho^A(\mathbf{r}_2)}{|\mathbf{r}_1 - \mathbf{r}_2|} d\mathbf{r}_2 + v_{\text{xc}}(\mathbf{r}_1). \tag{4}$$

The right terms are the external potential, usually due to the $A$ nuclei, the Coulomb contribution and the exchange/correlation potential due to the electron density of the atom $A$, respectively. The electron density is written as a superposition of atom-like densities centered on the nuclei $A$, as described by the Eq. 5:

$$\rho(\mathbf{r}_1) = \sum_A \rho^A(\mathbf{r}_A), \ \mathbf{r}_A = \mathbf{r}_1 - \mathbf{R}_A. \tag{5}$$

From the LCAO approach, the following secular problem is obtained:

$$\sum_\nu C_{\nu i}(F_{\mu\nu}^{\text{DFTB}} - \varepsilon_i S_{\mu\nu}) = 0 \quad \forall \mu\nu. \tag{6}$$

An extension of the DFTB, taking into account the first order density fluctuations in a simple but efficient way has been proposed by Elstner et al. (1998) and allows for the explicit treatment of charge-transfer effects. The charges are transferred between the different atoms according to the chemical hardness, which is related to the Hubbard parameter $U_\alpha$, in a self-consistent manner. The SCC-DFTB total energy is defined according to the Eq. 7:

$$E_{\text{SCC}} = E_{\text{bnd}} + \frac{1}{2} \sum_{A,B} \gamma_{AB} \Delta q_A \Delta q_B + E_{\text{rep}}. \tag{7}$$

The $\gamma_{AB} = \gamma_{AB}(U_A, U_B, |\mathbf{R}_A - \mathbf{R}_B|)$ and $\Delta q_A = (q_A^0 - q_A)$, where $q_A^0$ is the valence number of electrons of the isolated atom $A$ and $q_A$ is the Mulliken charge calculated using the KS orbitals.

The three center terms in the Hamiltonian are neglected and the two-center contributions can be considered of short range. The integrals arising in the Eq. 3 are tabulated for each pair of atoms normally using the local density approximation (LDA), which are called Slater–Koster parameters. The repulsion energy does not decay to 0 for long interatomic distances. At the SCC-DFTB method, the $E_{\text{rep}}$ is fitted to the difference between the DFT energy and $E_{\text{bnd}}$ as a function of the interatomic distance using a suitable reference structure. The $E_{\text{bnd}}$ is defined by the Eq. 8:

$$E_{\text{bnd}} = \sum_i n_i \varepsilon_i. \tag{8}$$

Slater–Koster parameters for describing the imogolite (Guimaraes et al., 2007) were used. It has been shown that reliable structural, energetic, and electronic properties are obtained for nanostructured clay minerals (Guimaraes et al., 2007, 2010; Lourenco et al., 2012, 2014; da Silva et al., 2013). "A posteriori" correction to the van der Waals interactions (SCC-DFTB-D) improves the results for organic molecules or non-bonding interactions of organic residues. We have used the approach proposed by Zhechkov et al. (2005). A model study for the intermolecular interactions of methoxy groups showed in Figure S4 in Supplementary Material indicated that the SCC-DFTB-D method provides interaction energies in good agreement with the PBE/aug-cc-pVTZ, with difference of <1 kJ·mol$^{-1}$ as shown in Table S1 in Supplementary Material.

Initial configurations of the modified NTs were based on the optimized structure of the zigzag (12,0) imogolite NT. As it was shown by Guimaraes et al. (2007), this NT is the most stable structure with 13.2 Å of inner diameter and the $b$ lattice parameter of 8.60 Å, as it is shown in **Figure 1**. We consider a one-dimensional periodic approximation for all systems, setting the periodicity of the systems along $y$ axis and applying vacuum of 100 Å in the other directions, $x$ and $z$. A converged Monkhorst–Pack sampling $1 \times 4 \times 1$ $k$-point grid was used in our calculations for all systems.

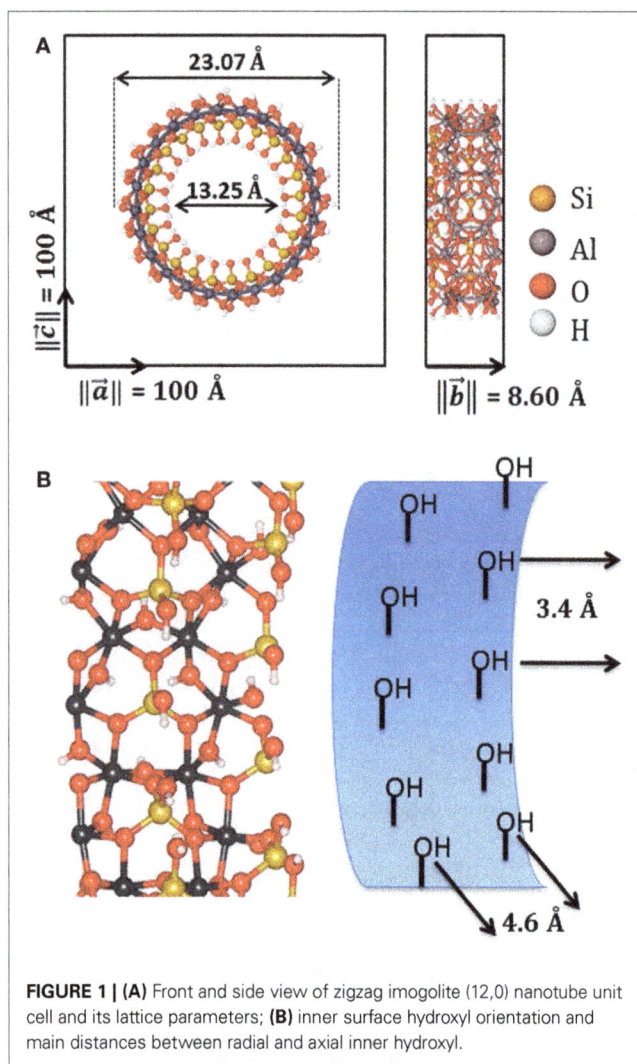

FIGURE 1 | **(A)** Front and side view of zigzag imogolite (12,0) nanotube unit cell and its lattice parameters; **(B)** inner surface hydroxyl orientation and main distances between radial and axial inner hydroxyl.

## YOUNG MODULUS

The elastic mechanical behavior of the NTs was investigated by calculating the Young modulus ($Y$) in the direction of the periodic axis. The total electronic SCC-DFTB-D energy ($E$) of a NT system must be related to the strain ($\varepsilon$) applied to this system along the periodic direction, Eqs 9 and 10 and **Figure 2**. In Eq. 9, $l_0$ is the equilibrium unit cell lattice vector size in the direction of the periodic axis ($y$), $l$ is the new size of the unit cell lattice vector after a small perturbation, and the strain ($\varepsilon$) is a dimensionless unit that reflects how much a system is compressed or stretched in the direction of the periodic axis. In Eq. 10, the total electronic SCC-DFTB-D energy ($E$) is expanded by Maclaurin series around the equilibrium position of $l_0$ or in this case $\varepsilon = 0$. To guarantee the extensivity of the elastic property, the total energy ($E$) is divided by the equilibrium unit cell volume ($V_{eq}$), Eq. 10:

$$\varepsilon = \frac{l - l_0}{l_0} \tag{9}$$

$$E(\varepsilon) = \frac{1}{V_{eq}} \sum_{n=0}^{M} \frac{1}{n!} \left( \frac{d^n E(\varepsilon)}{d\varepsilon^n} \right)_{\varepsilon=0} \varepsilon^n. \tag{10}$$

In the equilibrium position, Eq. 10 can be approximated by Eq. 11 if considering $E(\varepsilon = 0)$ as the 0 potential reference for the system and taking into account that in the equilibrium position the first derivative of the energy related to the strain must be 0. Hence, the second derivative will be related to the elastic force constant of the NT along the periodic direction ($k_{axi}$), Eq. 12. Taking strain transformations that maintain the unit cell volume ($V_{eq}$) constant, the Young modulus for the NT was defined by the Eqs 12 and 13:

$$E(\varepsilon) \approx \frac{1}{V_{eq}} \frac{1}{2} \left( \frac{d^2 E(\varepsilon)}{d\varepsilon^2} \right)_{\varepsilon=0} \varepsilon^2 \tag{11}$$

$$E(\varepsilon) \approx \frac{1}{V_{eq}} \frac{1}{2} K_{axi} \varepsilon^2 \tag{12}$$

$$E(\varepsilon) \approx \frac{1}{2} Y \varepsilon^2. \tag{13}$$

The protocol used in this work was to apply different strain factors from −0.5 to 0.5% in the unit cell periodic vector along the axis, relaxing the atomic position in each strain, and then obtaining a polynomial parabolic energetic curve, Eq. 14 and **Figure 3**. Comparing Eqs 13 and 14, the $Y$ is related to the third polynomial coefficient ($a_2$) of a polynomial fit curve, Eq. 15. By our approximation, the coefficients $a_0$ and $a_1$ were very close to 0:

$$f(\varepsilon) = a_0 + a_1 \varepsilon + a_2 \varepsilon^2 \tag{14}$$

$$Y = 2 \times a_2. \tag{15}$$

Systems like carbon or imogolite NT is quite simple to define the equilibrium unit cell volume ($V_{eq}$) using a transversal sector area of the NT ($S_0$) and the unit cell periodic vector length, **Figure 2**. Besides, in a simple NT the volume ($V_{eq}$) is $S_0$ times the lattice periodic vector length. This approach has been used to calculate the Young moduli of the clay mineral NTs (Guimaraes et al., 2007, 2010, 2013; Lourenco et al., 2014). However, in the present work, due to the inner modifications, the unit cell volume cannot be approached by this simply assumption. The volume of the unit cell was computed by numerical integration of the molecular region defined inside the unit cell. Monte Carlo integration strategies are the most powerful and less time consuming algorithm used to get molecular volumes compared to discrete numerical volume integrator. The van der Waals radius is used to establish a region of molecular volume. More details about the molecular volume estimates can be found in the Supplementary Material.

## RESULTS AND DISCUSSION
### IMOGOLITE STRUCTURES

The (12,0) imogolite NT has been predicted to be the most stable structure at SCC-DFTB-D level of theory with internal diameter about 13.25 Å (Guimaraes et al., 2007; Lourenco et al., 2014). DFT calculations have been performed at the B3LYP level indicating that (10,0) imogolite is the most stable (Demichelis et al., 2010). It is important to highlight that it has been shown recently that the diameter of the imogolite can be controlled in some extent depending of the electrolyte used in the synthesis (Yucelen et al., 2012). The **Figure 1B** shows the structural parameters of the inner

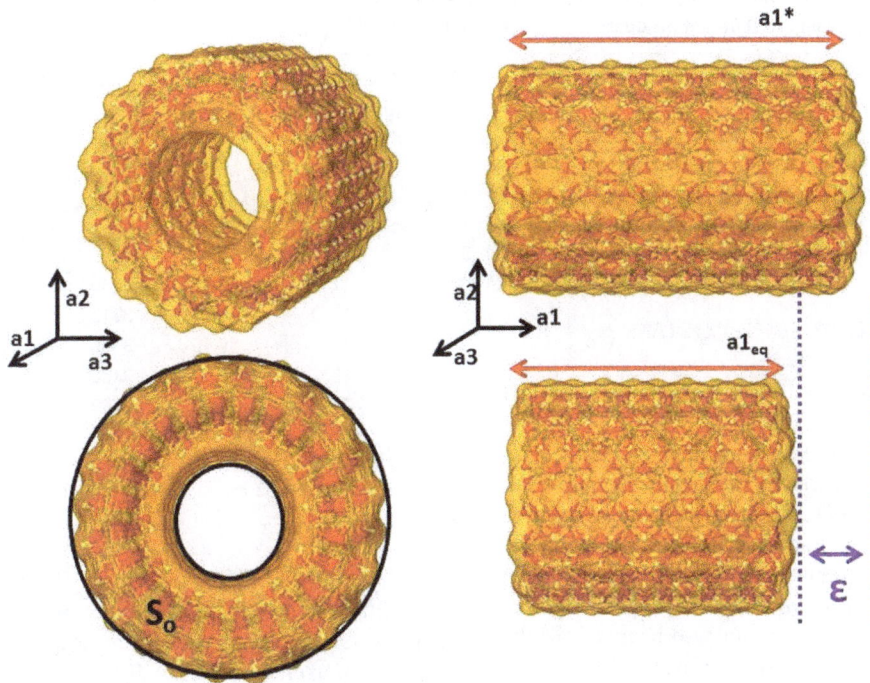

**FIGURE 2 | The unit cell of imogolite (12,0) showing the transversal area ($S_o$), lattice vector ($a_{1eq}$), periodic axis ($a_1$, $a_2$, $a_3$), stressed periodic vector ($a_{1*}$), and the deformation parameter ($\varepsilon$).**

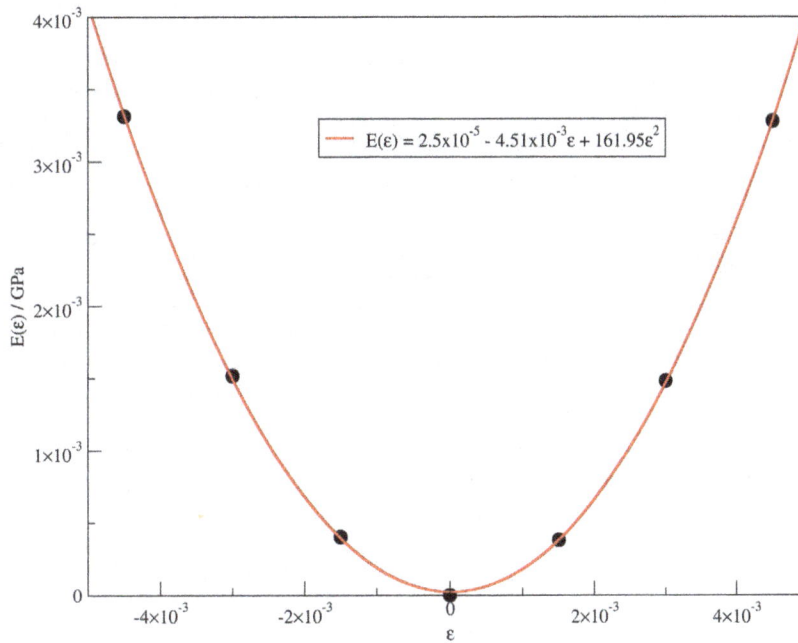

$$E(\varepsilon) = 2.5 \times 10^{-5} - 4.51 \times 10^{-3}\varepsilon + 161.95\varepsilon^2$$

**FIGURE 3 | Polynomial quadratic fit to the energy vs. strain data for imogolite (12,0) and fitted equation.**

side of the NTs. There are 24 silanol groups in each unit cell of (12,0), arranged in two alternate radial lines. Hydroxyls in the same radial line have distances of 3.4 Å and, in between different radial lines, of 4.6 Å. The silanol groups are easily modified by reaction with organosilanes or dehydroxylation by heat treatment releasing water molecules.

**Table 1 | Reaction energies for the condensation of $CH_3Si(OCH_3)_3$ with the inner surface silanols leading to different products.**

| Substituted product | $\Delta_r E$ [kJ·mol$^{-1}$(unit cell)$^{-1}$] |
|---|---|
| $\eta^1$-img | −118.2 |
| Radial $\eta^2$-img | −127.3 |
| Axial $\eta^2$-img | 47.8 |
| $\eta^3$-img | 95.2 |

**Table 2 | Mean values and SD of the reaction energy per number of substitution ($N_{sub}$).**

| $N_{sub}$ | $\Delta_r E$ [kJ·mol$^{-1}$ (unit cell)$^{-1}$] | BG (eV) | $V_{eq}$ (Å$^3$) | $k_{axi}$ (kN·m$^{-1}$) | Y (GPa) |
|---|---|---|---|---|---|
| 0 | – | 9.11 | 3007 | 1.342 | 330 |
| 1 | −127.3 | 9.18 | 3068 | 1.347 | 324 |
| 2 | −127.2 | 9.03 | 3127 | 1.348 | 319 |
| 3 | −121.3 | 8.43 | 3189 | 1.350 | 313 |
| 4 | −121.3 | 9.22 | 3249 | 1.345 | 306 |
| 5 | −120.8 | 8.77 | 3312 | 1.343 | 300 |
| 6 | −129.3 | 9.15 | 3389 | 1.348 | 294 |
| 7 | −122.1 | 9.02 | 3460 | 1.350 | 289 |
| 8 | −118.6 | 9.33 | 3485 | 1.361 | 289 |

*Electronic band gap (BG), equilibrium unit cell volume ($V_{eq}$), NT unit axial elastic constant ($k_{axi}$), and Young's modulus calculated for the $\eta^2$-img products for the most stable configuration.*

## FUNCTIONALIZATION WITH METHYLTRIMETHOXYSILANE [$CH_3Si(OCH_3)_3$]

**Figure 4** shows the representation of the reaction that was carried out by Kang et al. (2011). The $CH_3Si(OCH_3)_3$ molecule reacts with the silanol groups leading to methanol formation and the functionalized inner surface. Initially, the unit cell was modified with only one $CH_3Si(OCH_3)_3$ to form monosubstituted ($\eta^1$-img), bisubstituted ($\eta^2$-img), and trisubstituted ($\eta^3$-img). For the $\eta^2$-img, silanol groups from the same radial line or from neighboring radial lines (axial) were calculated, respectively. The $\eta^3$-img was calculated taking into account two hydroxyls in the same radial line and one from the neighbor hydroxyl radial line (see **Figure 5**). **Table 1** shows the reaction energy per unit cell for the condensation of $CH_3Si(OCH_3)_3$ with the inner surface silanols leading to the four different products. The distance of 4.6 Å between the hydroxyls in different radial lines leads to bonding stress for the axial $\eta^2$-img and $\eta^3$-img. The $\eta^1$-img and radial $\eta^2$-img are energetically more favored with reaction energy of −118.2 and −127.3 kJ·mol$^{-1}$(unit cell)$^{-1}$. The difference is about 10 kJ·mol$^{-1}$(unit cell)$^{-1}$ favoring the radial $\eta^2$-img. It is important to note that these calculations were performed at gas phase. The solvation energy and the entropy change, due to the formation of methanol, are expected to modify the reaction energy estimates. However, the relative stability is well characterized by the SCC-DFTB-D method.

Kang et al. (2011) reported solid $^{29}$Si NMR data for the studied reaction and suggested that three products are obtained when the imogolite NTs are treated with the $CH_3Si(OCH_3)_3$ leading to about 24–38% of the substituted silanols. Our results indicate that radial $\eta^2$-img is the most stable (hereafter only $\eta^2$-img) and it will be used in the subsequent calculations.

The condensation reaction was investigated for up to 8 $\eta^2$-img substitutions, i.e., up to 16 (a fraction of 66%) silanol groups per unit cell was modified. Initially, for two substitutions all possible arrangements were calculated. The orientation of the methyl and methoxy groups in the silane can be arranged in different manner with respect to each methylmethoxysilane ($CH_3SiOCH_3$). Table S2 in Supplementary Material shows the relative energies of the different configurations. The difference of the energy is not larger than 3.5 kJ·mol$^{-1}$(unit cell)$^{-1}$ and the most stable was found for methoxy–methoxy, i.e., the two methoxy oriented toward each other. This result is probably due to the possible hydrogen bond interaction between the two groups.

Concerning the configuration of the silane groups in the unit cell, different possibilities can be envisaged. For the two substitutions, it was calculated all possibilities and the relative energies

are very small, not larger than 3 kJ·mol$^{-1}$(unit cell)$^{-1}$. The most stable configuration is the one that the two groups are the most distant as possible from each other. This is expected since the methyl groups can contribute to the van der Walls repulsion destabilizing the system. We have also calculated all possibilities for the three substitutions and this is also the case, the configuration that keeps the $CH_3SiOCH_3$ groups the most distant is the most stable. For the other systems, we have calculated the most stable structures keeping in mind that the methoxy groups must be oriented to each other and the organosilanes are in a configuration that leads to the largest distance from each other.

**Table 2** shows the condensation reaction energies for the different number of $\eta^2$-img substitutions, their band gaps (BGs) and mechanical properties. The subsequent condensation reactions increasing the fraction of silanols substituted increase slightly the reaction energy, in about 8 kJ·mol$^{-1}$(unit cell)$^{-1}$. The optimized geometries of the most stable substituted imogolite are shown in **Figure 5** indicating that the NT is just slightly deformed upon the functionalization with organosilane. These results indicate the presence of the organosilane does not affect the reaction energy of the neighbor silanols. It is well known that SCC-DFTB-D calculations overestimate the BG energies; however, it can offer the general tendency of the system studied. It indicates that the BG is slightly decreased by 1 eV with the presence of the substituents. This is expected since at the Fermi level, the electronic states correspond to the lone electron pairs of the oxygen. The equilibrium volume of the unit cell is increased by about 61 Å$^3$ per substitution. This indicates that the modification of the inner part of the NTs does not change their main structural characteristics. The axial elastic constant and the Young's modulus are related to the stiffness of the NT with respect to the axial deformations. The estimated values indicate that these two properties are just slightly modified leading to a small decrease of the Young's modulus.

## DEHYDROXYLATION OF IMOGOLITE

Kang et al. (2010) also showed that upon heat treatment above 300°C imogolite undergoes dehydroxylation, which is only partly

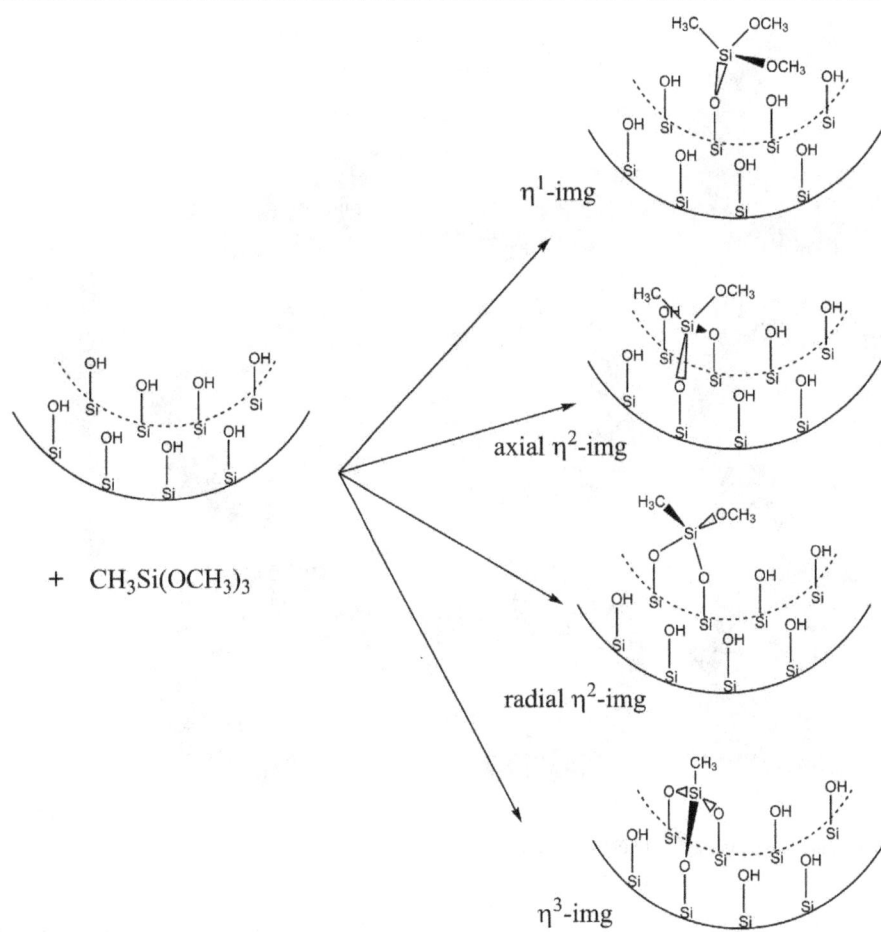

**FIGURE 4 | Inner surface representations for the modified zigzag (12,0) imogolite NT.**

reversed upon rehydrating conditions. The $^{29}$Si and $^{27}$Al NMR experiments have been carried out to show that about 73% of the silanol groups were dehydroxylated to form $SiO_2$ as shown in **Figure 6**. The gibbsite structure in the outer surface is not damaged with partial dehydroxylation. Therefore, the modification of the inner surface of the imogolite in a selective manner was successfully achieved by Kang et al. (2010).

The dehydroxylation will lead to the formation of the Si–O–Si bonds. The Si–Si distance in the inner side of the imogolite is about 4.27 Å between silicon atoms of the same radial silanols and 4.79 Å between silicon atoms of different radial silanols. Therefore, dehydroxylation takes place between the radial silicon atoms. The Si–OH bond distance is about 1.67 Å and the strain in the Si–O–Si bonding will be very large between silicon atoms of different radial silanols. Actually, even for the radial silicon atoms, it is expected that the strain in the Si–O–Si bonding will cause deformation of the NT. It was investigated the dehydroxylation of up to 50% of the silanol groups, which means about six dehydroxylations per unit cell of the (12,0) imogolite. The optimized structure for the first dehydroxylation ($n = 1$) presents Si–O–Si bond distance of about 1.80 Å, which is about 0.12 Å larger than the Si–O–Al bond distance. Furthermore, the Si–O–Si angle is

estimated to be about 118.6°, elongated from the ideal tetrahedral angle of 109.5° but 9° smaller than the mean value for the Si–O–Al angle. **Figure 6** shows the structure of the first dehydroxylation of the imogolite.

The second dehydroxylation can occur in different position of the unit cell. An extensive study has been performed to verify, which is the most favored site for the second ($n = 2$) dehydroxylation. Table S3 in Supplementary Material shows the relative energies of the different sites. The most stable is the one that leads to the most symmetric structure with the second dehydroxylation occurring in the opposite side of the NT but in the same radial silanols line as shown in **Figure 7**.

**Table 3** shows the reaction energy for the dehydroxylations in the same radial line and the respective BGs and Young moduli. As it can be seen in **Figure 7**, the respective hydroxylations lead to the deformation of the NT. The cylindrical structure is recovered with the subsequent dehydroxylations. The reaction energy of the dehydroxylation varies from 484 to 538 kJ·mol$^{-1}$ per hydroxylation. One could argue that the dehydroxylations in the different radial silanols would lead to a more favored structure. Table S3 in Supplementary Material shows a comparison of the two possibilities. The dehydroxylations in the same radial silanols lead to a structure

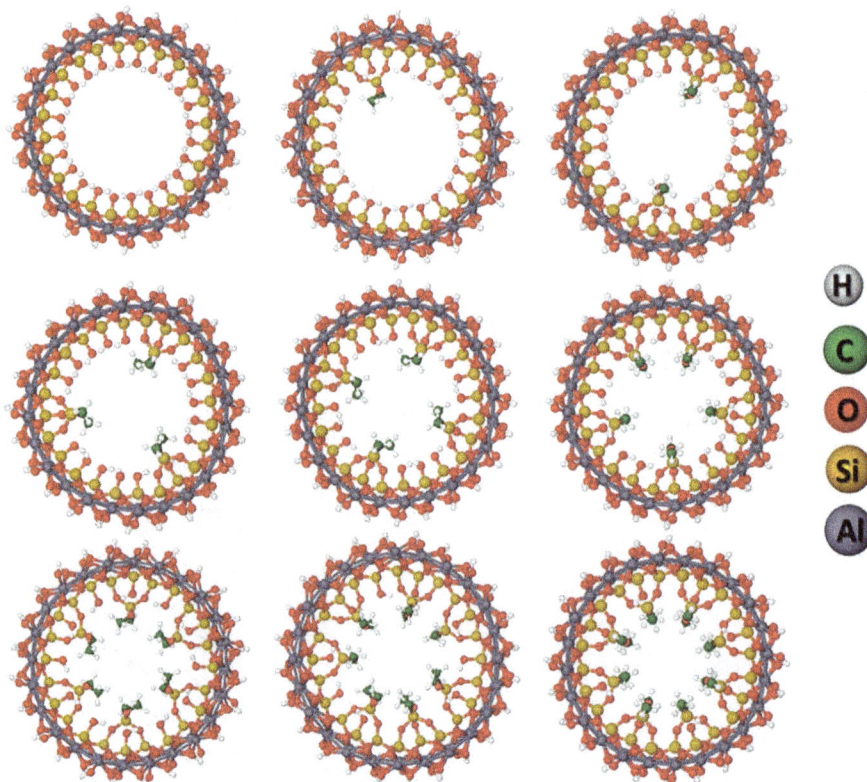

**FIGURE 5 | The optimized structures of the substituted (12,0) imogolite.**

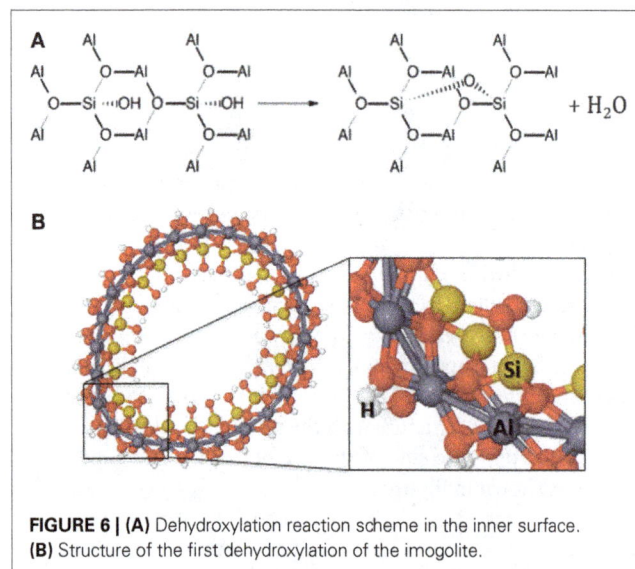

**FIGURE 6 | (A)** Dehydroxylation reaction scheme in the inner surface.
**(B)** Structure of the first dehydroxylation of the imogolite.

that is more favored. For the structure with six dehydroxylations, the difference is about 40 kJ·mol$^{-1}$ per dehydroxylation favoring the structure shown in **Figure 7**.

The Young modulus is increased from 324 to 371 GPa, as expected, since the Si–O–Si bound lead to a more rigid structure. However, the BG is drastically decreased from 9.1 to about 4.4 eV. The band structures of the different dehydroxylated structures are shown in Supplementary Material. It is important to highlight that this is an upper bound for the BG. At the B3LYP/86-21G* level of theory using helical symmetry, the BG energy was predicted for the imogolite to be about 7.2 eV (Demichelis et al., 2010). This value must be contrasted with the 9.1 eV of the SCC-DFTB-D calculations. This means that with the heat treatment of the imogolite, one can control the BG energy and, eventually, produce a semiconductor material with well-defined NT structure.

One could argue that the favored dehydroxylations in the same radial silanols of the unit cell are an artifact. It is expected that the dehydroxylations occur in alternate radial silanols in such way that the inner side structure of the NT will be more homogeneous. Therefore, a supercell containing three unit cells was constructed with alternate dehydroxylations as shown in **Figure 8**. The results are indicated in **Table 3**. The helical hydrogen bonding inside of the structure contributes to its stabilization. The BG energy is decreased to about 3.80 eV indicating that the material can be actually a semiconductor if one takes into account the SCC-DFTB-D method overestimates the BG of about 2–4 eV with respect to the DFT.

**FINAL REMARKS**

Imogolite is becoming a target material for developing advanced materials with tuned properties. It has been reported that

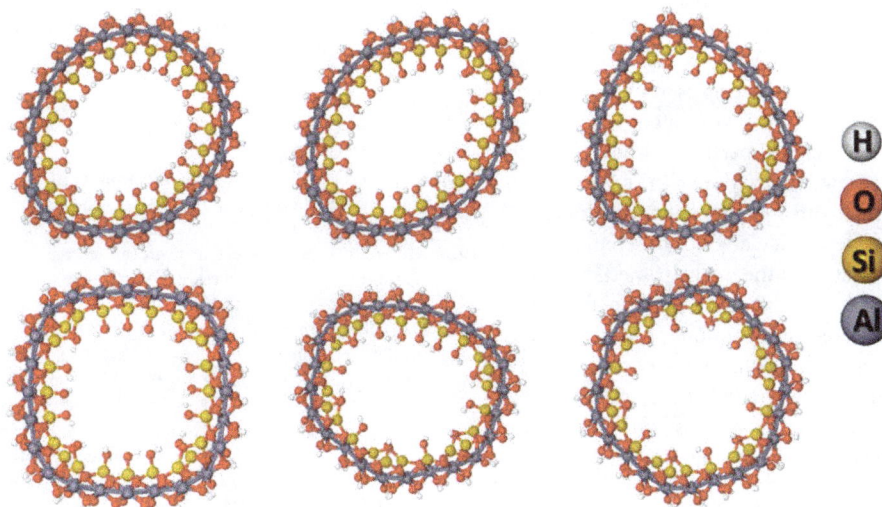

FIGURE 7 | Optimized structures of the dehydroxylated (12,0) imogolite.

Table 3 | Reaction energy of the dehydroxylation of imogolite in the same radial line.

| n | $\Delta_r E$ [kJ·mol$^{-1}$(unit cell)$^{-1}$] | $\Delta_r E/n$ [kJ·mol$^{-1}$(unit cell)$^{-1}$] | BG (eV) | $V_{eq}$ (Å$^3$) | $k_{axi}$ (kNm$^{-1}$) | Y (GPa) |
|---|---|---|---|---|---|---|
| 0 | – | – | 9.11 | 3004 | 1.317 | 324 |
| 1 | 484.1 | 484.1 | 5.59 | 2984 | 1.335 | 331 |
| 2 | 935.2 | 467.6 | 5.95 | 2966 | 1.330 | 332 |
| 3 | 1424.7 | 474.8 | 5.60 | 2947 | 1.361 | 341 |
| 4 | 1962.7 | 490.7 | 5.52 | 2926 | 1.397 | 351 |
| 5 | 2625.3 | 525.1 | 5.48 | 2906 | 1.387 | 358 |
| 6 | 3226.6 | 537.8 | 4.42 | 2901 | 1.417 | 371 |
| 6-Supercell[a] | 9551.6 | 530.6 | 3.80 | 8255 | 1.272[b] | 341 |

[a]This is a [3×(12,0) imogolite NT] supercell.
[b]Corrected to one unit cell.
The number of the dehydroxylation in the unit cell is given by n.

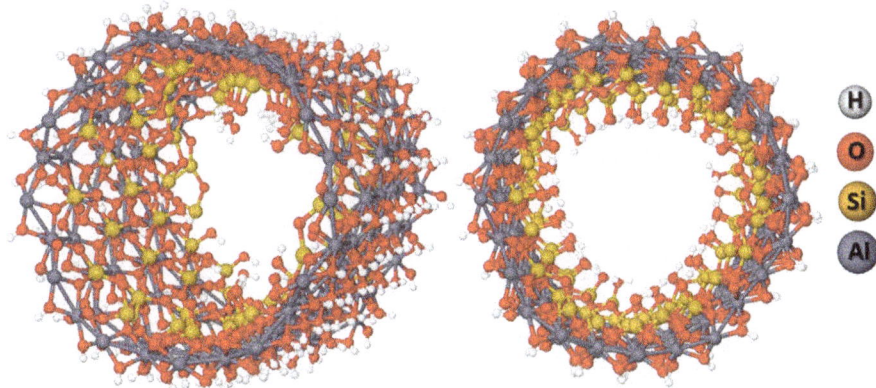

FIGURE 8 | (12,0) NT supercell with 50% of dehydroxylation leading to the helical hydrogen bonding network.

functionalization of the inner and outer surface have been successfully achieved by modifying its properties toward gas adsorption, reactivity, and as component of composites. However, the electronic, structural, and mechanical properties of such functionalized materials have not been detailed investigated. The present study provides insights about these properties and how they are modified with respect to the functionalization of the inner surface of the imogolite NTs. The methyltrimethoxysilane functionalization of the inner surface has been calculated for different level of substitution. The results indicate that the methyltrimethoxysilane prefers the $\eta^2$-img sites involving silanols that are in the same radial line in the unit cell. Different substitutions occur to provide the most symmetrical structure. The arrangement of the methoxy groups between silanes in the different radial lines is more favored when they are directed to each other leading to hydrogen bonding. The BGs and the Young moduli are slightly decreased with the functionalization.

The thermal treatment of the imogolite leads to its dehydroxylation releasing water. This is a more severe modification since the oxo groups bridging the silicon atoms of the imogolite structure change the nature of the bonding of the system. The different dehydroxylations are favored leading to the most symmetrical structure as expected. The Young moduli are increased by 50 GPa with the dehydroxylations and the reaction energy increases with the presence of dehydroxylated silanols. The dehydroxylation reaction energy varies from 484 to 538 kJ·mol$^{-1}$ indicating that the degree of the dehydroxylation can, in principle, be controlled by heat treatment. The dehydroxylation leads to a drastic decrease of the BG energy from 9.1 eV (imogolite) to 4.4 eV (with six dehydroxylations per unit cell). It is important to note that the SCC-DFTB-D provides an upper bound of the BG energy since it overestimates it. Indeed, it is expected for this system the BG is at least 3 eV smaller, taking into account the comparison between the SCC-DFTB-D and B3LYP calculations for the ideal imogolite. A supercell containing three unit cells with six dehydroxylations each were calculated leading to a helical hydrogen bonding network inside of the NT and a BG of 3.8 eV. Therefore, it seems the dehydroxylation can, eventually, lead to a semiconductor material. Actually, depending on the level of the dehydroxylation one could tune the BG energy.

## ACKNOWLEDGMENTS

MS would like to thank Prof. Dr. Dennis R. Salahub and Prof. Dr. Sergei Noskov of the University of Calgary for the facilities during the writing of the present manuscript. We would like to thank the Brazilian funding agencies: Conselho Nacional para o Desenvolvimento Cientifico e Tecnológico (CNPq); Coordenação de Aperfeiçoamento de Pessoal do Ensino Superior (CAPES); and, Fundação de Amparo a Pesquisa do Estado de Minas Gerais (FAPEMIG). This work is also supported by the National Institute of Science and Technology for Mineral Resources, Water and Biodiversity (INCT-ACQUA).

## REFERENCES

Alhuthali, A., and Low, I. M. (2013). Water absorption, mechanical, and thermal properties of halloysite nanotube reinforced vinyl-ester nanocomposites. *J. Sci. Mater.* 48, 4260–4273. doi:10.1007/s10853-013-7240-x

Aradi, B., Hourahine, B., and Frauenheim, T. (2007). DFTB+, a sparse matrix-based implementation of the DFTB method. *J. Phys. Chem. A* 111, 5678–5684. doi:10.1021/jp070186p

Assima, G. P., Larachi, F., Beaudoin, G., and Molson, J. (2013). Dynamics of carbon dioxide uptake in chrysotile mining residues – effect of mineralogy and liquid saturation. *Int. J. Greenh. Gas Control* 12, 124–135. doi:10.1016/j.ijggc.2012.10.001

Azzouz, A. (2012). Achievement in hydrogen storage on adsorbents with high surface-to-bulk ratio – prospects for Si-containing matrices. *Int. J. Hydrogen Energy* 37, 5032–5049. doi:10.1016/j.ijhydene.2011.12.024

Bates, T. F., Hildebrand, F. A., and Swineford, A. (1950a). Morphology and structure of endellite and halloysite. *Am. Mineral.* 35, 463–484.

Bates, T. F., Sand, L. B., and Mink, J. F. (1950b). Tubular crystals of chrysotile asbestos. *Science* 111, 512–513. doi:10.1126/science.111.2889.512

Cavallaro, G., Lazzara, G., Milioto, S., Palmisano, G., and Parisi, F. (2014). Halloysite nanotube with fluorinated lumen: non-foaming nanocontainer for storage and controlled release of oxygen in aqueous media. *J. Colloid Interface Sci.* 417, 66–71. doi:10.1016/j.jcis.2013.11.026

Cradwick, P. D., Wada, K., Russell, J. D., Yoshinag, N., Masson, C. R., and Farmer, V. C. (1972). Imogolite, a hydrated aluminum silicate of tubular structure. *Nat. Phys. Sci.* 240, 187. doi:10.1038/physci240187a0

da Silva, M. C., dos Santos, E. C., Lourenco, M. P., and Duarte, H. A. (2013). Structural, mechanical and electronic properties of nano-fibriform silica and its organic functionalization by dimethyl silane: a SCC-DFTB approach. *J. Mol. Model.* 19, 1995–2005. doi:10.1007/s00894-012-1583-0

Demichelis, R., Noel, Y., D'Arco, P., Maschio, L., Orlando, R., and Dovesi, R. (2010). Structure and energetics of imogolite: a quantum mechanical ab initio study with B3LYP hybrid functional. *J. Mater. Chem.* 20, 10417–10425. doi:10.1039/c0jm00771d

Elstner, M., Porezag, D., Jungnickel, G., Elsner, J., Haugk, M., Frauenheim, T., et al. (1998). Self-consistent-charge density-functional tight-binding method for simulations of complex materials properties. *Phys. Rev. B* 58, 7260–7268. doi:10.1103/PhysRevB.58.7260

Gomez, L., Hueso, J. L., Ortega-Liebana, M. C., Santamaria, J., and Cronin, S. B. (2014). Evaluation of gold-decorated halloysite nanotubes as plasmonic photocatalysts. *Catal. Commun.* 56, 115–118. doi:10.1016/j.catcom.2014.07.017

Guimaraes, L., Enyashin, A. N., Frenzel, J., Heine, T., Duarte, H. A., and Seifert, G. (2007). Imogolite nanotubes: stability, electronic, and mechanical properties. *ACS Nano* 1, 362–368. doi:10.1021/nn700184k

Guimaraes, L., Enyashin, A. N., Seifert, G., and Duarte, H. A. (2010). Structural, electronic, and mechanical properties of single-walled halloysite nanotube models. *J. Phys. Chem. C* 114, 11358–11363. doi:10.1021/jp100902e

Guimaraes, L., Pinto, Y. N., Lourenco, M. P., and Duarte, H. A. (2013). Imogolite-like nanotubes: structure, stability, electronic and mechanical properties of the phosphorous and arsenic derivatives. *Phys. Chem. Chem. Phys.* 15, 4303–4309. doi:10.1039/c3cp44250k

Kang, D.-Y., Brunelli, N. A., Yucelen, G. I., Venkatasubramanian, A., Zang, J., Leisen, J., et al. (2014). Direct synthesis of single-walled aminoaluminosilicate nanotubes with enhanced molecular adsorption selectivity. *Nat. Commun.* 5, 3342. doi:10.1038/ncomms4342

Kang, D. Y., Zang, J., Jones, C. W., and Nair, S. (2011). Single-walled aluminosilicate nanotubes with organic-modified interiors. *J. Phys. Chem. C* 115, 7676–7685. doi:10.1021/jp2010919

Kang, D. Y., Zang, J., Wright, E. R., McCanna, A. L., Jones, C. W., and Nair, S. (2010). Dehydration, dehydroxylation, and rehydroxylation of single-walled aluminosilicate nanotubes. *ACS Nano* 4, 4897–4907. doi:10.1021/nn101211y

Levard, C., Masion, A., Rose, J., Doelsch, E., Borschneck, D., Dominici, C., et al. (2009). Synthesis of imogolite fibers from decimolar concentration at low temperature and ambient pressure: a promising route for inexpensive nanotubes. *J. Am. Chem. Soc.* 131, 17080–17081. doi:10.1021/ja9076952

Levard, C., Masion, A., Rose, J., Doelsch, E., Borschneck, D., Olivi, L., et al. (2011). Synthesis of Ge-imogolite: influence of the hydrolysis ratio on the structure of the nanotubes. *Phys. Chem. Chem. Phys.* 13, 14516–14522. doi:10.1039/c1cp20346k

Levard, C., Rose, J., Masion, A., Doelsch, E., Borschneck, D., Olivi, L., et al. (2008). Synthesis of large quantities of single-walled aluminogermanate nanotube. *J. Am. Chem. Soc.* 130, 5862. doi:10.1021/ja801045a

Levard, C., Rose, J., Thill, A., Masion, A., Doelsch, E., Maillet, P., et al. (2010). Formation and growth mechanisms of imogolite-like aluminogermanate nanotubes. *Chem. Mater.* 22, 2466–2473. doi:10.1021/cm902883p

Li, C. P., Wang, J. Q., Luo, X., and Ding, S. J. (2014). Large scale synthesis of Janus nanotubes and derivative nanosheets by selective etching. *J. Colloid Interface Sci.* 420, 1–8. doi:10.1016/j.jcis.2013.12.062

Lourenco, M. P., de Oliveira, C., Oliveira, A. F., Guimaraes, L., and Duarte, H. A. (2012). Structural, electronic, and mechanical properties of single-walled chrysotile nanotube models. *J. Phys. Chem. C* 116, 9405–9411. doi:10.1021/jp301048p

Lourenco, M. P., Guimaraes, L., da Silva, M. C., de Oliveira, C., Heine, T., and Duarte, H. A. (2014). Nanotubes with well-defined structure: single- and double-walled imogolites. *J. Phys. Chem. C* 118, 5945–5953. doi:10.1021/jp411086f

Lun, H. L., Ouyang, J., and Yang, H. M. (2014). Natural halloysite nanotubes modified as an aspirin carrier. *RSC Adv.* 4, 44197–44202. doi:10.1039/c4ra09006c

Lvov, Y., Aerov, A., and Fakhrullin, R. (2014). Clay nanotube encapsulation for functional biocomposites. *Adv. Colloid Interface Sci.* 207, 189–198. doi:10.1016/j.cis.2013.10.006

Lvov, Y. M., Shchukin, D. G., Mohwald, H., and Price, R. R. (2008). Halloysite clay nanotubes for controlled release of protective agents. *ACS Nano* 2, 814–820. doi:10.1021/nn800259q

Ma, W., Kim, J., Otsuka, H., and Takahara, A. (2011). Surface modification of individual imogolite nanotubes with alkyl phosphate from an aqueous solution. *Chem. Lett.* 40, 159–161. doi:10.1246/cl.2011.159

Ma, W., Yah, W. O., Otsuka, H., and Takahara, A. (2012). Surface functionalization of aluminosilicate nanotubes with organic molecules. *Beilstein J. Nanotechnol.* 3, 82–100. doi:10.3762/bjnano.3.10

Machado, G. S., Ucoski, G. M., de Lima, O. J., Ciuffi, K. J., Wypych, F., and Nakagaki, S. (2013). Cationic and anionic metalloporphyrins simultaneously immobilized onto raw halloysite nanoscrolls catalyze oxidation reactions. *Appl. Catal. A Gen.* 460, 124–131. doi:10.1016/j.apcata.2013.04.014

Murali, R. S., Padaki, M., Matsuura, T., Abdullah, M. S., and Ismail, A. F. (2014). Polyaniline in situ modified halloysite nanotubes incorporated asymmetric mixed matrix membrane for gas separation. *Sep. Purif. Technol.* 132, 187–194. doi:10.1016/j.seppur.2014.05.020

Oliveira, A. F., Seifert, G., Heine, T., and Duarte, H. A. (2009). Density-functional based tight-binding: an approximate DFT method. *J. Braz. Chem. Soc.* 20, 1193–1205. doi:10.1590/S0103-50532009000700002

Peng, Q., Liu, M. X., Zheng, J. W., and Zhou, C. R. (2015). Adsorption of dyes in aqueous solutions by chitosan-halloysite nanotubes composite hydrogel beads. *Microporous Mesoporous Mater.* 201, 190–201. doi:10.1016/j.micromeso.2014.09.003

Porezag, D., Frauenheim, T., Kohler, T., Seifert, G., and Kaschner, R. (1995). Construction of tight-binding-like potentials on the basis of density-functional theory: application to carbon. *Phys. Rev. B Condens. Matter* 51, 12947–12957. doi:10.1103/PhysRevB.51.12947

Price, R. R., Gaber, B. P., and Lvov, Y. (2001). In-vitro release characteristics of tetracycline HCl, khellin and nicotinamide adenine dineculeotide from halloysite; a cylindrical mineral. *J. Microencapsul.* 18, 713–722. doi:10.1080/02652040010019532

Qi, X., Yoon, H., Lee, S.-H., Yoon, J., and Kim, S.-J. (2008). Surface-modified imogolite by 3-APS-OsO4 complex: synthesis, characterization and its application in the dihydroxylation of olefins. *J. Ind. Eng. Chem.* 14, 136–141. doi:10.1016/j.jiec.2007.08.010

Rao, K. M., Nagappan, S., Seo, D. J., and Ha, C. S. (2014). pH sensitive halloysite-sodium hyaluronate/poly(hydroxyethyl methacrylate) nanocomposites for colon cancer drug delivery. *Appl. Clay Sci.* 9, 33–42. doi:10.1016/j.clay.2014.06.002

Seifert, G., Porezag, D., and Frauenheim, T. (1996). Calculations of molecules, clusters, and solids with a simplified LCAO-DFT-LDA scheme. *Int. J. Quantum Chem.* 58, 185–192. doi:10.1002/(SICI)1097-461X(1996)58:2<185::AID-QUA7>3.3.CO;2-B

Shchukin, D. G., and Mohwald, H. (2007). Surface-engineered nanocontainers for entrapment of corrosion inhibitors. *Adv. Funct. Mater.* 17, 1451–1458. doi:10.1002/adfm.200601226

Shchukin, D. G., Sukhorukov, G. B., Price, R. R., and Lvov, Y. M. (2005). Halloysite nanotubes as biomimetic nanoreactors. *Small* 1, 510–513. doi:10.1002/smll.200400120

Shchukin, D. G., Zheludkevich, M., Yasakau, K., Lamaka, S., Ferreira, M. G. S., and Mohwald, H. (2006). Layer-by-layer assembled nanocontainers for self-healing corrosion protection. *Adv. Mater. Weinheim* 18, 1672. doi:10.1002/adma.200502053

Tenne, R., Margulis, L., Genut, M., and Hodes, G. (1992). Polyhedral and cylindrical structures of tungsten disulfide. *Nature* 360, 444–446. doi:10.1038/360444a0

Tham, W. L., Poh, B. T., Ishak, Z. A. M., and Chow, W. S. (2014). Thermal behaviors and mechanical properties of halloysite nanotube-reinforced poly(lactic acid) nanocomposites. *J. Therm. Anal. Calorim.* 118, 1639–1647. doi:10.1007/s10973-014-4062-2

Thill, A., Guiose, B., Bacia-Verloop, M., Geertsen, V., and Belloni, L. (2012a). How the diameter and structure of (OH)(3)Al2O3SixGe1-xOH imogolite nanotubes Are controlled by an adhesion versus curvature cornpetition. *J. Phys. Chem. C* 116, 26841–26849. doi:10.1021/jp310547k

Thill, A., Maillet, P., Guiose, B., Spalla, O., Belloni, L., Chaurand, P., et al. (2012b). Physico-chemical control over the single- or double-wall structure of aluminogermanate imogolite-like nanotubes. *J. Am. Chem. Soc.* 134, 3780–3786. doi:10.1021/ja209756j

Tu, J. X., Cao, Z., Jing, Y. H., Fan, C. J., Zhang, C., Liao, L. Q., et al. (2013). Halloysite nanocomposite hydrogels with tunable mechanical properties and drug release behavior. *Compos. Sci. Technol.* 85, 126–130. doi:10.1016/j.compscitech.2013.06.011

Veerabadran, N. G., Price, R. R., and Lvov, Y. M. (2007). Clay nanotubes for encapsulation and sustained release of drugs. *Nano* 2, 115–120. doi:10.1142/S1793292007000441

Wada, S., and Wada, K. (1982). Effects of substitution of germanium for silicon in imogolite. *Clays Clay Miner.* 30, 123–128. doi:10.1346/ccmn.1982.0300206

Wang, B., and Huang, H.-X. (2013). Effects of halloysite nanotube orientation on crystallization and thermal stability of polypropylene nanocomposites. *Polym. Degrad. Stab.* 98, 1601–1608. doi:10.1016/j.polymdegradstab.2013.06.022

Wang, Q., Zhang, J. P., and Wang, A. Q. (2013). Alkali activation of halloysite for adsorption and release of ofloxacin. *Appl. Surf. Sci.* 287, 54–61. doi:10.1016/j.apsusc.2013.09.057

Whittaker, E. J. W. (1956). The structure of chrysotile.2. Clino-chrysotile. *Acta Crystallogr.* 9, 855–862. doi:10.1107/S0365110X56002473

Xu, W., Luo, B. H., Wen, W., Xie, W. J., Wang, X. Y., Liu, M. X. (2015). Surface modification of halloysite nanotubes with L-lactic acid: an effective route to high-performance poly(L-lactide) composites. *J. Appl. Polym. Sci.* 132:41451. doi:10.1002/app.41451

Yucelen, G. I., Kang, D. Y., Guerrero-Ferreira, R. C., Wright, E. R., Beckham, H. W., and Nair, S. (2012). Shaping single-walled metal oxide nanotubes from precursors of controlled curvature. *Nano Lett.* 12, 827–832. doi:10.1021/nl203880z

Zhechkov, L., Heine, T., Patchkovskii, S., Seifert, G., and Duarte, H. A. (2005). An efficient a posteriori treatment for dispersion interaction in density-functional-based tight binding. *J. Chem. Theory Comput.* 1, 841–847. doi:10.1021/ct050065y

Zhong, S., Zhou, C. Y., Zhang, X. N., Zhou, H., Li, H., Zhu, X. H., et al. (2014). A novel molecularly imprinted material based on magnetic halloysite nanotubes for rapid enrichment of 2,4-dichlorophenoxyacetic acid in water. *J. Hazard. Mater.* 276, 58–65. doi:10.1016/j.jhazmat.2014.05.013

**Conflict of Interest Statement:** The authors declare that the research was conducted in the absence of any commercial or financial relationships that could be construed as a potential conflict of interest.

# 10

# A bottom-up approach for the synthesis of highly ordered fullerene-intercalated graphene hybrids

**Antonios Kouloumpis[1], Konstantinos Spyrou[1,2], Konstantinos Dimos[1], Vasilios Georgakilas[3], Petra Rudolf[2] and Dimitrios Gournis[1]***

[1] Department of Materials Science and Engineering, University of Ioannina, Ioannina, Greece
[2] Zernike Institute for Advanced Materials, University of Groningen, Groningen, Netherlands
[3] Department of Materials Science, University of Patras, Rio, Greece

**Edited by:**
*Emilia Morallon, Universidad de Alicante, Spain*

**Reviewed by:**
*Junyi Ji, Sichuan University, China
Piotr Gauden, Nicholaus Copernicus University, Poland*

**\*Correspondence:**
*Dimitrios Gournis, Department of Materials Science and Engineering, University of Ioannina, Ioannina 45110, Greece
e-mail: dgourni@cc.uoi.gr*

Much of the research effort on graphene focuses on its use as a building block for the development of new hybrid nanostructures with well-defined dimensions and properties suitable for applications such as gas storage, heterogeneous catalysis, gas/liquid separations, nanosensing, and biomedicine. Toward this aim, here we describe a new bottom-up approach, which combines self-assembly with the Langmuir–Schaefer deposition technique to synthesize graphene-based layered hybrid materials hosting fullerene molecules within the interlayer space. Our film preparation consists in a bottom-up layer-by-layer process that proceeds via the formation of a hybrid organo-graphene oxide Langmuir film. The structure and composition of these hybrid fullerene-containing thin multilayers deposited on hydrophobic substrates were characterized by a combination of X-ray diffraction, Raman and X-ray photoelectron spectroscopies, atomic force and scanning electron microscopies, and conductivity measurements. The latter revealed that the presence of $C_{60}$ within the interlayer spacing leads to an increase in electrical conductivity of the hybrid material as compared to the organo-graphene matrix alone.

Keywords: Langmuir–Blodgett, graphene, fullerene, thin films, hybrids

## INTRODUCTION

The outstanding mechanical, thermal, and electrical properties of graphene have attracted a lot of scientific effort aimed at exploiting them in the development of new hybrid nanostructures with well-defined dimensions and behavior, which contain graphene as building block (Wang et al., 2010; Yin et al., 2012). In fact, graphene sheets can be used as templates for the synthesis of novel intercalated carbon-based materials suitable for various applications in gas storage, gas/liquid separation, heterogeneous catalysis, energy storage, Li-ion batteries, supercapacitors, nanosensing, and biomedicine (Patil et al., 2009; Yu and Dai, 2009; Choi et al., 2010; Wang et al., 2010; Chen et al., 2011; Lei et al., 2011). Combining graphene's properties with the extraordinary properties of fullerenes (Iijima, 1991; Guldi and Prato, 2000) by incorporating the latter into well-defined graphene-based hybrid thin multilayers continues to be a challenging new field for developing novel hybrid nanocomposites.

Toward this aim, here we describe a new bottom-up layer-by-layer approach for the production of graphene hybrid materials where graphene acts as the structure directing interface and reaction media. This method, based on combining self-assembly with the Langmuir–Schaefer (LS) deposition technique, uses the graphene nanosheets as a template for incorporating $C_{60}$ molecules in a bi-dimensional array, and allows for perfect layer-by-layer growth with control at the molecular level (Gengler et al., 2012a). Similar synthetic protocols have been reported during the last decade for the development of hybrid multilayers and monolayers using the Langmuir–Blodgett (LB) technique. Layered

materials like clay minerals (Umemura et al., 2001, 2002, 2009; Yamamoto et al., 2004b; Junxiang et al., 2005; Yoshida et al., 2006; Gengler, 2010; Gengler et al., 2010, 2012a,b; Toma et al., 2010), graphene, and/or graphene oxide (GO) (Cote et al., 2009; Gengler et al., 2010; Zheng et al., 2011b; Michopoulos et al., 2014), or other carbon-based nanostructures (Bourlinos et al., 2012a,b, 2013) have been used to produce hybrid thin multilayers with unique properties. Clay minerals with amphiphilic species and other complexes were combined to fabricate monolayers and multilayers used as photoprobes (Hagerman et al., 2002), sensors (Junxiang et al., 2005), catalysts (Yoshida et al., 2006), and photomagnetic films (Yamamoto et al., 2004a,b). Moreover, the optoelectronic and mechanical properties of graphene can be modified, tuned, or enhanced by layer-by-layer assembly techniques such as the LB method, as has been reported in various studies over the last 3–4 years (for a review, see Kouloumpis et al., 2014). High-performance dye-sensitized solar cells (Roh et al., 2014) and transparent conductive films (Zheng et al., 2011a) are some examples of the uses of hybrid graphene-based thin multilayers. Finally, the advantages of the precise control and the homogeneous deposition over large areas makes the LB technique promising for preventing carbon-based nanostructures from agglomerating during the film synthesis, as demonstrated for fullerene derivatives or carbon dots (Bourlinos et al., 2012a,b, 2013).

Our film preparation approach involves a bottom-up layer-by-layer process that starts with the formation of a hybrid organo-graphene Langmuir film and proceeds via two self-assembly steps to create a layered structure hosting fullerene molecules within its interlayer space (a schematic representation of the synthetic procedure is illustrated in **Scheme 1**). The composition, structure, and transport properties of the fullerene-containing hybrid thin multilayers deposited on hydrophobic substrates were characterized by X-ray diffraction (XRD), Raman and X-ray photoelectron spectroscopies, atomic force and scanning electron microscopies (SEM), and electrical conductivity measurements.

## EXPERIMENTAL SECTION
### MATERIALS
Octadecylamine [$CH_3(CH_2)_{17}NH_2$, ODA], acetone, methanol, and ethanol were purchased from Sigma-Aldrich while fullerene ($C_{60}$, powder, 99% C) was obtained from Alfa Aesar. Ultrapure deionized water (18.2 MOhm) produced by a Millipore Simplicity® system was used throughout. The Si-wafer (P/Bor, single side polished) substrates were cleaned prior to use by 15 min ultrasonication in water, acetone, and ethanol. All reagents were of analytical grade and were used without further purification.

### SYNTHESIS OF GRAPHENE OXIDE
Graphene oxide was synthesized using a modified Staudenmaier's method (Staudenmaier, 1898; Gengler et al., 2010; Stergiou et al., 2010): 10 g of powdered graphite (purum, powder ≤0.2 mm, Fluka) were added to a mixture of 400 mL of 95–97% $H_2SO_4$ (Riedel-de Haën) and 200 mL of 65% $HNO_3$ (Fluka), while cooling in an ice–water bath. Two hundred grams of powdered $KClO_3$ (Fluka) were added to the mixture in small portions under vigorous stirring while cooling in the ice–water bath. The reaction was quenched after 18 h by pouring the mixture into ultrapure water; the oxidation product was washed until the pH reached 6.0, and finally dried at room temperature.

## PREPARATION OF HYBRID GRAPHENE/FULLERENE MULTILAYERS
A KSV 2000 Nima Technology LB trough was used for the preparation and deposition of multilayers at a temperature of $21 \pm 0.5°C$. The surface pressure in the LB trough was monitored with a Pt Wilhelmy plate. The deposition protocol is schematically illustrated in **Scheme 1**. A GO suspension in ultrapure water ($0.02 \, mg \, mL^{-1}$) was used as subphase. Two hundred microliters of a $0.2 \, mg \, mL^{-1}$ ODA solution in chloroform/methanol 9/1 (v/v) were spread onto the water surface using a microsyringe to achieve the hybridization of GO sheets by covalent bonding via the amide functionality. After a waiting time of 20 min to allow for solvent evaporation and GO-surfactant functionalization to occur, the hybrid ODA–GO layer was compressed at a rate of $5 \, mm \, min^{-1}$ until the chosen stabilization pressure of $20 \, mN \, m^{-1}$ was reached. As in any classical LB experiment, the applied pressure pushes the surfactant molecules along the water surface; the grafted GO sheets will simply follow that movement and therefore become packed, depending on the surface tension established in the trough (Gengler et al., 2010). This pressure was maintained throughout the deposition process. Layers were transferred onto the hydrophobic Si–wafer substrates by horizontal dipping [this way of transferring is known as Langmuir–Schaefer (LS) technique], with downward and lifting speeds of 10 and $5 \, mm \, min^{-1}$, respectively (Cote et al., 2009) – first step in **Scheme 1**. After the horizontal lift of the substrate, the second step of the deposition protocol consisted in a surface modification of the GO nanosheets, induced by bringing the surface of the transferred Langmuir film (ODA–GO) in contact with a solution of ODA surfactant (self-assembly) dissolved in methanol ($0.2 \, mg \, mL^{-1}$) (Gengler et al., 2010) as illustrated in **Scheme 1**. In the third and final step, the hybrid organo-GO film was lowered into a solution of $C_{60}$ in toluene ($0.2 \, mg \, mL^{-1}$)

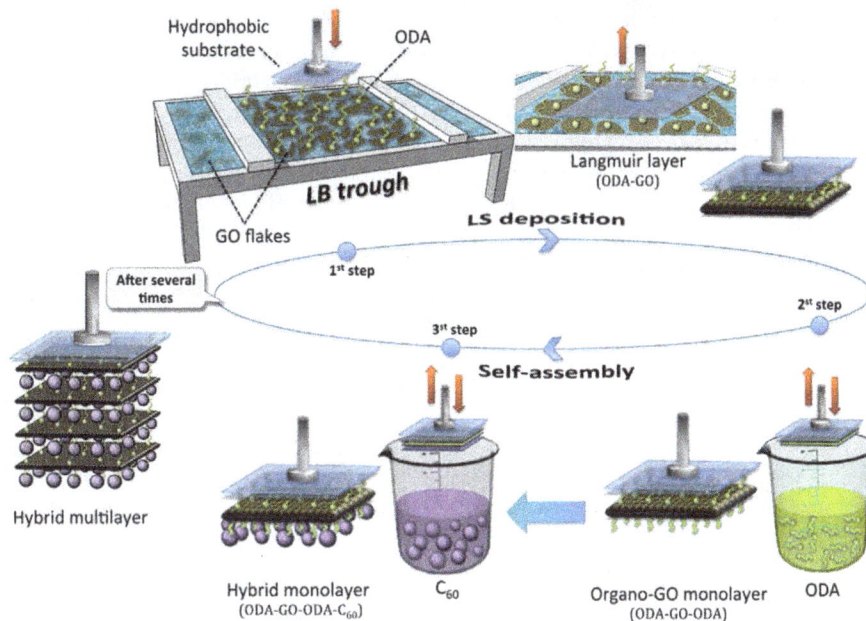

**SCHEME 1 | Schematic representation of the synthetic procedure followed for the synthesis of the hybrid GO/$C_{60}$ multilayer film consisting in a Langmuir–Schaefer deposition (first step) combined with two self-assembly steps (second and third step).**

for the formation of a hybrid graphene/fullerene monolayer by self-assembly. A hybrid multilayer film was constructed by repeating this three-step-dipping cycle 40 times, as shown in **Scheme 1** (sample denoted as ODA–GO–ODA–$C_{60}$). Each time when the substrate was lowered toward the liquid surface, it was allowed to touch the air–water interface or the solution surface in a very gentle dip of max 0.5 mm below the surface for 90 s. After each deposition step, the sample was rinsed several times by dipping into ultrapure water to eliminate any weakly attached molecules that remained from the deposition step and dried with nitrogen flow to avoid contaminating the Langmuir film air–water interface or the solution employed in the following step (Gengler et al., 2012a; Michopoulos et al., 2014). For comparison, an organo-GO hybrid multilayer (40 layers) was also fabricated under the same experimental conditions without the third step of $C_{60}$ incorporation (sample denoted as ODA–GO–ODA). Moreover, a reduction and an annealing step (fourth step) were also performed in both multilayers in order to convert the deposited GO to graphene and thus to increase the conductivity. For this, the deposited multilayers were immersed into an aqueous solution of $NaBH_4$ (1 mg mL$^{-1}$) for 10 min and annealed at 700°C for 1 h under argon (samples denoted as ODA–rGO–ODA–$C_{60}$ and ODA–rGO–ODA). This treatment results in a partial reconstruction of the graphene mesh and consequently causes a drastic increase in conductivity (Gengler et al., 2010).

## CHARACTERIZATION TECHNIQUES

Topographic atomic force microscopy (AFM) images were recorded in tapping mode with a Bruker Multimode 3D Nanoscope, using a silicon microfabricated cantilever type TAP-300G, with a tip radius <10 nm and force constant range ~20–75 N m$^{-1}$. Raman spectra were collected with a Micro-Raman system RM 1000 RENISHAW using a laser excitation line at 532 nm (laser diode). A 0.5–1 mW laser power was used with a 1 μm focus spot in order to avoid photodecomposition of the sample. X-ray photoelectron spectroscopy (XPS) measurements were performed in ultrahigh vacuum at a base pressure of $2 \times 10^{-10}$ mbar with a SPECS GmbH spectrometer equipped with a monochromatic Mg Kα source (hv = 1253.6 eV) and a Phoibos-100 hemispherical analyzer. The spectra were collected in normal emission and energy resolution was set to 1.16 eV to minimize measuring time. All binding energies were referenced to the C1s core level at 285.0 eV. Spectral analysis included a Shirley background subtraction and a peak deconvolution employing mixed Gaussian-Lorentzian functions, in a least squares curve-fitting program (WinSpec) developed at the Laboratoire Interdisciplinaire de Spectroscopie Electronique, University of Namur, Belgium. The XRD patterns were collected on a D8 Advance Bruker diffractometer by using a Cu Kα (λ = 1.5418 Å) radiation source (40 kV, 40 mA) and a secondary beam graphite monochromator. The patterns were recorded in the 2-theta (2Θ) range from 2 to 80°, in steps of 0.02° and with a counting time of 2 s per step. SEM images were recorded using a JEOL SEM-6510LV SEM equipped with an EDX analysis system xx-Act from Oxford Instruments. For the electrical conductivity, four-probe measurements were performed using an AFX DC 9660SB power supply and a Keithley 2000 multimeter. Thin films deposited on Si–wafers were measured without any treatment as well as after the reduction and annealing step.

## RESULTS AND DISCUSSION
### STRUCTURE CONTROL OF HYBRID ODA–GO LAYER

The surface pressure versus molecular area (Π–a) isotherms of ODA monolayers in pure water and on GO dispersion are shown in **Figure 1** (right). The curves show the change in the slope corresponding to the phase transitions of ODA–GO sheets from gas to condensed-liquid and then to solid state during the compression process (Michopoulos et al., 2014). In the absence of GO, the Π–a isotherm is a smoothly increasing curve with a lift-off area of 32.8 Å$^2$. When adding a small amount of GO (0.02 mg mL$^{-1}$) to the aqueous subphase, the lift-off area increased to 52 Å$^2$, verifying that GO flakes cause a stabilization effect on the ODA layer (Michopoulos et al., 2014) through chemical grafting (covalent bonding) of the terminal amine groups of ODA to the epoxide groups on the top side of the GO sheets. The amine end groups interact via a ring opening reaction (nucleophilic substitution reactions) of the epoxide groups of GO (Bourlinos et al., 2003; Dreyer et al., 2010; Gengler et al., 2010). At surface pressures above 40 mN m$^{-1}$, the monolayer collapses due to the formation of bilayers at approximately 26 Å$^2$ (Michopoulos et al., 2014).

The successful transfer of the hybrid Langmuir films onto the Si–wafer substrate can be deduced from the plot of the pressure measured at the surface of the subphase in the LB trough versus time (Gengler et al., 2012a). **Figure 1** (right) shows the time dependence of the total trough area of the hybrid Langmuir film. As the substrate is dipped into the subphase, this area reduces due to the transfer of one hybrid layer from the air–water interface to the substrate during each dip. This transfer is visible as a sharp step in the curve (Toma et al., 2010; Gengler et al., 2012a). If the step height (which gives an area value) is equal to the substrate surface area, the transfer ratio is 1 and the substrate surface is 100% covered by the hybrid layer each time it is lowered into the subphase. A different transfer ratio suggests a multilayer transfer or an incomplete coverage of the substrate (Gengler et al., 2012a). The curve shown in **Figure 1** is a typical one recorded during the deposition of a 10-layer hybrid ODA–GO–ODA–$C_{60}$ film. The transfer ratio was very close to 1 throughout the deposition, testifying to the successful transfer of the hybrid Langmuir film at each dip of substrate into the LB trough.

Representative AFM images of hybrid GO sheets (ODA–GO) deposited on Si–wafer with the LS method (at surface pressure 20 mN m$^{-1}$) during the first dip into the LB trough (first step in **Scheme 1**) are shown in **Figure 2**. In these images, one recognizes high quality GO flakes with well-defined edges and a relatively low amount of cracks and wrinkles. The surface coverage is quite high; the GO platelets in the transferred layer are contacting each other, with hardly any overlap but only small voids between them, forming a nearly continuous, close-packed array. Because of the ODA layer below the GO sheet, the average height of the flakes of 0.9–1.5 nm, as derived from topographical height profile, is larger than the value of 0.61 nm predicted for a single GO layer (Dekany et al., 1998). However, it is difficult to conclude anything on the orientation of the ODA molecules (straight or inclined) from these images because the height of single GO layers without ODA has also been reported in the literature to be larger than 0.61 nm, namely of the order of 1.1 ± 0.2 nm [see, for example, Schniepp et al. (2006) and

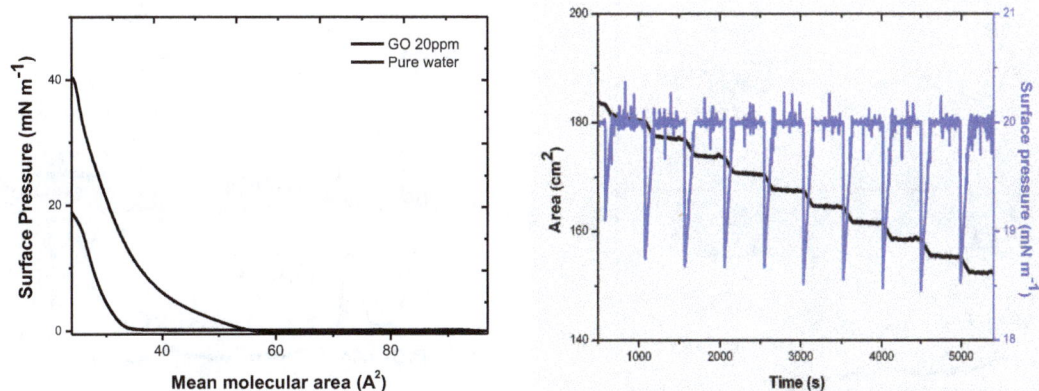

**FIGURE 1 | (Left panel)** Π–a isotherms of ODA monolayers in pure water and on an aqueous dispersion of 0.02 mg mL⁻¹ GO. **(Right panel)** The black curve corresponds to the trough area covered by the hybrid ODA–GO Langmuir film at the air-water surface, and the blue curve shows the surface pressure throughout (around 20 mN m⁻¹) the deposition of a 10-layer ODA–GO–ODA–$C_{60}$ hybrid film.

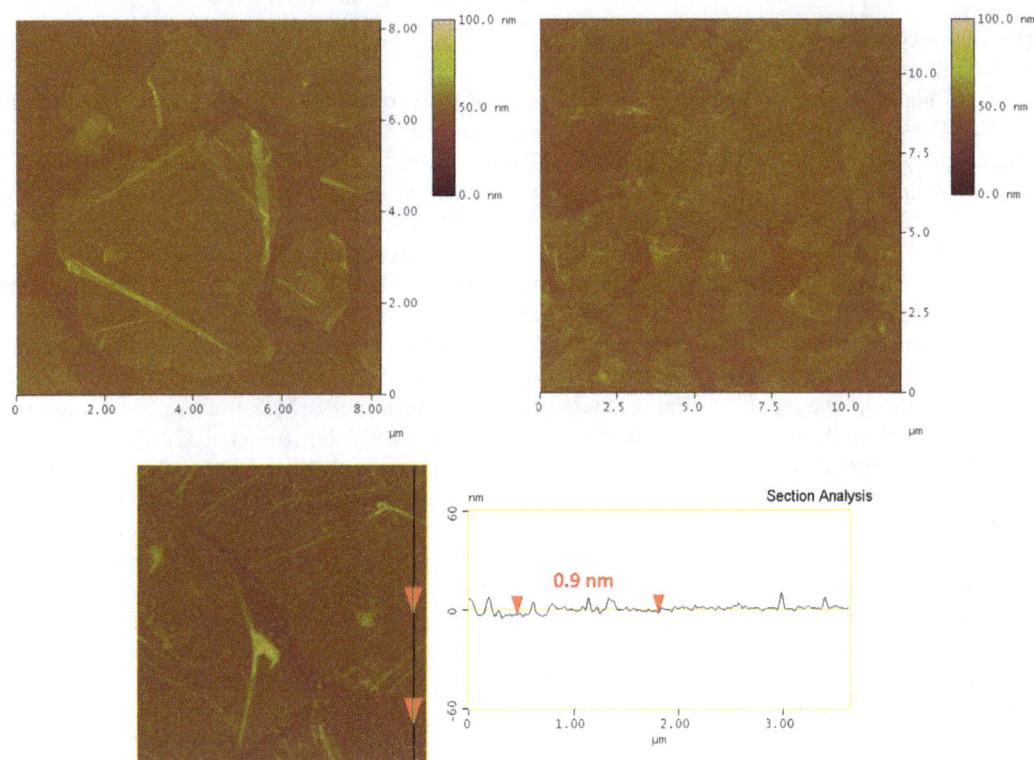

**FIGURE 2 | Tapping mode AFM micrographs (and line profile) of a ODA–GO monolayer deposited with the Langmuir–Schaefer technique at a surface pressure of 20 mN m⁻¹.**

Gomez-Navarro et al. (2007)] probably due to the presence of adsorbed water molecules.

## CHARACTERIZATION OF HYBRID GRAPHENE/FULLERENE MULTILAYERS

The XRD pattern of an ODA–GO–ODA–$C_{60}$ hybrid multilayer (40 layers) is shown in **Figure 3** in comparison with a ODA–GO–ODA hybrid multilayer (40 layers) that was synthesized under the same experimental conditions. The ODA–GO–ODA–$C_{60}$ hybrid multilayer shows a 001 diffraction peak at $2\theta = 2.2°$, which corresponds to a $d_{001}$-spacing of 40 Å. This value is higher compared with the corresponding value of the ODA–GO–ODA multilayer ($d_{001} = 37.6$ Å) testifying to the successful intercalation of the fullerene between the organo-graphene nanosheets. This increment does not correspond to the size of $C_{60}$ (~7 Å) implying that

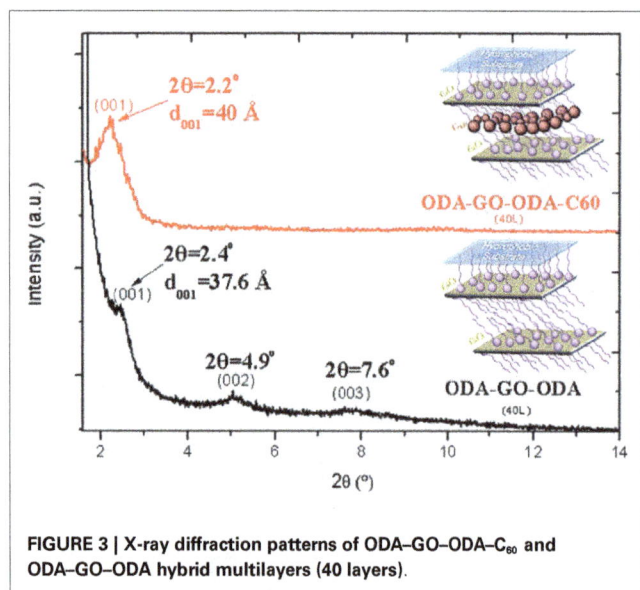

**FIGURE 3 | X-ray diffraction patterns of ODA–GO–ODA–C$_{60}$ and ODA–GO–ODA hybrid multilayers (40 layers).**

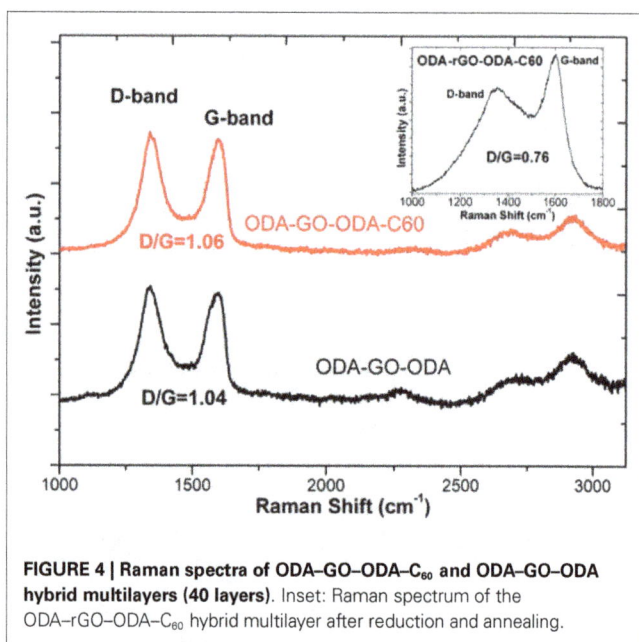

**FIGURE 4 | Raman spectra of ODA–GO–ODA–C$_{60}$ and ODA–GO–ODA hybrid multilayers (40 layers).** Inset: Raman spectrum of the ODA–rGO–ODA–C$_{60}$ hybrid multilayer after reduction and annealing.

the fullerene molecules are accommodated between the double alkyl chains of the surfactant (see inset sketch) and not on top of it. Moreover, the absence of higher order (00l) reflections in the XRD pattern of ODA–GO–ODA–C$_{60}$ suggests that the GO layers are not stacked in perfect registry but have slipped sideways and are turbostratically disordered. A similar behavior has been observed upon intercalation of fullerene derivatives in aluminosilicate clay minerals (Gournis et al., 2006). This hypothesis is further supported by calculating the size of the coherently diffracting domains along the c axis (also called mean crystalline thickness, t) that give rise to the 001 diffraction peak, given by the Debye–Scherrer equation, $t = K\lambda/\beta\cos\theta$, where K is a constant near unity ($K = 0.91$), $\lambda$ the X-ray wavelength ($\lambda = 1.5418$ Å), $\theta$ the angular position of the first diffraction peak, and $\beta$ the full width at half maximum of the 001 peak expressed in radians. The crystalline thickness (t) along c axis of the ODA–GO–ODA–C$_{60}$ and ODA–GO–ODA multilayers is found to be 114 and 153 Å, respectively. Dividing these values with the corresponding $d_{001}$-spacings, the number of stacked layers can be calculated to be ~3 for ODA–GO–ODA–C$_{60}$ and ~4 for ODA–GO–ODA, confirming the high degree of ordering of the produced multilayers.

Raman spectra of the ODA–GO–ODA–C$_{60}$ and the ODA–GO–ODA multilayer samples are shown in **Figure 4**. Spectra are typical of GO materials, in that both exhibit the characteristic first-ordered G- and D-bands at around 1600 and 1350 cm$^{-1}$, respectively. The G-band is associated with sp$^2$-hybridized carbon atoms and originates from the doubly degenerate zone center $E_{2g}$ mode. The D-band is correlated with sp$^3$ hybridized carbon atoms as it requires a defect for its activation by double resonance, thus indicating the presence of lattice defects and distortions (Ferrari et al., 2006; Torrisi et al., 2012). The ratio of the D- to G-band intensities (I$_D$/I$_G$) is indicative of the quality of the graphitic lattice. This ratio is equal to 1.06 for the ODA–GO–ODA–C$_{60}$ multilayer and 1.04 for the ODA–GO–ODA multilayer, implying that fullerene intercalation into the interlayer space of the organo-graphene nanosheets does not influence the GO structure. In addition, the

shape of the two spectra in the 2D region is alike; both exhibit two broad peaks at ~2700 and ~2930 cm$^{-1}$. The low intensity 2700 cm$^{-1}$ peak is attributed to the 2D (or else G′) vibrational mode, which is an overtone (second order) of the D peak. The second peak at ~2930 cm$^{-1}$ is assigned to the mode D + D′, originating from the combination of phonons with different momenta, thus requiring defects for its activation; its intensity agrees with the defective nature of the GO lattice revealed by the high I$_D$/I$_G$ ratios of both samples (Ferrari et al., 2006; Martins Ferreira et al., 2010; Cançado et al., 2011; Torrisi et al., 2012). However, the typical pentagonal pinch mode [A$_g$(2)] of C$_{60}$ that is expected between 1440 and 1470 cm$^{-1}$ (Spyrou et al., 2013) is not visible since it is superimposed on the broad D-band of GO. Moreover, upon reduction and annealing (see **Figure 4** inset), the GO sheets show a noticeable decrease in the D/G ratio from 1.06 to 0.76. This observation suggests that while most of the oxygenated groups are removed (in the form of CO or CO$_2$), the relative amount of disordered sp$^2$-hybridized atoms is still high. The later could be explained either by the fact that amine molecules covalently bound to graphene survive the reduction and annealing procedures (contributing to an enhanced sp$^3$ hybridization) or that vacancies produced during production of GO remain unchanged by the reduction process and ultimately define the intact graphene regions (Gomez-Navarro et al., 2007; Lomeda et al., 2008; Gengler et al., 2010).

The C1s core level X-ray photoelectron spectra of the graphene/fullerene hybrid multilayers before (ODA–GO–ODA–C$_{60}$) and after reduction and annealing (ODA–rGO–ODA–C$_{60}$) in comparison with pristine GO and the organo-GO multilayers (ODA–GO–ODA) are shown in **Figure 5**. The spectrum of ODA–GO–ODA displays four main contributions at 285.0, 286.1, 287.5, and 289.3 eV corresponding to different carbon environments while for ODA–GO–ODA–C$_{60}$ an additional contribution appear at 290.5 eV. The peak at 285.0 eV is attributed to the C–C bonds

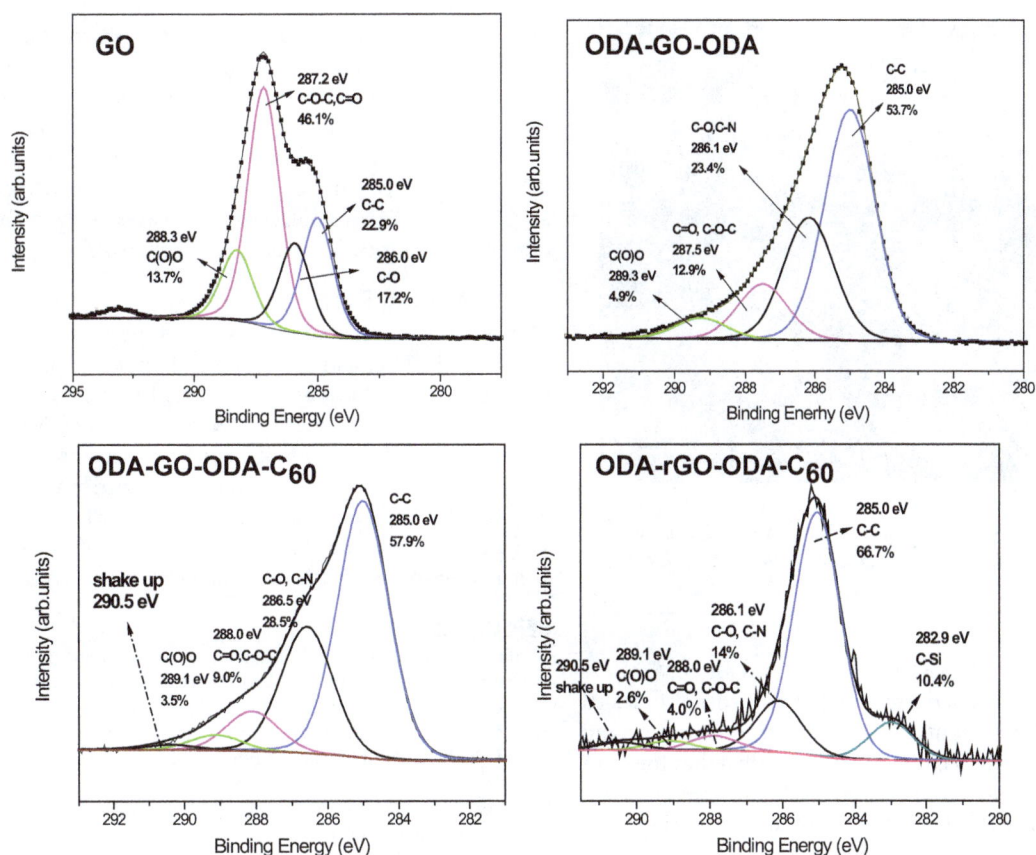

FIGURE 5 | X-ray photoelectron spectra of the C 1s core level region of GO, ODA–GO–ODA and graphene/fullerene hybrid multilayers (40 layers) before (ODA–GO–ODA–C$_{60}$) and after reduction and annealing (ODA–rGO–ODA–C$_{60}$).

of the aromatic ring of GO as well the organic carbon–carbon groups of ODA attached to graphene and represents 53.7% of the total carbon 1s intensity of ODA–GO–ODA (**Figure 5** left). This contribution is more intense in the case of ODA–GO–ODA–C$_{60}$ (**Figure 5** right) because in addition to photoemission signal due to the C–C bonds of ODA and GO as also that from the C$_{60}$ cage appears at this binding energy. The increased intensity of this peak to 57.9% of the total C1s signal testifies therefore to the successful incorporation of C$_{60}$ into the layered structure. C–O and C–N both contribute to the photoelectron peak centered at 286.1 eV for ODA–GO–ODA and 286.5 eV for ODA–GO–ODA–C$_{60}$; in both spectra its intensity is higher than in the corresponding spectrum of GO starting material due to the chemical grafting of the amino groups of ODA with the epoxy groups of GO. Similarly, to the contribution at 287.5 eV in the ODA–GO–ODA spectrum (at 288.0 eV for ODA–GO–ODA–C$_{60}$) assigned to the epoxy groups of GO is decreased as compared to GO starting material, attesting that the epoxy groups have reacted with the amines via chemical grafting. The peak at 289.3 eV in the ODA–GO–ODA spectrum (at 289.1 eV for ODA–GO–ODA–C$_{60}$) is attributed to carboxyl groups created during the oxidation of GO. The additional peak at 290.5 eV in the ODA–GO–ODA–C$_{60}$ spectrum is due to the C1s shake up features of C$_{60}$ (Maxwell et al., 1994), resulting from

$\pi$–$\pi^*$ transitions excited in the photoemission process, and therefore gives additional support to the presence of fullerenes in the hybrid multilayer (Felicissimo et al., 2009; Gengler et al., 2012a). Finally, upon reduction and subsequent annealing at 700°C the C1s spectral intensity due to oxygen-containing functional groups of graphite oxide (**Figure 5**) is significantly lower as compared to before the treatment, indicating a partial reconstruction of the graphene network expected to result in a better electrical conductivity (see below). It is noteworthy to mention that a new photoemission peak appears in the same spectrum at 282.9 eV; we attribute this peak to C–Si bonds formed as a result of the annealing procedure. Finally, no N1s photoelectron peak was observed for both the ODA–rGO–ODA and ODA–rGO–ODA–C$_{60}$ samples, indicating that annealing did not cause any nitrogen doping of the graphene layers.

The sheet resistance measurements of the ODA–GO–ODA and ODA–GO–ODA–C$_{60}$ multilayers gave values of about 15–20 and 10 MOhm, respectively. These values were remarkably reduced to 20 and 4 kOhm, respectively, after the reduction and annealing treatment was applied in order to convert the deposited GO to graphene. These values are mean values of all measurements performed in different spots on both samples. By taking into account the thickness (as estimated from the AFM analysis) of

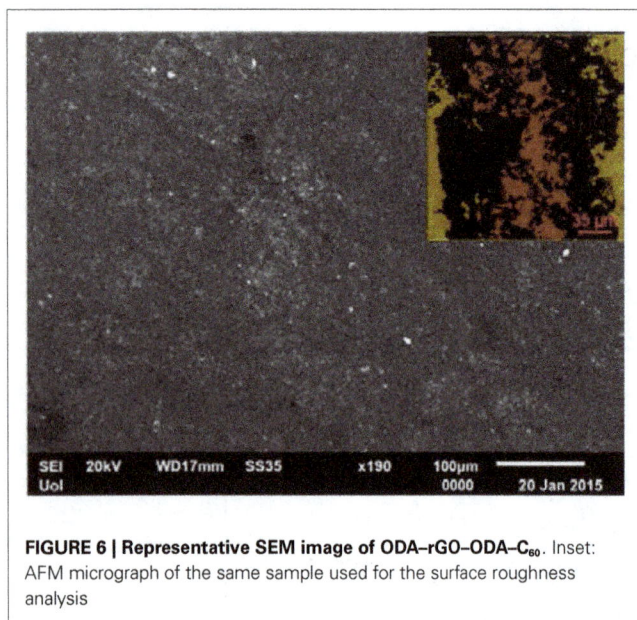

**FIGURE 6 | Representative SEM image of ODA–rGO–ODA–C$_{60}$.** Inset: AFM micrograph of the same sample used for the surface roughness analysis

the films (70 nm for ODA–rGO–ODA and 90 nm for ODA–rGO–ODA–C$_{60}$), the electrical conductivity of the reduced and annealed samples was found to amount to 714 and 2800 S m$^{-1}$ without and with the C$_{60}$, respectively. It is important to note that the films have a quite uniform and smooth surface as revealed by SEM measurements (**Figure 6**). The roughness analysis of the surface shown on the AFM micrograph in the inset of **Figure 6**, gave a mean roughness (RMS) of around 3 nm. The poor electrical conductivity of both the ODA–GO–ODA and ODA–GO–ODA–C$_{60}$ mulitilayers is due to the presence of both ODA and the oxygen-containing groups decorating the GO sheets. However, the reduction and annealing steps improved substantially the electrical conductivity by removing effectively these moieties. The difference in sheet resistance between the two multilayers indicates a strong influence of the presence of C$_{60}$. It is clear that the presence of C$_{60}$ entrapped between graphene layers increases the electrical conductivity by an order of magnitude.

## CONCLUSION

In conclusion, a highly controllable layer-by-layer synthetic approach for the production of a new class of hybrid carbonaceous structures was discussed. A 40-layer thick film consisting of organo-modified GO layers accommodating pure fullerene molecules (C$_{60}$) in the interlayer space was successfully fabricated by using a combination of the LS deposition method with two self-assembly steps from solution. Initially, the effectiveness of this method in terms of coverage, uniformity, and single-layer-level control of the assembly of the first organo-modified GO layer was confirmed by Π–α isotherms and AFM micrographs. For the hybrid, multilayer sample XRD measurements revealed the presence of the fullerene molecules within the interlayer space between the turbostratically layered organo-GO nanosheets and confirmed the high degree of ordering of the produced structure. The existence of fullerenes in the hybrid multilayer system was confirmed by XPS while Raman spectroscopy showed that

the intercalation of the fullerene within the interlayer space of the organo-GO nanosheets did not affect the structure of GO itself. Finally, a considerably improvement (an order of magnitude) in the electrical conductivity of the hybrid multilayer, compared to the organo-graphene analog, was observed due to the presence of C$_{60}$ entrapped between graphene layers. This graphene/fullerene hybrid material constitutes a novel hybrid system that could ideally be used in diverse applications such as transparent electrodes, thin film transistors, supercapacitors, or lubricants.

## ACKNOWLEDGMENTS

This work was co-financed by the European Union (European Social Fund – ESF) and Greek national funds through the Operational Program "Education and Lifelong Learning" of the National Strategic Reference Framework (NSRF) – Research Funding Program: THALES. Investing in knowledge society through the European Social Fund. This work was also supported by the "Graphene-based electronics" research program of the Stichting voor Fundementeel Onderzoek der Materie (FOM), part of the Nederlandse Organisatie voor Wetenschappelijk Onderzoek (NWO).

## REFERENCES

Bourlinos, A. B., Bakandritsos, A., Kouloumpis, A., Gournis, D., Krysmann, M., Giannelis, E. P., et al. (2012a). Gd(III)-doped carbon dots as a dual fluorescent-MRI probe. *J. Mater. Chem.* 22, 23327–23330. doi:10.1039/c2jm35592b

Bourlinos, A. B., Georgakilas, V., Bakandritsos, A., Kouloumpis, A., Gournis, D., and Zboril, R. (2012b). Aqueous-dispersible fullerol-carbon nanotube hybrids. *Mater. Lett.* 82, 48–50. doi:10.1016/j.matlet.2012.05.026

Bourlinos, A. B., Gournis, D., Petridis, D., Szabo, T., Szeri, A., and Dekany, I. (2003). Graphite oxide: chemical reduction to graphite and surface modification with primary aliphatic amines and amino acids. *Langmuir* 19, 6050–6055. doi:10.1021/la026525h

Bourlinos, A. B., Karakassides, M. A., Kouloumpis, A., Gournis, D., Bakandritsos, A., Papagiannouli, I., et al. (2013). Synthesis, characterization and non-linear optical response of organophilic carbon dots. *Carbon N. Y.* 61, 640–643. doi:10.1016/j.carbon.2013.05.017

Cançado, L. G., Jorio, A., Ferreira, E. H. M., Stavale, F., Achete, C. A., Capaz, R. B., et al. (2011). Quantifying defects in graphene via Raman spectroscopy at different excitation energies. *Nano Lett.* 11, 3190–3196. doi:10.1021/nl201432g

Chen, W., Li, S., Chen, C., and Yan, L. (2011). Self-assembly and embedding of nanoparticles by in situ reduced graphene for preparation of a 3D graphene/nanoparticle aerogel. *Adv. Mater.* 23, 5679–5683. doi:10.1002/adma.201102838

Choi, B. G., Park, H., Park, T. J., Yang, M. H., Kim, J. S., Jang, S.-Y., et al. (2010). Solution chemistry of self-assembled graphene nanohybrids for high-performance flexible biosensors. *ACS Nano* 4, 2910–2918. doi:10.1021/nn100145x

Cote, L. J., Kim, F., and Huang, J. X. (2009). Langmuir-Blodgett assembly of graphite oxide single layers. *J. Am. Chem. Soc.* 131, 1043–1049. doi:10.1021/ja806262m

Dekany, I., Kruger-Grasser, R., and Weiss, A. (1998). Selective liquid sorption properties of hydrophobized graphite oxide nanostructures. *Colloid Polym. Sci.* 276, 570–576. doi:10.1007/s003960050283

Dreyer, D. R., Park, S., Bielawski, C. W., and Ruoff, R. S. (2010). The chemistry of graphene oxide. *Chem. Soc. Rev.* 39, 228–240. doi:10.1039/b917103g

Felicissimo, M. P., Jarzab, D., Gorgoi, M., Forster, M., Scherf, U., Scharber, M. C., et al. (2009). Determination of vertical phase separation in a polyfluorene copolymer: fullerene derivative solar cell blend by X-ray photoelectron spectroscopy. *J. Mater. Chem.* 19, 4899–4901. doi:10.1039/b906297a

Ferrari, A. C., Meyer, J. C., Scardaci, V., Casiraghi, C., Lazzeri, M., Mauri, F., et al. (2006). Raman spectrum of graphene and graphene layers. *Phys. Rev. Lett.* 97, 187401. doi:10.1103/PhysRevLett.97.187401

Gengler, R. Y. N. (2010). *A Modified Langmuir Schaefer Method for the Creation of Functional Thin Films.* Ph.D. thesis, University of Groningen, Groningen, 161.

Gengler, R. Y. N., Gournis, D., Aimon, A. H., Toma, L. M., and Rudolf, P. (2012a). The molecularly controlled synthesis of ordered bi-dimensional C$_{60}$ arrays. *Chem. Eur. J.* 18, 7594–7600. doi:10.1002/chem.201103528

Gengler, R. Y. N., Toma, L. M., Pardo, E., Lloret, F., Ke, X. X., Van Tendeloo, G., et al. (2012b). Prussian blue analogues of reduced dimensionality. *Small* 8, 2532–2540. doi:10.1002/smll.201200517

Gengler, R. Y. N., Veligura, A., Enotiadis, A., Diamanti, E. K., Gournis, D., Jozsa, C., et al. (2010). Large-yield preparation of high-electronic-quality graphene by a Langmuir-Schaefer approach. *Small* 6, 35–39. doi:10.1002/smll.200901120

Gomez-Navarro, C., Weitz, R. T., Bittner, A. M., Scolari, M., Mews, A., Burghard, M., et al. (2007). Electronic transport properties of individual chemically reduced graphene oxide sheets. *Nano Lett.* 7, 3499–3503. doi:10.1021/nl072090c

Gournis, D., Jankovic, L., Maccallini, E., Benne, D., Rudolf, P., Colomer, J. F., et al. (2006). Clay-fulleropyrrolidine nanocomposites. *J. Am. Chem. Soc.* 128, 6154–6163. doi:10.1021/ja0579661

Guldi, D. M., and Prato, M. (2000). Excited-state properties of C60 fullerene derivatives. *Acc. Chem. Res.* 33, 695–703. doi:10.1021/ar990144m

Hagerman, M. E., Salamone, S. J., Herbst, R. W., and Payeur, A. L. (2002). Tris(2,2′-bipyridine)ruthenium(II) cations as photoprobes of clay tactoid architecture within hectorite and laponite films. *Chem. Mater.* 15, 443–450. doi:10.1021/cm0209160

Iijima, S. (1991). Helical microtubules of graphitic carbon. *Nature* 354, 56–58. doi:10.1038/354056a0

Junxiang, H., Sato, H., Umemura, Y., and Yamagishi, A. (2005). Sensing of molecular chirality on an electrode modified with a clay-metal complex hybrid film. *J. Phys. Chem. B* 109, 4679–4683. doi:10.1021/jp0451086

Kouloumpis, A., Zygouri, P., Dimos, K., and Gournis, D. (2014). "Layer-by-layer assembly of graphene-based hybrid materials," in *Functionalization of Graphene*, ed. V. Georgakilas (Weinheim: Wiley-VCH Verlag GmbH & Co. KGaA), 359–400.

Lei, Z., Christov, N., and Zhao, X. S. (2011). Intercalation of mesoporous carbon spheres between reduced graphene oxide sheets for preparing high-rate supercapacitor electrodes. *Energy Environ. Sci.* 4, 1866–1873. doi:10.1039/c1ee01094h

Lomeda, J. R., Doyle, C. D., Kosynkin, D. V., Hwang, W. F., and Tour, J. M. (2008). Diazonium functionalization of surfactant-wrapped chemically converted graphene sheets. *J. Am. Chem. Soc.* 130, 16201–16206. doi:10.1021/ja806499w

Martins Ferreira, E. H., Moutinho, M. V. O., Stavale, F., Lucchese, M. M., Capaz, R. B., Achete, C. A., et al. (2010). Evolution of the Raman spectra from single-, few-, and many-layer graphene with increasing disorder. *Phys. Rev. B* 82, 125429. doi:10.1103/PhysRevB.82.125429

Maxwell, A. J., Bruhwiler, P. A., Nilsson, A., Martensson, N., and Rudolf, P. (1994). Photoemission, autoionization, and X-ray adsorption spectroscopy of ultrathin-film C60 on Au(110). *Phys. Rev. B* 49, 10717–10725. doi:10.1103/PhysRevB.49.10717

Michopoulos, A., Kouloumpis, A., Gournis, D., and Prodromidis, M. I. (2014). Performance of layer-by-layer deposited low dimensional building blocks of graphene-prussian blue onto graphite screen-printed electrodes as sensors for hydrogen peroxide. *Electrochim. Acta* 146, 477–484. doi:10.1016/j.electacta.2014.09.031

Patil, A. J., Vickery, J. L., Scott, T. B., and Mann, S. (2009). Aqueous stabilization and self-assembly of graphene sheets into layered bio-nanocomposites using DNA. *Adv. Mater.* 21, 3159–3164. doi:10.1002/adma.200803633

Roh, K.-M., Jo, E.-H., Chang, H., Han, T. H., and Jang, H. D. (2014). High performance dye-sensitized solar cells using graphene modified fluorine-doped tin oxide glass by Langmuir-Blodgett technique. *J. Solid State Chem.* doi:10.1016/j.jssc.2014.04.022

Schniepp, H. C., Li, J. L., McAllister, M. J., Sai, H., Herrera-Alonso, M., Adamson, D. H., et al. (2006). Functionalized single graphene sheets derived from splitting graphite oxide. *J. Phys. Chem. B* 110, 8535–8539. doi:10.1021/jp060936f

Spyrou, K., Kang, L., Diamanti, E. K., Gengler, R. Y., Gournis, D., Prato, M., et al. (2013). A novel route towards high quality fullerene-pillared graphene. *Carbon N. Y.* 61, 313–320. doi:10.1016/j.carbon.2013.05.010

Staudenmaier, L. (1898). Verfahren zur Darstellung der Graphitsaure. *Ber. Deut. Chem. Ges.* 31, 1481. doi:10.1002/cber.18980310237

Stergiou, D. V., Diamanti, E. K., Gournis, D., and Prodromidis, M. I. (2010). Comparative study of different types of graphenes as electrocatalysts for ascorbic acid. *Electrochem. Commun.* 12, 1307–1309. doi:10.1016/j.elecom.2010.07.006

Toma, L. M., Gengler, R. Y. N., Prinsen, E. B., Gournis, D., and Rudolf, P. (2010). A Langmuir-Schaefer approach for the synthesis of highly ordered organoclay thin films. *Phys. Chem. Chem. Phys.* 12, 12188–12197. doi:10.1039/c0cp00286k

Torrisi, F., Hasan, T., Wu, W., Sun, Z., Lombardo, A., Kulmala, T. S., et al. (2012). Inkjet-printed graphene electronics. *ACS Nano* 6, 2992–3006. doi:10.1021/nn2044609

Umemura, Y., Shinohara, E., and Schoonheydt, R. A. (2009). Preparation of Langmuir-Blodgett films of aligned sepiolite fibers and orientation of methylene blue molecules adsorbed on the film. *Phys. Chem. Chem. Phys.* 11, 9804–9810. doi:10.1039/b817635c

Umemura, Y., Yamagishi, A., Schoonheydt, R., Persoons, A., and De Schryver, F. (2001). Fabrication of hybrid films of alkylammonium cations (CnH2n+1NH3+; n=4-18) and a smectite clay by the Langmuir-Blodgett method. *Langmuir* 17, 449–455. doi:10.1021/la0011376

Umemura, Y., Yamagishi, A., Schoonheydt, R., Persoons, A., and De Schryver, F. (2002). Langmuir-Blodgett films of a clay mineral and ruthenium(II) complexes with a noncentrosymmetric structure. *J. Am. Chem. Soc.* 124, 992–997. doi:10.1021/ja016005t

Wang, D., Kou, R., Choi, D., Yang, Z., Nie, Z., Li, J., et al. (2010). Ternary self-assembly of ordered metal oxide-graphene nanocomposites for electrochemical energy storage. *ACS Nano* 4, 1587–1595. doi:10.1021/nn901819n

Yamamoto, T., Umemura, Y., Sato, O., and Einaga, Y. (2004a). Photomagnetic Co-Fe Prussian blue thin films fabricated by the modified Langmuir-Blodgett technique. *Chem. Lett.* 33, 500–501. doi:10.1246/cl.2004.500

Yamamoto, T., Umemura, Y., Sato, O., and Einaga, Y. (2004b). Photoswitchable magnetic films: Prussian blue intercalated in Langmuir-Blodgett films consisting of an amphiphilic azobenzene and a clay mineral. *Chem. Mater.* 16, 1195–1201. doi:10.1021/cm035223d

Yin, S., Niu, Z., and Chen, X. (2012). Assembly of graphene sheets into 3D macroscopic structures. *Small* 8, 2458–2463. doi:10.1002/smll.201102614

Yoshida, J., Saruwatari, K., Kameda, J., Sato, H., Yamagishi, A., Sun, L., et al. (2006). Electron transfer through clay monolayer films fabricated by the Langmuir-Blodgett technique. *Langmuir* 22, 9591–9597. doi:10.1021/la061668f

Yu, D., and Dai, L. (2009). Self-assembled graphene/carbon nanotube hybrid films for supercapacitors. *J. Phys. Chem. Lett.* 1, 467–470. doi:10.1021/jz9003137

Zheng, Q., Ip, W. H., Lin, X., Yousefi, N., Yeung, K. K., Li, Z., et al. (2011a). Transparent conductive films consisting of ultralarge graphene sheets produced by Langmuir-Blodgett assembly. *ACS Nano* 5, 6039–6051. doi:10.1021/nn2018683

Zheng, Q. B., Ip, W. H., Lin, X. Y., Yousefi, N., Yeung, K. K., Li, Z. G., et al. (2011b). Transparent conductive films consisting of ultra large graphene sheets produced by Langmuir-Blodgett assembly. *ACS Nano* 5, 6039–6051. doi:10.1021/nn2018683

**Conflict of Interest Statement:** The authors declare that the research was conducted in the absence of any commercial or financial relationships that could be construed as a potential conflict of interest.

# 11

# Graphene-based transparent electrodes for hybrid solar cells

*Pengfei Li, Caiyun Chen, Jie Zhang, Shaojuan Li, Baoquan Sun and Qiaoliang Bao\**

*Jiangsu Key Laboratory for Carbon-Based Functional Materials and Devices, Collaborative Innovation Center of Suzhou Nano Science and Technology, Institute of Functional Nano and Soft Materials (FUNSOM), Soochow University, Suzhou, China*

**Edited by:**
Xiaobin Fan, Tianjin University, China

**Reviewed by:**
Thiagarajan Soundappan, Washington University in St. Louis, USA
Qiang Zhang, Tsinghua University, China
Junyi Ji, Sichuan University, China

**\*Correspondence:**
Qiaoliang Bao, Jiangsu Key Laboratory for Carbon-Based Functional Materials and Devices, Collaborative Innovation Center of Suzhou Nano Science and Technology, Institute of Functional Nano and Soft Materials (FUNSOM), Soochow University, Suzhou 215123, China
e-mail: qlbao@suda.edu.cn, qiaoliang.bao@gmail.com

The graphene-based transparent and conductive films were demonstrated to be cost-effective electrodes working in organic–inorganic hybrid Schottky solar cells. Large area graphene films were produced by chemical vapor deposition on copper foils and transferred onto glass as transparent electrodes. The hybrid solar cell devices consist of solution-processed poly (3,4-ethlene-dioxythiophene): poly (styrenesulfonate) (PEDOT: PSS), which is sandwiched between silicon wafer and graphene electrode. The solar cells based on graphene electrodes, especially those doped with $HNO_3$, have comparable performance to the reference devices using commercial indium tin oxide (ITO). Our work suggests that graphene-based transparent electrode is a promising candidate to replace ITO.

**Keywords: graphene, hybrid Schottky solar cell, transparent electrode, indium tin oxide**

## INTRODUCTION

Solar energy is a potential substitute for fossil fuels in the future because it is a renewable energy source and inexhaustible. Over a few decades, the photovoltaic industry has been grown rapidly following the improvements in the efficiency and the demand for alternative energy resources. Commercial silicon photovoltaic with the power conversion efficiency (PCE) of above 20% plays a dominate role but the high-manufacturing cost is still a major issue for large-scale implementation. Many efforts have been made to reduce the cost of photovoltaic devices. The combination of organic and inorganic materials to form a hybrid solar cell is very promising approach as it marries lower process cost of organic materials with the good performance of silicon. However, the conventional anode for organic materials, indium tin oxide (ITO), is becoming expensive because of the dwindling supplies of indium and the growth of the environmental costs (Wang et al., 2008; Novoselov et al., 2012; Son et al., 2012). The silver electrodes demonstrated in the laboratory are also not suitable for large-scale commercial usage. There are huge demands for developing new electrode materials with lower cost and comparable performance.

Graphene has been proposed to be an effective transparent electrode to replace ITO in solar cell (Bao et al., 2009; Park et al., 2011; Wang et al., 2011a; Miao et al., 2012) as graphene exhibits excellent properties such as low-sheet resistance, high transmittance, good mechanical property, and good thermal and chemical stability (Yin et al., 2014). Graphene has very high-carrier mobility as charge carriers in it are delocalized over large areas, resulting in an unencumbered platform for electron/hole transport. High Fermi-velocity and the ability to be doped chemically contribute to extremely high in-plane conductivities. As early as 2007, Wang et al. (2008) fabricated polymer solar cells using reduced graphene oxide as transparent electrode and achieved a PCE of 0.26%. Afterwards, chemical vapor deposition (CVD) approach has been used by many groups to synthesize single- or few-layer graphene films with large area for energy harvesting applications, which is a significant advance in this field. Gomez De Arco et al. (2010) reported the continuous, highly flexible, and transparent graphene films produced by CVD as transparent conductive electrodes in organic solar cells. The efficiency of organic solar cells with graphene electrode was 1.18%, which is close to that of organic solar cells with ITO electrode (~1.27%). In 2011, Wang et al. (2011a) used layer-by-layer transfer method to fabricate multilayer CVD graphene films with less defects and lower sheet resistance. The organic solar cells with the electrode of four layers graphene have an improved PCE up to 2.5%, which is 83.3% of the PCE of ITO-based devices. For the hybrid solar cell, Wu et al. (2013) demonstrated the use of graphene as transparent conductive electrodes with the structure of graphene/organic/silicon, which has a PCE of 10%. However, the solar cell has a very small device area of about 0.1 cm$^2$.

In this work, we used graphene films as transparent electrodes to work in organic–inorganic hybrid Schottky solar cells with a relatively large device area. The hybrid solar cells based on solution-processed poly (3,4-ethlene-dioxythiophene): poly (styrenesulfonate) (PEDOT: PSS) in combination with silicon wafer are fabricated through a simple and low-cost process (Li et al., 2010; Song et al., 2012; Liu et al., 2013; Shen et al., 2013; Zhu et al., 2013). It was found that the surface doping of graphene film can effectively improve the device performance. This work suggests that the application of graphene-based transparent electrode can be extended to a wide range of new optoelectronic devices.

## MATERIALS AND METHODS
### GROWTH AND TRANSFER OF GRAPHENE FILMS
Large area graphene films were produced using CVD on the copper foils as the method described in the literature (Yu et al., 2010; Manu et al., 2011). We used a modified transfer method to prepared the graphene-based transparent electrode and it includes the following steps: (1) deposition of PMMA and curing; (2) etching of copper foil in FeCl$_3$ solution and rinsing in deionized (DI) water for three times; (3) rinsing PMMA/graphene in DI water three times; (4) fishing of PMMA/graphene onto glass substrate; (5) re-deposition of one drop of PMMA solution and curing; and (6) removal of PMMA with acetone. The as-transferred graphene films were doped by putting them in the vapor of HNO$_3$ (concentration of 69%) for 10 s.

### MATERIALS CHARACTERIZATIONS
The morphology of the graphene film was checked by optical microscopy (Leica DM2700 M). The domain of the graphene film was indicated by scanning electron microscope (SEM Quanta 200 FEG). The quality of the graphene film was verified by Raman spectroscopy (HORIBA JOBIN-YVON LABRAM HR800). The optical property was investigated by UV–visible spectrometer (Perkin Elmer, Lambda 750). The work function of intrinsic and doped graphene film is measured using ultraviolet photoelectron spectroscopy (UPS, KRATOS Analytical).

### PREPARATION OF SILICON WAFER AND PEDOT: PSS
The organic–inorganic hybrid solar cells were fabricated by *n*-doped Si and PEDOT: PSS. The Si wafer was first cleaned for half an hour using acetone, ethanol, and DI water. Following, the wafer was treated by chlorination and alkylation. Highly conductive PEDOT: PSS (CLEVIOS PH 1000) was incorporated with 5% wt DMSO to increase conductivity and 1% wt Triton was used as surface activator. The resulting solution was stirred thoroughly to ensure that the different chemicals are uniformly mixed (Shen et al., 2010; Zhang et al., 2011).

### FABRICATION OF SOLAR CELLS
Using physics vapor deposition (PVD), 0.6 nm of LiF and 200 nm of Al electrode were deposited on the back of the silicon wafer. Then PEDOT: PSS was spin-coated onto the silicon wafer and the graphene/glass. (Speed: 4000 rpm, time: 1 min) The resulted organic films (about 70–80 nm thick) was then annealed at 125°C for 30 min in a glove box (Shen et al., 2013; Zhu et al., 2013). The last step is the encapsulation of the solar cell devices, in which a clamp and AB glue were used to firmly stick the graphene/glass and silicon wafer together.

## RESULTS
### PREPARATION AND CHARACTERIZATIONS OF GRAPHENE FILMS
**Figure 1** schematically shows the procedures to prepare graphene transparent electrode. Graphene grown on copper was transferred onto glass substrate using PMMA as a supporting host. The PMMA layer can be further cleaned by annealing the resulted sample in inert gas at above 350°C. Multilayer graphene films can be prepared by repeating the procedures in **Figures 1A–D** and stacking the PMMA supported graphene onto graphene covered glass.

**Figure 2A** shows the optical image of single layer graphene transferred onto SiO$_2$ wafer. It is found that the graphene film is generally uniform with fewer cracks. **Figure 2B** shows the SEM image of graphene film grown on copper substrate, indicating very large graphene domains on copper grains. In order to further check the quality of as-produced graphene film, Raman spectrum was measured, as shown in **Figure 2C**. We can clearly see two characteristic peaks: 2D peak at 2687 cm$^{-1}$, which is originated from the phonon resonance and G peak at 1579 cm$^{-1}$, which is correlated to the crystallinity of graphene (Ferrari et al., 2006). The intensity of 2D peak is about two times that of G peak, suggesting that the graphene synthesized on the copper foils is single layer. There is no obvious D peak at around 1350 cm$^{-1}$, indicating very less defects or disorders. **Figure 2D** displays the Raman mapping image of intensity ratio of G/2D over an area of 5 μm × 5 μm. It is revealed that G/2D ratio is almost below 0.5 over the whole area, indicating the nature of single atomic layer (Bao et al., 2011).

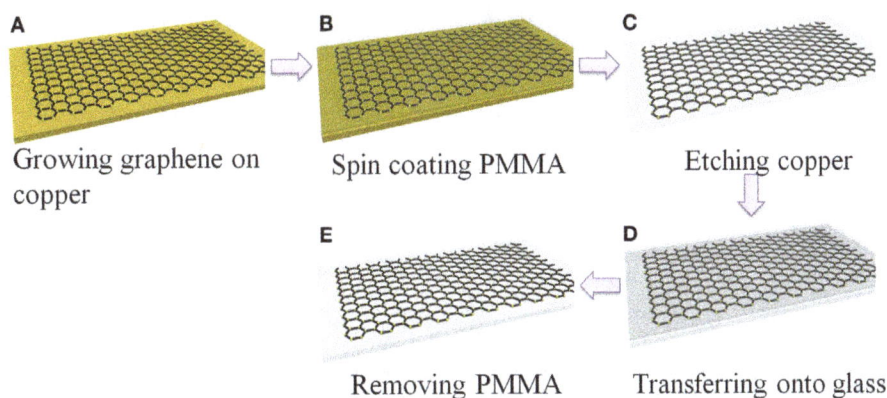

**FIGURE 1 | Schematic of transfer graphene onto glass substrate**. (A) Synthesis of graphene on copper foil, (B) spin-coating of PMMA onto the Cu foil to protect graphene, (C) etching of Cu, (D) transfer of graphene/PMMA film onto glass, (E) removal of PMMA in acetone.

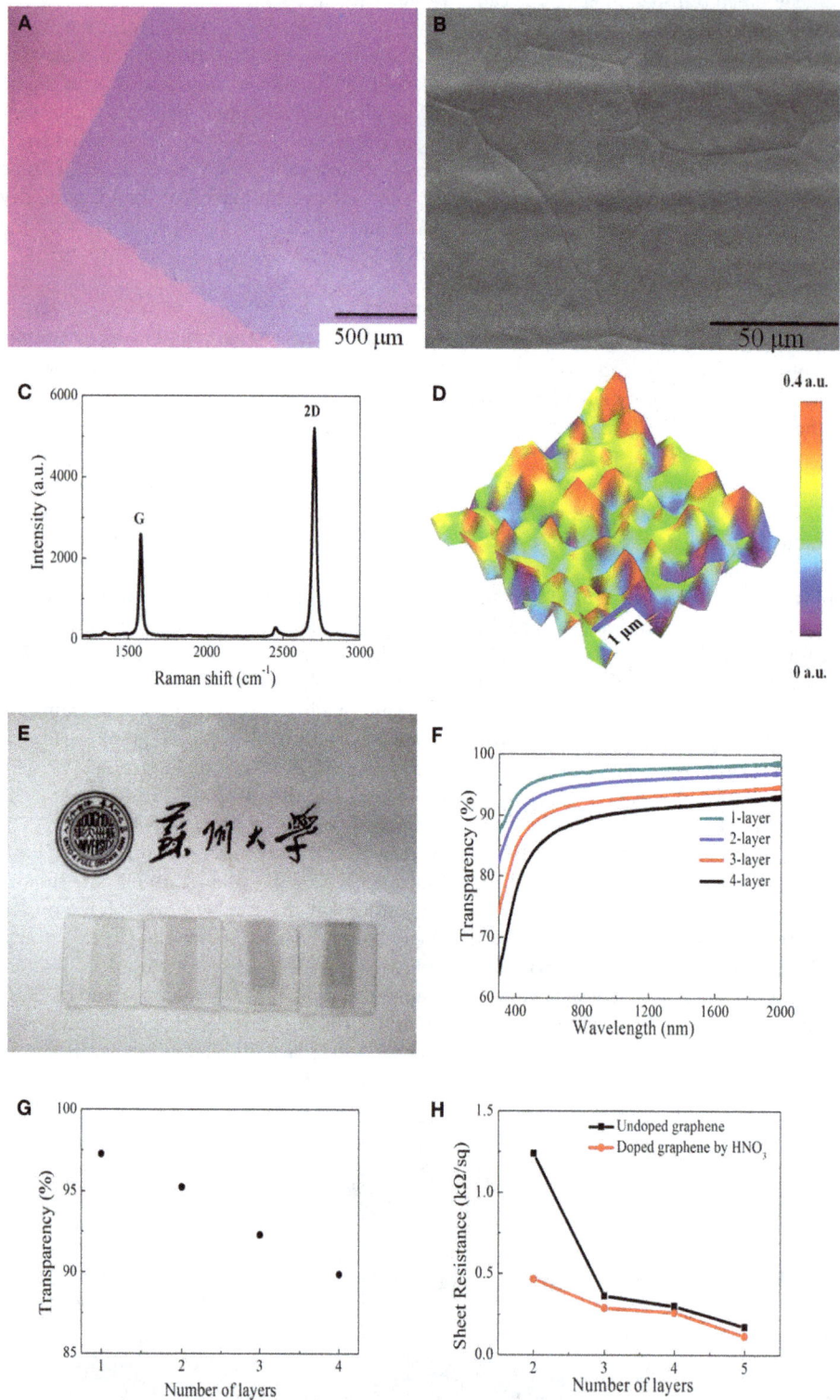

**FIGURE 2 | (A)** Optical image of monolayer graphene on SiO₂ substrate. **(B)** SEM of graphene film grown on copper substrate. **(C)** Raman spectrum of monolayer graphene measured on silicon substrate. **(D)** G/2D peak Raman map of monolayer graphene. **(E)** Photograph of one to four layers graphene films covering on quartz substrates. **(F)** Transmittance spectra of graphene with different numbers of layers. **(G)** Optical transparency of graphene film as a function of number of layers. **(H)** Sheet resistance of the undoped and doped graphene films with different numbers of layers.

Figure 2E shows the photograph of one to four layers of graphene on quartz substrates. With the increasing of the number of layers, the contrast becomes darker from left to right. Figure 2F shows the transmittance spectra from visible to near-infrared range. The smoothness of the transmittance spectrum for each graphene sample suggests that the films are generally uniform. The transmittance decreases as the number of layers (see Figure 2G) in graphene increases, i.e., each additional layer results in a decrease of about 2.3% in the transmittance, which is due to the universal absorption of graphene film shows in Figure 2F. We also found that the sheet resistance of both undoped and doped graphene is reduced while increasing the number of layers, as shown in Figure 2H.

## CHARACTERIZATIONS OF THE SOLAR CELLS

Figure 3A shows the schematic structure of the hybrid Schottky solar cells with graphene electrode. The cross-section of the device

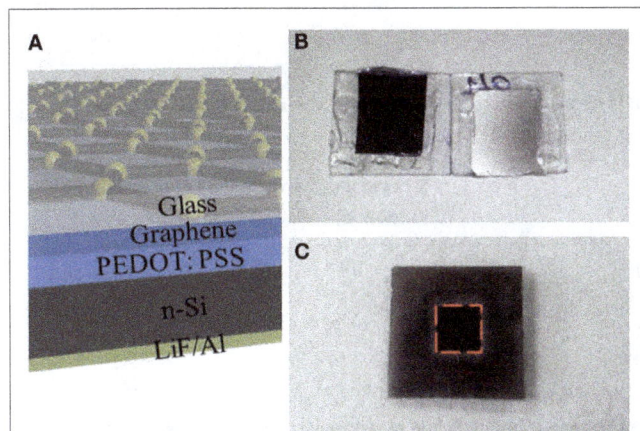

FIGURE 3 | (A) Schematic diagrams of hybrid solar cells using graphene as electrode. The cross-section of solar cell shows different functional layers. (B) Photograph of the real hybrid solar cell device after encapsulation. Left: top view; right: bottom view. (C) Photograph of solar cell device covered with a mask.

reveals a vertical structure where PEDOT: PSS is sandwiched between and $n$-type silicon. Figure 3B shows the photographs of real solar cell device. Both the top view and bottom view are presented. In order to standardize the photovoltaic measurements, a mask is used to expose an area of $0.5\,cm \times 0.5\,cm$ for light illumination, as shown in Figure 3C.

Figure 4 shows the output characteristics and external quantum efficiency (EQE) of doped graphene-based hybrid solar cells under AM1.5G illumination. The $J_{SC}$ of the devices is above $22\,mA/cm^2$ and the average circuit voltage ($V_{oc}$) is about $0.55 \pm 0.2\,V$. The detailed performance results are summarized in Table 1. For 2-layer and 3-layer graphene films, the same fill factor is obtained; however, 3-layer graphene gives higher PCE and the highest $J_{SC}$ ($25.09\,mA/cm^2$). Overall, 4-layer graphene gives the highest fill factor and PCE. It may result from a lower sheet resistance in the 3-layer and 4-layer graphene. It is generally found that the fill factor and PCE are increased while increasing the number of graphene layers. This may be attributed to the reduction of the sheet resistance as well as the improved contact between graphene layers and PEDOT: PSS. This trend agrees with the observed trend in literatures (Wang et al., 2011a). However, it should be noted that the performance of different batches of devices is also different due to many unpredictable factors during the device fabrication. By optimizing the graphene film quality, balancing the optical transparency with conductivity and the fill factor, it is possible to obtain a relatively high efficiency from these hybrid Schottky solar cells.

In order to further study the effect of sheet resistance and work function of graphene on the solar cell performance, we compare the devices based on undoped graphene with those based on doped graphene. It is well known that $HNO_3$ could cause $p$-type doping in graphene, which improves the conductivity and charge transfer efficiency (Wang et al., 2011b; Feng et al., 2012; Miao et al., 2012; Cui et al., 2013). Figure 5A shows the photovoltaic characteristics and Table 2 lists the detailed performance parameters. Figure 5B shows the schematic energy diagram of the hybrid solar cell device. The down-shift of Fermi level in graphene ($p$-type doping) will facilitate the collection of

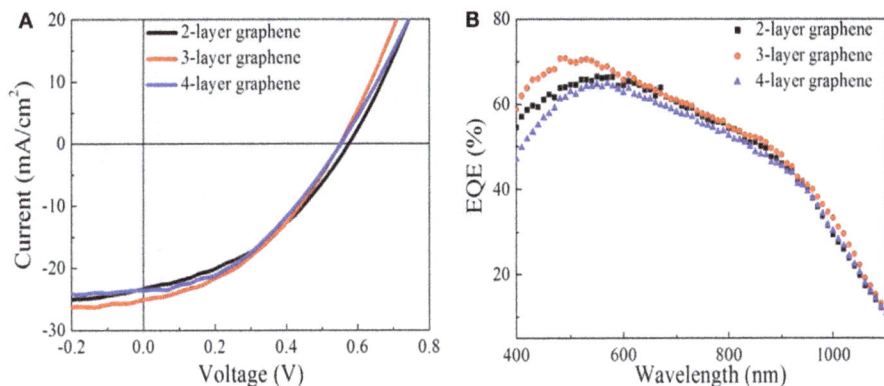

FIGURE 4 | Characterizations of graphene-based hybrid solar cells. (A) The current density versus voltage (J–V) curve of solar cells with two to four layers graphene electrodes. (B) External quantum efficiency of the solar cells.

photo-excited holes from PEDOT: PSS/Si hetero-junction. Interestingly, the PCE of graphene-based solar cell is improved by 40% after doping with $HNO_3$, as illustrated in **Table 2**. This is understandable because the doping treatment causes the sheet resistance to decrease from 320 to 247 $\Omega/\square$ (see **Table 3**). Another consequence of doping is that the short-circuit current ($J_{sc}$) of the graphene-based solar cell increases from 22.88 to 25.08 $mA/cm^2$ (see **Table 2**).

The engineering of work function of graphene by doping also plays an important role to affect the final device performance. As shown in **Table 3**, the measured work function of 3-layer graphene is increased from 4.36 to 4.81 eV. The $V_{oc}$ of the solar cell before the doping of graphene is lower than that of ITO-based solar cell, however, it is increased from 0.48 to 0.55 V but after the doping, which is even higher than that of ITO-based solar cell. As a result, the PCE of graphene-based solar cell is enhanced from 3.92 to 5.48%. Using ITO-based hybrid solar cell as a reference, it is found that doped graphene electrode can deliver comparable energy conversion performance (Wang et al., 2011a), 3-layer graphene electrode gives a PCE of 5.48%, which is about 87% of that of ITO. There is still much space to improve for graphene-based electrode as the sheet resistance of ITO is ~100 $\Omega/\square$ or less. However, considering rising price of ITO because of the dwindling supplies of indium, graphene films, with additional chemical tunability, could be a very promising candidate to replace ITO as transparent electrode in solar cells.

The degradation in solar cells is also a very important figure of merit to evaluate the device performance. We tested the ITO and graphene electrode solar cells after 1 month storage in dry cabinet. **Figure 6** and **Table 4** show the results before and after 1 month. It is found that both the solar cell with graphene and that with ITO experience a decrease in PCE after a month. The fill factor of both solar cells is decreased by almost 6.3%. The $J_{sc}$ of ITO-based solar cell is decreased from 25.64 to 22.4 $mA/cm^2$ and the $V_{oc}$ is decreased from 0.53 to 0.50 V. In contrast, the $V_{oc}$ and $J_{sc}$ of the graphene-based device have a relatively smaller decrease, from 0.55 to 0.54 V and from 25.09 to 23.49 $mA/cm^2$, respectively. Consequently, the PCE of ITO-based solar cell is decreased by 22.7% and that of graphene-based solar cell is decreased by 11.7% only. Therefore, it is concluded that graphene-based hybrid solar cell device degrade slower than ITO-based hybrid solar cell. It may correlate to good chemical compatibility of graphene with organic molecular such as PEDOT: PSS and this is an interesting topic in demand of further investigation.

**Table 1 | The parameters of solar cells made from doped graphene electrodes with different thicknesses.**

|  | 2-Layer graphene | 3-Layer graphene | 4-Layer graphene |
| --- | --- | --- | --- |
| $V_{oc}$ (V) | 0.57 | 0.55 | 0.55 |
| $J_{sc}$ ($mA/cm^2$) | 23.32 | 25.08 | 22.24 |
| Fill factor | 0.40 | 0.39 | 0.47 |
| PCE (%) | 5.35 | 5.48 | 5.76 |

**Table 2 | The parameters of the different solar cells based on graphene and ITO electrodes.**

|  | ITO | 3-Layer graphene (undoped) | 3-Layer graphene (doped) |
| --- | --- | --- | --- |
| $V_{oc}$ (V) | 0.52 | 0.48 | 0.55 |
| $J_{sc}$ ($mA/cm^2$) | 26.38 | 22.88 | 25.08 |
| Fill factor | 0.46 | 0.36 | 0.39 |
| PCE (%) | 6.33 | 3.92 | 5.48 |

**Table 3 | The work function and sheet resistance of 3-layer graphene before and after doped by $HNO_3$.**

| 3-Layer graphene | Undoped graphene | Doped graphene |
| --- | --- | --- |
| Work function (eV) | 4.36 | 4.81 |
| Sheet resistance ($\Omega/\square$) | 320 $\Omega/\square$ | 247 $\Omega/\square$ |

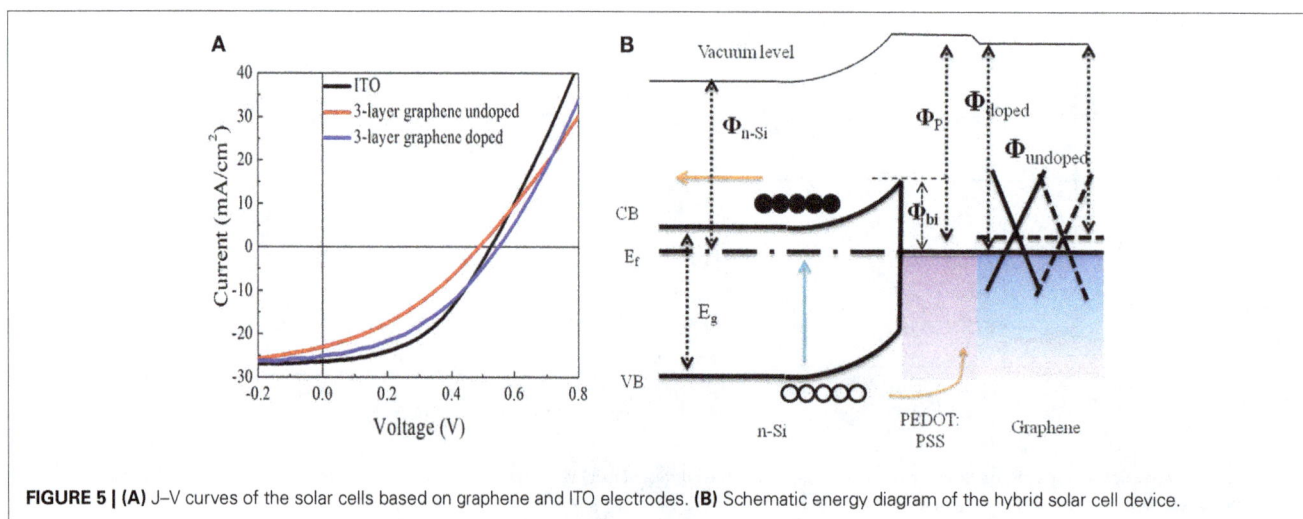

**FIGURE 5 | (A)** J–V curves of the solar cells based on graphene and ITO electrodes. **(B)** Schematic energy diagram of the hybrid solar cell device.

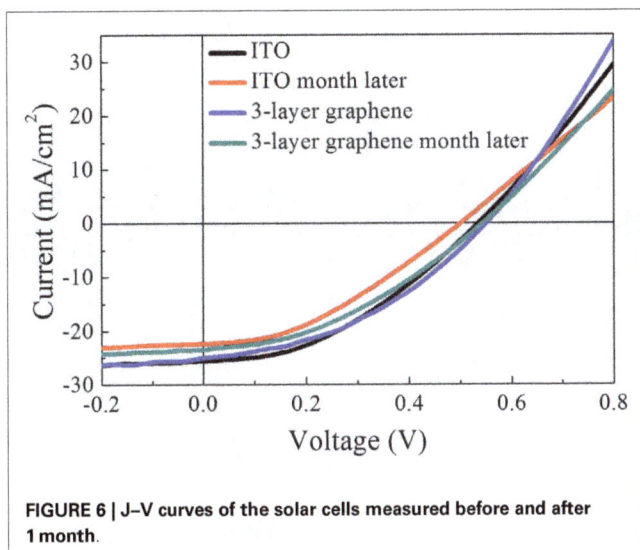

**FIGURE 6 | J–V curves of the solar cells measured before and after 1 month.**

**Table 4 | The degradation of different solar cells with ITO and graphene electrodes.**

|  | ITO | ITO after 1 month | 3-Layer graphene | 3-Layer graphene after 1 month |
|---|---|---|---|---|
| $V_{OC}$ (V) | 0.53 | 0.5 | 0.55 | 0.54 |
| $J_{SC}$ (mA/cm$^2$) | 25.64 | 22.4 | 25.09 | 23.49 |
| Fill factor | 0.40 | 0.37 | 0.40 | 0.38 |
| PCE (%) | 5.38 | 4.16 | 5.48 | 4.84 |

## CONCLUSION

The graphene-based thin films were successfully applied as transparent electrodes working in organic–inorganic hybrid Schottky solar cells. In comparison to similar solar cell devices using ITO as electrodes, graphene-based solar cells can deliver comparable photovoltaic performance. It is found that the chemical doping by $HNO_3$ can effectively increase the work function, reduce the sheet resistance, which in turn improve the solar cell performance. The lifetime study suggests that graphene-based solar cell experiences smaller degradation than ITO-based device. This work indicates that graphene-based electrodes have the potential to replace ITO in a wide range of optoelectronic devices.

## ACKNOWLEDGMENTS

This work was supported by the National High Technology Research and Development Program of China (863 Program) (Grant No. 2013AA031903), the youth 973 program (2015CB932700), the National Natural Science Foundation of China (Grant No. 51222208, 51290273), the Doctoral Fund of Ministry of Education of China (Grant No. 20123201120026).

## REFERENCES

Bao, Q., Zhang, H., Ni, Z., Wang, Y., Polavarapu, L., Shen, Z., et al. (2011). Monolayer graphene as a saturable absorber in a mode-locked laser. *Nano Res.* 4, 297–307. doi:10.1364/OL.38.001745

Bao, Q., Zhang, H., Wang, Y., Ni, Z., Yan, Y., Shen, Z. X., et al. (2009). Atomic-layer graphene as a saturable absorber for ultrafast pulsed lasers. *Adv. Funct. Mater.* 19, 3077–3083. doi:10.1002/adfm.200901007

Cui, T., Lv, R., Huang, Z.-H., Chen, S., Zhang, Z., Gan, X., et al. (2013). Enhanced efficiency of graphene/silicon heterojunction solar cells by molecular doping. *J. Mater. Chem. A* 1, 5736–5740. doi:10.1039/c3ta01634j

Feng, T., Xie, D., Lin, Y., Zhao, H., Chen, Y., Tian, H., et al. (2012). Efficiency enhancement of graphene/silicon-pillar-array solar cells by HNO3 and PEDOT-PSS. *Nanoscale* 4, 2130–2133. doi:10.1039/c2nr12001a

Ferrari, A. C., Meyer, J. C., Scardaci, V., Casiraghi, C., Lazzeri, M., Mauri, F., et al. (2006). Raman spectrum of graphene and graphene layers. *Phys. Rev. Lett.* 97, 187401. doi:10.1103/physrevlett.97.187401

Gomez De Arco, L., Zhang, Y., Schlenker, C. W., Ryu, K., Thompson, M. E., and Zhou, C. (2010). Continuous, highly flexible, and transparent graphene films by chemical vapor deposition for organic photovoltaics. *ACS Nano* 4, 2865–2873. doi:10.1021/nn901587x

Li, X., Zhu, H., Wang, K., Cao, A., Wei, J., Li, C., et al. (2010). Graphene-on-silicon Schottky junction solar cells. *Adv. Mater. Weinheim* 22, 2743–2748. doi:10.1002/adma.200904383

Liu, D., Zhang, Y., Fang, X., Zhang, F., Song, T., and Sun, B. (2013). An 11%-power-conversion-efficiency organic-inorganic hybrid solar cell achieved by facile organic passivation. *Electron Dev. Lett. IEEE* 34, 345–347. doi:10.1109/LED.2013.2239255

Manu, J., Xuan, L. C. H. Y., Qiaoliang, B., Tat, T. C., Ping, L. K., and Barbaros, O. (2011). Controlled hydrogenation of graphene sheets and nanoribbons. *ACS Nano* 5, 888–896. doi:10.1021/nn102034y

Miao, X., Tongay, S., Petterson, M. K., Berke, K., Rinzler, A. G., Appleton, B. R., et al. (2012). High efficiency graphene solar cells by chemical doping. *Nano Lett.* 12, 2745–2750. doi:10.1021/nl204414u

Novoselov, K. S., Fal, V., Colombo, L., Gellert, P., Schwab, M., and Kim, K. (2012). A roadmap for graphene. *Nature* 490, 192–200. doi:10.1038/nature11458

Park, H., Brown, P. R., Bulović, V., and Kong, J. (2011). Graphene as transparent conducting electrodes in organic photovoltaics: studies in graphene morphology, hole transporting layers, and counter electrodes. *Nano Lett.* 12, 133–140. doi:10.1021/nl2029859

Shen, X., Sun, B., Yan, F., Zhao, J., Zhang, F., Wang, S., et al. (2010). High-performance photoelectrochemical cells from ionic liquid electrolyte in methyl-terminated silicon nanowire arrays. *ACS Nano* 4, 5869–5876. doi:10.1021/nn101980x

Shen, X., Zhu, Y., Song, T., Lee, S.-T., and Sun, B. (2013). Hole electrical transporting properties in organic-Si Schottky solar cell. *Appl. Phys. Lett.* 103, 013504. doi:10.1063/1.4812988

Son, D. I., Kwon, B. W., Park, D. H., Seo, W.-S., Yi, Y., Angadi, B., et al. (2012). Emissive ZnO-graphene quantum dots for white-light-emitting diodes. *Nat. Nanotechnol.* 7, 465–471. doi:10.1038/nnano.2012.71

Song, T., Lee, S.-T., and Sun, B. (2012). Silicon nanowires for photovoltaic applications: the progress and challenge. *Nano Energy* 1, 654–673. doi:10.1016/j.nanoen.2012.07.023

Wang, X., Zhi, L., and Müllen, K. (2008). Transparent, conductive graphene electrodes for dye-sensitized solar cells. *Nano Lett.* 8, 323–327. doi:10.1021/nl072838r

Wang, Y., Tong, S. W., Xu, X. F., Özyilmaz, B., and Loh, K. P. (2011a). Interface engineering of layer-by-layer stacked graphene anodes for high-performance organic solar cells. *Adv. Mater. Weinheim* 23, 1514–1518. doi:10.1002/adma.201190044

Wang, Y., Tong, S. W., Xu, X. F., Ozyilmaz, B., and Loh, K. P. (2011b). Interface engineering of layer-by-layer stacked graphene anodes for high-performance organic solar cells. *Adv. Mater.* 23, 1514–1518. doi:10.1002/adma.201190044

Wu, Y., Zhang, X., Jie, J., Xie, C., Zhang, X., Sun, B., et al. (2013). Graphene transparent conductive electrodes for highly efficient silicon nanostructures-based hybrid heterojunction solar cells. *J. Phys. Chem. C* 117, 11968–11976. doi:10.1021/jp402529c

Yin, Z., Zhu, J., He, Q., Cao, X., Tan, C., Chen, H., et al. (2014). Graphene-based materials for solar cell applications. *Adv. Energy Mater.* 4, 1–19. doi:10.1002/aenm.201300574

Yu, W., Xiangfan, X., Jiong, L., Ming, L., Qiaoliang, B., Barbaros, Ö., et al. (2010). Toward high throughput interconvertible graphane-to-graphene growth and patterning. *ACS Nano* 4, 6146–6152. doi:10.1021/nn1017389

Zhang, F., Sun, B., Song, T., Zhu, X., and Lee, S. (2011). Air stable, efficient hybrid photovoltaic devices based on poly(3-hexylthiophene) and silicon nanostructures. *Chem. Mater.* 23, 2084–2090. doi:10.1021/cm103221a

Zhu, Y., Song, T., Zhang, F., Lee, S.-T., and Sun, B. (2013). Efficient organic-inorganic hybrid Schottky solar cell: the role of built-in potential. *Appl. Phys. Lett.* 102, 113504. doi:10.1063/1.4796112

**Conflict of Interest Statement:** The authors declare that the research was conducted in the absence of any commercial or financial relationships that could be construed as a potential conflict of interest.

# 12

# Direct growth of $Ge_{1-x}Sn_x$ films on Si using a cold-wall ultra-high vacuum chemical-vapor-deposition system

*Aboozar Mosleh[1,2]\*, Murtadha A. Alher[2,3], Larry C. Cousar[1,4], Wei Du[2], Seyed Amir Ghetmiri[1,2], Thach Pham[2], Joshua M. Grant[5], Greg Sun[6], Richard A. Soref[6], Baohua Li[4], Hameed A. Naseem[2] and Shui-Qing Yu[2]*

[1] Microelectronics-Photonics Graduate Program (μEP), University of Arkansas, Fayetteville, AR, USA
[2] Department of Electrical Engineering, University of Arkansas, Fayetteville, AR, USA
[3] Mechanical Engineering Department, University of Karbala, Karbala, Iraq
[4] Arktonics, LLC, Fayetteville, AR, USA
[5] Engineering-Physics Department, Southern Arkansas University, Magnolia, AR, USA
[6] Department of Engineering, University of Massachusetts Boston, Boston, MA, USA

**Edited by:**
Jifeng Liu, Dartmouth College, USA

**Reviewed by:**
Fabio Iacona, National Research Council, Italy
Christophe Labbé, Ecole Nationale Supérieure d'Ingénieurs de Caen, France

**\*Correspondence:**
Aboozar Mosleh, Engineering Research Center (ENRC), 700 Research Center Boulevard, Fayetteville, AR 72701, USA
e-mail: amosleh@gmail.com

Germanium–tin alloys were grown directly on Si substrate at low temperatures using a cold-wall ultra-high vacuum chemical-vapor-deposition system. Epitaxial growth was achieved by adopting commercial gas precursors of germane and stannic chloride without any carrier gases. The X-ray diffraction analysis showed the incorporation of Sn and that the $Ge_{1-x}Sn_x$ films are fully epitaxial and strain relaxed. Tin incorporation in the Ge matrix was found to vary from 1 to 7%. The scanning electron microscopy images and energy-dispersive X-ray spectra maps show uniform Sn incorporation and continuous film growth. Investigation of deposition parameters shows that at high flow rates of stannic chloride the films were etched due to the production of HCl. The photoluminescence study shows the reduction of band-gap from 0.8 to 0.55 eV as a result of Sn incorporation.

**Keywords: chemical-vapor-deposition, Si photonics, Ge alloys, photoluminescence, Ge–Sn**

## INTRODUCTION

The discovery and development of $Ge_{1-x}Sn_x$ epitaxy technology has enabled silicon photonics to be explored in a different scope of a material platform. The ability of band-gap engineering by varying Sn mole fraction, along with its compatibility to the complementary metal–oxide–semiconductor (CMOS) process, has paved the way for highly competitive Si-based near and mid-infrared optoelectronic devices. Recent reports on the fabrication and characterization of high performance $Ge_{1-x}Sn_x$ devices such as modulators (Kouvetakis et al., 2005), photodetectors (Conley et al., 2014a,b), and light emitting diodes (LEDs) (Du et al., 2014a) show great potential for $Ge_{1-x}Sn_x$ being adopted by industry in the near future. Cutting-edge reports on $Ge_{1-x}Sn_x$, achieving a direct band-gap group IV alloy (Du et al., 2014b; Ghetmiri et al., 2014a; Li et al., 2014; Wirths et al., 2014), is a turning point for the technology to be pursued for the demonstration of an efficient group IV laser. In addition, due to the tunable lattice constant and formation of Lomer dislocations, $Ge_{1-x}Sn_x$ has been shown to work as a universal compliant buffer layer to grow high quality lattice mismatched materials, like III–Vs, on Si (Beeler et al., 2011a; Mosleh et al., 2014).

A variety of challenges exist for the growth of $Ge_{1-x}Sn_x$ alloys on Si such as large lattice mismatch between $Ge_{1-x}Sn_x$ and Si (more than 4.2%), low solid solubility of Sn in Ge (less than 0.5%), and instability of diamond lattice Sn (α-Sn) above 13°C. Therefore, growth can only possibly be done under non-equilibrium conditions. Different growth methods have been demonstrated for $Ge_{1-x}Sn_x$ growth in which molecular beam epitaxy (MBE) and chemical-vapor-deposition (CVD) have obtained device quality material and high Sn incorporation. For the MBE method,

both gas source and solid source MBE have been used by different groups to grow $Ge_{1-x}Sn_x$ films (Gurdal et al., 1998; Takeuchi et al., 2007; Chen et al., 2011; Werner et al., 2011; Stefanov et al., 2012; Bhargava et al., 2013; Oehme et al., 2013; Wang et al., 2013).

The other parallel approach of $Ge_{1-x}Sn_x$ growth is CVD. The early results of CVD growth by Kouvetakis and Chizmeshya (2007) at Arizona State University (ASU) showed the ability to grow $Ge_{1-x}Sn_x$ film directly on Si using a hot-wall ultra-high vacuum CVD (UHV-CVD) system with deuterated Stannane ($SnD_4$) as the Sn precursor along with different chemistries of germanium. Due to the high cost and instability of $SnD_4$, other precursors such as tetramethyl tin [$Sn(CH_3)_4$] and stannic chloride ($SnCl_4$) have been explored to grow $Ge_{1-x}Sn_x$ alloys. Vincent et al. (2011) (from IMEC using atmospheric pressure CVD) and Kim et al. (Chen et al., 2013) [from Applied Materials/Stanford University using reduced pressure-CVD (RP-CVD)] have reported successful growth of $Ge_{1-x}Sn_x$ by using $SnCl_4$ and a high cost Ge precursor digermane ($Ge_2H_6$) and carrier gases on a Ge-buffered Si substrate. Using the same $SnCl_4$ and $Ge_2H_6$ precursors and carrier gases, Mantl et al. (Wirths et al., 2013) (from PGI9-IT) demonstrated direct growth of $Ge_{1-x}Sn_x$ on Si using showerhead technology in an RP-CVD chamber. In the recent report, Tolle et al. (Margetis et al., 2014; Mosleh et al., 2014a) (ASM company) have achieved $Ge_{1-x}Sn_x$ growth using an industry prevail RP-CVD reactor in collaboration with University of Arkansas (UA). Low-cost Germane ($GeH_4$) and $SnCl_4$ with carrier gasses of $N_2/H_2$ were used to grow $Ge_{1-x}Sn_x$. A Ge buffer was deposited between the Si substrate and the $Ge_{1-x}Sn_x$ layer in order to compensate the lattice mismatch between the layers. **Table 1** lists the different research groups that have grown $Ge_{1-x}Sn_x$ using CVD. Different

**Table 1 | A summary of reports on $Ge_{1-x}Sn_x$ growth using CVD methods by different research groups.**

| Growth team | Deposition system | Deposition gas precursors | | | | Carrier gas | Buffer layer |
|---|---|---|---|---|---|---|---|
| | | Ge | Cost | Sn | Cost | | |
| ASU (Kouvetakis and Chizmeshya, 2007) | UHV-CVD | Different chemistries | High | $SnD_4$ | High | Yes | No |
| IMEC (Vincent et al., 2011) | AP-CVD | $Ge_2H_6$ | High | $SnCl_4$ | Low | Yes | Ge |
| Applied materials (Chen et al., 2013) | RP-CVD | $Ge_2H_6$ | High | $SnCl_4$ | Low | Yes | Ge |
| PGI9-IT (Wirths et al., 2013) | RP-CVD | $Ge_2H_6$ | High | $SnCl_4$ | Low | Yes | No |
| ASM/UA (Margetis et al., 2014; Mosleh et al., 2014a) | RP-CVD | $GeH_4$ | Low | $SnCl_4$ | Low | Yes | Ge |
| UA (this work) | UHV-CVD | $GeH_4$ | Low | $SnCl_4$ | Low | No | No |

growth methods and the cost effectiveness of the gas precursors are compared.

In this paper, we report direct growth of strain-relaxed $Ge_{1-x}Sn_x$ films on Si substrates with Sn mole fractions up to 7% using a cold-wall UHV-CVD system. Stannic chloride and germane were chosen as the precursors which are low-cost and commercially available. The growth of $Ge_{1-x}Sn_x$ films was achieved without using any carrier gases and buffer layers. In order to investigate the material quality, the X-ray diffraction (XRD), high-resolution transmission electron microscopy (TEM), energy-dispersive X-ray spectroscopy (EDX), Raman spectroscopy, and photoluminescence (PL) measurements have been conducted.

## EXPERIMENT
### GROWTH METHOD
A cold-wall UHV-CVD system was adopted to grow $Ge_{1-x}Sn_x$ films (see **Figure 1** for machine schematic). The system composes a load-lock chamber with a base pressure of $10^{-6}$ Pa and a process chamber whose base pressure reaches $10^{-8}$ Pa using the turbo-molecular and cryogenic pumps, respectively. Due to low-temperature growth of the films, removal of oxygen and water vapor is critical which was achieved by using a cryogenic pump. The turbo-molecular pumps are backed by mechanical pumps. The heating stage consisted of a pyrolytic graphite heater with a thermocouple placed at the same distance away from the heater as the wafer. The sample holder rotates up to 80 rpm for uniform film growth. The gas flow is through a side entry port, controlled by mass flow controllers (MFCs). Stannic chloride is a volatile liquid with vapor pressure of 2.4 kPa at one atmospheric pressure. Therefore, the evaporation could produce enough pressure to be passed through the MFC.

Germanium–tin films were grown on 4″ (001) p-type Si substrates with 5–10 $\Omega$ cm resistivity. Prior to loading the samples, they were cleaned in a two-step process: (1) Piranha etch solution [$H_2SO_4$:$H_2O_2$ (1:1)], (2) oxide strip HF dipping [$H_2O$:HF (10:1) using 48% pure HF] followed by nitrogen blow drying. The final oxide strip step was not followed by a water rinse as it reduces the life-time of hydrogen passivation and exposes the surface to ambient oxygen (Mosleh et al., 2013, 2014b). The experiments were carried out at reduced pressures of 13, 40, 65, 95, 130, 200, and 260 Pa and at temperatures as low as 300°C. Germane ($GeH_4$) and stannic chloride ($SnCl_4$) were used as the precursors for $Ge_{1-x}Sn_x$ growth. The gas flow ratio ($GeH_4$/$SnCl_4$) was set to 5, 3.3, 2.5, and 1.6. Depending on the growth parameters such as gas flow

ratio and deposition pressure, a growth rate of 20–3.3 nm/min was achieved.

## CHARACTERIZATION METHOD
Analysis of Sn mole fraction, lattice constant, growth quality, and strain in the $Ge_{1-x}Sn_x$ films were conducted using a high-resolution X-ray diffractometer. High-resolution TEM (TITAN) with an accelerating voltage of 300 kV was used to investigate crystal orientation and defects in the grown epi-layers as well as determining the thicknesses of the samples. Surface morphology of the samples was investigated by a scanning electron microscope equipped with EDX. Room temperature PL measurements were carried out using a 690-nm excitation laser. The PL signal was collected by a grating-based spectrometer equipped with a thermoelectric-cooled PbS detector (cut-off at 3 μm) for spectral analysis.

## RESULTS AND DISCUSSION
### MATERIAL CHARACTERIZATION
The $2\theta$-$\omega$ XRD scan was performed from the symmetric (004) plane to obtain the out-of-plane lattice constant of the $Ge_{1-x}Sn_x$ films. **Figure 2A** shows the peak at 69° corresponding to a satisfaction of the Bragg condition by Si (001) substrate, and the peaks at lower angles of 66–65° due to larger lattice size of the $Ge_{1-x}Sn_x$ layers. The difference in the position of $Ge_{1-x}Sn_x$ peaks is due to the difference in the Sn mole fractions of $Ge_{1-x}Sn_x$ layers. Different compositions were achieved from 1 to 7% with desirable crystal quality. The $Ge_{1-x}Sn_x$ peaks are broadened for two reasons: (1) thin film thickness of the layers and (2) presence of mosaicity in the Ge–Sn crystal and formation of defects as a result of strain relaxation. The full width at half maximum (FWHM) of the $Ge_{1-x}Sn_x$ peaks are between 0.28 for 1% Sn film and 0.36 for 7% Sn film. The change in FWHM depends on various factors such as film thickness, relaxation, and quality and there is no trend showing that the FWHM of the peaks change as the Sn composition increases.

In order to calculate the total lattice constant and the strain in the film, an asymmetric reciprocal space mapping (RSM) from (−2, −2, 4) plane was performed. The RSM scans provide measurement of the in-plane ($a_{\parallel}$) and out-of-plane ($a_{\perp}$) lattice constant of $Ge_{1-x}Sn_x$ alloys. The total lattice constant $a_0^{GeSn}$ was calculated by taking into account the elastic constants of $Ge_{1-x}Sn_x$ (Beeler et al., 2011b). Knowing the total lattice constant, the Sn mole fractions is calculated through Vegard's law with the bowing factor of $b = 0.0166$ Å (Moontragoon et al., 2012). **Figure 2B**

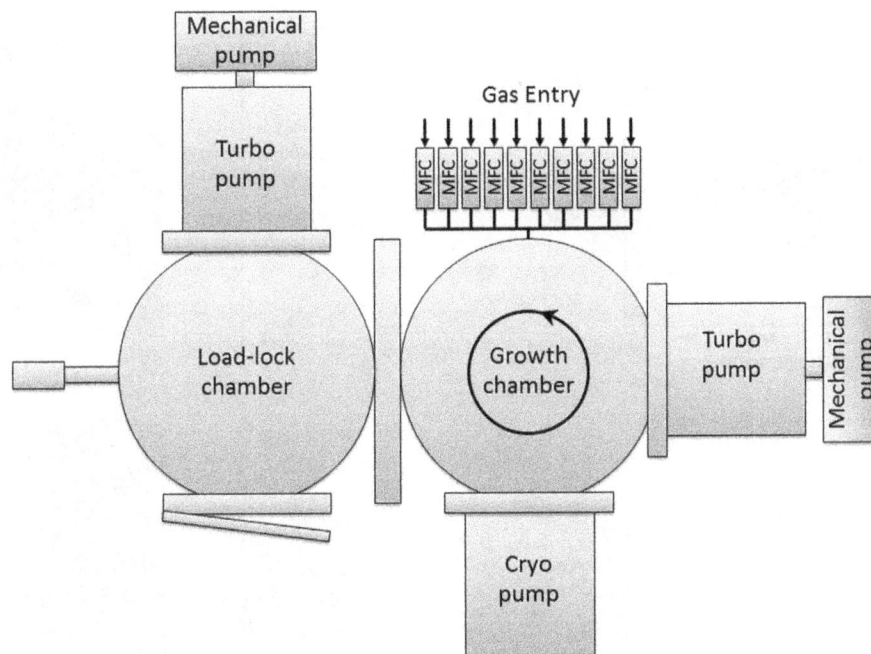

**FIGURE 1 | Cold-wall UHV-CVD system with a substrate rotation**. Samples are transferred through a load-lock chamber equipped with a turbo-molecular pump. The growth chamber is equipped with a turbo-molecular pump and a cryogenic pump. Side entry of the gases is controlled by mass flow controllers.

shows the RSM of 6% Sn sample. The $x$-axis shows $Q_z$ in reciprocal lattice unit (rlu) which is related to the out-of-plane lattice constant (L) and the $y$-axis shows $Q_x$ which is related to the in-plane lattice constant (H or K). Direction of the spread in the $Ge_{0.94}Sn_{0.06}$ peak does not show a compositional gradient in the sample because it is related to the relaxation of the lattice on Si substrate. Large lattice mismatch between Sn and Ge is the main reason for a large spread in the omega direction. The relaxation line in **Figure 2B** shows that the films which are grown above are tensile strained and the films grown underneath are compressively strained. The $Ge_{0.94}Sn_{0.06}$ peak is observed to be on the relaxation line and the relaxation is measured to be 97%.

Calculation of total strain in other samples shows that all the films are more than 95% relaxed. **Table 2** shows the lattice constants of the $Ge_{1-x}Sn_x$ alloys, their Sn mole fraction, and strain relaxation percentage. $Ge_{1-x}Sn_x$ films were almost fully relaxed mainly due to large lattice mismatch between Si (5.431 Å) and $Ge_{1-x}Sn_x$ (above 5.658 Å) and small critical thickness (Mosleh et al., 2014a). The other reason for relaxation of Ge (and similarly $Ge_{1-x}Sn_x$) films on Si is the thermal mismatch between these two materials. High temperature growth (above 500°C) and rapid cool down has been the main method for achieving tensile strained Ge on Si (Conley et al., 2014a). The $Ge_{1-x}Sn_x$ samples were grown at 300°C for 30 min and we have not achieved tensile strained films; however, the thermal mismatch between Si and $Ge_{1-x}Sn_x$ has helped relaxing the compressive strain. The strain has mainly relieved through formation of misfit dislocations including Lomer misfit dislocation. The cross-sectional TEM image in **Figure 2C** shows formation of such dislocations

at the $Ge_{1-x}Sn_x$/Si interface. In addition, **Figure 2B** shows that strain relaxation occurred by formation of misfit dislocations at the interface. The TEM image shows that the grown film was fully epitaxial. Film thickness of the samples is listed in **Table 2**.

The SEM scan/EDX spectra of the samples show surface morphology of the sample as well as Sn incorporation in the Ge matrix. The EDX spectra in **Figure 2D** show the presence of Ge, Si, and Sn in the $Ge_{0.94}Sn_{0.06}$ film. Due to the high count collection of secondary electrons from the substrate, the ratio of Sn and Ge cannot exactly reveal the percentage of Sn in Ge. The presence of carbon and oxygen in the EDX spectra is mainly due to the contamination and oxidation of the film after exposure to ambient air. The EDX maps for Ge (**Figure 2E**) and Sn (**Figure 2F**) display uniform incorporation of Sn. The SEM image shows continuous growth of $Ge_{1-x}Sn_x$ without observation of locally crystalline patches. No segregation and precipitation of Sn was observed on the films which indicates robust and stable growth of the films.

**GROWTH MECHANISM**

Growth of $Ge_{1-x}Sn_x$ on a Si substrate requires considering the reaction of byproducts and reduction of activation energy by introducing carrier gases. Stannic chloride has a tendency to etch Ge due to the presence of chlorine in the chemistry of the molecule. The byproduct of $GeH_4 + SnCl_4$ reaction is HCl which is an etchant gas for germanium and silicon (Bogumilowicz et al., 2005). Following reactions show different mechanisms of film deposition as well as HCl production in the chamber:

$$GeH_4 \rightarrow Ge + 2H_2 \qquad (1)$$

FIGURE 2 | (A) Symmetric (004) 2θ-ω scan of $Ge_{1-x}Sn_x$ films which are grown on a Si substrate. The peak at 69° shows the Si substrate peak and the peaks between 66° and 65° belong to $Ge_{1-x}Sn_x$ films. (B) Reciprocal space map from asymmetrical plane (−2, −2, 4) for $Ge_{0.94}Sn_{0.06}$ grown on a Si substrate. The x-coordinate shows out-of-plane lattice constant and the y-coordinate shows in-plane lattice constant in units of reciprocal lattice unit. The relaxation line shows that the films grown above are tensile strained and below are compressively strained. Presence of the $Ge_{0.94}Sn_{0.06}$ on the relaxation line shows that the film is strain relaxed. (C) Transmission electron microscopy images of $Ge_{0.94}Sn_{0.06}$ film shows epitaxial growth of Ge–Sn on a Si substrate. Arrows show misfit dislocations formed at the $Ge_{1-x}Sn_x$/Si interface. (D) The EDX spectrum of $Si/Ge_{0.94}Sn_{0.06}$ film shows the presence of Si (substrate), Ge and Sn (film), O (native oxide), and C (carbon contamination from the ambient). (E) The EDX surface maps of Ge and (F) Sn taken from scanning electron micrographs for $Ge_{0.94}Sn_{0.06}$ film shows uniform growth of $Ge_{1-x}Sn_x$ alloy.

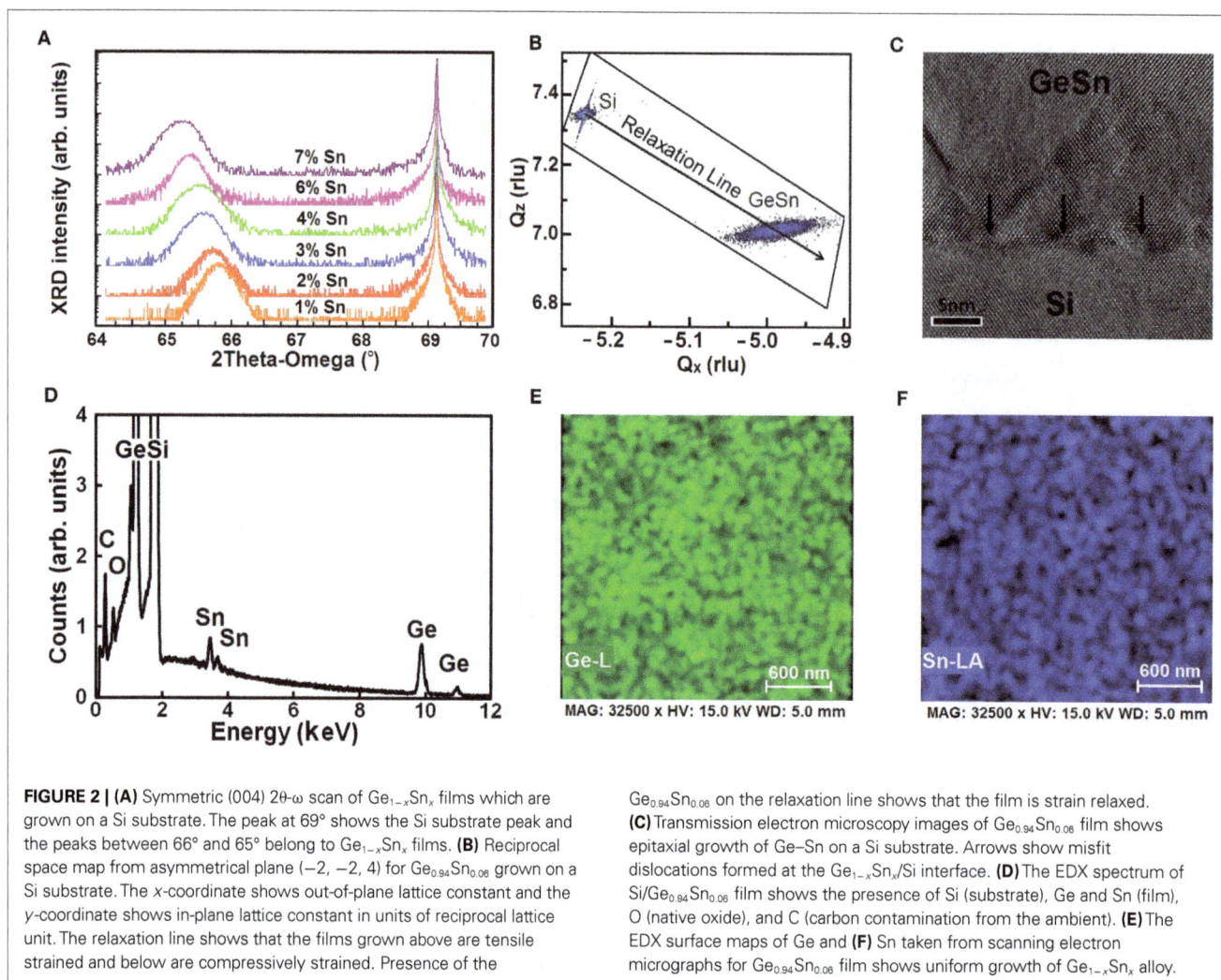

**Table 2 | Tin mole fraction calculation, lattice constant, and relaxation percentage of the grown samples.**

| Sample no. | Sn (%) | $a_{\parallel\Pi}$ (nm) | $a_{\perp}$ (nm) | $a$ (nm) | Relaxation (%) | Thickness (nm) |
|---|---|---|---|---|---|---|
| 1 | 1.2 | 5.666 | 5.671 | 5.668 | 98 | 615 |
| 2 | 2.1 | 5.673 | 5.679 | 5.676 | 98 | 423 |
| 3 | 2.9 | 5.678 | 5.687 | 5.682 | 97 | 295 |
| 4 | 4.2 | 5.689 | 5.695 | 5.692 | 98 | 207 |
| 5 | 5.8 | 5.699 | 5.712 | 5.706 | 97 | 108 |
| 6 | 7.0 | 5.715 | 5.719 | 5.717 | 99 | 532 |

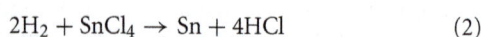

$$2H_2 + SnCl_4 \rightarrow Sn + 4HCl \quad (2)$$

$$GeH_4 + SnCl_4 \rightarrow Ge + Sn + 4HCl \quad (3)$$

Higher temperature of the substrate results in higher density of depositing ad-atoms (Ge and Sn); however, it will result in production of HCl at a higher rate. In addition, higher flow rate of $SnCl_4$ increases the production rate of HCl as well. Controlling the temperature and flow rate of the gases could control the process so that growth is the dominant process in the chamber. The Ge/Sn film will be etched by HCl through the following reactions:

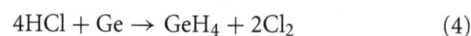

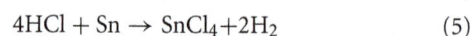

$$4HCl + Ge \rightarrow GeH_4 + 2Cl_2 \quad (4)$$

$$4HCl + Sn \rightarrow SnCl_4 + 2H_2 \quad (5)$$

Domination of etching over growth is the main mechanism that prevents direct growth of $Ge_{1-x}Sn_x$ on Si.

By controlling the flow through MFCs, we have grown $Ge_{1-x}Sn_x$ films on Si at different pressures with a fixed flow ratio of $GeH_4/SnCl_4 = 1.6$. Growth was observed at 13 Pa of deposition pressure and continued until the deposition pressure increased to 130 Pa. **Figure 3A** shows the thickness of $Ge_{1-x}Sn_x$ films versus deposition pressure of the chamber as well as Sn incorporation percentage. Incorporation of Sn in the Ge lattice is increased by raising the pressure due to the higher residence time of the precursors in the chamber. The residence time of the gases has increased from 2 s at 13 Pa to 19 s at 130 Pa. Meanwhile, HCl etched more of the $Ge_{1-x}Sn_x$ films after deposition at higher pressures. This trend has continued to 130 Pa and no growth has been observed at 200 and 265 Pa. The increase in Sn composition from 1 to 6%

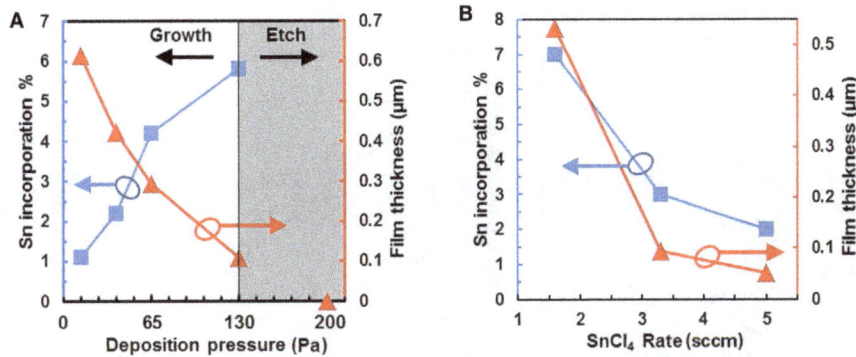

**FIGURE 3 | (A)** Variation of Sn incorporation percentage versus deposition pressure. Films were etched away for deposition pressures higher than 130 Pa. The secondary axis on the right shows the reduction of film thickness as a result of increase in the deposition pressure. **(B)** Tin incorporation and film thickness of the samples grown at 65 Pa growth pressure versus GeH₄/SnCl₄ flow ratios.

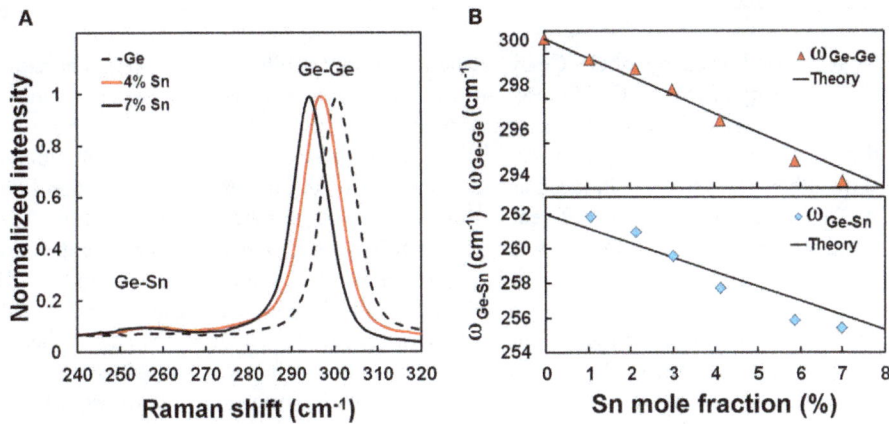

**FIGURE 4 | (A)** Raman spectra of the Ge₁₋ₓSnₓ film grown on a Si substrate. The shift in the Ge–Ge peak is due to the incorporation of Sn in Ge lattice. The shoulder on the left side of the Ge–Ge peak is due to the Ge–Sn peak at 285 cm⁻¹. The Ge–Sn peak is shown at lower wavenumber of 250–260 cm⁻¹. **(B)** Ge–Ge and Ge–Sn peak shifts versus Sn mole fraction. The solid symbols are experimental data and the curves are theoretical predictions for relaxed films. The Ge–Ge peak is expected to shift 0.8310 cm⁻¹ for every 1% Sn incorporation in relaxed films. The expected shift (0.8311 cm⁻¹) for Ge–Sn peak is very close to that of Ge–Ge.

has been accompanied with reduction in the thickness from 615 to 108 nm. Films that were expected to have higher than 6% Sn content were totally etched off. Therefore, in order to grow higher Sn content films, growth mechanism under fixed pressure and changing the SnCl₄ flow was studied. Higher film thickness and higher Sn incorporation was achieved as a result of domination of growth over etching. **Figure 3B** shows Sn incorporation in Ge₁₋ₓSnₓ films versus SnCl₄ flow rate at 95 Pa deposition pressure. The secondary axis of **Figure 3B** shows film thicknesses of the samples. Due to the dominance of etching for higher SnCl₄ flow rate, the films were mostly etched and the film thickness was less than 100 nm.

Introduction of carrier gases has different effects on the growth of Ge₁₋ₓSnₓ films. Hydrogen changes the balance in the reaction to produce more HCl. Consequently, the GeH₄/SnCl₄ ratio at which the Ge₁₋ₓSnₓ films were depositing will not result in growth when hydrogen is introduced in the chamber. In addition, introduction of nitrogen and argon as carrier gases will reduce the activation energy of the growth (Wirths et al., 2013). Although reduction of activation energy enables easier breakdown of the molecules on the surface and enhances the growth quality and growth rate, it would prepare the conditions for easier etch due to the presence of an etchant agent. Therefore, the presence of carrier gases pushes the competition between growth and etching toward etching, resulting in film etching at even lower flow rates of carrier gases when the flow rate of SnCl₄ is of the same order of GeH₄.

## OPTICAL CHARACTERIZATION
### Raman spectroscopy

The Ge₁₋ₓSnₓ films were further investigated by Raman spectroscopy in order to analyze the crystal structure. Room temperature Raman spectra of the grown samples as well as a Ge reference sample are plotted in **Figure 4A**. The Ge–Ge longitudinal optical (LO) peak was observed at 300 cm⁻¹ for the Ge reference sample while the Ge–Ge peak in the Ge₁₋ₓSnₓ films was shifted to lower

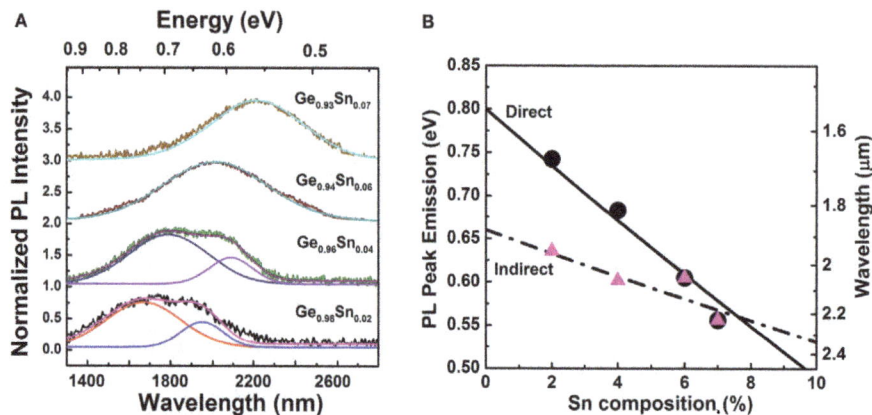

**FIGURE 5 | (A)** Photoluminescence spectra of the $Ge_{1-x}Sn_x$ films with 2, 4, 6, and 7% Sn mole fraction showing a red-shift in the band-gap of the films. Incorporation of Sn has shifted both direct band-gap and indirect band-gap toward lower energies. **(B)** The bowed Vegard's law interpolation for the direct (solid line) and indirect band-gap (dash line) of $Ge_{1-x}Sn_x$ alloy is plotted for different Sn compositions and is overlaid with experimental data (solid symbols).

wavenumbers due to the change in bonding energy of Ge–Ge by incorporation of Sn atoms. The intensity of the Ge–Ge LO peak at $300\,cm^{-1}$ is normalized for all the samples for comparison of the peak positions. In addition to the main Ge–Ge peak, Raman spectra of $Ge_{1-x}Sn_x$ films show other peaks that are induced as a result of Sn incorporation. The Ge–Sn LO peaks for different Sn mole fractions were observed at $250–260\,cm^{-1}$ in the films. A second peak of Ge–Sn is observed at $285\,cm^{-1}$, which can be seen as a shoulder of Ge–Ge main peak.

The peak positions are obtained by Lorentzian fitting to find the exact position for further analysis. The shift in the Ge–Ge LO peak depends on both strain and Sn composition of the films. Theoretical calculations for $\Delta\omega$ are different for strain-relaxed films and strained films for different Sn $(x)$ content $[\Delta\omega_{Ge-Ge}(x) = bx\,cm^{-1}]$. The Ge–Ge peak is expected to shift by a factor of $b = -30.30$ for a strained alloy while this factor varies to $b = -83.10$ for a strain-relaxed film (Cheng et al., 2013). **Figure 4B** shows the experimental data obtained for Ge–Ge and Ge–Sn Raman shift from the sample compared with the theoretical calculations. The peak shifts match well with the theoretical calculations for strain-relaxed films.

*Photoluminescence*

Germanium has an indirect band-gap in the L valley with the energy of 0.644 eV and a direct band-gap at the $\gamma$ point with 0.8 eV energy at room temperature. Incorporation of Sn in Ge lattice lowers the conduction band edge at the $\gamma$-point at a faster rate than that at the L-point. PL measurements on $Ge_{1-x}Sn_x$ samples allow determination of the band-gap edge for various Sn compositions.

**Figure 5** depicts room temperature PL intensity spectra for as-grown $Ge_{1-x}Sn_x$ films with 2, 4, 6, and 7% Sn mole fractions. As indicated in **Figure 5A**, increase of the Sn mole fraction results in a band-gap reduction. Both direct and indirect PL peaks exhibit red-shift with Sn compositions increase from 2 to 7%. A Gaussian fitting function was employed to extract the PL peak positions of both direct and indirect transitions as described in Ghetmiri et al. (2014b). In $Ge_{0.94}Sn_{0.06}$ and $Ge_{0.93}Sn_{0.07}$ samples, the energies difference between direct and indirect transitions are very small, therefore the PL emissions from these indirect and direct transitions cannot be identified. A temperature-dependent study is needed to differentiate the direct and indirect peak positions which will be reported in the future. The PL peaks from the samples with 2, 4, 6, and 7% Sn compositions are shown in **Figure 5B** as solid symbols. The solid and the dashed lines show the direct and indirect band-gap energies based on bowed Vegard's law for the relaxed $Ge_{1-x}Sn_x$ alloy (Ghetmiri et al., 2014b), respectively. Since the $Ge_{1-x}Sn_x$ films are almost strain-free, as confirmed by XRD measurements, the experimental results closely follow the predicted values from Vegard's law.

## CONCLUSION

Direct growth of $Ge_{1-x}Sn_x$ layers on Si substrates was achieved using a cold-wall UHV-CVD system. The films were grown by employing low-cost commercial available $GeH_4$ and $SnCl_4$ precursors without using any carrier gases and buffer layers. Characterizations of the samples with XRD showed successful incorporation of Sn up to 7%. The TEM images show fully epitaxial growth of the samples without any precipitation of Sn from the Ge lattice. The Raman results verified the Sn incorporation and PL measurements showed reduction of the band-gap to 0.55 eV for 7% Sn sample. The low-cost and CMOS compatible growth method and the performance of the samples indicate a promising future for $Ge_{1-x}Sn_x$ applications in Si photonics. Moreover, the samples were grown strain-relaxed enabling this material to be a universal compliant buffer layer which can be used in hybrid integration.

## ACKNOWLEDGMENTS

The work at the UA was supported by NSF (EPS-1003970), the Arkansas Bioscience Institute, the Arktonics, LLC (Air Force SBIR, FA9550-14-C-0044, Dr. Gernot Pomrenke, Program Manager), and DARPA (W911NF-13-1-0196, Dr. Dev Palmer, Program Manager). Drs. RS and GS acknowledge support from AFOSR

(FA9550-14-1-0196, Dr. Gernot Pomrenke, Program Manager). JG acknowledges the support of NSF REU Program under Grant number EEC-1359306.

# REFERENCES

Beeler, R. T., Grzybowski, G. J., Roucka, R., Jiang, L., Mathews, J., Smith, D. J., et al. (2011a). Synthesis and materials properties of Sn/P-doped Ge on Si (100): photoluminescence and prototype devices. *Chem. Mater.* 23, 4480–4486. doi:10.1021/cm201648x

Beeler, R., Roucka, R., Chizmeshya, A., Kouvetakis, J., and Menéndez, J. (2011b). Nonlinear structure-composition relationships in the Ge1−ySnySi (100) (y< 0.15) system. *Phys. Rev. B* 84, 035204. doi:10.1103/PhysRevB.84.035204

Bhargava, N., Coppinger, M., Gupta, J. P., Wielunski, L., and Kolodzey, J. (2013). Lattice constant and substitutional composition of GeSn alloys grown by molecular beam epitaxy. *Appl. Phys. Lett.* 103, 041908. doi:10.1063/1.4816660

Bogumilowicz, Y., Hartmann, J. M., Truche, R., Campidelli, Y., Rolland, G., and Billon, T. (2005). Chemical vapour etching of Si, SiGe and Ge with HCl; applications to the formation of thin relaxed SiGe buffers and to the revelation of threading dislocations. *Semicond. Sci. Technol.* 20, 127. doi:10.1088/0268-1242/20/2/004

Chen, R., Huang, Y., Gupta, S., Lin, A. C., Sanchez, E., Kim, Y., et al. (2013). Material characterization of high Sn-content, compressively-strained GeSn epitaxial films after rapid thermal processing. *J. Cryst. Growth* 365, 29–34. doi:10.1016/j.jcrysgro.2012.12.014

Chen, R., Lin, H., Huo, Y., Hitzman, C., Kamins, T. I., and Harris, J. S. (2011). Increased photoluminescence of strain-reduced, high-Sn composition Ge1−xSnx alloys grown by molecular beam epitaxy. *Appl. Phys. Lett.* 99, 181125. doi:10.1063/1.3658632

Cheng, R., Wang, W., Gong, X., Sun, L., Guo, P., Hu, H., et al. (2013). Relaxed and strained patterned germanium-tin structures: a Raman scattering study. *ECS J. Solid State Sci. Technol.* 2, 138–145. doi:10.1149/2.013304jss

Conley, B. R., Mosleh, A., Ghetmiri, S. A., Du, W., Soref, R. A., Sun, G., et al. (2014a). Temperature dependent spectral response and detectivity of GeSn photoconductors on silicon for short wave infrared detection. *Opt. Express* 22, 15639–15652. doi:10.1364/OE.22.015639

Conley, B. R., Margetis, J., Du, W., Tran, H., Mosleh, A., Ghetmiri, S. A., et al. (2014b). Si based GeSn photoconductors with a 1.63 A/W peak responsivity and a 2.4 μm long-wavelength cutoff. *App. Phys. Lett.* 105, 221117. doi:10.1063/1.4903540

Du, W., Zhou, Y., Ghetmiri, S. A., Mosleh, A., Conley, B. R., Nazzal, A., et al. (2014a). Room-temperature electroluminescence from Ge/Ge1−xSnx/Ge diodes on Si substrates. *Appl. Phys. Lett.* 104, 241110. doi:10.1063/1.4884380

Du, W., Ghetmiri, S. A., Conley, B. R., Mosleh, A., Nazzal, A., Soref, R. A., et al. (2014b). Competition of optical transitions between direct and indirect bandgaps in Ge1−xSnx. *Appl. Phys. Lett.* 105, 051104. doi:10.1063/1.4892302

Ghetmiri, S. A., Du, W., Margetis, J., Mosleh, A., Cousar, L., Conley, B. R., et al. (2014a). Direct-bandgap GeSn grown on Silicon with 2230 nm photoluminescence. *Appl. Phys. Lett.* 105, 151109. doi:10.1063/1.4898597

Ghetmiri, S. A., Du, W., Conley, B. R., Mosleh, A., Naseem, H. A., Yu, S., et al. (2014b). Shortwave-infrared photoluminescence from Ge1−xSnx thin films on silicon. *J. Vac. Sci. Technol. B* 32, 060601. doi:10.1116/1.4897917

Gurdal, O., Desjardins, P., Carlsson, J., Taylor, N., Radamson, H., Sundgren, J., et al. (1998). Low-temperature growth and critical epitaxial thicknesses of fully strained metastable Ge1−xSnx (x≲0.26) alloys on Ge (001) 2 × 1. *J. Appl. Phys.* 83, 162–170. doi:10.1063/1.366690

Kouvetakis, J., and Chizmeshya, A. V. G. (2007). New classes of Si-based photonic materials and device architectures via designer molecular routes. *J. Mater. Chem.* 17, 1649–1655. doi:10.1039/b618416b

Kouvetakis, J., Menendez, J., and Soref, R.A. *Strain-Engineered Direct-Gap Ge/SnxGe1-x Heterodiode and Multi-Quantum-Well Photodetectors, Laser, Emitters and Modulators Grown on SnySizGe1-y-z Buffered Silicon*. United States patent US 6897471 B1 (2005).

Li, H., Brouillet, J., Salas, A., Chaffin, I., Wang, X., and Liu, J. (2014). Low temperature geometrically confined growth of pseudo single crystalline GeSn on amorphous layers for advanced optoelectronics. *ECS Trans.* 64, 819–827. doi:10.1149/06406.0819ecst

Margetis, J., Ghetmiri, S. A., Du, W., Conley, B. R., Mosleh, A., Soref, R., et al. (2014). Growth and characterization of epitaxial Ge1−xSnx alloys and heterostructures

using a commercial CVD system. *ECS Trans.* 64, 1830–1830. doi:10.1149/06406. 0711ecst

Moontragoon, P., Soref, R., and Ikonic, Z. (2012). The direct and indirect bandgaps of unstrained Six Ge1-xySny and their photonic device applications. *J. Appl. Phys.* 112, 073106–073106–8. doi:10.1063/1.4757414

Mosleh, A., Benamara, M., Ghetmiri, S. A., Conley, B. R., Alher, M. A., Du, W., et al. (2014). Investigation on the formation and propagation of defects in GeSn thin films. *ECS Trans.* 64, 1845–1845. doi:10.1149/06406.0895ecst

Mosleh, A., Ghetmiri, S. A., Conley, B. R., Abu-Safe, H., Waqar, Z., Benamara, M., et al. (2013). "Nucleation-step study of silicon homoepitaxy for low-temperature fabrication of Si solar cells," in *Photovoltaic Specialists Conference (PVSC), IEEE 39th*. Tampa, FL.

Mosleh, A., Ghetmiri, S. A., Conley, B. R., Hawkridge, M., Benamara, M., Nazzal, A., et al. (2014a). Material characterization of Ge1−xSnx alloys grown by a commercial CVD system for optoelectronic device applications. *J. Electron. Mater.* 43, 938–946. doi:10.1007/s11664-014-3089-2

Mosleh, A., Ghetmiri, S. A., Conley, B. R., Abu-Safe, H., Benamara, M., Waqar, Z., et al. (2014b). Investigation of growth mechanism and role of H2 in very low temperature Si epitaxy. *ECS Trans.* 64, 967–975. doi:10.1149/06406.0967ecst

Oehme, M., Buca, D., Kostecki, K., Wirths, S., Holländer, B., Kasper, E., et al. (2013). Epitaxial growth of highly compressively strained GeSn alloys up to 12.5% Sn. *J. Cryst. Growth* 384, 71–76. doi:10.1016/j.jcrysgro.2013.09.018

Stefanov, S., Conde, J., Benedetti, A., Serra, C., Werner, J., Oehme, M., et al. (2012). Laser synthesis of germanium tin alloys on virtual germanium. *Appl. Phys. Lett.* 100, 104101. doi:10.1063/1.3692175

Takeuchi, S., Sakai, A., Yamamoto, K., Nakatsuka, O., Ogawa, M., and Zaima, S. (2007). Growth and structure evaluation of strain-relaxed Ge1−xSn buffer layers grown on various types of substrates. *Semicond. Sci. Technol.* 22, S231. doi:10.1088/0268-1242/22/1/S54

Vincent, B., Gencarelli, F., Bender, H., Merckling, C., Douhard, B., Petersen, D. H., et al. (2011). Undoped and in-situ B doped GeSn epitaxial growth on Ge by atmospheric pressure-chemical vapor deposition. *Appl. Phys. Lett.* 99, 152103–152103–3. doi:10.1063/1.3645620

Wang, L., Su, S., Wang, W., Gong, X., Yang, Y., Guo, P., et al. (2013). Strained germanium–tin (GeSn) p-channel metal-oxide-semiconductor field-effect-transistors (p-MOSFETs) with ammonium sulfide passivation. *Solid State Electron.* 83, 66–70. doi:10.1016/j.sse.2013.01.031

Werner, J., Oehme, M., Schmid, M., Kaschel, M., Schirmer, A., Kasper, E., et al. (2011). Germanium-tin pin photodetectors integrated on silicon grown by molecular beam epitaxy. *Appl. Phys. Lett.* 98, 061108–061108–3. doi:10.1063/1.3555439

Wirths, S., Buca, D., Mussler, G., Tiedemann, A., Holländer, B., Bernardy, P., et al. (2013). Reduced pressure CVD growth of Ge and Ge1−xSnx alloys. *ECS J. Solid State Sci. Technol.* 2, N99–N102. doi:10.1149/2.006305jss

Wirths, S., Geiger, R., Scherrer, P., Ikonic, Z., Tiedemann, A. T., Mussler, G., et al. (2014). "Epitaxy and photoluminescence studies of high quality GeSn heterostructures with Sn concentrations up to 13 at.%," in *11th International Conference on Group IV Photonics 27–29 August*. Paris.

**Conflict of Interest Statement:** The authors declare that the research was conducted in the absence of any commercial or financial relationships that could be construed as a potential conflict of interest.

# 13

# Chalcogenide glass hollow-core microstructured optical fibers

*Vladimir S. Shiryaev* [1,2] *

[1] G.G. Devyatykh Institute of Chemistry of High-Purity Substances of the Russian Academy of Sciences, Nizhny Novgorod, Russia
[2] N.I. Lobachevski Nizhny Novgorod State University, Nizhny Novgorod, Russia

**Edited by:**
Petra Granitzer, University of Graz, Austria

**Reviewed by:**
Tatiana S. Perova, Trinity College Dublin, Ireland
Roman Golovchak, Austin Peay State University, USA

**\*Correspondence:**
Vladimir S. Shiryaev, G.G. Devyatykh Institute of Chemistry of High-Purity Substances of the Russian Academy of Sciences, 49 Tropinin Street, Nizhny Novgorod 603950, Russia
e-mail: shiryaev@ihps.nnov.ru

The recent developments on chalcogenide glass hollow-core microstructured optical fibers (HC-MOFs) are presented. The comparative analysis of simulated optical properties for chalcogenide HC-MOFs of negative curvature with different size and number of capillaries is given. The technique for the manufacture of microstructured chalcogenide preforms, which includes the assembly of the substrate glass tube and 8–10 capillaries, is described. Further trends to improve the optical transmission in chalcogenide negative curvature hollow-core photonic crystal fibers are considered.

Keywords: chalcogenide glass, microstructured fibers, optical loss, mid-IR, preform

## INTRODUCTION

The hollow-core microstructured optical fibers (HC-MOFs) (Yablonovitch, 1987; Knight et al., 1996) have attracted much attention in optical applications over the past decade as they exhibit many unique optical properties such as controlled dispersion, endlessly single-mode operation, supercontinuum generation, and soliton propagation (Birks et al., 1997; Monro et al., 1999; Revathi et al., 2014; Skryabin and Wadsworth, 2010; Wadsworth et al., 2001) over a wide range of wavelengths that cannot be realized in conventional step-index fibers. The light guidance in HC-MOF fiber is based on the mechanism of photonic band gap (PBG) guidance. The PBG mechanism has the capabilities to control the guidance of light with a certain frequency band (Knight et al., 1998; Cregan et al., 1999). In HC-MOFs, core localized modes can exist with mode effective indices $b/k_0$ (where $b$ is the propagation constant and $k_0$ is the wavenumber in free space) that are lower than the mean index of the microstructured cladding (Cregan et al., 1999), so it is possible to guide the light in air.

The theoretical losses in HC-MOFs can be very low, because the material absorption and Rayleigh scattering in air are negligible compared to glass (Knight et al., 1996). Special potential is accorded to HC-MOFs with negative curvature of refractive index. Experiments on silica fibers of that type have shown not only the possibility of achieving low optical losses, but also a considerable extension of the transmission region to longer wavelengths in comparison with conventional step-index optical fibers (Knight et al., 1996; Pryamikov et al., 2011).

There are many reasons for interest in guiding light in a hollow-core. For example, the effective single-mode guidance of light was demonstrated in HC-MOF with six-layer cladding structure made from a lead-silicate "soft" glass (Jiang et al., 2011). Also, in gas-filled HC-MOFs, there is a strong interaction between the light and the gas which fills the core, giving a lower energy threshold for

Raman amplification (Benabid et al., 2005). HC-MOFs are also of interest for fiber sensors and applications requiring single-mode guidance in a large mode area, such as power delivery, fluid and gas-filled devices, and particle transport (Benabid, 2006).

In recent years, great attention has been paid to development of the mid-infrared (IR) region, which is due to numerous potential applications in analytical IR spectroscopy, pyrometers, and transmission of IR lasers. Among the wide variety of IR glasses, for mid-IR MOFs, non-silica compound glasses like chalcogenide glasses have the best prospects. Chalcogenide glasses are based on sulfur, selenium, tellurium, and the addition of other elements such as arsenic, germanium, antimony, gallium, etc. They are characterized by a number of significant advantages, such as a wide transmittance range (1–12 μm), low intrinsic losses in the mid-IR, low phonon energy, and the absence of free-carrier effects (Snopatin et al., 2009b). The large refractive index of chalcogenide glasses (compared to other glasses) of 2.4–3.0 opens up the possibility of achieving compact non-linear devices (Hu et al., 2014; Zhang et al., 2014; He et al., 2012). The non-linear refractive index $n_2$ in chalcogenide glasses is by (100–1000) times higher than in silica glass. These two properties are more suitable for fiber-based photonic wire devices (He et al., 2012).

There are a large number of technical problems in fabrication of HC-MOFs based on chalcogenide glasses. Several research groups from different countries are involved in tackling these problems. The main efforts are focused on the development of optimum design and geometric parameters of MOF structure, the increase in the chemical and phase purity of bases of chalcogenide glass, the improvement of drawing technique. The results of these studies are not widely presented in the literature, since the development of solutions to the abovementioned problems is at a very early stage. This review presents the modern status of developments on chalcogenide glass HC-MOFs.

## THE FIRST EXPERIMENTS ON PREPARING THE CHALCOGENIDE HC-MOFs

The development of near- and mid-IR hollow-core chalcogenide MOFs is an important and urgent task because of the theoretical possibility to achieve optical losses, which are lower than material losses, an expansion of transmission range, as well as to transmit high-power CO and $CO_2$-laser radiation (Pryamikov et al., 2011; Temelkuran et al., 2002).

Therefore, after the first publication on MOF of Ga–La–S system in 1998 (Monro et al., 2000), an interest in the world in the preparation and study of such fibers of different chalcogenide glasses, as a solid-core and hollow-core MOFs, has increased significantly. Published data on chalcogenide MOFs with a hollow-core is limited because the production of such fibers is a rather difficult technical challenge.

In published papers on preparing chalcogenide MOFs with a solid-core, the following techniques of preform manufacturing have been used: "stack and draw" (assembly of capillaries inside the substrate tube) (Brilland et al., 2006; Desevedavy et al., 2008), molding (with use of silica glass pattern of the specified photonic crystal design) (Coulombier et al., 2010; Conseil et al., 2011), and drilling (El-Amraoui et al., 2010). Usually, the thin capillaries are drawn from the chalcogenide glass tube prepared by centrifugal casting. However, the additional heat treatment can provoke crystallization of the glass and increase the excess optical loss in the resulting fibers. The main disadvantage of the drilling method to make holes in the substrate chalcogenide glass tube is the formation of broken surface layers of the holes.

There are one-dimensional (1-D) (hybrid Bragg photonic crystal fiber) and two-dimensional (2-D) hollow fibers made of chalcogenide glasses.

One-dimensional MOF is a hybrid structure consisting of dielectric mirrors to guide light inside the fibers. A dielectric mirror is basically simultaneous layers of quarter wave stacks, made of thermo mechanically compatible materials with high-refractive-index difference. The modification of thin dielectric mirror layers designed at the inner surface of a hollow fiber allows this fiber to guide a certain spectrum of light through air efficiently.

The first HC-MOFs were hybrid Bragg fibers (Yeh et al., 1978) composed of successive circular layers presenting two different refractive indices. An example of such hollow-core Bragg fiber composed of layers of $As_2Se_3$ chalcogenide glass with a refractive index of 2.8 and high glass-transition temperature thermoplastic polymer poly(ether sulfone) (PES) having a refractive index of 1.55 was described in paper (Temelkuran et al., 2002) (**Figure 1**). A hollow preform from these two materials was made by vapor depositing an $As_2Se_3$ coating on a thin sheet of PES, with the subsequent rolling the $As_2Se_3$/PES sheet into a fiber preform. Confinement of light in the hollow-core is provided by the large photonic bandgaps established by the multiple alternating submicrometer-thick layers of a high-refractive-index glass and a low-refractive-index polymer. The fundamental and high-order transmission windows are determined by the layer dimensions and can be change in wavelength range from 0.75 to 10.6 μm. The transmission losses are found to be less than 1.0 dB/m. The large photonic bandgaps result in very short electromagnetic penetration depths within the layer structure, significantly reducing radiation and absorption losses. The maximum laser power density coupled into these fibers was approximately 300 W/$cm^2$.

Another method to prepare a 1-D photonic bandgap structure was proposed in paper (Gibson and Harrington, 2004). The authors extruded a stack of alternating plates of $As_2Se_3$/PSU (polysulfone) through a die into a hollow-core preform with dielectric multi-layers.

The most common design of 2-D MOF introduced by Birks et al. (1997) can be described as a hollow-core fiber in which the cross-section is a periodic array of air holes placed along the length of the fiber. The origin of the band gap for 2-D MOFs is similar to the 2-D band gaps investigated for planar light-wave circuits. The most common geometry for 2-D MOFs is presented in **Figure 2A**.

The first all-chalcogenide hollow-core photonic crystal fibers (HC-MOF) was reported in paper (Desevedavy et al., 2010). These fibers composed of six rings of holes and regular microstructures with Kagome and hexagonal lattices were fabricated by the "stack and draw" technique from $Te_{20}As_{30}Se_{50}$ glass (**Figure 2B**). Unfortunately, the targeted HC-MOF profile was not reached and

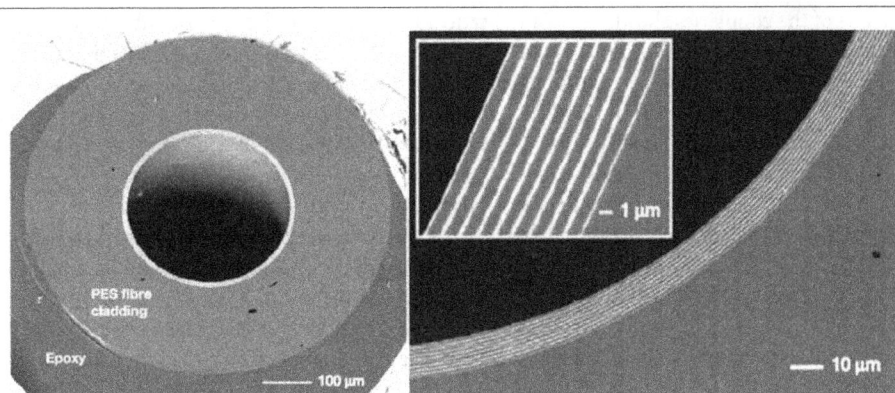

**FIGURE 1 | Cross-sectional SEM micrographs at various magnifications of hollow cylindrical multi-layer fiber mounted in epoxy.** The hollow-core appears black, the PES layers and cladding gray, and the $As_2Se_3$ layers bright white (Temelkuran et al., 2002).

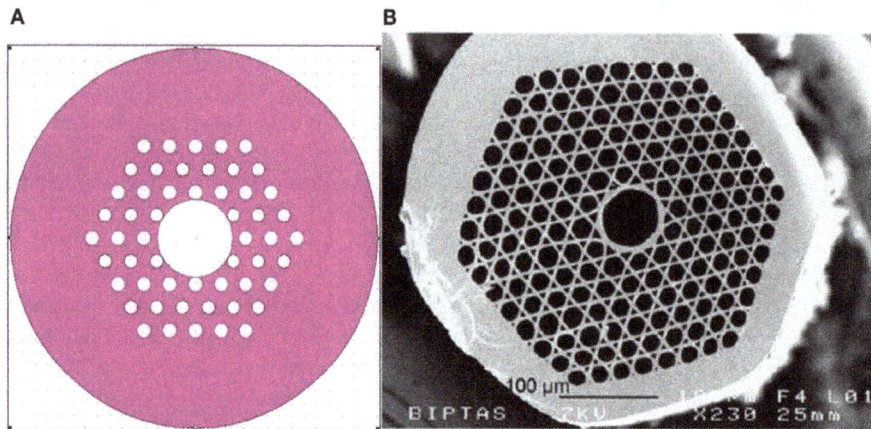

**FIGURE 2 | (A)** Geometry for 2-D PCFs; **(B)** $Te_{20}As_{30}Se_{50}$ hollow-core PCF profiles with Kagome lattice cladding (Desevedavy et al., 2010).

no propagation was observed in the core fibers between 2 and 20 μm.

By now, the optical loss in step-index chalcogenide glass fibers has been reduced to 1.5–2 dB/m at $\lambda = 10.6$ μm due to a decrease of the oxide impurity content causing absorption in 7.5–15 spectral region (Nishii et al., 1992; Katsuyama and Matsumura, 1986; Shiryaev et al., 2005). The maximum power transmitted through a 1-m length of such a fiber was 10.7 W, when an anti-reflection coating and water cooling were used (Nishii et al., 1992). Better transmission results have been achieved with hollow-core fibers, in which radiation power is transmitted through the air. A hollow-core polymer fiber with a cladding in the form of a multi-layer Bragg mirror demonstrated a loss below 1 dB/m at $\lambda = 10.6$ μm (Temelkuran et al., 2002).

## CHALCOGENIDE NEGATIVE CURVATURE HOLLOW-CORE PHOTONIC CRYSTAL FIBERS. THEORY

To develop a model chalcogenide HC-MOF (Kosolapov et al., 2011), a relatively simple design of photonic crystal structure of eight capillaries within the substrate tube was used (**Figure 3**), as tested previously for silica glass (Pryamikov et al., 2011). The ratio of the inner and outer diameters of the capillaries should be from 0.8 to 0.9. These negative curvature hollow-core photonic crystal fibers (NCHCFs) with a negative curvature of the core boundary belongs to the type of HC-MOFs that do not support PBGs. Such NCHCFs have relatively high transmission losses in comparison with PBG HC-MOFs, but possess a larger bandwidth, because the air-core localized modes are coupled only weakly with the cladding modes in the low loss wavelength regions. The high loss wavelengths in the transmission spectrum correspond to the avoided crossing of the air-core modes and the cladding modes.

The simulation of the optimal design of the NCHCF was carried out by the FemLab 3.1 software, by means of the finite element method. **Figure 4** gives the theoretical losses of the fundamental $HE_{11}$ air-core mode for $As_2S_3$ and $As_{30}Se_{50}Te_{20}$ glass fibers for the structure illustrated in **Figure 3**, with air-core diameter $D_{core}$ of 223 μm, the outer diameter of capillaries of 178 μm, the

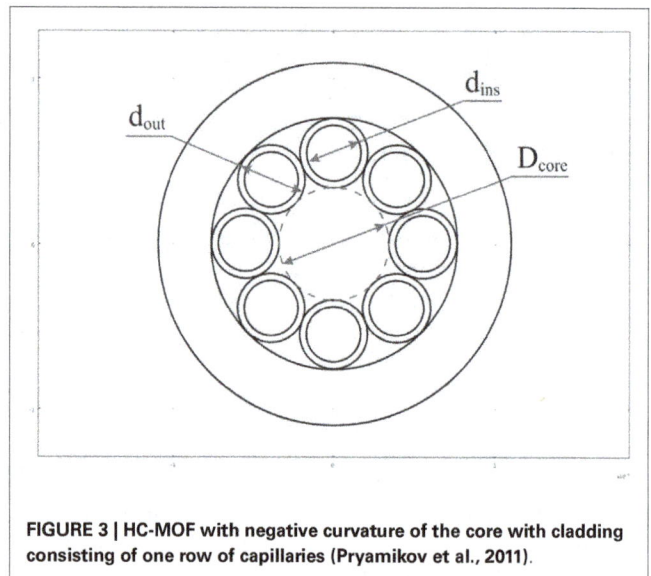

**FIGURE 3 | HC-MOF with negative curvature of the core with cladding consisting of one row of capillaries (Pryamikov et al., 2011).**

ratio of the inner and outer diameters of the capillaries of 0.8 (Shiryaev et al., 2014a). It can be seen that the spectrum of ideal fiber with negative curvature of the hollow-core is irregular, the optical losses change in the transmission windows from a minimum value of 0.001 dB/m to a maximum value of 100 dB/m. Such irregular behavior of optical losses can be explained by the weak coupling of the core modes with the dielectric modes having a high Azimuthal dependence. Theoretical spectra of optical losses in ideal NCHCFs based on $As_2S_3$ and $As_{30}Se_{50}Te_{20}$ glasses are similar, but transmittance bands for $As_2S_3$ are wider.

The calculated bending losses for fiber of arsenic sulfide were calculated to be about 3–4 times lower as compared with $As_{30}Se_{50}Te_{20}$ (**Figure 5**) (Shiryaev et al., 2014a). The losses of the fundamental $HE_{11}$ air-core mode for small bending radii are characterized by the resonance behavior. The theoretical losses in $As_2S_3$ NCHCF in the field of $CO_2$-laser radiation at a wavelength of 10.6 μm were about 0.1 dB/m.

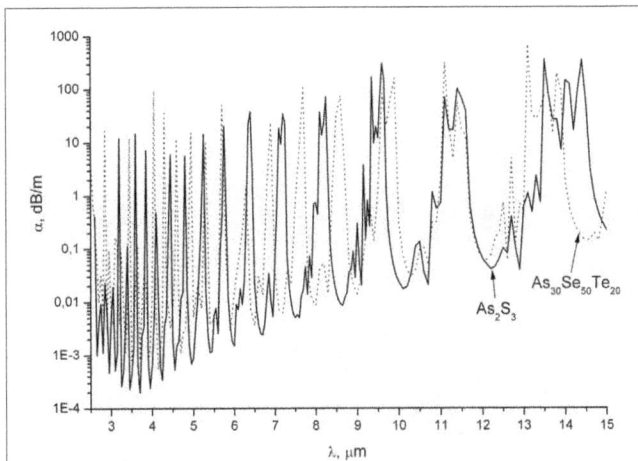

FIGURE 4 | The calculated transmission bands for ideal NCHCF (eight capillaries; $D_{core} = 223\,\mu m$, $d_{in}/d_{out} = 0.8$) for $As_2S_3$ (solid line) and $As_{30}Se_{50}Te_{20}$ (dotted line) glasses (Shiryaev et al., 2014b).

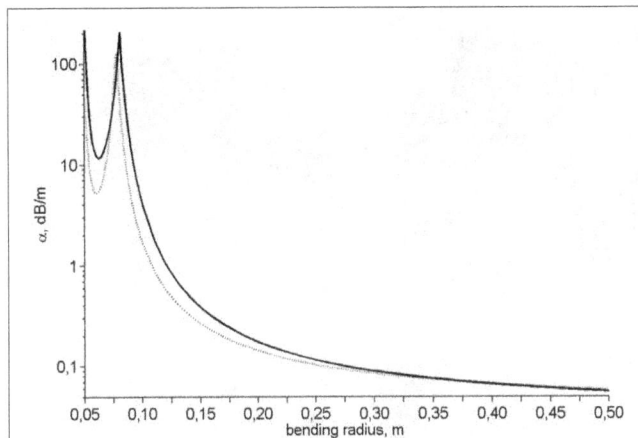

FIGURE 5 | Calculated dependence of the bending losses of the fundamental $HE_{11}$ air-core mode on bending radius at $\lambda = 10.3\,\mu m$ for $As_2S_3$ (dotted line) and $As_{30}Se_{50}Te_{20}$ (solid line) glasses (Shiryaev et al., 2014b).

In paper (Pryamikov et al., 2012), four types of HC-MOFs with the claddings consisting of 6, 8, 10, and 12 capillaries were considered. The calculations were carried out for two values of the glass refractive indices $n = 2.4$, 2.8 and with three values of $d_{in}/d_{out} = 0.8$, 0.85, 0.9. All calculations were made in the narrow spectral region near $\lambda = 10.6\,\mu m$. It has been shown that the achievement of a low loss waveguide regime for HC MFs with the cladding consisting of capillaries is complicated multi parameter task. All the parameters characterizing the HC MFs such as $D_{core}$ (hollow-core diameter), $d_{in}$ (inner diameter of capillaries), $d_{out}$ (outside diameter of capillaries), $n$, $N$ (number of the capillaries in the cladding) have an effect on the waveguide regime in the considered spectral range. In this way, two main factors affect the loss level of the HC-MOFs: the density of eigen states of the individual capillary and the discrete rotational symmetry of the core boundary. The first factor is determined by geometry parameters

of a capillary and the value of a glass refractive index. The second factor is connected to the symmetry of the capillary arrangement in the cladding. It was established that a balance between the number of capillaries and the air-core diameter should be found.

The authors of paper (Wei et al., 2014) have carried out the simulating analysis for chalcogenide negative curvature hollow-core MOFs with different capillary wall thickness ($t$) to diameter ratios, $t/d_{out}$ and different number of capillaries, $N$, of 8, 10, 12, and 14. They have found that leakage loss and power ratio in the glass decrease as the number of capillaries increases or tube wall thickness to diameter ratio decreases.

## PREPARATION OF PREFORMS FOR CHALCOGENIDE NEGATIVE CURVATURE FIBERS

The description of preparation of fiber preforms for chalcogenide negative curvature fibers is given in papers (Kosolapov et al., 2011) and (Shiryaev et al., 2014a). These performs were manufactured by the "stack and draw" technique from a substrate tube and 8 or 10 capillaries of the same chalcogenide glass. For this, high-purity chalcogenide glasses with low tendency to crystallization such as $As_2S_3$, $As_{40}S_{30}Se_{30}$, and $As_{30}Se_{50}Te_{20}$ were used.

The $As_2S_3$ glass was produced by melting purified arsenic monosulfide ($As_4S_4$) as the arsenic-containing component with the required amount of elementary sulfur in a sealed silica ampoule (Snopatin et al., 2009a). The preparation method of high-purity $As_{40}S_{30}Se_{30}$ and $As_{30}Se_{50}Te_{20}$ glasses included the purification of glass-forming chalcogenides by chemical distillation methods, the melting of the purified fractions up to homogeneous state, and the melt solidification (Shiryaev et al., 2004; Shiryaev and Balda, 2006).

The obtained $As_{40}S_{30}Se_{30}$ and $As_{30}Se_{50}Te_{20}$ glasses had a low content of the limiting impurities: hydrogen – <0.06 ppm wt, oxygen – 0.2 ppm wt, carbon – 1 ppm wt, silicon – 0.5 ppm wt (as follows from laser mass spectroscopy and IR spectroscopy). The content of limiting impurities in $As_2S_3$ glass was as follows: carbon – <0.1 ppm wt, hydrogen as SH group – <0.5 ppm wt, silicon – $\leq$0.5 ppm wt, metals – <0.1 ppm wt. The optical loss spectra of mono-index fibers drawn from these glasses are given in the **Figures 6A,B**.

The substrate tubes with the outer diameter of 16 mm, the inner diameter of 11 mm, and the length of 140–170 mm (**Figure 7A**) were fabricated by centrifugal casting of the chalcogenide glass melt inside an evacuated silica tube (Kosolapov et al., 2011; Shiryaev and Churbanov, 2013; Shiryaev et al., 2014a).

The capillaries with the outer diameter of 2.4–3.2 mm were prepared by the double crucible method from the chalcogenide glass melt. For this, the chalcogenide glass was placed in the cladding crucible, while the core crucible remained empty. The capillary wall thickness was determined by the geometrical dimensions of the double crucible die, the excess inert gas pressure over the melt, and the pressure inside the capillary. The cross-sections of the capillaries had a coaxial geometry with a ratio of internal and outside diameters of 0.8–0.9. Concentricity of obtained capillaries was over 80%.

The final preform consisted of the substrate tube and 8 or 10 capillaries forming a complete azimuthally symmetric layer of holes on the inner surface of the tube (**Figures 7B** and **8A**).

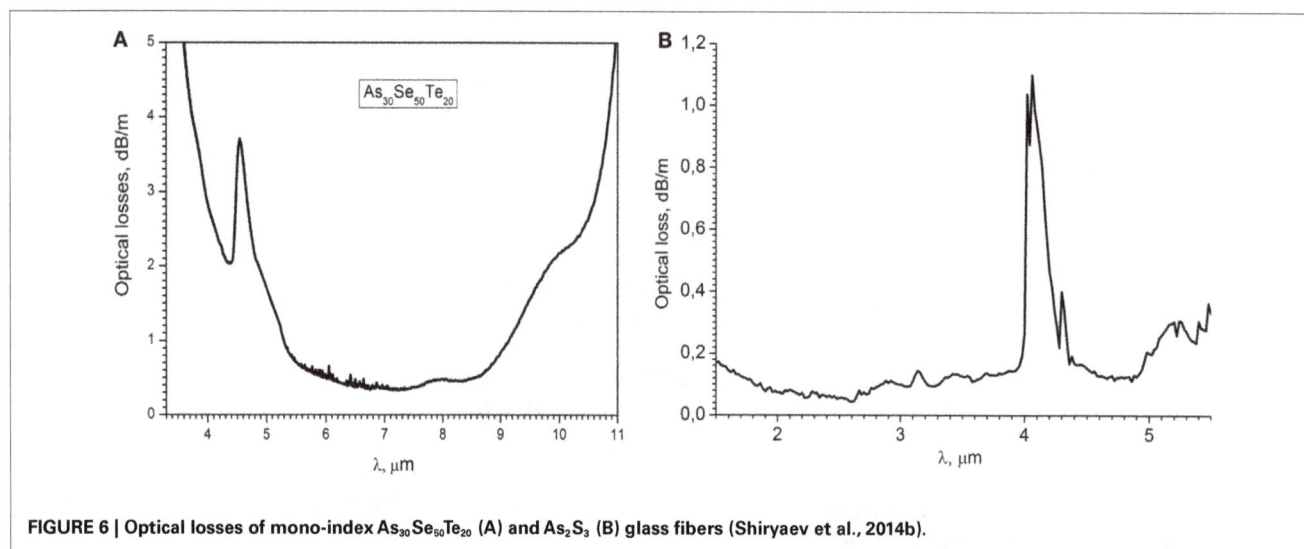

**FIGURE 6 | Optical losses of mono-index $As_{30}Se_{50}Te_{20}$ (A) and $As_2S_3$ (B) glass fibers (Shiryaev et al., 2014b).**

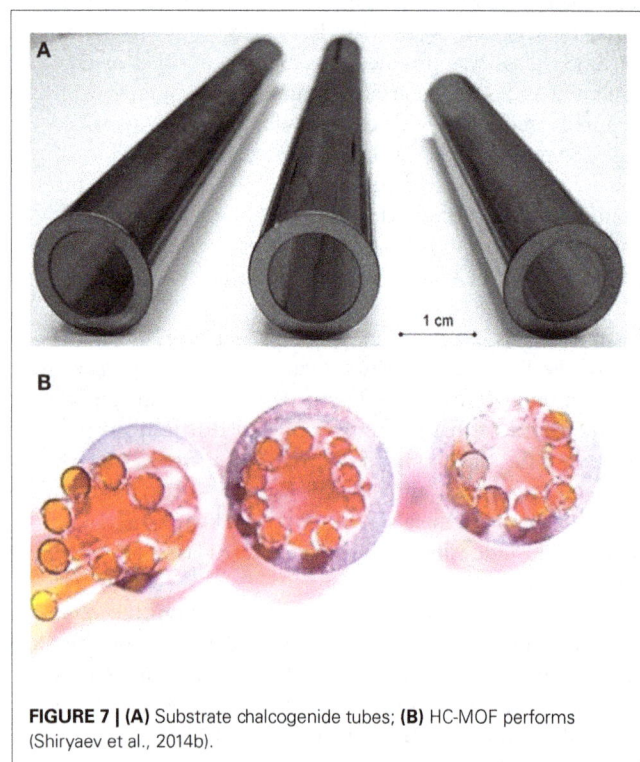

FIGURE 7 | (A) Substrate chalcogenide tubes; (B) HC-MOF performs (Shiryaev et al., 2014b).

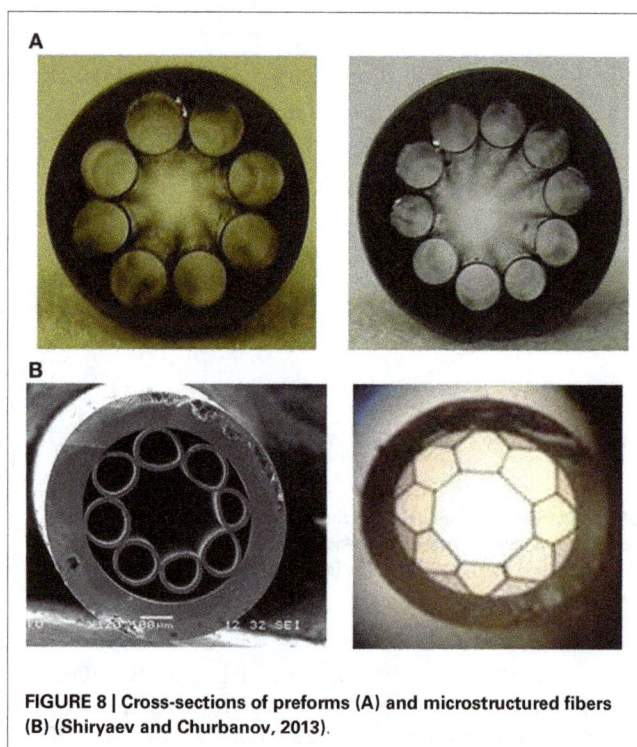

**FIGURE 8 | Cross-sections of preforms (A) and microstructured fibers (B) (Shiryaev and Churbanov, 2013).**

## ADHESION OF CHALCOGENIDE GLASSES TO SILICA GLASS

The realization of the centrifugal casting method of chalcogenide melt inside a silica ampoule is associated with the need to determine the time-temperature conditions to obtain the perfect substrate tube. However, the substrate chalcogenide tubes, especially based on arsenic sulfide glass, were found to crack due to the strong adhesion of chalcogenide glass to silica glass at annealing temperatures. Therefore, to determine the optimal conditions for manufacturing the perfect substrate chalcogenide tubes, the investigation of temperature dependence of chalcogenide glass adhesion to silica glass was carried out (Shiryaev et al., 2014b; Shiryaev

et al., 2015). It was carried out by the steady detachment method. This method is used to measure the force required to separate the adhesive from the substrate simultaneously in all areas of contact.

Results of the temperature dependence of separation tension of six chalcogenide glasses from silica glass are shown in **Figure 9** (Shiryaev et al., 2015). These dependences show that the adhesion strength has the maximum near $T_g \pm 10°C$ of chalcogenide glasses. Adhesion strength exceeds the value of 2500 kPa for $As_{40}S_{60}$ and $As_{40}S_{30}Se_{30}$ glasses at temperatures above (170–185)°C.

The adhesive strength was determined to increase with contact temperature, as well as with the exposure time of assembly up to the stationary value, depending on the contact temperature. The investigation of chalcogenide glass composition on the value of adhesion to the silica glass showed, that the adhesion strength of chalcogenide glasses to silica glass increases with decrease in chalcogen molar mass and with increase in chalcogen content in the glass composition. The obtained results made it possible to explain the observed experimental data for forming chalcogenide glass tube preforms and to optimize the process for their preparation.

Adhesion is the critical parameter during the annealing of the samples in the form of rods and tubes near the glass-transition temperature (185°C). To prevent the destruction of preforms during the cooling process, a separation of the surface of the chalcogenide glass sample from the walls of the silica ampoule, executing a forced local cooling of interface of chalcogenide and silica glasses, must be used.

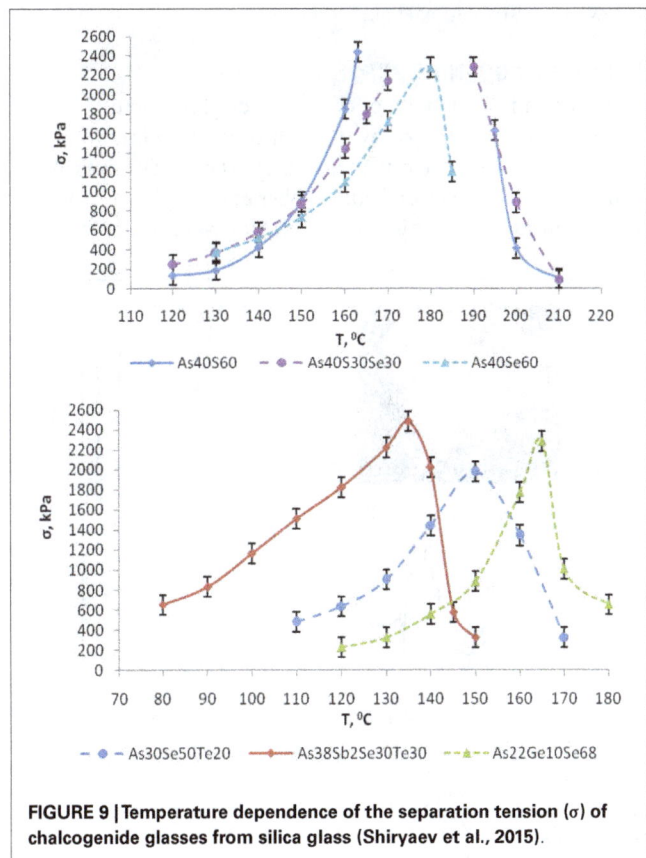

FIGURE 9 | Temperature dependence of the separation tension (σ) of chalcogenide glasses from silica glass (Shiryaev et al., 2015).

## PREPARATION OF CHALCOGENIDE NEGATIVE CURVATURE FIBERS

The prepared preforms were used to draw NCHCFs (Kosolapov et al., 2011; Shiryaev and Churbanov, 2013; Shiryaev et al., 2014a). For this, the overpressure of dry inert gas (argon) was established in each capillary and inside of the substrate tube. The preforms with an outside diameter of 16 mm were drawn in fibers with diameter of 700–900 μm. To prepare a defined structure of NCHCF, the conditions of fiber drawing were determined by experiment. A high temperature of fiber drawing resulted in the random distribution of capillaries inside the substrate tube (**Figures 10A–C**). In case of $As_2S_3$ glass, the NCHCF had a structure close to the defined one, at a drawing temperature of 310°C and an excess argon pressure in capillaries of 4000 Pa. The capillary rings had no discontinuities, and the capillaries themselves had deviations from the original spherical shape due to conditions of fiber drawing (**Figures 8B and 10D,E**).

## OPTICAL TRANSMISSION OF CHALCOGENIDE NEGATIVE CURVATURE FIBERS

The fiber loss spectra of as-drawn NCHCFs (Kosolapov et al., 2011; Shiryaev and Churbanov, 2013; Shiryaev et al., 2014a) were

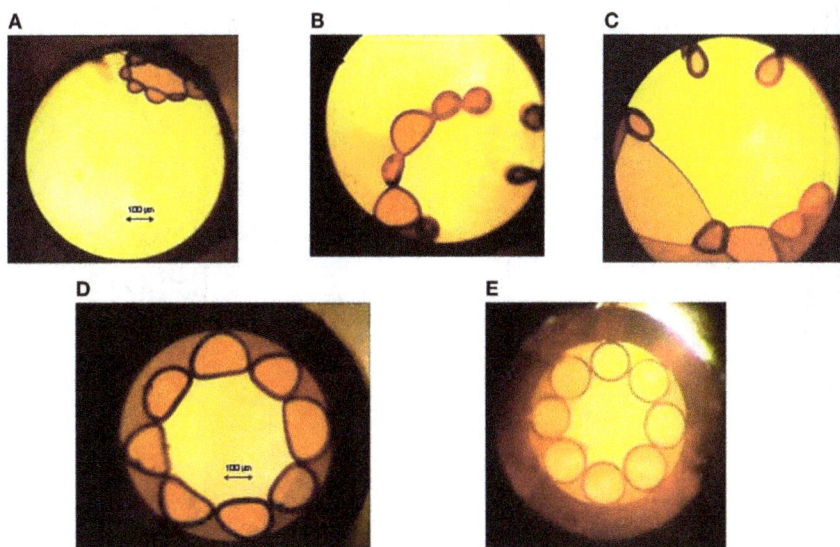

FIGURE 10 | The cross-sections of $As_2S_3$ MOFs obtained under different drawing conditions. The temperature: (**A–C**) 330–340°C, (**D**) 320°C, (**E**) 310°C. The excess argon pressure in capillaries: (**A**) 0, (**B**) 100 Pa, (**C**) 600 Pa, (**D**) 1000 Pa, (**E**) 4000 Pa (Shiryaev et al., 2014b).

measured by the cut-back method on a Bruker IFS-113v Fourier transform IR spectrometer. The measured length of optical fibers was from 1 to 2 m. In this experiment, a radiation propagated not only through the air-core but also through the substrate tube and the cladding capillaries. To remove the influence of cladding modes, a liquid gallium–indium immersing alloy was applied on the surface of input and output ends of fiber.

The spectra of optical losses of prepared $As_2S_3$ NCHCFs with diameter of 700 μm illustrated in **Figures 10D,E** are given in **Figure 11** (Shiryaev et al., 2014b). The optical losses in fibers without Ga–In immersion were higher than in Ga–In immersed fibers. Our best fiber was transparent in the 1.5–8.2 μm wavelength range and has minimum loss about 3 dB/m at 4.8 μm. Unfortunately, the obtained MOFs contained the absorption bands of impurity associated with arsenic sulfide. There are some bands due to S–H bonds at 6.8; 4.1; 3.7; 3.1 μm; due to $CO_2$ impurity at 4.31 and 4.34 μm; due to OH groups at 2.92 μm; due to COS at 4.9 μm; and due to molecular $H_2O$ at 6.33 μm.

The spectrum of optical losses of prepared $As_2S_3$ NCHCF with 10 capillaries is given in **Figure 12**. The minimum loss in this fiber was about 1.2 dB/m in the range of 2.7–3.4 μm.

Investigation of the as-prepared hollow-core microstructured fibers based on $As_{30}Se_{50}Te_{20}$ and $As_{40}S_{30}Se_{30}$ glasses has shown their transparency in the range of 2–11 μm and possibility to transfer of the $CO_2$-laser radiation. Minimum optical loss in As–Se–Te glass hollow-core PCF was about 4 dB/m at 9.5 μm (6.5 dB/m at 10.6 μm) (**Figure 13**) (Kosolapov et al., 2011), and one in As–S–Se glass hollow-core PCF was 30 dB/m at 10.6 μm and 10 dB/m at 2 μm) (Shiryaev and Churbanov, 2013).

To investigate the distribution of the $CO_2$-laser radiation intensity over the fiber cross-section, a ZnSe lens was used to launch the radiation into the fiber, while the intensity distribution over the fiber output endface was observed with the help of an Electrophysics PV320 thermal imaging camera. It was found that a much greater power can be transmitted through the core, which means that the light propagating through the glass owing to the total internal reflection experiences much stronger attenuation. The distribution in **Figure 14** obtained under proper excitation conditions shows that the $CO_2$-laser radiation is well confined in the hollow-core (Kosolapov et al., 2011).

## AGING OF CHALCOGENIDE PCFs

The efficiency and the stability of chalcogenide glass microstructured optical fibers are limited by the shift of their optical properties that occurs over time due to an aging process. The optical aging on $As_2S_3$ microstructured optical fiber upon exposure to air was reported in papers (Toupin et al., 2014; Mouawad et al., 2014).

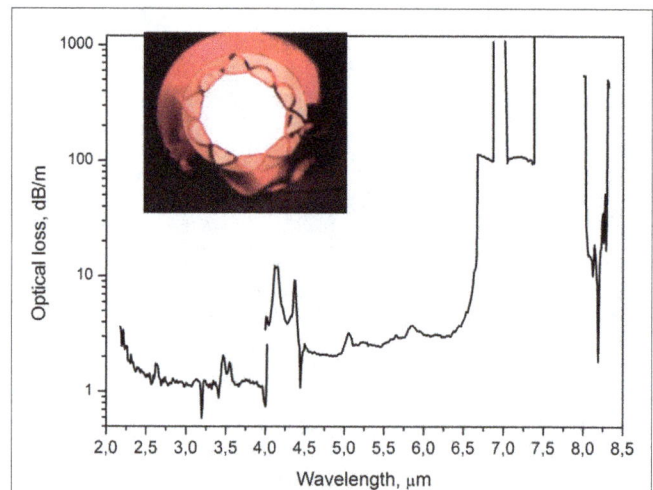

FIGURE 12 | Spectrum of optical losses of as-prepared $As_2S_3$ NCHCF with 10 capillaries.

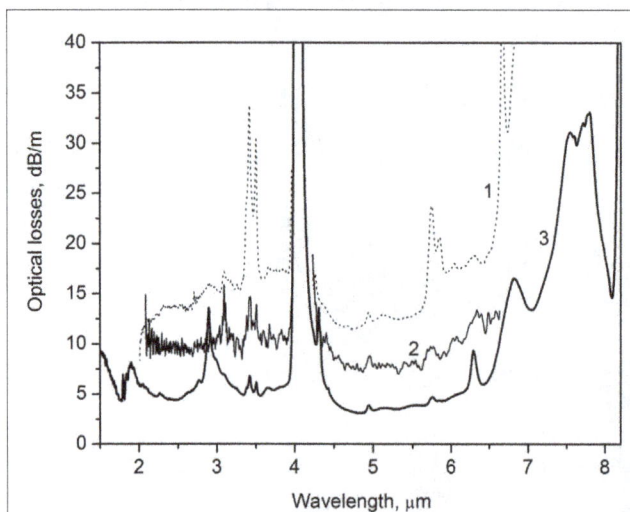

FIGURE 11 | The spectra of optical losses in prepared $As_2S_3$ NCHCFs with diameter of 700 μm: (1) fiber illustrated in Figure 10D without Ga immersion; (2) fiber illustrated in Figure 10D with Ga–In immersion; (3) fiber illustrated in Figure 10E with Ga–In immersion (Shiryaev et al., 2014b).

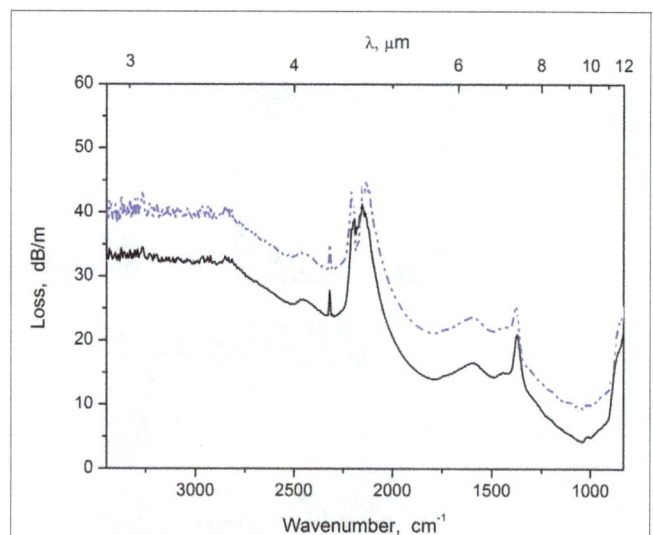

FIGURE 13 | Measured optical loss spectra of the As–Se–Te microstructured hollow-core fibers (two different pieces) (Kosolapov et al., 2011).

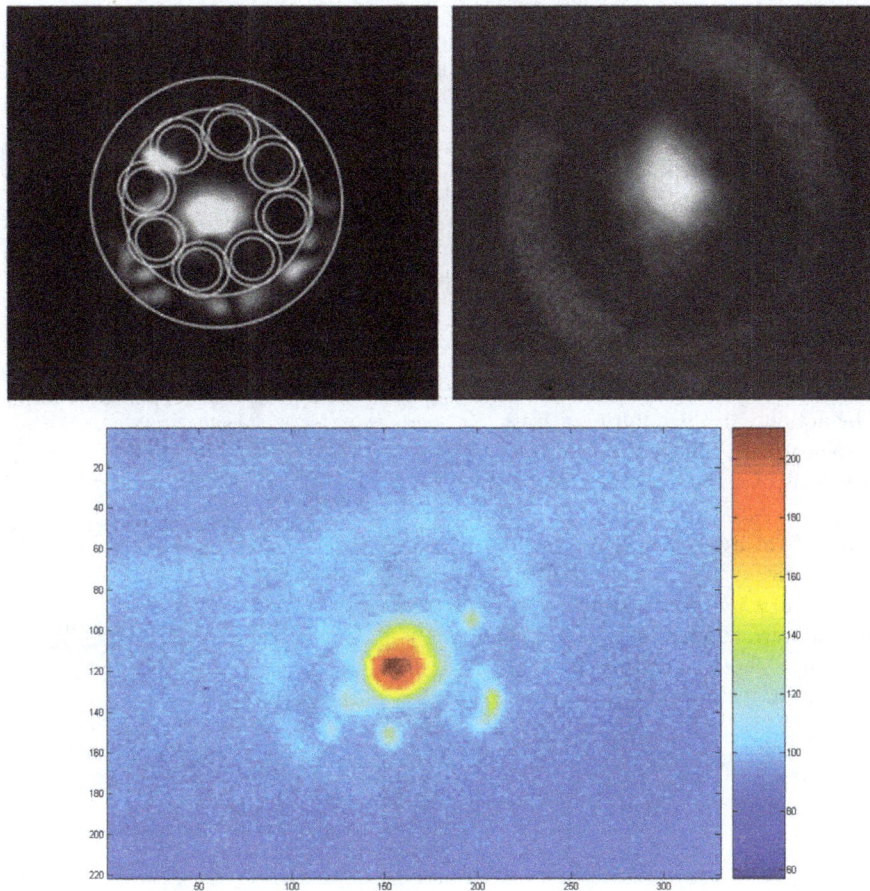

**FIGURE 14 | Intensity distribution of CO₂-laser radiation over the fiber cross-section (Kosolapov et al., 2011).**

The optical aging was established to associate with a dynamic grow of OH and $H_2O$ attributed absorption bands.

Paper (Toupin et al., 2014) describes the optical aging of MOFs prepared from four glass compositions ($As_{30}Se_{50}Te_{20}$, $As_{38}Se_{62}$, $Ge_{10}As_{22}Se_{68}$, and $As_{40}S_{60}$) and stored in air. The results have shown that the same absorption bands associated with the presence of OH hydroxyl groups and molecular water appear whatever the glass composition of the fiber. The continuous evolution of the concentrations of hydroxyl groups in the MOFs as a function of exposure period was observed. These changes upon exposure period have shown that the harmful effect of atmospheric moisture in the holes of the MOF occurs rapidly over the first few hours and is the most important in the early centimeters of the MOF. The content of hydroxyl groups decreases exponentially with distance away from the MOF extremity. However, the growing rate of absorption peaks depends on the glass composition. Optical aging as a function of fiber composition was summarized as: $As_{40}S_{60} < Ge_{10}As_{22}Se_{68} < As_{38}Se_{62} < As_{30}Se_{50}Te_{20}$. The value of "optical contamination pollution rate" at 6.3 μm wavelength was estimated to be 0.16, 0.02, 0.001, and 0.0003 dB/m/day for $As_{40}S_{60}$, $Ge_{10}As_{22}Se_{68}$, $As_{38}Se_{62}$, and $As_{30}Se_{50}Te_{20}$, respectively.

The authors of paper (Mouawad et al., 2014) have demonstrated the deleterious time evolution of $As_2S_3$ MOFs upon exposure of the MOFs core to atmospheric moisture. FTIR experiments have shown that an increase in the OH and SH content in the MOFs causes extra losses. The exposure of $As_2S_3$ glass to atmospheric conditions was determined to induce the formation and the growth of pyramidal defects on the glass surface. These surface structures are $As_2O_3$ crystals according to the XDS results and arise from reaction with the atmosphere. Substantial increase in the intensity of OH absorption bands occurring over the first hours requires the specific storage of $As_2S_3$ MOFs in dry conditions, immediately after the drawing process. To decrease the chemical aging process of the MOF core, a protection of the MOF holes from diffusion of atmospheric steam is required. For that purpose, the authors (Mouawad et al., 2014) have airproofed the fiber ends by means of a methacrylate-based polymer. The fiber ends were soaked in the liquid polymeric solution, and then let to polymerize in free atmosphere at room temperature.

## SUMMARY

The review on the recent developments on chalcogenide glass HC-MOFs demonstrates their attractive perspectives for mid-IR applications, as well as the trends and challenges of the technique of manufacturing.

The calculation of the structural design of chalcogenide NCHCF has shown the possibility of achieving optical losses less than 1 dB/km in the mid-IR spectral region. NCHCFs with a hollow-core are promising especially for the transmission of high optical power.

Preparation of chalcogenide glass capillaries by extrusion from a double crucible, in contrast to commonly used capillary drawing of a tubular preform, eliminates the additional step of heat treatment of the sample to reduce the crystallization and to obtain the smooth outer and inner surfaces of the capillaries.

The prepared NCHCFs of chalcogenide glass were suitable for the mid-IR radiation transmission, although the optical losses were higher than the theoretically predicted level.

The achievement of optical losses close to the theoretically calculated level is determined by not only the high purity of glass on impurities and its optical homogeneity, but also the high accuracy of the geometrical dimensions of a given photonic crystal structure.

Thus, to prepare the hollow-core microstructured fibers with low optical losses in the mid-IR range (special interest in fibers transmitted at wavelengths of 5–6 and 9.3–10.6 $\mu$m), further optimization of their design and fiber drawing conditions is required to prevent the deformation of their photonic crystal structure, which is also one of the reasons for the increase in optical losses.

# REFERENCES

Benabid, F. (2006). Hollow-core photonic bandgap fibre: new light guidance for new science and technology. *Phil. Trans. R. Soc. A* 364, 3439–3462. doi:10.1098/rsta.2006.1908

Benabid, F., Couny, F., Knight, J. C., Birks, T. A., and Russell, P. (2005). Compact, stable and efficient all-fibre gas cells using hollow-core photonic crystal fibres. *Nature* 434, 488–491. doi:10.1038/nature03349

Birks, T. A., Knight, J. C., and Russell, P. (1997). Endlessly single-mode photonic crystal fiber. *Opt. Lett.* 22, 961–963. doi:10.1364/OL.22.000961

Brilland, L., Smektala, F., Renversez, G., Chartier, T., Troles, J., Nguyen, T., et al. (2006). Fabrication of complex structures of holey fibers in chalcogenide glass. *Opt. Express* 14, 1280–1285. doi:10.1364/OE.14.001280

Conseil, C., Coulombier, Q., Boussard-Pledel, C., Troles, J., Brilland, L., Renversez, G., et al. (2011). Chalcogenide step index and microstructured single mode fibers. *J. Non Cryst. Solids* 357, 2480–2483. doi:10.1016/j.jnoncrysol.2010.11.090

Coulombier, Q., Brilland, L., Houizot, P., Chartier, T., N'Guyen, T. N., Smektala, F., et al. (2010). Casting method for producing low-loss chalcogenide microstructured optical fibers. *Opt. Express* 18, 9107–9112. doi:10.1364/OE.18.009107

Cregan, R. F., Mangan, B. J., Knight, J. C., Birks, T. A., Russell, P. J., Roberts, P. J., et al. (1999). Single-mode photonic band gap guidance of light in air. *Science* 285, 1537–1539. doi:10.1126/science.285.5433.1537

Desevedavy, F., Renversez, G., Brilland, L., Houizot, P., Troles, J., Coulombier, Q., et al. (2008). Small-core chalcogenide microstructured fibers for the infrared. *Appl. Opt.* 47, 6014–6021. doi:10.1364/AO.47.006014

Desevedavy, F., Renversez, G., Troles, J., Houizot, P., Brilland, L., Vasilief, I., et al. (2010). Chalcogenide glass hollow core photonic crystal fibers. *Opt. Mater.* 32, 1532–1539. doi:10.1016/j.optmat.2010.06.016

El-Amraoui, M., Fatome, J., Jules, J. C., Kibler, B., Gadret, G., Fortier, C., et al. (2010). Strong infrared spectral broadening in low-loss As-S chalcogenide suspended core microstructured optical fibers. *Opt. Express* 18, 4547–4556. doi:10.1364/OE.18.004547

Gibson, D. J., and Harrington, J. A. (2004). Extrusion of hollow waveguide performs with a one-dimensional photonic bandgap structure. *J. Appl. Phys.* 95, 3895–3900. doi:10.1063/1.1667277

He, J., Xiong, C., Clark, A., Collins, M., Gai, X., Choi, D., et al. (2012). Effect of low-Raman window position on correlated photon-pair generation in a chalcogenide $Ge_{11.5}As_{24}Se_{64.5}$ nanowire. *J. Appl. Phys.* 112, 1–5. doi:10.1063/1.4769740

Hu, K., Kabakova, I., Buttner, T., Lefrancois, S., Hudson, D., He, S., et al. (2014). Low-threshold Brillouin laser at 2 mum based on suspended-core chalcogenide fiber. *Opt. Lett.* 39, 4651–4654. doi:10.1364/OL.39.004651

Jiang, X., Euser, T. G., Abdolvand, A., Babic, F., Tani, F., Joly, N. Y., et al. (2011). Single-mode hollow-core photonic crystal fiber made from soft glass. *Opt. Express* 19, 15438–15444. doi:10.1364/OE.19.015438

Katsuyama, T., and Matsumura, H. (1986). Low loss Te-based chalcogenide glass optical fibers. *Appl. Phys. Lett.* 49, 22–23. doi:10.1063/1.97089

Knight, J. C., Birks, T. A., Russell, P. J., and Atkin, D. M. (1996). All-silica single-mode fiber with photonic crystal cladding. *Opt. Lett.* 21, 1547–1549. doi:10.1364/OL.21.001547

Knight, J. C., Birks, T. A., Russell, P. J., and Sandro, J. P. (1998). Properties of photonic crystal fiber and the effective index model. *J. Opt. Soc. Am. A* 15, 748–752. doi:10.1364/JOSAA.15.000748

Kosolapov, A. F., Pryamikov, A. D., Biriukov, A. S., Shiryaev, V. S., Astapovich, M. S., Philippovsky, D. V., et al. (2011). Demonstration of $CO_2$ laser power delivery through chalcogenide glass fiber with hollow negative-curvature core. *Opt. Express* 19, 25723–25728. doi:10.1364/OE.19.025723

Monro, T. M., Richardson, D. J., Broderick, N. G. R., and Bennett, P. J. (1999). Holey optical fibers: an efficient modal model. *J. Lightwave Technol.* 17, 1093–1101. doi:10.1109/50.769313

Monro, T. M., West, Y. D., Hewak, D. W., Broderick, N. G. R., and Richardson, D. J. (2000). Chalcogenide holey fibres. *Electron. Lett.* 36, 1998–2000. doi:10.1049/el:20001394

Mouawad, O., Strutynski, C., Picot-Clémente, J., and Désévédavy, F. (2014). Optical aging behaviour naturally induced on $As_2S_3$ microstructured optical fibres. *Opt. Mater. Express* 4, 2190–2203. doi:10.1364/OME.4.002190

Nishii, J., Morimoto, S., Inagawa, I., Iizuka, R., Yamashita, T., and Yamagishi, T. (1992). Recent advances and trends in chalcogenide glass fiber technology: a review. *J. Non Cryst. Solids* 140, 199–208. doi:10.1016/S0022-3093(05)80767-7

Pryamikov, A. D., Biriukov, A. S., Kosolapov, A. F., Plotnichenko, V. G., Semjonov, S. L., and Dianov, E. M. (2011). Demonstration of a waveguide regime for a silica hollow – core microstructured optical fiber with a negative curvature of the core boundary in the spectral region > 3.5 $\mu$m. *Opt. Express* 19, 1441–1448. doi:10.1364/OE.19.001441

Pryamikov, A. D., Kosolapov, A. F., Plotnichenko, V. G., and Dianov, E. M. (2012). "Transmission of CO2 laser radiation through glass hollow core microstructured fibers," in *CO2 Laser – Optimisation and Application*, ed. D. C. Dumitras (InTech). Available from: http://www.intechopen.com/books/co2-laser-optimisation-and-application/transmission-of-co2-laser-radiationthrough-glass-hollow-core-microstructured-fibers

Revathi, S., Inbathini, S. R., and Saifudeen, R. A. (2014). Highly nonlinear and birefringent spiral photonic crystal fiber. *Adv. Optoelectron.* 2014, 6. doi:10.1155/2014/464391

Shiryaev, V. S., Adam, J.-L., Zhang, X. H., Boussard-Plédel, C., Lucas, J., and Churbanov, M. F. (2004). Infrared fibers based on Te-As-Se glass system with low optical losses. *J. Non Cryst. Solids* 336, 113–119. doi:10.1016/j.jnoncrysol.2004.01.006

Shiryaev, V. S., and Balda, R. (2006). "Preparation of high purity chalcogenide glasses and fibers," in *Photonic Glasses*, ed. R. Balda (Kerala, India: Research Signpost), 151–195.

Shiryaev, V. S., and Churbanov, M. F. (2013). Trends and prospects of development of chalcogenide fibers for mid infrared transmission. *J. Non Cryst. Solids* 377, 225–230. doi:10.1016/j.jnoncrysol.2012.12.048

Shiryaev, V. S., Churbanov, M. F., Dianov, E. M., Plotnichenko, V. G., Adam, J.-L., and Lucas, J. (2005). Recent progress in preparation of chalcogenide As-Se-Te glasses with low impurity content. *J. Optoelectron. Adv. Mater.* 7, 1773–1779.

Shiryaev, V. S., Kosolapov, A. F., Pryamikov, A. D., Snopatin, G. E., Churbanov, M. F., Biriukov, A. S., et al. (2014a). Development of technique for preparation of $As_2S_3$ glass preforms for hollow core microstructured optical fibers. *J. Optoelectron. Adv. Mater.* 16, 1020–1025.

Shiryaev, V. S., Mishinov, S. V., and Churbanov, M. F. (2014b). Adhesion of glassy arsenic sulfide to quartz glass. *Inorg. Mater.* 50, 1157–1161. doi:10.1134/S0020168514110181

Shiryaev, V. S., Mishinov, S. V., and Churbanov, M. F. (2015). Investigation of adhesion of chalcogenide glasses to silica glass. *J. Non Cryst. Solids.* 408, 71–75. doi:10.1016/j.jnoncrysol.2014.10.010

Skryabin, D. V., and Wadsworth, W. J. (2010). "Nonlinear optics and solitons in photonic crystal fibres," in *Nonlinearities in Periodic Structures and Metamaterials*,

*Springer Series in Optical Sciences*, Vol. 150, eds C. Denz, S. Flach, and Y. S. Kivshar (Berlin: Springer), 37–54.

Snopatin, G. E., Churbanov, M. F., Pushkin, A. A., Gerasimenko, V. V., Dianov, E. M., and Plotnichenko, V. G. (2009a). High purity arsenic-sulfide glasses and fibers with minimum attenuation of 12 dB/km. *J. Opt. Adv. Mater. Rapid Commun.* 3, 669–671.

Snopatin, G. E., Shiryaev, V. S., Plotnichenko, G. E., Dianov, E. M., and Churbanov, M. F. (2009b). High-Purity Chalcogenide Glasses for Fiber Optics. *Inorg. Mater.* 45, 1439–1460. doi:10.1134/S0020168509130019

Temelkuran, B., Hart, S. D., Benoit, G., Joannopoulos, J. D., and Fink, Y. (2002). Wavelength-scalable hollow optical fibres with large photonic bandgaps for $CO_2$ laser transmission. *Nature* 420, 650–653. doi:10.1038/nature01275

Toupin, P., Brilland, L., Mechin, D., Adam, J.-L., and Troles, J. (2014). Optical aging of chalcogenide microstructured optical fibers. *J. Lightwave Technol.* 32, 2428–2432. doi:10.1109/JLT.2014.2326461

Wadsworth, W. J., Knight, J. C., Ortigosa-Blanch, A., Arriaga, J., Silvestre, E., and Russell, P. S. J. (2001). Soliton effects in photonic crystal fibres at 850 nm. *Electron. Lett.* 36, 53–55. doi:10.1049/el:20000134

Wei, C., Kuis, R., Chenard, F., and Hu, J. (2014). "Chalcogenide negative curvature hollow-core photonic crystal fibers with low loss and low power ratio in the glass," in *CLEO: 2014, Laser Science to Photonic Applications Technical Conference* (San Jose, CA: SM1N), 5.

Yablonovitch, E. (1987). Inhibited spontaneous emission in solid-state physics and electronics. *Phys. Rev. Lett.* 58, 2059–2062. doi:10.1103/PhysRevLett.58.2059

Yeh, P., Yariv, A., and Marom, E. (1978). Theory of Bragg fiber. *J. Opt. Soc. Am.* 68, 1196–1201. doi:10.1364/JOSA.68.001196

Zhang, Y., Schroeder, J., Husko, C., Lefrancois, S., Choi, D., Madden, S., et al. (2014). Pump degenerate phase-sensitive amplification in chalcogenide waveguides. *J. Opt. Soc. Am. B Opt. Phys.* 31, 780–787. doi:10.1364/JOSAB.31.000780

**Conflict of Interest Statement:** The author declares that the research was conducted in the absence of any commercial or financial relationships that could be construed as a potential conflict of interest.

# 14

# Strain localization and shear band propagation in ductile materials

**Nicola Bordignon, Andrea Piccolroaz, Francesco Dal Corso and Davide Bigoni ***

*Department of Civil, Environmental and Mechanical Engineering (DICAM), University of Trento, Trento, Italy*

*Edited by:*
*Simone Taioli, Bruno Kessler*
*Foundation, Italy*

*Reviewed by:*
*Gianni F. Royer Carfagni, University of Parma, Italy*
*Stefano Vidoli, Sapienza University of Rome, Italy*

*\*Correspondence:*
*Davide Bigoni, via Mesiano 77, Trento I-38123, Italy*
*e-mail: bigoni@ing.unitn.it*

A model of a shear band as a zero-thickness non-linear interface is proposed and tested using finite element simulations. An imperfection approach is used in this model where a shear band that is assumed to lie in a ductile matrix material (obeying von Mises plasticity with linear hardening), is present from the beginning of loading and is considered to be a zone in which yielding occurs before the rest of the matrix. This approach is contrasted with a perturbative approach, developed for a $J_2$-deformation theory material, in which the shear band is modeled to emerge at a certain stage of a uniform deformation. Both approaches concur in showing that the shear bands (differently from cracks) propagate rectilinearly under shear loading and that a strong stress concentration should be expected to be present at the tip of the shear band, two key features in the understanding of failure mechanisms of ductile materials.

Keywords: slip lines, plasticity, failure, stress concentration, rectilinear growth

## 1. INTRODUCTION

When a ductile material is brought to an extreme strain state through a uniform loading process, the deformation may start to localize into thin and planar bands, often arranged in regular lattice patterns. This phenomenon is quite common and occurs in many materials over a broad range of scales: from the kilometric scale in the earth crust (Kirby, 1985), down to the nanoscale in metallic glass (Yang et al., 2005), see the examples, reported in **Figure 1**.

After localization, unloading typically[1] occurs in the material outside the bands, while strain quickly evolves inside, possibly leading to final fracture (as in the examples shown in **Figure 2**, where the crack lattice is the signature of the initial shear band network[2]) or to a progressive accumulation of deformation bands (as, for instance, in the case of the drinking straws, or of the iron meteorite, or of the uPVC sample shown in **Figure 1**, or in the well-known case of granular materials, where fracture is usually absent and localization bands are made up of material at a different relative density, Gajo et al., 2004).

It follows from the above discussion that as strain localization represents a prelude to failure of ductile materials, its mechanical understanding paves the way to the innovative use of materials in extreme mechanical conditions. For this reason, shear bands have been the focus of a thorough research effort. In particular, research initiated with pioneering works by Hill (1962), Nadai (1950), Mandel (1962), Prager (1954), Rice (1977), Thomas (1961), and developed – from theoretical point of view – into two principal directions, namely, the dissection of the specific constitutive features responsible for strain localization in different materials (for instance, as related to the microstructure, Danas and Ponte Castaneda, 2012; Bacigalupo and Gambarotta, 2013; Tvergaard, 2014) and the struggle for the overcoming of difficulties connected with numerical approaches [reviews have been given by Needleman and Tvergaard (1983) and Petryk (1997)]. Although these problems are still not exhausted, surprisingly, the most important questions have only marginally been approached and are therefore still awaiting explanation. These are as follows: (i) Why are shear bands a preferred mode of failure for ductile materials? (ii) Why do shear bands propagate rectilinearly under mode II, while cracks do not? (iii) How does a shear band interact with a crack or with a rigid inclusion? (iv) Does a stress concentration exist at a shear band tip? (v) How does a shear band behave under dynamic conditions?

The only systematic[3] attempt to solve these problems seems to have been a series of works by Bigoni and co-workers, based on the perturbative approach to shear bands (Bigoni and Capuani, 2002, 2005; Piccolroaz et al., 2006; Argani et al., 2013, 2014). In fact, problems (i), (ii), and (iv) were addressed in Bigoni and Dal Corso (2008) and Dal Corso and Bigoni (2010), problem (iii) in Bigoni et al. (2008), Dal Corso et al. (2008), and Dal Corso and Bigoni (2009), and (v) in Bigoni and Capuani (2005).

The purpose of the present article is to present a model of a shear band as a zero-thickness interface and to rigorously motivate

---

[1] For granular materials, there are cases in which unloading occurs inside the shear band, as shown by Gajo et al. (2004).

[2] The proposed explanation for the crack patterns shown in **Figure 2** relies on the fact that the fracture network has formed during the plastic evolution of a ductile homogeneously deformed material. Other explanations may be related to bonding of an external layer to a rigid substrate (Peron et al., 2013), or to surface instability (Destrade and Merodio, 2011; Boulogne et al., 2015), or to instabilities occurring during shear (Destrade et al., 2008; Ciarletta et al., 2013).

[3] Special problems of shear band propagation in geological materials have been addressed by Puzrin and Germanovich (2005) and Rice (1973).

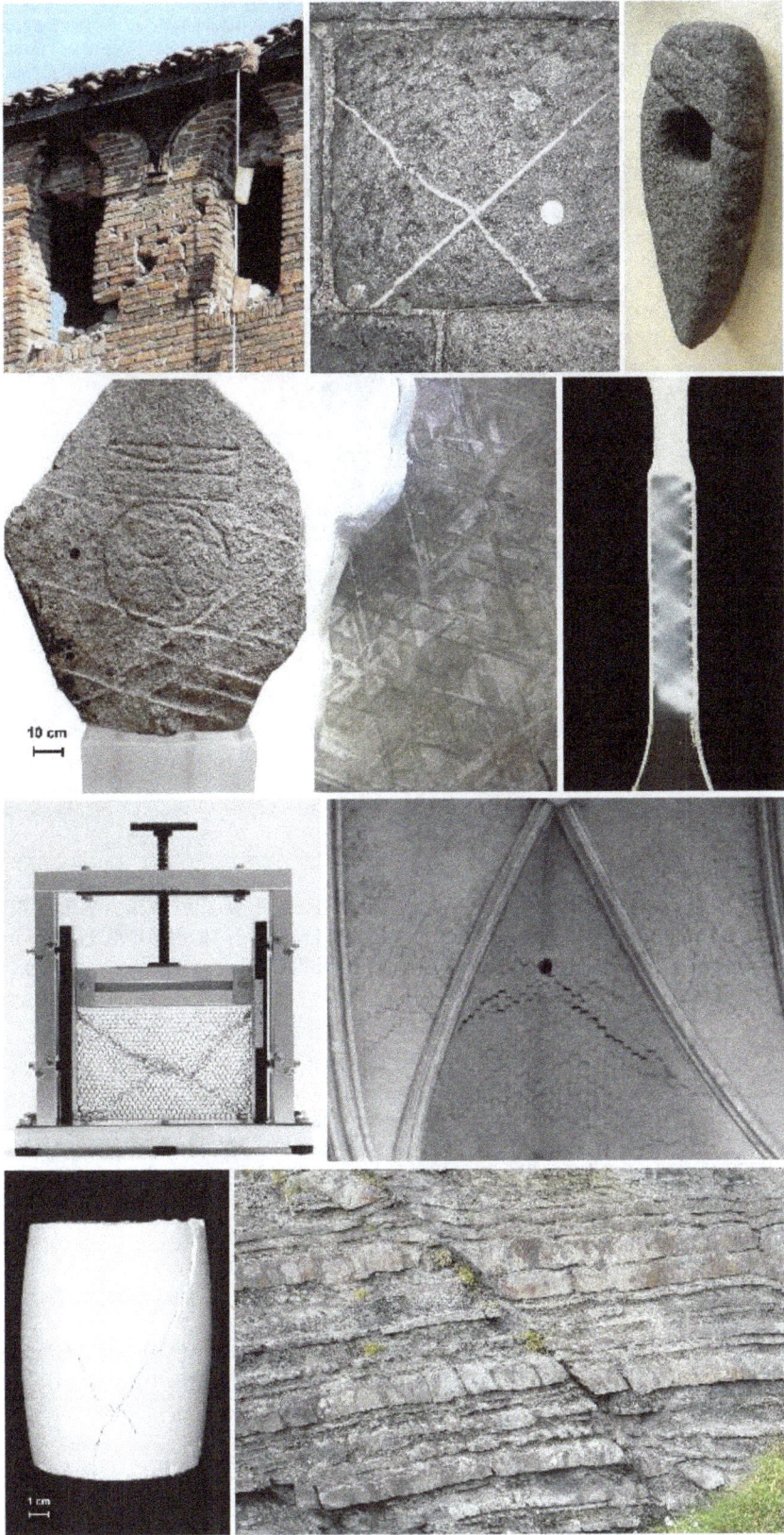

**FIGURE 1 | Examples of strain localization**.

*(Continued)*

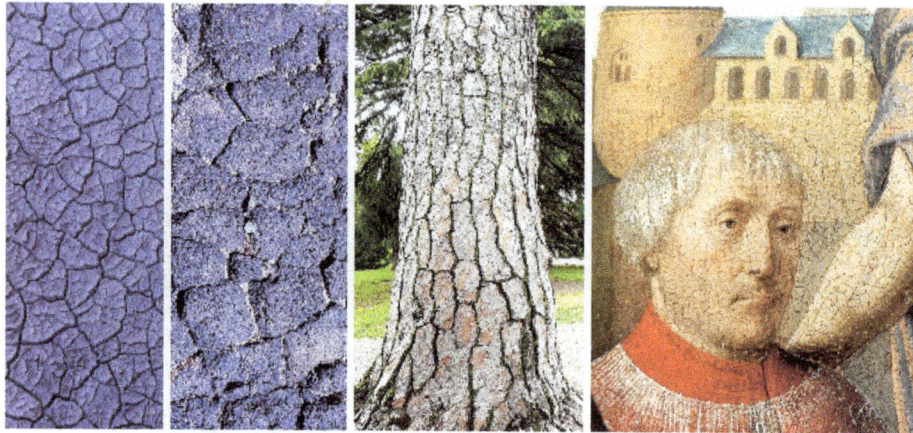

**FIGURE 2 | Regular patterns of localized cracks as the signature of strain localization lattices**. From left to right: Dried mud; Lava cracked during solidification (near Amboy crater); Bark of a maritime pine (*Pinus pinaster*); Cracks in a detail of a painting by J. Provost ("Saint Jean-Baptiste," Valenciennes, Musée des Beaux Arts).

this as the asymptotic behavior of a thin layer of material, which is extremely compliant in shear (Section 2). Once the shear band model has been developed, it is used (in Section 3) to demonstrate two of the above-listed open problems, namely (ii) that a shear band grows rectilinearly under mode II remote loading in a material deformed near to failure and (iv) to estimate the stress concentration at the shear band tip. In particular, a pre-existing shear band is considered to lie in a matrix as a thin zone of material with properties identical to the matrix, but lower yield stress. This is an imperfection, which remains neutral until the yield is reached in the shear band[4]. The present model is based on an imperfection approach and shares similarities to that pursued by Abeyaratne and Triantafyllidis (1981) and Hutchinson and Tvergaard (1981), so that it is essentially different from a perturbative approach, in which the perturbation is imposed at a certain stage of a uniform deformation process[5].

---

[4]A different approach to investigate shear band evolution is based on the exploitation of phase-field models (Zheng and Li, 2009), which has been often used for brittle fracture propagation (Miehe et al., 2010).

[5]To highlight the differences and the analogies between the two approaches, the incremental strain field induced by the emergence of a shear band of finite length (modeled as a sliding surface) is determined for a $J_2$-deformation theory material and compared with finite element simulations in which the shear band is modeled as a zero-thickness layer of compliant material.

## 2. ASYMPTOTIC MODEL FOR A THIN LAYER OF HIGHLY COMPLIANT MATERIAL EMBEDDED IN A SOLID

A shear band, inside a solid block of dimension $H$, is modeled as a thin layer of material (of semi-thickness $h$, with $h/H \ll 1$) yielding at a uniaxial stress $\sigma_Y^{(s)}$, which is lower than that of the surrounding matrix $\sigma_Y^{(m)}$, **Figure 3**. Except for the yield stress, the material inside and outside the layer is described by the same elastoplastic model, namely, a von Mises plasticity with associated flow rule and linear hardening, defined through the elastic constants, denoted by the Young modulus $E$ and Poisson's ratio $v$, and the plastic modulus $E_p$, see **Figure 3B**.

At the initial yielding, the material inside the layer [characterized by a low hardening modulus $E_{ep} = EE_p/(E + E_p)$] is much more compliant than the material outside (characterized by an elastic isotropic response $E$).

For $h/H \ll 1$, the transmission conditions across the layer imply the continuity of the tractions, $\boldsymbol{t} = [t_{21}, t_{22}]^T$, which can be expressed in the asymptotic form

$$[\![t_{21}]\!] = O(h), \quad [\![t_{22}]\!] = O(h), \tag{1}$$

where $[\![\cdot]\!]$ denotes the jump operator.

The jump in displacements, $[\![\boldsymbol{u}]\!] = [[\![u_1]\!], [\![u_2]\!]]^T$, across the layer is related to the tractions at its boundaries through the asymptotic relations (Mishuris et al., 2013; Sonato et al., 2015)

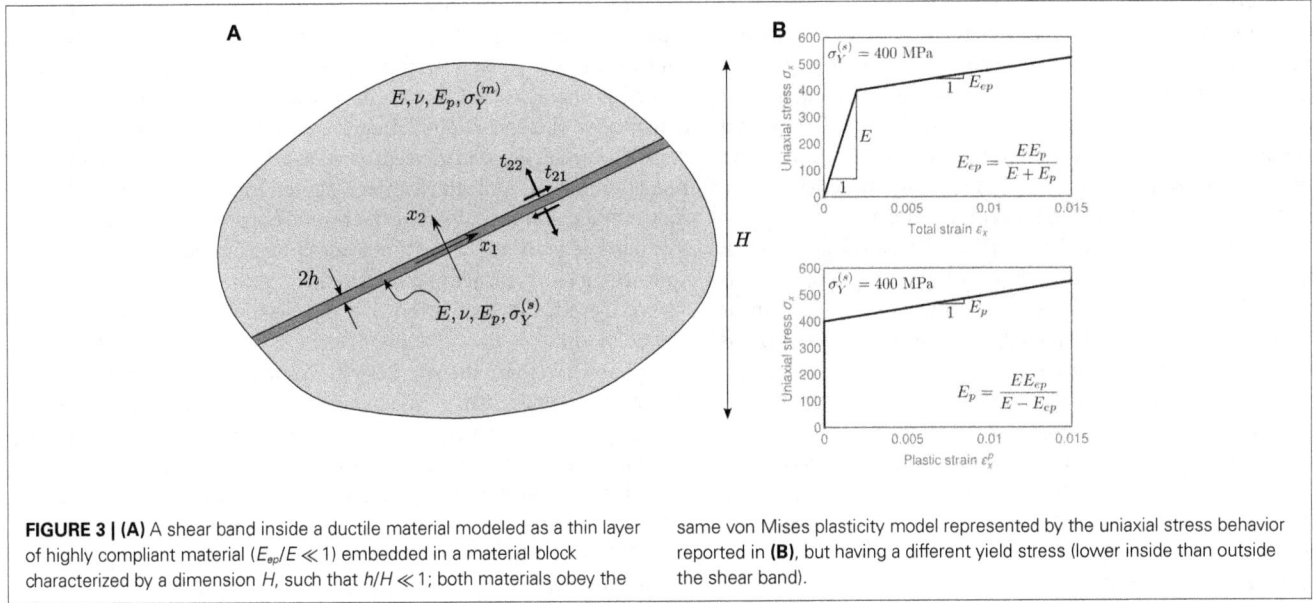

**FIGURE 3 | (A)** A shear band inside a ductile material modeled as a thin layer of highly compliant material ($E_{ep}/E \ll 1$) embedded in a material block characterized by a dimension $H$, such that $h/H \ll 1$; both materials obey the same von Mises plasticity model represented by the uniaxial stress behavior reported in **(B)**, but having a different yield stress (lower inside than outside the shear band).

$$t_{21}(\llbracket u_1 \rrbracket, \llbracket u_2 \rrbracket) = \frac{E_p\sqrt{3\llbracket u_1 \rrbracket^2 + 4\llbracket u_2 \rrbracket^2} + 6h\sigma_Y^{(s)}}{(3E + 2(1+\nu)E_p)\sqrt{3\llbracket u_1 \rrbracket^2 + 4\llbracket u_2 \rrbracket^2}} \frac{E\llbracket u_1 \rrbracket}{2h} + O(h), \qquad (2)$$

$$t_{22}(\llbracket u_1 \rrbracket, \llbracket u_2 \rrbracket) = \frac{(E + 2(1-\nu)E_p)\sqrt{3\llbracket u_1 \rrbracket^2 + 4\llbracket u_2 \rrbracket^2} + 8h(1-2\nu)\sigma_Y^{(s)}}{(1-2\nu)(3E + 2(1+\nu)E_p)\sqrt{3\llbracket u_1 \rrbracket^2 + 4\llbracket u_2 \rrbracket^2}} \frac{E\llbracket u_2 \rrbracket}{2h} + O(h), \qquad (3)$$

involving the semi-thickness $h$ of the shear band, which enters the formulation as a *constitutive parameter for the zero-thickness interface model* and introduces a *length scale*. Note that, by neglecting the remainder $O(h)$, equations (2) and (3) define non-linear relationships between tractions and jump in displacements.

The time derivative of equations (2) and (3) yields the following asymptotic relation between incremental quantities

$$\dot{\boldsymbol{t}} \sim \left[ \frac{1}{h}\boldsymbol{K}_{-1} + \boldsymbol{K}_0(\llbracket u_1 \rrbracket, \llbracket u_2 \rrbracket) \right] \llbracket \dot{\boldsymbol{u}} \rrbracket, \qquad (4)$$

where the two stiffness matrices $\boldsymbol{K}_{-1}$ and $\boldsymbol{K}_0$ are given by

$$\boldsymbol{K}_{-1} = \frac{E}{2(3E + 2(1+\nu)E_p)} \begin{bmatrix} E_p & 0 \\ 0 & \dfrac{E + 2(1-\nu)E_p}{1-2\nu} \end{bmatrix}, \qquad (5)$$

$$\boldsymbol{K}_0 = \frac{12E\sigma_Y^{(s)}}{(3E + 2(1+\nu)E_p)(3\llbracket u_1 \rrbracket^2 + 4\llbracket u_2 \rrbracket^2)^{3/2}}$$

$$\times \begin{bmatrix} \llbracket u_2 \rrbracket^2 & -\llbracket u_1 \rrbracket \llbracket u_2 \rrbracket \\ -\llbracket u_1 \rrbracket \llbracket u_2 \rrbracket & \llbracket u_1 \rrbracket^2 \end{bmatrix}, \qquad (6)$$

Assuming now a perfectly plastic behavior, $E_p = 0$, in the limit $h/H \rightarrow 0$ the condition

$$\llbracket u_2 \rrbracket = 0 \qquad (7)$$

is obtained, so that the incremental transmission conditions equation (4) can be approximated to the leading order as

$$\dot{\boldsymbol{t}} \sim \frac{1}{h}\boldsymbol{K}_{-1}\llbracket \dot{\boldsymbol{u}} \rrbracket. \qquad (8)$$

Therefore, when the material inside the layer is close to the perfect plasticity condition, the incremental conditions assume the limit value

$$\dot{t}_{21} = 0, \quad \llbracket \dot{u}_2 \rrbracket = 0, \qquad (9)$$

which, together with the incremental version of equation (1)$_2$, namely,

$$\llbracket \dot{t}_{22} \rrbracket = 0, \qquad (10)$$

correspond to the incremental boundary conditions proposed in Bigoni and Dal Corso (2008) to define a pre-existing shear band of null thickness.

The limit relations, equations (9) and (10), motivate the use of the imperfect interface approach (Bigoni et al., 1998; Antipov et al., 2001; Mishuris, 2001, 2004; Mishuris and Kuhn, 2001; Mishuris and Ochsner, 2005, 2007) for the modeling of shear band growth in a ductile material. A computational model, in which the shear bands are modeled as interfaces, is presented in the next section.

## 3. NUMERICAL SIMULATIONS

Two-dimensional plane-strain finite element simulations are presented to show the effectiveness of the above-described asymptotic model for a thin and highly compliant layer in modeling a shear band embedded in a ductile material. Specifically, we will show that the model predicts rectilinear propagation of a shear band under simple shear boundary conditions and it allows the investigation of the stress concentration at the shear band tip.

The geometry and material properties of the model are shown in **Figure 4**, where a rectangular block of edges $H$ and $L \geq H$ is subject to boundary conditions consistent with a simple shear deformation, so that the lower edge of the square domain is clamped, the vertical displacements are constrained along the vertical edges and along the upper edge, where a constant horizontal displacement $u_1$ is prescribed. The domain is made of a ductile material and contains a thin ($h/H \ll 1$) and highly compliant ($E_{ep}/E \ll 1$) layer of length $H/2$ and thickness $2h = 0.01$ mm, which models a shear band. The material constitutive behavior is described by an elastoplastic model based on linear isotropic elasticity ($E = 200000$ MPa, $v = 0.3$) and von Mises plasticity with linear hardening (the plastic modulus is denoted by $E_p$). The uniaxial yield stress $\sigma_Y^{(m)}$ for the matrix material is equal to 500 MPa, whereas the layer is characterized by a lower yield stress, namely, $\sigma_Y^{(s)} = 400$ MPa.

The layer remains neutral until yielding, but, starting from that stress level, it becomes a material inhomogeneity, being more compliant (because its response is characterized by $E_{ep}$) than the matrix (still in the elastic regime and thus characterized by $E$). The layer can be representative of a pre-existing shear band and can be treated with the zero-thickness interface model, equations (2) and (3). This zero-thickness interface was implemented in the ABAQUS finite element

software[6] through cohesive elements, equipped with the traction-separation laws, equations (2) and (3), by means of the user subroutine UMAT. An interface, embedded into the cohesive elements, is characterized by two dimensions: a geometrical and a constitutive thickness. The latter, $2h$, exactly corresponds to the constitutive thickness involved in the model for the interface equations (2) and (3), while the former, denoted by $2h_g$, is related to the mesh dimension in a way that the results become independent of this parameter, in the sense that a mesh refinement yields results converging to a well-defined solution.

We consider two situations. In the first, we assume that the plastic modulus is $E_p = 150000$ MPa (both inside and outside the shear band), so that the material is in a state far from a shear band instability (represented by loss of ellipticity of the tangent constitutive operator, occurring at $E_p = 0$) when at yield. In the second, we assume that the material is prone to a shear band instability, though still in the elliptic regime, so that $E_p$ (both inside and outside the shear band) is selected to be "sufficiently small", namely, $E_p = 300$ MPa. The pre-existing shear band is therefore employed as an imperfection triggering shear strain localization when the material is still inside the region, but close to the boundary, of ellipticity.

### 3.1. DESCRIPTION OF THE NUMERICAL MODEL

With reference to a square block ($L = H = 10$ mm) containing a pre-existing shear band with constitutive thickness $h = 0.005$ mm, three different meshes were used, differing in the geometrical thickness of the interface representing the pre-existing shear band (see **Figure 5** where the shear band is highlighted with a black line), namely, $h_g = \{0.05; 0.005; 0.0005\}$ mm corresponding to coarse, fine, and ultra-fine meshes.

The three meshes were generated automatically using the mesh generator available in ABAQUS. In order to have increasing mesh refinement from the exterior (upper and lower parts) to the interior (central part) of the domain, where the shear band is located, and to ensure the appropriate element shape and size according to the geometrical thickness $2h_g$, the domain was partitioned into rectangular subdomains with increasing mesh seeding from the exterior to the interior. Afterwards, the meshes were generated by employing a free meshing technique with quadrilateral elements and the advancing front algorithm.

The interface that models the shear band is discretized using 4-node two-dimensional cohesive elements (COH2D4), while the matrix material is modeled using 4-node bilinear, reduced integration with hourglass control (CPE4R).

It is important to note that the constitutive thickness used for traction-separation response is always equal to the actual size of the shear band $h = 0.005$ mm, whereas the geometric thickness $h_g$, defining the height of the cohesive elements, is different for the three different meshes. Consequently, all the three meshes used in the simulations correspond to the same problem in terms of both material properties and geometrical dimensions (although the geometric size of the interface is different), so that

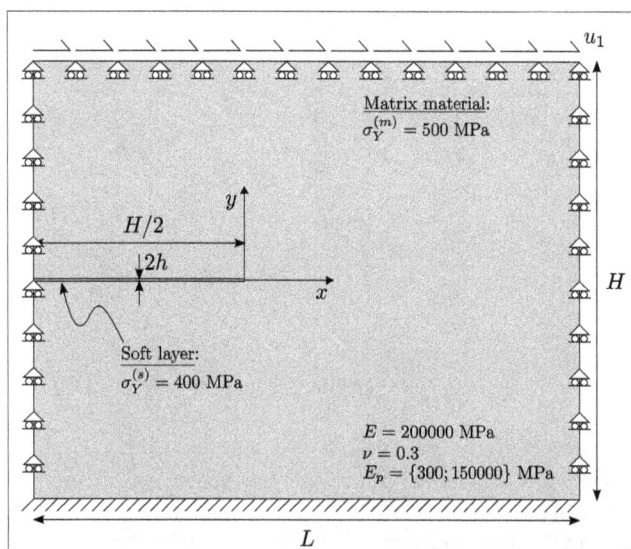

**FIGURE 4 | Geometry of the model, material properties, and boundary conditions (which would correspond to a simple shear deformation in the absence of the shear band).** The horizontal displacement $u_1$ is prescribed at the upper edge of the domain.

---

[6] ABAQUS Standard Ver. 6.13 has been used, available on the AMD Opteron cluster Stimulus at UniTN.

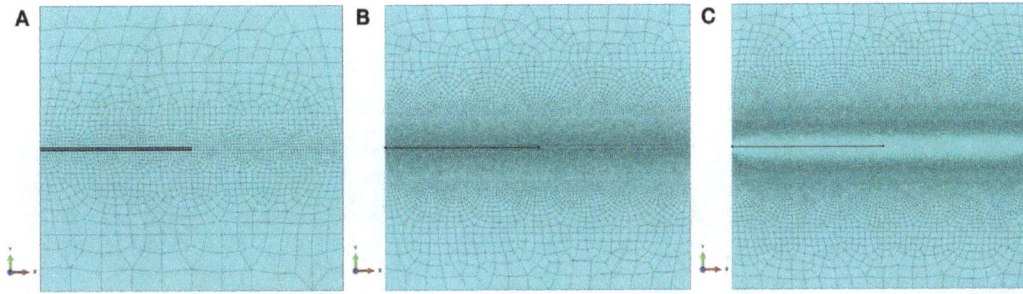

**FIGURE 5 | The three meshes used in the analysis to simulate a shear band (highlighted in black) in a square solid block ($L = H = 10$ mm).** The shear band is represented in the three cases as an interface with the same constitutive thickness $h = 0.005$ mm, but with decreasing geometric thickness $h_g$; **(A)** coarse mesh (1918 nodes, 1874 elements, $h_g = 0.05$ mm); **(B)** fine mesh (32,079 nodes, 31,973 elements, $h_g = 0.005$ mm); **(C)** ultra-fine mesh (1,488,156 nodes, 1,487,866 elements, $h_g = 0.0005$ mm).

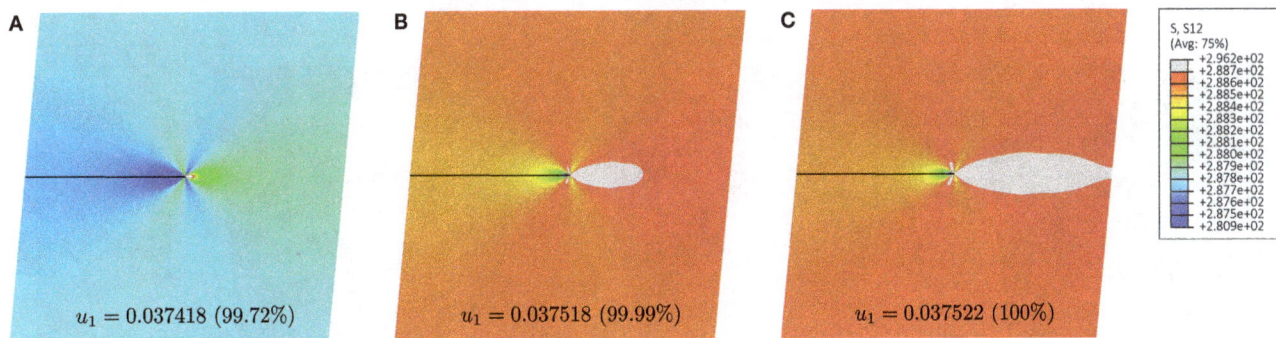

$u_1 = 0.037418$ (99.72%)     $u_1 = 0.037518$ (99.99%)     $u_1 = 0.037522$ (100%)

**FIGURE 6 | Contour plots of the shear stress $\sigma_{12}$ for the case of material far from shear band instability ($E_p = 150000$ MPa).** The gray region corresponds to the material at yielding $\sigma_{12} \geq 500/\sqrt{3} = 288.68$ MPa. Three different stages of deformation are shown, corresponding to a prescribed displacement at the upper edge of the square domain $u_1 = 0.037418$ mm **(A)**, $u_1 = 0.037518$ mm **(B)**, $u_1 = 0.037522$ mm **(C)**. The displacements in the figures are amplified by a deformation scale factor of 25 and the percentages refer to the final displacement.

the results have to be, and indeed will be shown to be, mesh independent[7].

## 3.2. NUMERICAL RESULTS

Results (obtained using the fine mesh, **Figure 5B**) in terms of the shear stress component $\sigma_{12}$ at different stages of a deformation process for the boundary value problem sketched in **Figure 4** are reported in **Figures 6** and **7**.

In particular, **Figure 6** refers to a matrix with high plastic modulus, $E_p = 150000$ MPa, so that the material is far from the shear band formation threshold. The upper limit of the contour levels was set to the value $\sigma_{12} = 500/\sqrt{3} \simeq 288.68$ MPa, corresponding to the yielding stress of the matrix material. As a result, the gray zone in the figure represents the material at yielding, whereas the material outside the gray zone is still in the elastic regime. Three stages of deformation are shown, corresponding to: the initial yielding of the matrix material (left), the yielding zone occupying

approximately one-half of the space between the shear band tip and the right edge of the domain (center), and the yielding completely linking the tip of the shear band to the boundary (right). Note that the shear band, playing the role of a material imperfection, produces a stress concentration at its tip. However, the region of high stress level rapidly grows and diffuses in the whole domain. At the final stage, shown in **Figure 6C**, almost all the matrix material is close to yielding.

**Figure 7** refers to a matrix with low plastic modulus, $E_p = 300$ MPa, so that the material is close (but still in the elliptic regime) to the shear band formation threshold ($E_p = 0$). Three stages of deformation are shown, from the condition of initial yielding of the matrix material near the shear band tip (left), to an intermediate condition (center), and finally to the complete yielding of a narrow zone connecting the shear band tip to the right boundary (right). In this case, where the material is prone to shear band localization, the zone of high stress level departs from the shear band tip and propagates toward the right. This propagation occurs in a highly concentrated narrow layer, rectilinear, and parallel to the pre-existing shear band. At the final stage of deformation, shown in **Figure 7C**, the layer of localized shear has reached the boundary of the block.

[7]Note that, in the case of null hardening, mesh dependency may occur in the simulation of shear banding nucleation and propagation (Needleman, 1988; Loret and Prevost, 1991, 1993). This numerical issue can be avoided by improving classical inelastic models through the introduction of characteristic length-scales (Lapovok et al., 2009; Dal Corso and Willis, 2011).

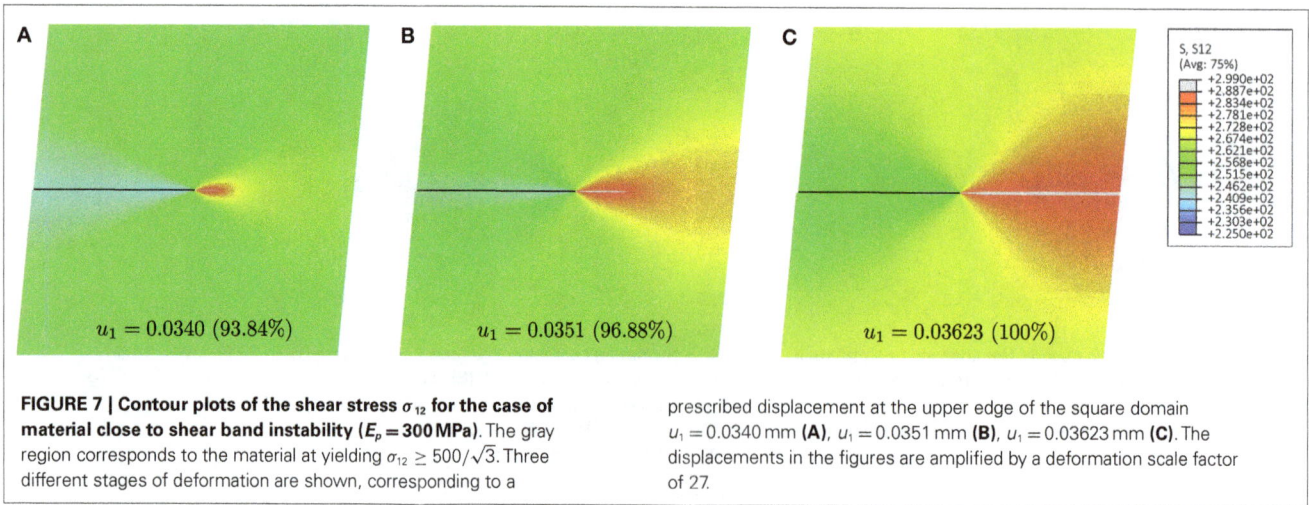

**FIGURE 7 | Contour plots of the shear stress $\sigma_{12}$ for the case of material close to shear band instability ($E_p = 300\,\text{MPa}$).** The gray region corresponds to the material at yielding $\sigma_{12} \geq 500/\sqrt{3}$. Three different stages of deformation are shown, corresponding to a prescribed displacement at the upper edge of the square domain $u_1 = 0.0340\,\text{mm}$ **(A)**, $u_1 = 0.0351\,\text{mm}$ **(B)**, $u_1 = 0.03623\,\text{mm}$ **(C)**. The displacements in the figures are amplified by a deformation scale factor of 27.

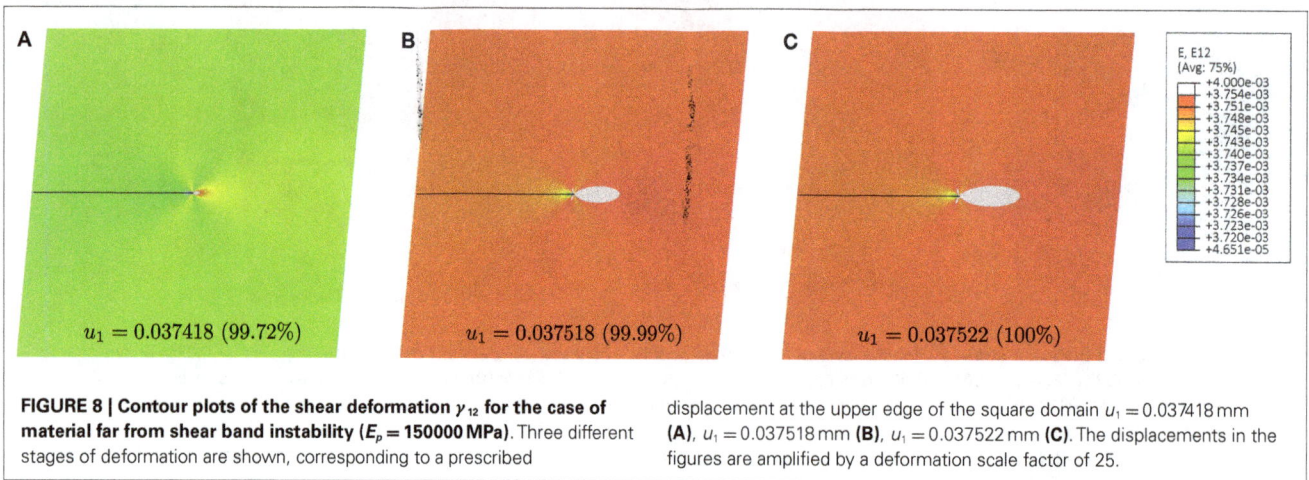

**FIGURE 8 | Contour plots of the shear deformation $\gamma_{12}$ for the case of material far from shear band instability ($E_p = 150000\,\text{MPa}$).** Three different stages of deformation are shown, corresponding to a prescribed displacement at the upper edge of the square domain $u_1 = 0.037418\,\text{mm}$ **(A)**, $u_1 = 0.037518\,\text{mm}$ **(B)**, $u_1 = 0.037522\,\text{mm}$ **(C)**. The displacements in the figures are amplified by a deformation scale factor of 25.

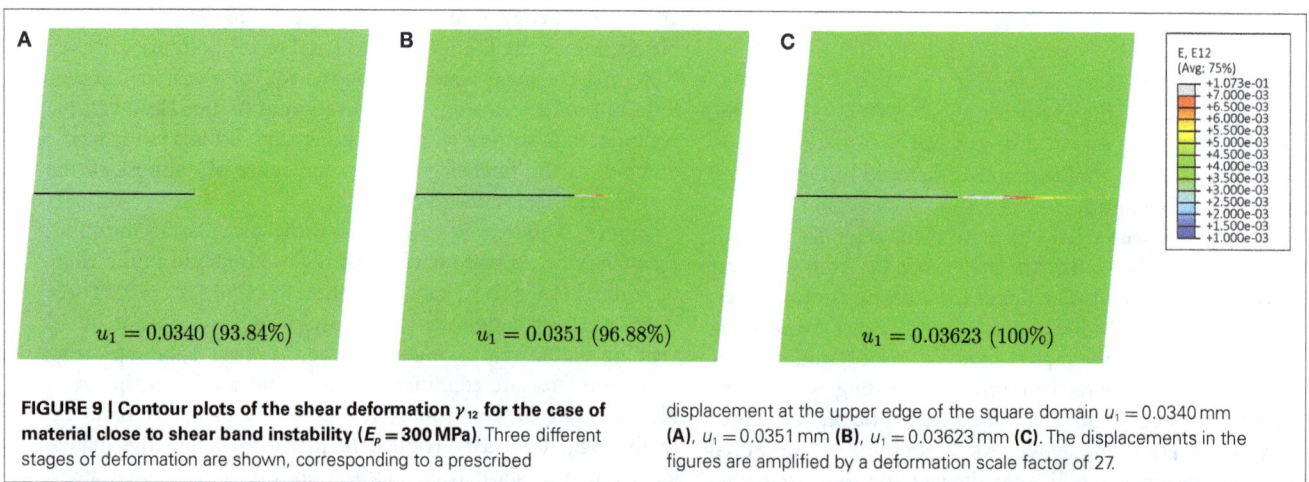

**FIGURE 9 | Contour plots of the shear deformation $\gamma_{12}$ for the case of material close to shear band instability ($E_p = 300\,\text{MPa}$).** Three different stages of deformation are shown, corresponding to a prescribed displacement at the upper edge of the square domain $u_1 = 0.0340\,\text{mm}$ **(A)**, $u_1 = 0.0351\,\text{mm}$ **(B)**, $u_1 = 0.03623\,\text{mm}$ **(C)**. The displacements in the figures are amplified by a deformation scale factor of 27.

Results in terms of the shear strain component $\gamma_{12}$, for both cases of material far from, and close to shear band instability are reported in **Figures 8** and **9**, respectively. In particular, **Figure 8** shows contour plots of the shear deformation $\gamma_{12}$ for the case of a material far from the shear band instability ($E_p = 150000\,\text{MPa}$) at the same three stages of deformation as those reported in **Figure 6**. Although the tip of the shear band acts as a strain raiser, the contour plots show that the level of shear deformation is high and remains diffused in the whole domain.

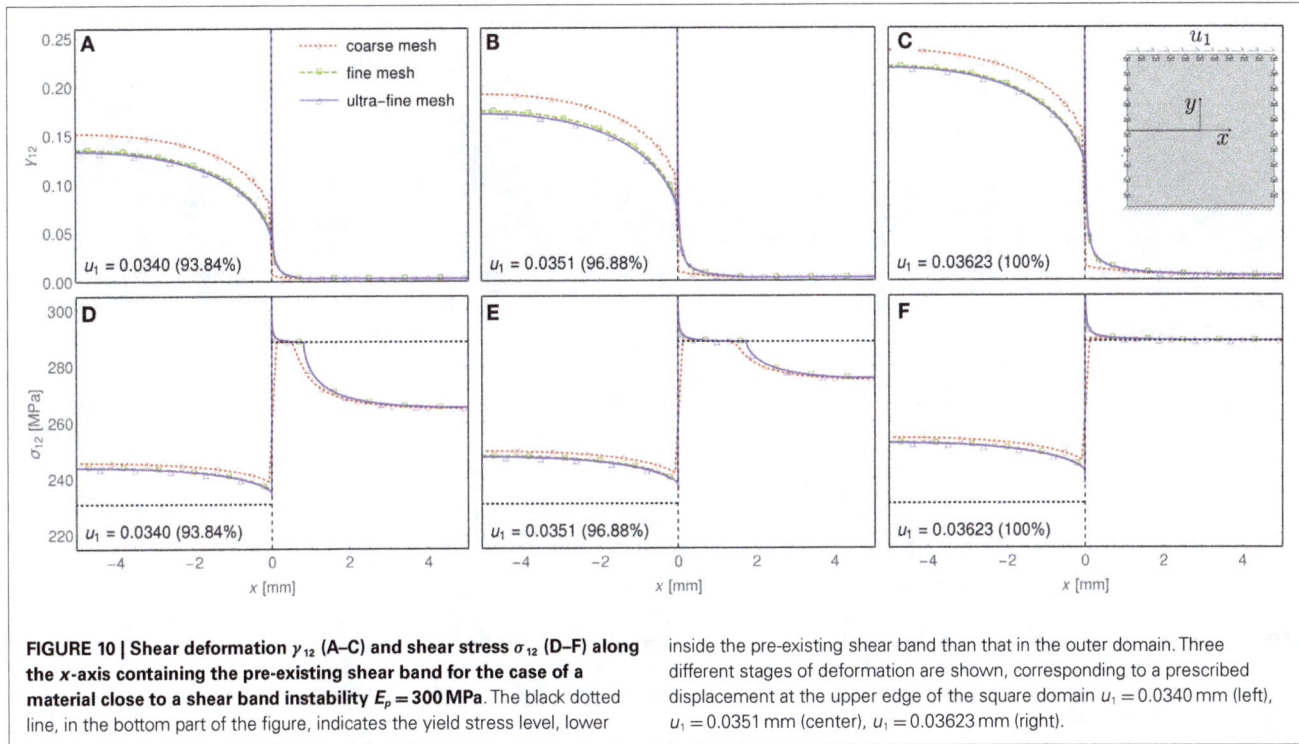

FIGURE 10 | Shear deformation $\gamma_{12}$ (A–C) and shear stress $\sigma_{12}$ (D–F) along the x-axis containing the pre-existing shear band for the case of a material close to a shear band instability $E_p = 300$ MPa. The black dotted line, in the bottom part of the figure, indicates the yield stress level, lower inside the pre-existing shear band than that in the outer domain. Three different stages of deformation are shown, corresponding to a prescribed displacement at the upper edge of the square domain $u_1 = 0.0340$ mm (left), $u_1 = 0.0351$ mm (center), $u_1 = 0.03623$ mm (right).

**Figure 9** shows contour plots of the shear deformation $\gamma_{12}$ for the case of a material close to the shear band instability ($E_p = 300$ MPa), at the same three stages of deformation as those reported in **Figure 7**. It is noted that the shear deformation is localized along a rectilinear path ahead of the shear band tip, confirming results that will be reported later with the perturbation approach (Section 4).

The shear deformation $\gamma_{12}$ and the shear stress $\sigma_{12}$ along the x-axis containing the pre-existing shear band for the case of a material close to strain localization, $E_p = 300$ MPa, are shown in **Figure 10**, upper and lower parts, respectively. Results are reported for the three meshes, coarse, fine and ultra-fine (**Figure 5**) and at the same three stages of deformation as those shown in **Figures 7** and **9**. The results appear to be mesh independent, meaning that the solution converges as the mesh is more and more refined.

The deformation process reported in **Figures 7, 9,** and **10** can be described as follows. After an initial homogeneous elastic deformation (not shown in the figure), in which the shear band remains neutral (since it shares the same elastic properties with the matrix material), the stress level reaches $\sigma_{12} = 400/\sqrt{3} \simeq 230.9$ MPa, corresponding to the yielding of the material inside the shear band. Starting from this point, the pre-existing shear band is activated, which is confirmed by a high shear deformation $\gamma_{12}$ and a stress level above the yield stress inside the layer, $-5$ mm $< x < 0$ (left part of **Figure 10**). The activated shear band induces a strain localization and a stress concentration at its tip, thus generating a zone of material at yield, which propagates to the right (central part of **Figure 10**) until collapse (right part of **Figure 10**).

In order to appreciate the strain and stress concentration at the shear band tip, a magnification of the results shown in **Figure 10** in the region $-0.2$ mm $< x < 0.2$ mm is presented in **Figure 11**. Due to the strong localization produced by the shear band, only the ultra-fine mesh is able to capture accurately the strain and stress raising (blue solid curve), whereas the coarse and fine meshes smooth over the strain and stress levels (red dotted and green dashed curves, respectively). The necessity of an ultra-fine mesh to capture details of the stress/strain fields is well-known in computational fracture mechanics, where special elements (quarter-point or extended elements) have been introduced to avoid the use of these ultra-fine meshes at corner points.

For the purpose of a comparison with an independent and fully numerical representation of the shear band, a finite element simulation was also performed, using standard continuum elements (CPE4R) instead of cohesive elements (COH2D4) inside the layer. This simulation is important to assess the validity of the asymptotic model of the layer presented in Section 2. In this simulation, reported in **Figure 12**, the layer representing the shear band is a "true" layer of a given and finite thickness, thus influencing the results (while these are independent of the geometrical thickness $2h_g$ of the cohesive elements, when the constitutive thickness $2h$ is the same). Therefore, only the fine mesh, shown in **Figure 5B**, was used, as it corresponds to the correct size of the shear band. The coarse mesh (**Figure 5A**) and the ultra-fine mesh (**Figure 5C**) would obviously produce different results, corresponding, respectively, to a thicker or thinner layer. Results pertaining to the asymptotic model, implemented into the traction-separation law for the cohesive elements COH2D4, are also reported in the figure (red solid curve) and are spot-on with the results obtained with a fully numerical solution employing standard continuum elements CPE4R (blue dashed curve).

A mesh of the same size as that previously called "fine" was used to perform a simulation of a rectangular block ($H = 10$ mm,

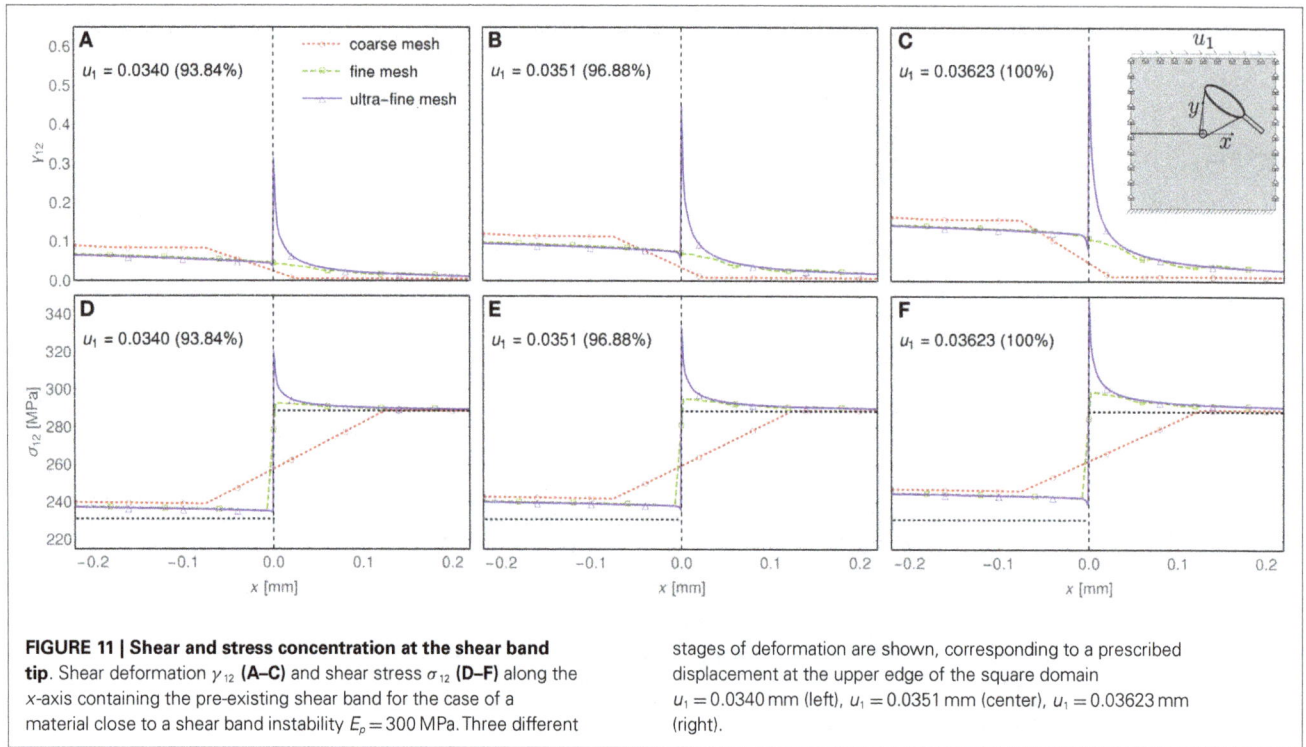

**FIGURE 11 | Shear and stress concentration at the shear band tip**. Shear deformation $\gamma_{12}$ **(A–C)** and shear stress $\sigma_{12}$ **(D–F)** along the $x$-axis containing the pre-existing shear band for the case of a material close to a shear band instability $E_p = 300$ MPa. Three different stages of deformation are shown, corresponding to a prescribed displacement at the upper edge of the square domain $u_1 = 0.0340$ mm (left), $u_1 = 0.0351$ mm (center), $u_1 = 0.03623$ mm (right).

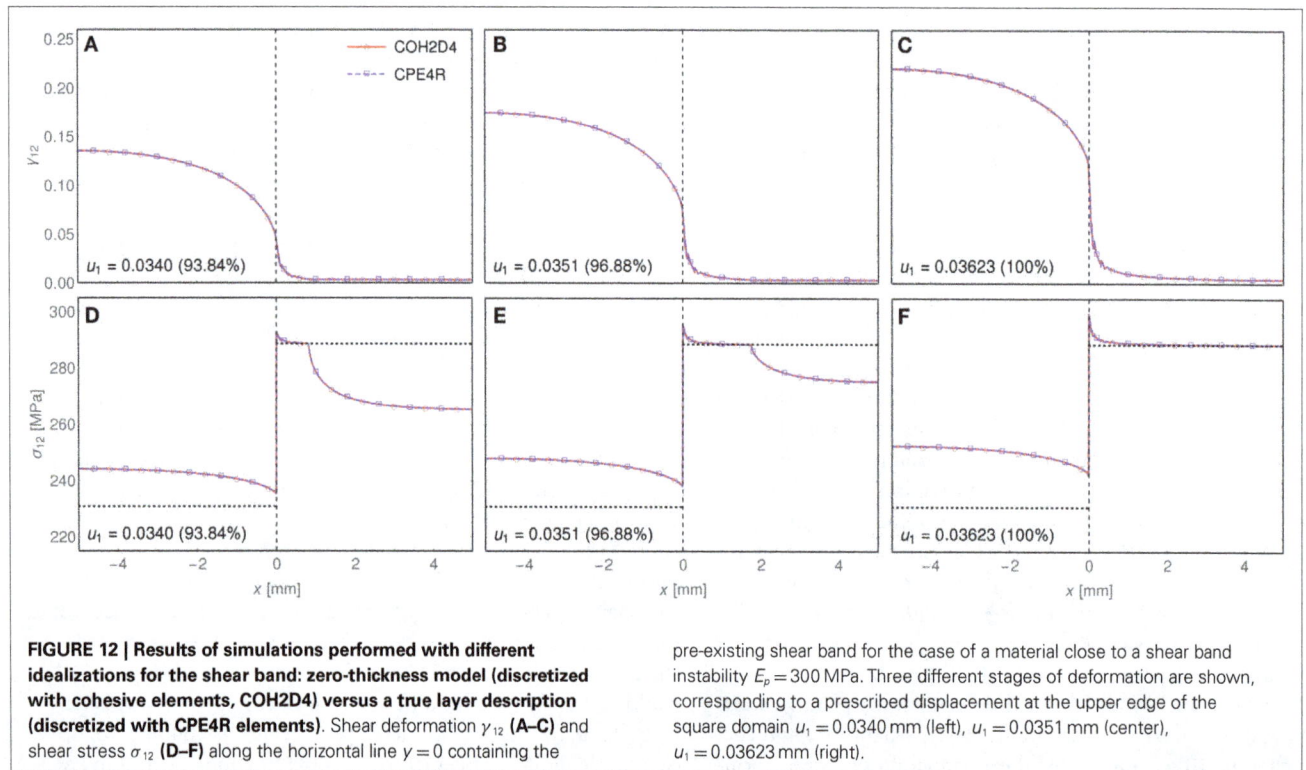

**FIGURE 12 | Results of simulations performed with different idealizations for the shear band: zero-thickness model (discretized with cohesive elements, COH2D4) versus a true layer description (discretized with CPE4R elements)**. Shear deformation $\gamma_{12}$ **(A–C)** and shear stress $\sigma_{12}$ **(D–F)** along the horizontal line $y = 0$ containing the pre-existing shear band for the case of a material close to a shear band instability $E_p = 300$ MPa. Three different stages of deformation are shown, corresponding to a prescribed displacement at the upper edge of the square domain $u_1 = 0.0340$ mm (left), $u_1 = 0.0351$ mm (center), $u_1 = 0.03623$ mm (right).

$L = 4$ $H = 40$ mm) made up of a material close to shear band instability ($E_p = 300$ MPa) and containing a shear band (of length $H/2 = 5$ mm and constitutive thickness $2h = 0.01$ mm). Results are presented in **Figure 13**. In parts (**Figures 13A,B**) (the latter is a detail of part **Figure 13A**) of this figure the overall response curve

is shown of the block in terms of average shear stress $\bar{\sigma}_{12} = T/L$ ($T$ denotes the total shear reaction force at the upper edge of the block) and average shear strain $\bar{\gamma}_{12} = u_1/H$. In part (**Figure 13C**) of the figure contour plots of the shear deformation $\gamma_{12}$ are reported at different stages of deformation. It is clear that the

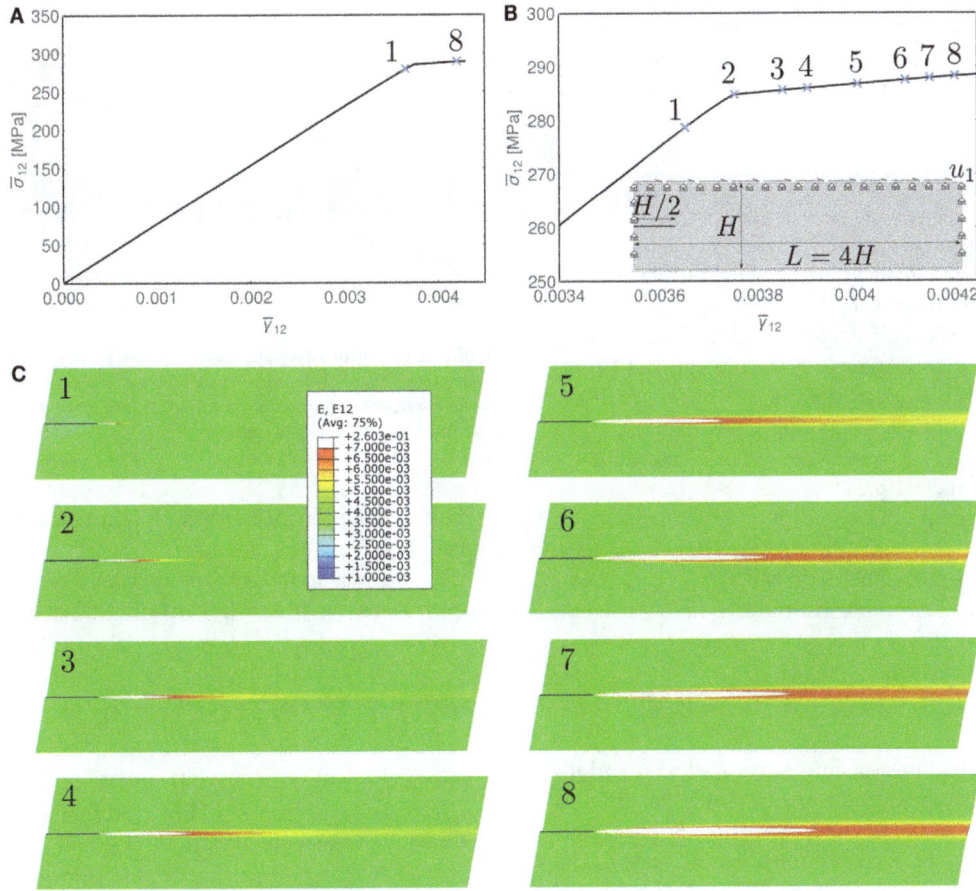

**FIGURE 13 | Results for a rectangular domain ($L = 40$ mm, $H = 10$ mm) of material close to shear band instability ($E_p = 300$ MPa) and containing a pre-existing shear band (of length $H/2 = 5$ mm and constitutive thickness $2h = 0.01$ mm). (A)** Overall response curve of the block in terms of average shear stress $\bar{\sigma}_{12} = T/L$, where $T$ is the total shear reaction force at the upper edge of the block, and average shear strain $\bar{\gamma}_{12} = u_1/H$. **(B)** Magnification of the overall response curve $\bar{\sigma}_{12} - \bar{\gamma}_{12}$ around the stress level corresponding to the yielding of the shear band. **(C)** Contour plots of the shear deformation $\gamma_{12}$ at different stages of deformation, corresponding to the points along the overall response curve shown in **(B)** of the figure. The deformation is highly focused along a rectilinear path emanating from the shear band tip. The displacements in the figures are amplified by a deformation scale factor of 50.

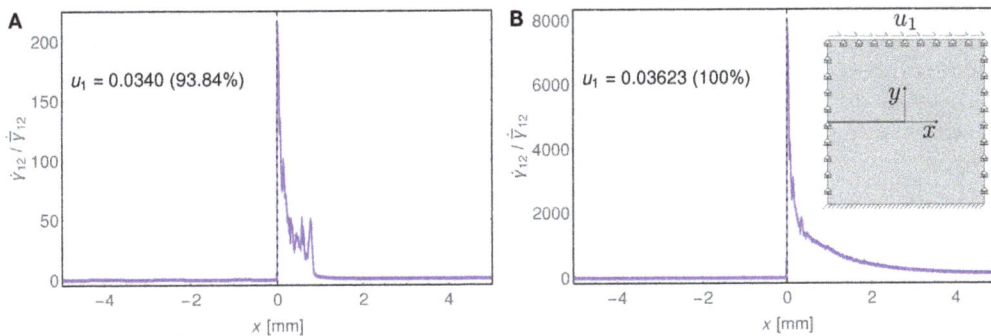

**FIGURE 14 | The incremental shear strain $\dot{\gamma}_{12}$ (divided by the mean incremental shear strain $\dot{\bar{\gamma}}_{12}$) along the x-axis at the two stages of deformation, (A) $u_1 = 0.0340$ mm and (B) $u_1 = 0.03623$ mm, reported in Figure 10 and labeled there as**

**(Figures 10A,C)**. It is clear that a strong strain concentration develops at the tip of the shear band, which becomes similar to the square-root singularity that is found with the perturbative approach (Section 4 and **Figure 16**).

deformation is highly focused along a rectilinear path emanating from the shear band tip, thus demonstrating the tendency of the shear band toward rectilinear propagation under shear loading.

Finally, the incremental shear strain (divided by the mean incremental shear strain) has been reported along the x-axis in **Figure 14**, at the two stages of deformation considered in **Figure 10** and referred there as (**Figures 10A,C**). These results, which have been obtained with the fine mesh, show that a strong incremental strain concentration develops at the shear band tip and becomes qualitatively similar to the square-root singularity found in the perturbative approach.

## 4. THE PERTURBATIVE VERSUS THE IMPERFECTION APPROACH

With the perturbative approach, a perturbing agent acts at a certain stage of uniform strain of an infinite body, while the material is subject to a uniform prestress. Here, the perturbing agent is a pre-existing shear band, modeled as a planar slip surface, emerging at a certain stage of a deformation path (Bigoni and Dal Corso, 2008), in contrast with the imperfection approach in which the imperfection is present from the beginning of the loading.

With reference to a $x_1 - x_2$ coordinate system (inclined at 45° with respect to the principal prestress axes $x_I - x_{II}$), where the incremental stress $\dot{t}_{ij}$ and incremental strain $\dot{\varepsilon}_{ij}$ are defined ($i, j = 1$, 2), the incremental orthotropic response under plane-strain conditions ($\dot{\varepsilon}_{i3} = 0$) for incompressible materials ($\dot{\varepsilon}_{11} + \dot{\varepsilon}_{22} = 0$) can be expressed through the following constitutive equations (Bigoni, 2012)[8].

$$\dot{t}_{11} = 2\mu\dot{\varepsilon}_{11} + \dot{p}, \quad \dot{t}_{22} = -2\mu\dot{\varepsilon}_{11} + \dot{p}, \quad \dot{t}_{12} = \mu_*\dot{\gamma}_{12}, \quad (11)$$

where $\dot{p}$ is the incremental in-plane mean stress, while $\mu$ and $\mu_*$ describe the incremental shear stiffness, respectively, parallel and inclined at 45° with respect to prestress axes.

The assumption of a specific constitutive model leads to the definition of the incremental stiffness moduli $\mu$ and $\mu_*$. With reference to the $J_2$-deformation theory of plasticity (Bigoni and Dal Corso, 2008), particularly suited to model the plastic branch of the constitutive response of ductile metals, the in-plane deviatoric stress can be written as

$$t_I - t_{II} = k\varepsilon_I|\varepsilon_I|^{(N-1)}. \quad (12)$$

In equation (12), $k$ represents a stiffness coefficient and $N \in (0, 1]$ is the strain hardening exponent, describing perfect plasticity (null hardening) in the limit $N \to 0$ and linear elasticity in the limit $N \to 1$. For the $J_2$-deformation theory, the relation between the two incremental shear stiffness moduli can be obtained as

$$\mu_* = N\mu, \quad (13)$$

so that a very compliant response under shear ($\mu_* \ll \mu$) is described in the limit of perfect plasticity $N \to 0$.

The perturbative approach (Bigoni and Dal Corso, 2008) can now be exploited to investigate the growth of a shear band within a solid. To this purpose, an incremental boundary value problem is formulated for an infinite solid, containing a zero-thickness pre-existing shear band of finite length $2l$ parallel to the $x_1$ axis

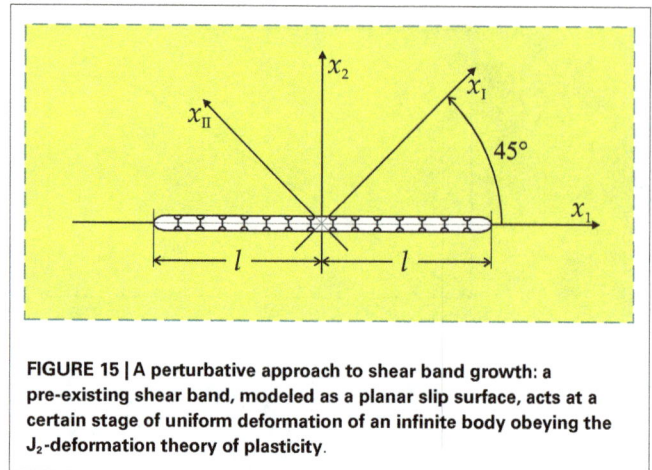

**FIGURE 15 | A perturbative approach to shear band growth: a pre-existing shear band, modeled as a planar slip surface, acts at a certain stage of uniform deformation of an infinite body obeying the $J_2$-deformation theory of plasticity.**

(see **Figure 15**) and loaded at infinity through a uniform shear deformation $\dot{\gamma}_{12}^{\infty}$.

The incremental boundary conditions introduced by the presence of a pre-existing shear band can be described by the following equations:

$$\dot{t}_{21}(x_1, 0^{\pm}) = 0, \quad [\![\dot{t}_{22}(x_1, 0)]\!] = 0, \quad [\![\dot{u}_2(x_1, 0)]\!] = 0, \quad \forall |x_1| < l. \quad (14)$$

A stream function $\psi(x_1, x_2)$ is now introduced, automatically satisfying the incompressibility condition and defining the incremental displacements $\dot{u}_j$ as $\dot{u}_1 = \psi_{,2}$, and $\dot{u}_2 = -\psi_{,1}$. The incremental boundary value problem is therefore solved as the sum of $\psi^{\circ}(x_1, x_2)$, solution of the incremental homogeneous problem, and $\psi^p(x_1, x_2)$, solution of the incremental perturbed problem.

The incremental solution is reported in **Figure 16** for a low hardening exponent, $N = 0.01$, as a contour plot (left) and as a graph (along the $x_1$-axis, right) of the incremental shear deformation $\dot{\gamma}_{12}$ (divided by the applied remote shear $\dot{\gamma}_{12}^{\infty}$). Note that, similarly to the crack tip fields in fracture mechanics, the incremental stress and deformation display square-root singularities at the tips of the pre-existing shear band. Evaluation of the solution obtained from the perturbative approach analytically confirms the conclusions drawn from the imperfection approach (see the numerical simulations reported in **Figures 9** and **13**), in particular:

- It can be noted from **Figure 16** (left) that the incremental deformation is highly focused along the $x_1$ direction, confirming that the shear band grows rectilinearly;
- The blow-up of the incremental deformation observed in the numerical simulations near the shear band tip (**Figure 14**) is substantiated by the theoretical square-root singularity found in the incremental solution (**Figure 16**, right).

We finally remark that, although the tendency toward rectilinear propagation of a shear band has been substantiated through the use of a von Mises plastic material, substantial changes are not expected when a different yield criterion (for instance, pressure-sensitive as Drucker–Prager) is employed.

---

[8]Note that the notation used here differs from that adopted in Bigoni and Dal Corso (2008), where the principal axes are denoted by $x_1$ and $x_2$ and the system inclined at 45° is denoted by $\hat{x}_1$ and $\hat{x}_2$.

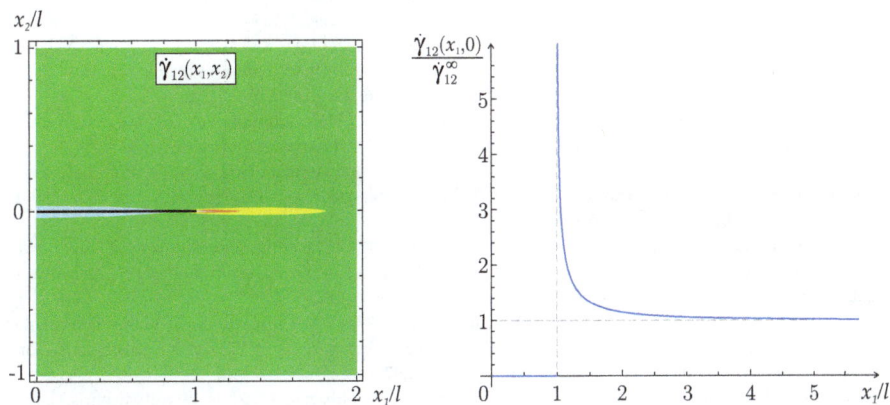

**FIGURE 16 | Incremental shear strain near a shear band obtained through the perturbative approach: level sets (left) and behavior along the $x_1$-axis (right).**

## 5. CONCLUSION

Two models of shear band have been described, one in which the shear band is an imperfection embedded in a material and another in which the shear band is a perturbation, which emerges during a homogeneous deformation of an infinite material. These two models explain how shear bands tend toward a rectilinear propagation under continuous shear loading, a feature not observed for fracture trajectories in brittle materials. This result can be stated in different words pointing out that, while crack propagation occurs following a maximum tensile stress criterion, a shear band grows according to a maximum Mises stress, a behavior representing a basic micromechanism of failure for ductile materials. The developed models show also a strong stress concentration at the shear band tip, which strongly concur to shear band growth.

## ACKNOWLEDGMENTS

DB, NB, and FDC gratefully acknowledge financial support from the ERC Advanced Grant "Instabilities and non-local multi-scale modeling of materials" FP7-PEOPLE-IDEAS-ERC-2013-AdG (2014-2019). AP thanks financial support from the FP7-PEOPLE-2013-CIG grant PCIG13-GA-2013-618375-MeMic.

## REFERENCES

Abeyaratne, R., and Triantafyllidis, N. (1981). On the emergence of shear bands in plane strain. *Int. J. Solids Struct.* 17, 1113–1134. doi:10.1016/0020-7683(81)90092-5

Antipov, Y. A., Avila-Pozos, O., Kolaczkowski, S. T., and Movchan, A. B. (2001). Mathematical model of delamination cracks on imperfect interfaces. *Int. J. Solids Struct.* 38, 6665–6697. doi:10.1016/S0020-7683(01)00027-0

Argani, L., Bigoni, D., Capuani, D., and Movchan, N. V. (2014). Cones of localized shear strain in incompressible elasticity with prestress: Green's function and integral representations. *Proc. Math. Phys. Eng. Sci.* 470, 20140423. doi:10.1098/rspa.2014.0423

Argani, L., Bigoni, D., and Mishuris, G. (2013). Dislocations and inclusions in prestressed metals. *Proc. R. Soc. A* 469, 20120752. doi:10.1098/rspa.2012.0752

Bacigalupo, A., and Gambarotta, L. (2013). A multi-scale strain-localization analysis of a layered strip with debonding interfaces. *Int. J. Solids Struct.* 50, 2061–2077. doi:10.1016/j.ijsolstr.2013.03.006

Bigoni, D. (2012). *Nonlinear Solid Mechanics Bifurcation Theory and Material Instability*. New York: Cambridge University Press.

Bigoni, D., and Capuani, D. (2002). Green's function for incremental nonlinear elasticity: shear bands and boundary integral formulation. *J. Mech. Phys. Solids* 50, 471–500. doi:10.1016/S0022-5096(01)00090-4

Bigoni, D., and Capuani, D. (2005). Time-harmonic Green's function and boundary integral formulation for incremental nonlinear elasticity: dynamics of wave patterns and shear bands. *J. Mech. Phys. Solids* 53, 1163–1187. doi:10.1016/j.jmps.2004.11.007

Bigoni, D., and Dal Corso, F. (2008). The unrestrainable growth of a shear band in a prestressed material. *Proc. R. Soc. A* 464, 2365–2390. doi:10.1098/rspa.2008.0029

Bigoni, D., Dal Corso, F., and Gei, M. (2008). The stress concentration near a rigid line inclusion in a prestressed, elastic material. Part II. Implications on shear band nucleation, growth and energy release rate. *J. Mech. Phys. Solids* 56, 839–857. doi:10.1016/j.jmps.2007.07.003

Bigoni, D., Serkov, S. K., Movchan, A. B., and Valentini, M. (1998). Asymptotic models of dilute composites with imperfectly bonded inclusions. *Int. J. Solids Struct.* 35, 3239–3258. doi:10.1016/S0020-7683(97)00366-1

Boulogne, F., Giorgiutti-Dauphine, F., and Pauchard, L. (2015). Surface patterns in drying films of silica colloidal dispersions. *Soft Matter* 11, 102. doi:10.1039/c4sm02106a

Ciarletta, P., Destrade, M., and Gower, A. L. (2013). Shear instability in skin tissue. *Q. J. Mech. Appl. Math.* 66, 273–288. doi:10.1097/TA.0b013e3182092e66

Dal Corso, F., and Bigoni, D. (2009). The interactions between shear bands and rigid lamellar inclusions in a ductile metal matrix. *Proc. R. Soc. A* 465, 143–163. doi:10.1098/rspa.2008.0242

Dal Corso, F., and Bigoni, D. (2010). Growth of slip surfaces and line inclusions along shear bands in a softening material. *Int. J. Fract.* 166, 225–237. doi:10.1007/s10704-010-9534-1

Dal Corso, F., Bigoni, D., and Gei, M. (2008). The stress concentration near a rigid line inclusion in a prestressed, elastic material. Part I. Full field solution and asymptotics. *J. Mech. Phys. Solids* 56, 815–838. doi:10.1016/j.jmps.2007.07.002

Dal Corso, F., and Willis, J. R. (2011). Stability of strain-gradient plastic materials. *J. Mech. Phys. Solids* 59, 1251–1267. doi:10.1016/j.jmps.2011.01.014

Danas, K., and Ponte Castaneda, P. (2012). Influence of the Lode parameter and the stress triaxiality on the failure of elasto-plastic porous materials. *Int. J. Solids Struct.* 49, 1325–1342. doi:10.1016/j.ijsolstr.2012.02.006

Destrade, M., Gilchrist, M., Prikazchikov, D., and Saccomandi, G. (2008). Surface instability of sheared soft tissues. *J. Biomech. Eng.* 130, 061007. doi:10.1115/1.2979869

Destrade, M., and Merodio, J. (2011). Compression instabilities of tissues with localized strain softening. *Int. J. Appl. Mech.* 3, 69–83. doi:10.1142/S1758825111000877

Gajo, A., Bigoni, D., and Muir Wood, D. (2004). Multiple shear band development and related instabilities in granular materials. *J. Mech. Phys. Solids* 52, 2683–2724. doi:10.1016/j.jmps.2004.05.010

Hill, R. (1962). Acceleration waves in solids. *J. Mech. Phys. Solids* 10, 1–16. doi:10.1016/0022-5096(62)90024-8

Hutchinson, J. W., and Tvergaard, V. (1981). Shear band formation in plane strain. *Int. J. Solids Struct.* 17, 451–470. doi:10.1111/j.1365-2818.2009.03250.x

Kirby, S. H. (1985). Rock mechanics observations pertinent to the rheology of the continental lithosphere and the localization of strain along shear zones. *Tectonophysics* 119, 1–27. doi:10.1016/0040-1951(85)90030-7

Lapovok, R., Toth, L. S., Molinari, A., and Estrina, Y. (2009). Strain localisation patterns under equal-channel angular pressing. *J. Mech. Phys. Solids* 57, 122–136. doi:10.1016/j.jmps.2008.09.012

Loret, B., and Prevost, J. H. (1991). On the existence of solutions in layered elasto-(visco-)plastic solids with negative hardening. *Eur. J. Mech. A Solids* 10, 575–586.

Loret, B., and Prevost, J. H. (1993). On the occurrence of unloading in 1D elasto-(visco-)plastic structures with softening. *Eur. J. Mech. Solids* 12, 757–772.

Mandel, J. (1962). Ondes plastiques dans un milieu indéfini à trois dimensions. *J. Mécanique* 1, 3–30.

Miehe, C., Hofacker, M., and Welschinger, F. (2010). A phase field model for rate-independent crack propagation: robust algorithmic implementation based on operator splits. *Comput. Methods Appl. Mech. Eng.* 199, 2765–2778. doi:10.1016/j.cma.2010.04.011

Mishuris, G. (2001). Interface crack and nonideal interface concept (Mode III). *Int. J. Fract.* 107, 279–296. doi:10.1023/A:1007664911208

Mishuris, G. (2004). Imperfect transmission conditions for a thin weakly compressible interface. 2D problems. *Arch. Mech.* 56, 103–115.

Mishuris, G., and Kuhn, G. (2001). Asymptotic behaviour of the elastic solution near the tip of a crack situated at a nonideal interface. *Z. Angew. Math. Mech.* 81, 811–826. doi:10.1002/1521-4001(200112)81:12<811::AID-ZAMM811>3.0.CO;2-I

Mishuris, G., Miszuris, W., Ochsner, A., and Piccolroaz, A. (2013). "Transmission conditions for thin elasto-plastic pressure-dependent interphases," in *Plasticity of Pressure-Sensitive Materials*, eds H. Altenbach and A. Ochsner (Berlin: Springer-Verlag), 205–251.

Mishuris, G., and Ochsner, A. (2005). Transmission conditions for a soft elasto-plastic interphase between two elastic materials. Plane Strain State. *Arch. Mech.* 57, 157–169.

Mishuris, G., and Ochsner, A. (2007). 2D modelling of a thin elasto-plastic interphase between two different materials: plane strain case. *Compos. Struct.* 80, 361–372. doi:10.1016/j.compstruct.2006.05.017

Nadai, A. (1950). *Theory of Flow and Fracture of Solids.* New York, NY: McGraw-Hill.

Needleman, A. (1988). Material rate dependence and mesh sensitivity in localization problems. *Comput. Methods Appl. Mech. Eng.* 67, 69–75. doi:10.1016/0045-7825(88)90069-2

Needleman, A., and Tvergaard, V. (1983). "Finite element analysis of localization in plasticity", in *Finite Elements: Special Problems in Solid Mechanics, Vol. V*, eds J. T. Oden and G. F. Carey (Englewood Cliffs: Prentice-Hall), 94–157.

Peron, H., Laloui, L., Hu, L. B., and Hueckel, T. (2013). Formation of drying crack patterns in soils: a deterministic approach. *Acta Geotech.* 8, 215–221. doi:10.1007/s11440-012-0184-5

Petryk, H. (1997). Plastic instability: criteria and computational approaches. *Arch. Comput. Meth. Eng.* 4, 111–151. doi:10.1007/BF03020127

Piccolroaz, A., Bigoni, D., and Willis, J. R. (2006). A dynamical interpretation of flutter instability in a continuous medium. *J. Mech. Phys. Solids* 54, 2391–2417. doi:10.1016/j.jmps.2006.05.005

Prager, W. (1954). "Discontinuous fields of plastic stress and flow," in *2nd Nat. Congr. Appl. Mech* (Ann Arbor, MI: ASME), 21–32.

Puzrin, A. M., and Germanovich, L. N. (2005). The growth of shear bands in the catastrophic failure of soils. *Proc. R. Soc. A* 461, 1199–1228. doi:10.1098/rspa.2004.1378

Rice, J. R. (1973). "The initiation and growth of shear bands," in *Plasticity and Soil Mechanics*, ed. A. C. Palmer (Cambridge: Cambridge University Engineering Department), 263.

Rice, J. R. (1977). "The localization of plastic deformation," in *Theoretical and Applied Mechanics*, ed. W. T. Koiter (Amsterdam: North-Holland Publishing Co.), 207.

Sonato, M., Piccolroaz, A., Miszuris, W. and Mishuris, G. (2015). General transmission conditions for thin elasto-plastic pressure-dependent interphase between dissimilar materials. *Int. J. Solids Struct.* doi:10.1016/j.ijsolstr.2015.03.009

Thomas, T. Y. (1961). *Plastic Flows and Fracture of Solids.* New York, NY: Academic Press.

Tvergaard, V. (2014). Bifurcation into a localized mode from non-uniform periodic deformations around a periodic pattern of voids. *J. Mech. Phys. Solids* 69, 112–122. doi:10.1016/j.jmps.2014.05.002

Yang, B., Morrison, M. L., Liaw, P. K., Buchanan, R. A., Wang, G., Liu, C. T., et al. (2005). Dynamic evolution of nanoscale shear bands in a bulk-metallic glass. *Appl. Phys. Lett.* 86, 141904. doi:10.1063/1.1891302

Zheng, G. P., and Li, M. (2009). Mesoscopic theory of shear banding and crack propagation in metallic glasses. *Phys. Rev. B* 80, 104201. doi:10.1103/PhysRevB.80.104201

**Conflict of Interest Statement:** The authors declare that the research was conducted in the absence of any commercial or financial relationships that could be construed as a potential conflict of interest.

# 15

# High-throughput multiple dies-to-wafer bonding technology and III/V-on-Si hybrid lasers for heterogeneous integration of optoelectronic integrated circuits

*Xianshu Luo[1], Yulian Cao[2], Junfeng Song[1], Xiaonan Hu[2], Yuanbing Cheng[2], Chengming Li[2], Chongyang Liu[2], Tsung-Yang Liow[1], Mingbin Yu[1], Hong Wang[2], Qi Jie Wang[2] and Patrick Guo-Qiang Lo[1]\**

[1] Institute of Microelectronics, Agency for Science, Technology and Research (A*STAR), Singapore, Singapore
[2] Photonics Center of Excellence (OPTIMUS), School of Electrical and Electronic Engineering, Nanyang Technological University, Singapore, Singapore

**Edited by:**
*Laurent Vivien, Université Paris-Sud, France*

**Reviewed by:**
*Junichi Fujikata, Photonics Electronics Technology Research Association, Japan*
*Yasuhiko Ishikawa, The University of Tokyo, Japan*

**\*Correspondence:**
*Patrick Guo-Qiang Lo, Institute of Microelectronics, Agency for Science, Technology and Research (A*STAR), Singapore Science Park II, 11 Science Park Road, 117685 Singapore*
*e-mail: logq@ime.a-star.edu.sg*

Integrated optical light source on silicon is one of the key building blocks for optical interconnect technology. Great research efforts have been devoting worldwide to explore various approaches to integrate optical light source onto the silicon substrate. The achievements so far include the successful demonstration of III/V-on-Si hybrid lasers through III/V gain material to silicon wafer bonding technology. However, for potential large-scale integration, leveraging on mature silicon complementary metal oxide semiconductor (CMOS) fabrication technology and infrastructure, more effective bonding scheme with high bonding yield is in great demand considering manufacturing needs. In this paper, we propose and demonstrate a high-throughput multiple dies-to-wafer (D2W) bonding technology, which is then applied for the demonstration of hybrid silicon lasers. By temporarily bonding III/V dies to a handle silicon wafer for simultaneous batch processing, it is expected to bond unlimited III/V dies to silicon device wafer with high yield. As proof-of-concept, more than 100 III/V dies bonding to 200 mm silicon wafer is demonstrated. The high performance of the bonding interface is examined with various characterization techniques. Repeatable demonstrations of 16-III/V die bonding to pre-patterned 200 mm silicon wafers have been performed for various hybrid silicon lasers, in which device library including Fabry–Perot (FP) laser, lateral-coupled distributed-feedback laser with side wall grating, and mode-locked laser (MLL). From these results, the presented multiple D2W bonding technology can be a key enabler toward the large-scale heterogeneous integration of optoelectronic integrated circuits.

**Keywords: silicon photonics, hybrid lasers, heterogeneous integration, die-to-wafer bonding, optoelectronic integrated circuits**

## INTRODUCTION

In the future generation of datacom and computercom, which demand ever higher bandwidth and lower power, the conventional electrical interconnection routing the electronic signals becomes bandwidth-limited along with prohibitively high power consumption (Beausoleil et al., 2008). One solution to the challenge is the optical interconnect technology (Goodman et al., 1984; Miller, 2000, 2009; Ohashi et al., 2009), in which high bandwidth optical signals are routed by low-loss optical fiber and waveguides. In contrast to the electrical interconnection (i.e., the copper wire), optical interconnect has many merits, e.g., high speed, low crosstalk, immunity to electromagnetic interference, low overall power consumption (Alduino and Paniccia, 2007). Most importantly, with the up scaling potential, optical interconnect is expected to provide much higher transmission capacity and longer signal transmission distance than the electrical interconnect.

Although it was proposed initially 30 years ago (Goodman et al., 1984), there was no significant development progress with solid demonstrations of optical interconnect for very-large-scale integration (VLSI). The situation has changed since the concept of silicon photonics (Pavesi and Lockwood, 2004; Reed and Knights, 2004; Lipson, 2005; Guillot and Pavesi, 2006; Jalali and Fathpour, 2006; Soref, 2006; Poon et al., 2009a,b; Vivien and Pavesi, 2013; Xu et al., 2014), which utilizes low-cost silicon material along with leveraging on the advancement of silicon complementary metal oxide semiconductor (CMOS) process, integration, and mature infrastructure. Envisioned by Soref and Lorenzo (1985), silicon photonics has emerged and progressed steadily. Especially in the past decade, we have been witnessing rapid growth in research and development activities along with product development efforts exploiting silicon photonics technology for the optical interconnect (Pavesi and Lockwood, 2004; Reed and Knights, 2004; Lipson, 2005; Guillot and Pavesi, 2006; Jalali and Fathpour, 2006; Soref, 2006; Fedeli et al., 2008; Poon et al., 2009a,b; Michel et al., 2010; Reed et al., 2010; Feng et al., 2012; Liow et al., 2013; Vivien and Pavesi, 2013; Dong et al., 2014a,b; Lim et al., 2014; Xu et al., 2014). For instance, to minimize the small core silicon waveguide propagation losses, considerable research work has been devoted to minimize the waveguide sidewall roughness by using the deep ultra-violet (DUV) photolithography and optimized patterning technique (Dumon et al., 2004; Bogaerts et al., 2005) and sidewall smoothing technique [e.g., double thermal oxidations (Sparacin et al., 2005; Xia et al., 2006)]. Indeed, submicrometer-scale silicon wire waveguides have shown a propagation loss of 2 dB/cm and less (Xia et al., 2006). Furthermore, owing to the enabling CMOS fabrication technologies, we have seen the establishment and utilization of a myriad of essential silicon photonic passive and active components including optical filters (Xiao et al., 2007; Zhou and Poon, 2007; Guha et al., 2010; Fang et al., 2012), optical switches (Poon et al., 2009a,b; Van Campenhout et al., 2009; Luo et al., 2012; Song et al., 2013), low-power-consuming modulators with up to 50 Gb/s-speed operation (Dong et al., 2009; Reed et al., 2010; Tu et al., 2013, 2014), and Ge-on-Si photodetectors with bandwidth larger than 40 GHz (Michel et al., 2010; Liow et al., 2013). It is these demonstrated silicon photonic devices and technologies that make ultimate optical interconnection a viable solution to address the distance/bandwidth/cost and power-consumption challenges. To this end, silicon photonics provides nearly all key building blocks for optical interconnection. Furthermore, the CMOS-compatible fabrication processes make it possible to integrate both electronics and photonics either through monolithic or heterogeneous approach. Such significant progress has led the optical interconnect to become a much more practical technology.

However, silicon-based on-chip optical light source, which is one of the key components for the light generation for carrying information, has been the missing piece for optical interconnect. This is mainly because silicon is transparent in the telecommunication wavelengths (i.e., 1310 and 1550 nm wavelengths) due to the indirect bandgap, which prohibits efficient light emission from silicon. Thus, to solve the challenge, numerous research efforts have been devoted to explore various technologies for light source on silicon chips.

## REVIEW ON RESEARCH FOR LASERS ON SILICON

Historically, researchers worldwide have devoted many research efforts by exploring various possibilities for the development of lasers on silicon, which mainly focused in the following directions:

(1) silicon material engineering by introducing emissive centers to assist the efficient light emission (Pavesi et al., 2000; Han et al., 2001; Rotem et al., 2007a,b; Shainline and Xu, 2007),

(2) strained Ge (Liu et al., 2007, 2009, 2010; Cheng et al., 2009; Sun et al., 2009b,c; Camacho-Aguilera et al., 2012),

(3) silicon Raman laser (Boyraz and Jalali, 2004, 2005; Rong et al., 2005a,b, 2007), and

(4) heterogeneous integration of III/V gain materials through packaging (Chu et al., 2009; Fujioka et al., 2010; Urino et al., 2011) or wafer bonding (Park et al., 2005; Fang et al., 2007a,b; Liang et al., 2009a,b, 2010; Stanković et al., 2010; Grenouillet et al., 2012).

Here, we will limit our review to the heterogeneous integrated silicon lasers. With regard to the silicon laser through heterogeneous integration of III/V gain materials on silicon, there are two major types of integration strategies, namely the packaging scheme and the bonding scheme. Research groups from Japan devoted many efforts for the development of silicon lasers using packaging methods. Chu et al. (2009) demonstrated the first wavelength-tunable-laser fabricated with silicon photonic

technology, which comprised a semiconductor optical amplifier (SOA) chip and a silicon photonic chip, and were hybrid-integrated by using passive alignment technology. An SiON mode-size converter was adopted between the silicon waveguide and III/V SOA for low coupling loss. Later on, silicon photonic-based optical interconnects were also demonstrated by integrating lasers, silicon modulators, and Ge photodetectors on single silicon substrate (Urino et al., 2011). While such demonstrations have shown the advantage in principal of being capable to integrate various building blocks together for optical interconnection, the main issue is the complicated fabrication process. It typically requires precise alignment between the SOA and the silicon waveguide, even with assistance of the mode-size converter. Considering the III/V gain region of <200 nm in thickness, for instance, it became a difficult challenge for alignment with acceptable coupling loss. Such complicated fabrication process is a potential show stopper for future massive production demanding high yield, thus significantly increases the product cost.

Heterogeneous integration of III/V gain materials on silicon through wafer bonding technology is another major directional strategy for silicon lasers. UMR-CNRS and LETI initiated the research work of III/V laser on silicon wafers for photonic integration by using wafer bonding technology. In 2001, they demonstrated the first InP-based microdisk laser integrated on a silicon wafer through $SiO_2$–$SiO_2$ molecular bonding (Seassal et al., 2001). Although this work did not show complete integration of III/V optoelectronics with silicon photonics waveguide structures, it showed the potential of such wafer bonding technology for future heterogeneous integrated optoelectronic circuit. Following such demonstration, Hattori et al. (2006) demonstrated an integration scheme of III/V microdisk laser with silicon waveguide in 2006. By aligning the microdisk laser atop silicon waveguide, the laser emissions can be vertically coupled into the underneath silicon waveguide with 35% coupling efficiency. Such demonstration showed the capability of the hybrid photonic integration of III/V laser with silicon waveguide for photonic links application.

The so-called hybrid silicon laser was proposed and first demonstrated by Park et al. (2005) with optical injection. In this work, the III/V wafer with AlGaInAs quantum well structure is directly bonded to pre-patterned silicon wafer using low-temperature oxygen plasma-assisted wafer bonding. The laser cavity was defined by endface-polished silicon waveguide structure, while the III/V provides the optical gain. As the III/V optoelectronic structures are fabricated after the wafer bonding with best precise, only possibly achieved via lithographic process, alignment to the silicon device layer, thus there is no stringent alignment requirement to the bonding process, which significantly simplifies the fabrication process and makes the possibility of wafer-level-oriented manufacturing ability. Subsequently, Fang et al. (2006) demonstrated an electrically pumped AlGaInAs-silicon evanescent laser with continuous-wave (CW) operation in 2006. Subsequently, various hybrid lasers with different structures and also enhanced laser performances are demonstrated by various research groups using molecular wafer bonding technology, including Fabry–Pérot lasers (FP) (Ben Bakir et al., 2011; Dong et al., 2013), racetrack lasers (Fang et al., 2007a,b), distributed Bragg reflector (DBR) lasers (Fang et al., 2008a,b,c), distributed-feedback (DFB) lasers (Fang et al., 2008a,b,c), microring lasers (Liang et al., 2009a,b, 2012), wavelength tunable lasers (Keyvaninia et al., 2013a,b,c), multiple-wavelength lasers (Van Campenhout et al., 2008; Kurczveil et al., 2011), and mode-locked lasers (MLL) (Fang et al., 2008a,b,c).

Besides such direct bonding method, wafer bonding can also be realized through an adhesive material as the bonding interlayer. Among all kinds of adhesive bonding materials, divinylsiloxane-bisbenzocyclobutene (DVS-BCB or BCB) is the most popular one for hybrid silicon lasers due to the merits such as the high bonding strength and the sustainability in the subsequent III/V process. IMEC has used BCB-assisted adhesive bonding method for heterogeneous integration (Roelkens et al., 2005). In 2006, Roelkens et al. (2006) demonstrated the first electrically injected InP/InGaAsP laser integrated on silicon waveguide circuit using BCB-assisted adhesive bonding technology. Similar to Seassal et al. (2001), the optical laser is purely made with III/V layer with the laser facets being defined by dry etching. With optimized mode-size converter, the optical light can be vertically coupled down to the underneath silicon waveguide with high efficiency. By designing hybrid mode waveguide comprising silicon waveguide and III/V gain medium, they also demonstrated a hybrid FP laser (Stanković et al., 2011) and a DFB laser (Stanković et al., 2012; Keyvaninia et al., 2013a,b,c), and multiple-wavelength laser (Keyvaninia et al., 2013a,b,c) using such adhesive bonding technology.

Apart from BCB, some kinds of metal can also be adopted as the bonding interlayer for adhesive bonding. AuGeNi is one of the most popular metals for metal bonding as it not only functions as a bonding media but can also be used for the Ohmic contact to the InGaAsP structure. Tanabe et al. (2010) demonstrated a InAs/GaAs quantum-dot laser on Si substrate by metal-assisted wafer bonding with room temperature operation at 1.3 μm wavelength. Meanwhile, Hong et al. (2010) also demonstrated an FP laser through selective-area metal bonding using AuGeNi. The silicon waveguide in such demonstration is with 5 μm and 800 nm thickness. The demonstrated FP laser is with threshold current density of 1.7 kA/cm and a maximum output power of 3 mW. However, the drawback of the AuGeNi-assisted bonding is the Au contamination. Thus, Tatsumi et al. (2012) further demonstrated an Au-free metal-assisted wafer bonding for lasers on silicon chip. Besides, Creazzo et al. (2013) also demonstrated another type of silicon laser by using metal-assisted bonding of III/V epitaxial material directly onto the silicon substrate. The demonstrated that silicon laser had a threshold of ~50 mA and maximum optical power of ~8 mW. The benefit of such metal-assisted bonding is the advantage of effective thermal dissipation, which shows a thermal resistant of only 21°C/W.

Beyond these two major heterogeneous integration schemes, there are also other methods for III/V-on-Si lasers, including direct III/V epitaxy on silicon substrate (Liu et al., 2011; Lee et al., 2012) and III/V epitaxial layer transfer-printing to silicon wafers (Justice et al., 2012; Yang et al., 2012). However, while the direct epitaxy method faces major challenges of high-density dislocations due to the lattice mismatch between III/V material and silicon after many years of research, the transfer-printing method for hybrid silicon laser needs further demonstrations to show the repeatability and reliability.

From these analyses, it shows that among various approaches, the hybrid silicon laser through wafer bonding technology can be considered as the most successful and promising one for silicon photonic heterogeneous integration circuits due to the ever-demonstrated advanced performances and the fabrication process compatibility with silicon photonics. **Table 1** summarizes some of the representative demonstrations of hybrid silicon lasers through wafer bonding technology.

## WAFER BONDING TECHNOLOGIES FOR ON-CHIP SILICON LASERS

In general, there are two mainstreams of wafer bonding methods applying to heterogeneous integrated silicon photonics, namely the molecular bonding through interfacial bonds, and the adhesive bonding assisted with another adhesive material as bonding interface such as polymer or metal. Such wafer bonding technology is a mature process, which is widely applicable for SOI wafer fabrication, MEMS technology, and optoelectronic device fabrication. As a lot of review papers already exist (Lasky, 1986; Maszara, 1991; Tong and Goesele, 1999; Alexe and Gösele, 2004; Christiansen et al., 2006), we thus only focus the discussion on the application of hybrid silicon lasers. According to the existing demonstrations, we further summarize here the major bonding technologies as below:

(1) wafer-to-wafer (W2W) molecular bonding,
(2) die-to-wafer (D2W) molecular bonding,
(3) BCB-assisted D2W adhesive bonding, and
(4) metal-assisted adhesive bonding.

The W2W molecular bonding for hybrid silicon lasers is mainly driven by the UCSB group. Through such plasma-activated low-temperature W2W molecular bonding (Pasquariello and Hjort, 2002), they, together with their collaborators, have demonstrated various hybrid silicon lasers, starting from the first-hybrid FP laser (Fang et al., 2006), followed by racetrack-shaped laser (Fang et al., 2007a,b), DBR lasers (Fang et al., 2008a,b,c), DFB lasers (Fang et al., 2008a,b,c), MLL (Fang et al., 2008a,b,c), and multiwavelength arbitrary waveform generation (AWG) laser (Kurczveil et al., 2011). However, for the conventional III/V-to-Si W2W bonding without thick oxide interlayer, the generated gas by-products of $H_2$ and $H_2O$ are easily trapped inside the bonding interface and form the interfacial voids, which subsequently affect the bonding quality. In order to effectively remove such trapped gases, some proper outgas channels are designed, such as in-plane outgassing channels (IPOC) (Kissinger and Kissinger, 1993) or vertical outgassing channels (VOC) (Liang and Bowers, 2008). IPOS is formed by etching some lateral channels extended to the chip edges, so that the by-product gases can be directed to outside the bonding interface to the chip edge during post bonding annealing. However, for some close-loop structures, such as microrings, there is no way to design such IPOS. In order to solve such issue, VOC is proposed by etching some array of holes down to the BOX layer. The generated by-product gases can migrate to the closest VOC and are absorbed by SOI BOX. As both IPOS and VOC can be formed during the waveguide etching, there is no particular design requirement from the fabrication point of view. However, as the formation of such outgas channels affects the silicon layer

**Table 1 | Representative demonstrations of hybrid silicon lasers through wafer bonding technology.**

| Laser types | Bonding type | Waveguide scheme | Performances | | | | | | Reference |
|---|---|---|---|---|---|---|---|---|---|
| | | | $\lambda$ (nm) | $T$ (°C) | $I_{th}$ (mA) | $P_{out}$ (mW) | SE (mW/mA) | $Z_t$ (°C/W) | |
| Fabry–Perot laser | Molecular bonding | Hybrid WG (75 vs. 3% mode confinement within Si WG and QW) | 1577 | CW @ 15 | 65 | 1.8 | 0.013[a] | 40 | Fang et al. (2006) |
| DBR laser | Molecular bonding | Hybrid waveguide with inverse taper (66 vs. 4.4% mode confinement within Si WG and QW) | 1569 | CW @ 15 | 65 | 11 | 0.088[a] | 40 | Fang et al. (2008a,b,c) |
| | D2W molecular bonding with oxide interlayer | Hybrid waveguide with adiabatic mode transformer | 1570 | Pulse @ 20 | 100 | 7.2 | 0.021[a] | – | Ben Bakir et al. (2011) |
| | W2W molecular bonding with oxide interlayer | Hybrid waveguide with inverse taper, thermal tunable microring for wavelength tuning | 1553 | CW @ 20 | 40 | 4 | 0.025[a] | – | Keyvaninia et al. (2013a,b,c) |
| | Metal-assisted D2W bonding | III/V gain material butt-coupling with Si waveguide through a waveguide coupler | 1562 | CW @ 20 | 41 | 8 | 0.038 | 21 | Creazzo et al. (2013) |
| DFB laser | Molecular bonding | Hybrid waveguide with inverse taper (59.2 vs. 5.2% mode confinement within Si WG and QW) | 1600 | CW @ 10 | 25 | 5.4 | 0.072[a] | 132 | Fang et al. (2008a,b,c) |
| | D2D BCB adhesive bonding | Hybrid waveguide (70 vs. 3% mode confinement within Si WG and QW) | 1308 | CW @ 20 | 20 | 2.1 | 0.026 | – | Stanković et al. (2011) |
| | Selective-area metal bonding | Hybrid WG (94% mode confinement within Si) | 1554 | Pulse @ RT | 35 | 3 | 0.05 | – | Hong et al. (2010) |
| Microdisk laser | D2W molecular bonding | InP microdisk laser light vertically coupling to Si WG | 1590 | CW @ 20 | 0.9 | 0.012 | 0.008 | – | Van Campenhout et al. (2008) |
| Microring laser | Molecular bonding | Racetrack microring, hybrid waveguide | 1590 | CW @ 15 | 175 | 29 | 0.089 | – | Fang et al. (2007a,b) |
| | Molecular bonding | Hybrid microring with side coupled Si WG | – | CW @ 10 | 7.5 | 2.5 | 0.2 | – | Liang et al. (2011) |

[a] Data are extracted from the power–current curves.

pattern density, which will finally affect the bonding strength, it is desirable to take into account the design tradeoff between the bonding strength and the gas removal effectiveness.

Alternatively, plasma-assisted D2W molecular bonding has also been investigated for hybrid silicon lasers mainly by LETI. For large-scale manufacturability for potential massive production, the key enabling capability is the multiple dies to wafer bonding with high yield. Kostrzewa et al. (2006) first demonstrated a molecular bonding of multiple InP dies to a 200 mm silicon CMOS wafer with only 1 mm × 1 mm die size (Kostrzewa et al., 2006). For strong hydrophilic molecular bonding, both InP and silicon wafers were covered with oxide layer. Pick-and-place technology was used in order to align the InP dies to specific spots in silicon wafer, as well as to supply mechanical force to the dies through pick-and-place head. Using such D2W bonding, they demonstrated electrically pumped microdisk lasers integrated with a silicon waveguide circuit (Van Campenhout et al., 2007). However, in such D2W bonding, as the cleaning of the dies is performed ahead of the pick-and-place process, the bonding surface could be contaminated, and subsequently affecting the bonding quality and bonding yield. Furthermore, with the consideration of the pick-and-place time of 30 s/die, it takes approximately an hour for bonding 100 dies. Such long bonding time through pick-and-place process for individual die would cause the bonding surface deactivation for the molecular bonding with plasma activation.

BCB-assisted D2W adhesive bonding can address such issue with potential capability of bonding unlimited number of dies. Stanković et al. (2010) demonstrated such D2D adhesive bonding technology using BCB. The BCB is first spin-coated on silicon wafer with controlled thickness of <100 nm in order to ensure the vertical light coupling efficiency, followed with die attaching and subsequent curing at 240°C for 1 h in a nitrogen atmosphere at 1000 mbar. With the assistant of BCB adhesion, the stringent requirements of contamination-free and smooth bonding surfaces for molecular bonding are relieved significantly. Furthermore, there is, in principle, no limitation for multiple dies bonding by the assistance of BCB adhesive layer (Keyvaninia et al., 2012). Through such bonding method, various hybrid lasers, including FP laser (Stanković et al., 2011), DFB laser (Stanković et al., 2012), microring and AWG integrated multi-wavelength DBR lasers (Keyvaninia et al., 2013a,b,c), and microdisk laser (Mechet et al., 2013) have been demonstrated. However, although such adhesive bonding is with good robustness and bonding strength, the thermal dissipation could be a major problem due to the low thermal conductivity of the BCB layer. Besides, robust polymer coating process ensuring the controllable BCB thickness is also very important.

Apart from these three major bonding methods, metal-assisted adhesive bonding (Hong et al., 2010; Tanabe et al., 2010; Creazzo et al., 2013) is another one that can be used for hybrid laser integration. However, due to the potential metal contamination and the non-compatibility with the subsequent fabrication process, such as acid etching for substrate removal, the metal-assisted adhesive bonding method might not be an optimal choice for silicon heterogeneous optoelectronic integrated circuits.

In **Table 2**, we summarize and compare these four different bonding methods.

## OUTLINE OF THE MANUSCRIPT

The rest of the submission is organized as follows. In the Section "III/V-to-Si Wafer-to-Wafer (W2W) Bonding Technology," we show a demonstration of the wafer-to-wafer bonding by using low-temperature plasma-activated molecular bonding method with oxide as the bonding interlayer. In the Section "High-Throughput Multiple Dies-to-Wafer Bonding Technology," we propose and show the demonstration of an alternative bonding technology that can perform high-throughput D2W bonding for potential massive production of silicon hybrid lasers, which is based on a batch process to simultaneously bonding all the dies to the silicon wafer. In the Section "Design of III/V-on-Si Lasers," we provide some design guidelines of hybrid silicon laser, including the design of III/V multiple quantum wells (MQW) structures, the silicon waveguide thickness selection for hybrid laser, and the design of the vertical coupling structures. The Section "Demonstration of III/V-on-Si Hybrid Lasers" shows some hybrid silicon laser demonstrations using the bonded wafers from the proposed high-throughput D2W bonding, including FP laser, lateral-coupled distributed-feedback (LC-DFB) laser with side wall grating, and MLL. The Section "Summary and Future Outlook" summarizes this paper and addresses some of the future challenges.

## III/V-TO-Si WAFER-TO-WAFER BONDING TECHNOLOGY

We have started the development of wafer bonding technology for hybrid silicon photonics integration in 2011. Considering the complete integration with existing silicon photonic integrated circuit, which consisting various silicon passive waveguide devices, high-speed modulator, and photodetectors, and are normally

**Table 2 | The major bonding technology for hybrid integration.**

| Bonding methods | Process description | Fabrication tolerance | Manufacturing scalability | Comments |
|---|---|---|---|---|
| W2W molecular bonding | O₂ plasma-assisted direct bonding with 12 h annealing at 300°C | Small tolerance of contamination-free, smooth, and flat bonding surfaces | Difficult due to wafer size mismatch | Low utilization of both III/V and silicon wafers Sensitive to the wafer bowing |
| D2W molecular bonding | O₂ plasma-assisted bonding with oxide interlayer and 3 h annealing at 250°C | Small tolerance of contamination-free and smooth bonding surfaces | Difficult due to contamination and surface deactivation during pick-and-place process for large number of dies bonding | Difficult to ensure high yield with large number of dies bonding |
| BCB-assisted D2W adhesive bonding | BCB adhesive bonding with post curing of 1 h at 240°C | Large tolerance with low requirement on the bonding surface. Yet it requires controllable polymer coating regarding the thickness and flatness | Easy to be scalable with multiple dies and large-sized wafers | Thermal dissipation problem due to the low thermal conductivity of the BCB layer |
| Metal-assisted adhesive bonding | Metal-assisted bonding with annealing | Large tolerance with low requirement on the bonding surface. However, potential metal contamination, and process incompatibility. Potential coupling problem due to the metal absorption | Easy to be scalable with multiple dies and large-sized wafers | Enhanced thermal resistant due to the metal utilization |

with thick oxide cladding, we adopt the low-temperature plasma-activated molecular bonding method (Pasquariello and Hjort, 2002) with oxide as the bonding interlayer. Furthermore, such thick cladding oxide also serves as the diffusion and absorption medium for the bonding by-products gases, thus with enhanced bonding quality and bonding yield.

For the initial development, we deposit 1.1 μm PECVD oxide on top of silicon wafers, followed with chemical mechanic polishing (CMP) to remove 100 nm oxide in order to ensure the smooth bonding interface. For all the bonding process described hereafter, we will use the similar PECVD oxide as cladding followed with CMP to smooth the bonding surface. Thus, we characterize and compare the oxide properties in terms of wafer-level uniformity and surface roughness before and after CMP. **Figures 1A,B** show the wafer-level oxide thickness before and after CMP. The non-uniformity is only ~1% after CMP, which suggests a very flat surface. **Figures 1C,D** show the AFM results before and after CMP. As deposited, the surface is relatively rough, with RMS of ~2.5 nm, while after CMP, the surface roughness is reduced significantly with RMS of ~0.4 nm, which is more suitable for wafer molecular bonding (Christiansen et al., 2006).

The bonding process starts with the wafer cleaning, separately for silicon wafer and III/V wafer. First, standard SPM clean for 10 min is performed to the silicon wafer in order to remove any organic contaminants, followed with 5 min SC1 clean with megasonic to remove any particle on the surface. The III/V wafer is separately cleaned in the $NH_4OH$ solution ($NH_4OH$:DI water $= 1:15$) for 1 min. Second, $O_2$ plasma activation in a RIE chamber is performed for both silicon wafer and III/V wafer, subsequently followed with DI water rinse. These two bonded wafers are then physically contacted with each other immediately after drying and placed inside to the EVG 520 bonder for pre-bonding under $N_2$ for 3 min with 1000 N mechanic force. After that, post bonding annealing at 300°C in vacuum is applied to the bonded pair for 12 h. **Figure 2A** shows the optical image of a 50 mm InP wafer bonding to a 200 mm silicon wafer before unloading from the bonder track.

The bonded wafers are first characterized by scanning acoustic microscope (SAM) using Sonix HS3000. **Figure 2B** shows the typical CSAM image for the bonded wafer. We observe that larger than 98% of the 50 mm InP area is bonded to the silicon wafer, with only limited voids, which are attributed to the particles remaining on bonding surface. Besides, the bonding quality in the wafer

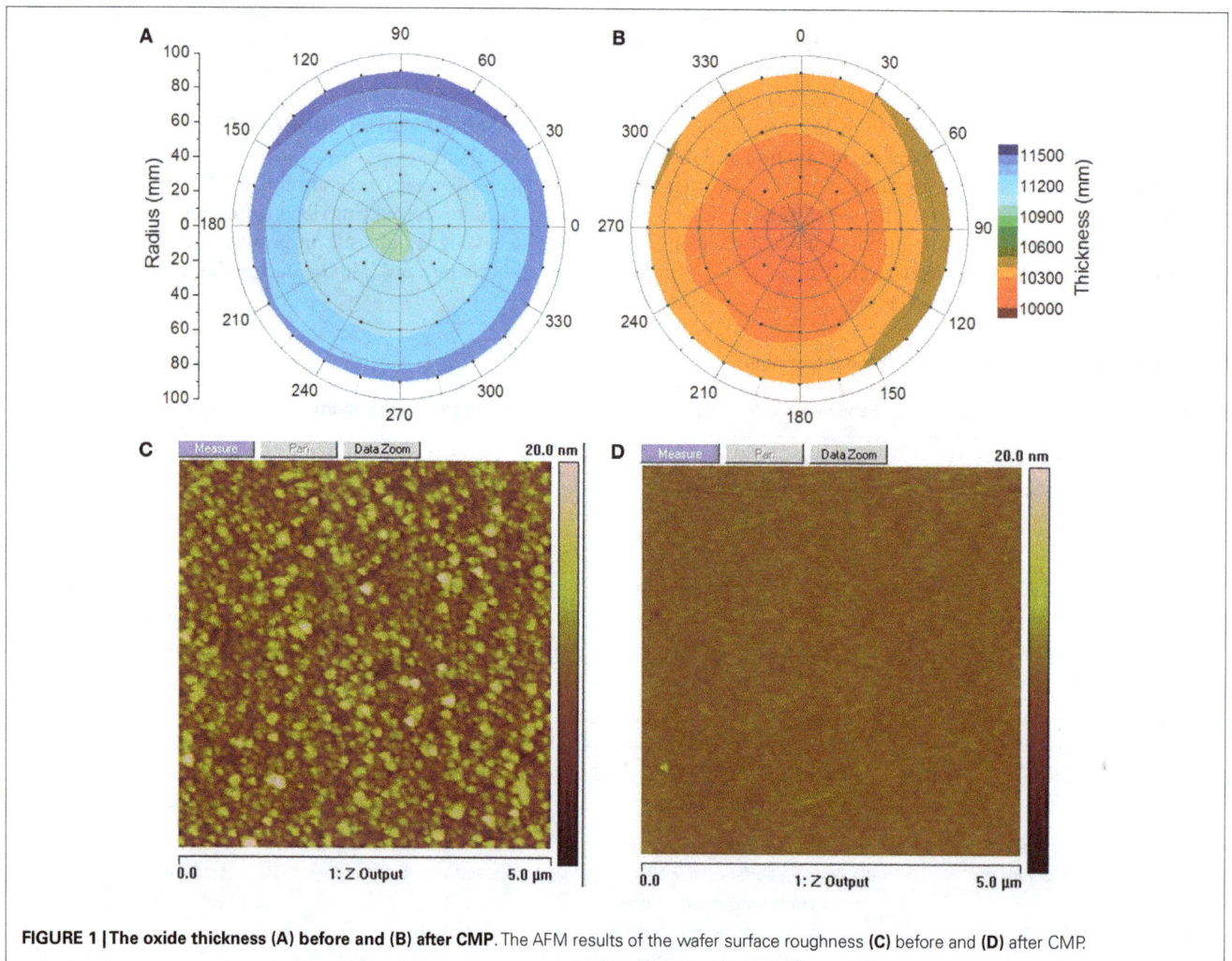

**FIGURE 1 | The oxide thickness (A) before and (B) after CMP.** The AFM results of the wafer surface roughness **(C)** before and **(D)** after CMP.

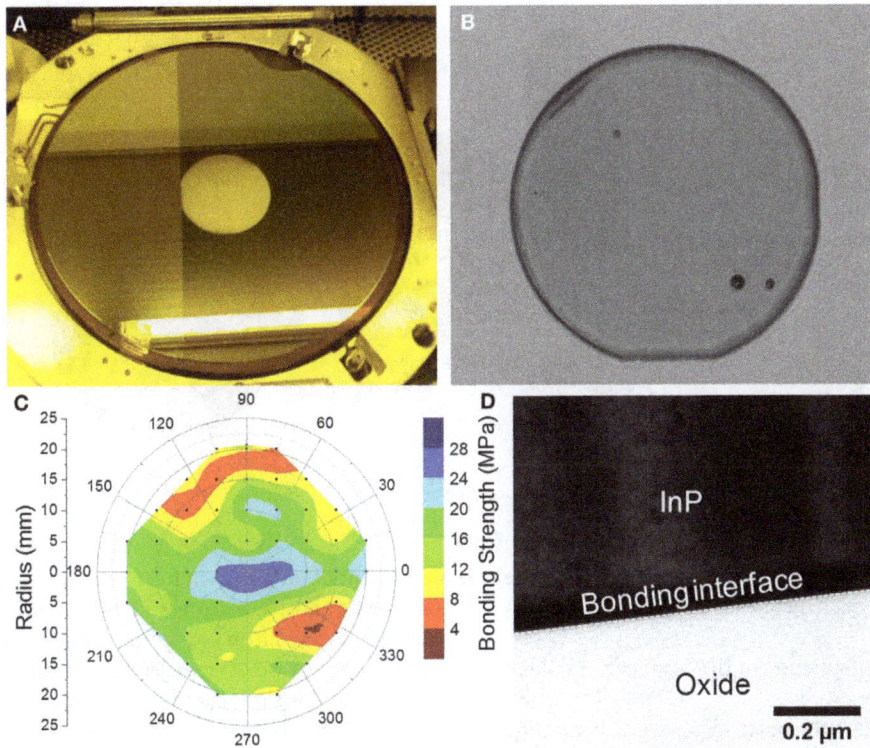

**FIGURE 2 | (A)** The bonded wafer before unloading from the EVG bonder. **(B)** The CSAM result. **(C)** The shear testing result. **(D)** The TEM results show the high-quality bonding interface.

periphery is also not good enough, which is due to the ring-shaped imperfection of the InP wafer.

The whole wafer is then diced into 5 mm × 5 mm dies for shear testing by using a Die Shear Tester (Dage Series 4000). **Figure 2C** shows the extracted bonding strength, with maximum bonding strength of ~30 MPa in the wafer center region, and the averaged value of 15 MPa. We believe such bonding strength is high enough for any of the post optoelectronic fabrication process. **Figure 2D** shows the TEM results of the bonded wafer, which indicate a very tight bonding between InP and oxide, again suggesting a high-quality bonding.

However, although such W2W bonding has been demonstrated with high quality, there are still existing big challenges, including:

(1) insufficient III/V wafer utilization,
(2) insufficient silicon wafer utilization due to wafer size mismatch,
(3) III/V wafer global stress-induced bonding voids.

First of all, for practical application of optical interconnection, only very small portion of the silicon waveguide area needs to be bonded with III/V material for optoelectronic fabrication to form optical lasers. With whole III/V wafer bonding, most of the III/V material will be subsequently etched away during post optoelectronic fabrication. Giving such precious III/V wafers, the insufficient utilization of the III/V material results in significantly

increased device cost and waste, which in turn makes it ineffectual to use the silicon photonics though it is of low cost. Second, the main stream silicon photonics has already adopted 200 mm silicon wafers. However, due to the brittleness of the InP wafers, it is very difficult to make large-sized wafers to match with silicon wafers. Although the largest available III/V epitaxial wafer can go with 150 mm, the commercially available largest-sized III/V epitaxial wafer is only 75 mm. Thus, such wafer size mismatching definitely results in the insufficient utilization of the silicon device wafer, which in turn increased the cost. Furthermore, InP wafers with multiple quantum well structures are normally with high global stress, which induces the wafer bowing. Such stress-induced wafer bowing will easily trap the air between the bonding interfaces with remained voids, thus reducing the bonding quality.

## HIGH-THROUGHPUT MULTIPLE DIES-TO-WAFER BONDING TECHNOLOGY

Based on the aforementioned W2W bonding method, we propose an alternative proprietary high-throughput multiple D2W bonding method, which is based on temporarily bonding III/V dies to a handle silicon wafer through pick-and-place process for simultaneous batch processing. Such high-throughput multiple D2W bonding method is the key enabling technique for potential manufacturability of large-scale hybrid optoelectronic integrated circuit (H-OEIC).

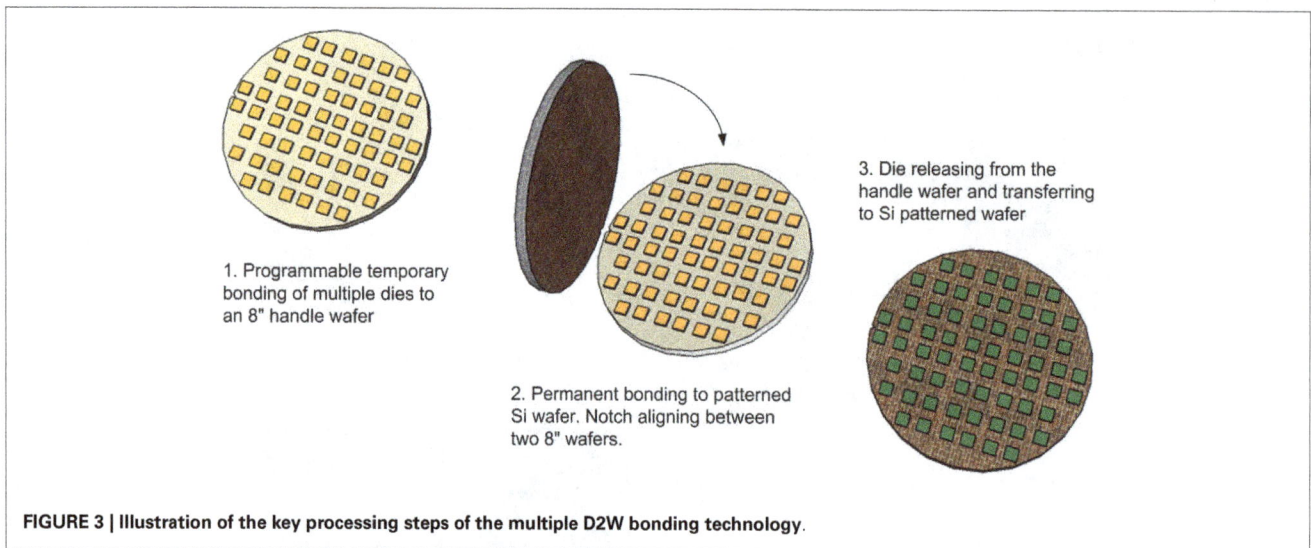

**FIGURE 3 | Illustration of the key processing steps of the multiple D2W bonding technology.**

**Figure 3** shows the key process steps of the proposed multiple D2W bonding technology, which includes:

(a) the programmable reconfiguration of III/V dies onto a handle wafer via pick-and-place process,

(b) the D2W bonding through the notch alignment between the two 8″ wafers, after batch processing of wafer cleaning and plasma activation, and

(c) the dies releasing from the handle wafer and transferring to silicon device wafer.

The most critical step here is the choice of the adhesion layer for the temporary III/V dies bonding to the handle wafer, which includes the following two trade-off considerations.

(1) The adhesion should be strong enough to stick the III/V dies on the handle wafer without peeling off during the subsequent III/V dies batch processing, including InGaAs cap layer wet etching, pre-clean, wafer drying, and plasma activation, etc.

(2) The adhesion should not be excessively strong so that the III/V dies can be successfully released and transferred to the Si device wafer after pre-bonding.

The programmable reconfiguration of the dies onto the handle wafer is realized through pick-and-place process by pre-determining the position coordinates of each die with considering the wafer-level silicon device die distribution. Unlike the pick-and-place process in flip-chip bonding, which directly bonds the dies to the actual wafer (Kostrzewa et al., 2006), the pick-and-place in our proposed method only helps to distribute the dies onto a handle wafer without flipping the chips. Thus, all the dies attached to the handle wafer can be simultaneously performed with different process steps for wafer bonding, such as InGaAs cap layer etching, wafer clean, and plasma activation.

The D2W bonding alignment accuracy is mainly determined by the notch alignment, which is performed manually and induces a relatively large misalignment of $\pm 500\,\mu m$, compared to the

misalignment of only $\pm 5\,\mu m$ from the programmable reconfiguration by pick-and-place process. However, as the alignment of the III/V devices to the silicon waveguide device is determined through photolithography during optoelectronic fabrication process after wafer bonding, such misalignment can easily be compensated by adopting relatively large-sized III/V dies.

**Figure 4** schematically illustrates the detailed bonding process flow starting from the preparations of silicon and III/V wafer. For either blanket silicon wafer or patterned silicon wafer with photonic devices, the wafers are cladded with PECVD oxide, followed with CMP process to smooth the bonding surface. As the hybrid laser performance is largely dependent on the vertical coupling efficiency, which is determined by the inter-layer oxide thickness, it is of very importance to control the oxide thickness by CMP process.

The preparation of the III/V dies includes the III/V wafer dicing into certain sized dies, the preparation of the adhesion layer to the handle wafer, and the programmable reconfiguration of the III/V dies temporary bonding to the handle wafer through pick-and-place process. Typically, the mechanical wafer dicing will result in edge roughness along the dicing lane, thus subsequently cause the low quality bonding near the die periphery. Besides, such dicing process may also introduce particles to the wafer surface cause contamination. Thus, a sacrificial InGaAs cap layer in order to protect the III/V bonding surface is designed in our demonstration. The applied mechanic force by the pick-and-place head also needs to be well controlled in order to ensure the successful die releasing from the handle wafer after pre-bonding. Due to the direct contact of the pick-and-place head, the die surface could be contaminated. However, owing to the sacrificial InGaAs layer, the bonding surface can be well protected without contamination or surface damage. We have checked and compared the surface condition of the III/V dies before and after the etching of the sacrificial InGaAs layer. We observe the particles on the chip surface after the wafer dicing and pick-and-place process, which is with relatively high RMS of 0.198 nm. In comparison, after the etching of InGaAs cap layer, the surface roughness is improved with reduced RMS of

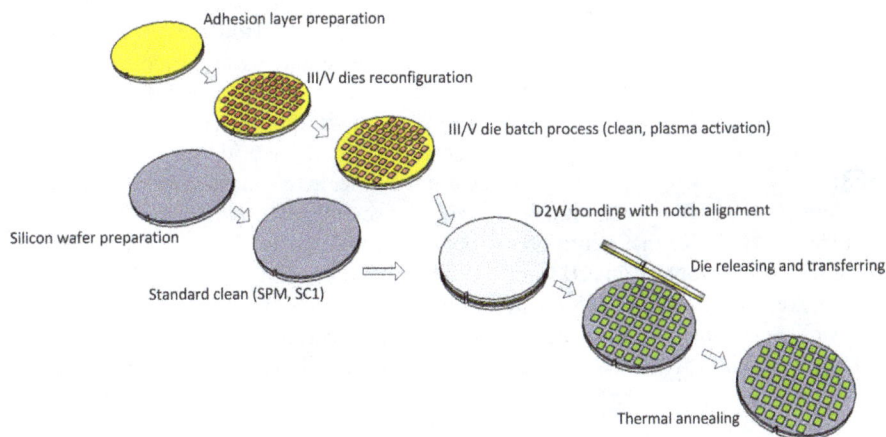

**FIGURE 4 | The fabrication process flow of the multiple D2W bonding technology**. The process includes two different folds, i.e., the bonding wafers preparation including the III/V dies adhesion to handle wafer for batch process and the silicon device wafer fabrication, and D2W bonding through dies releasing from handle wafer and transferring to silicon device wafer.

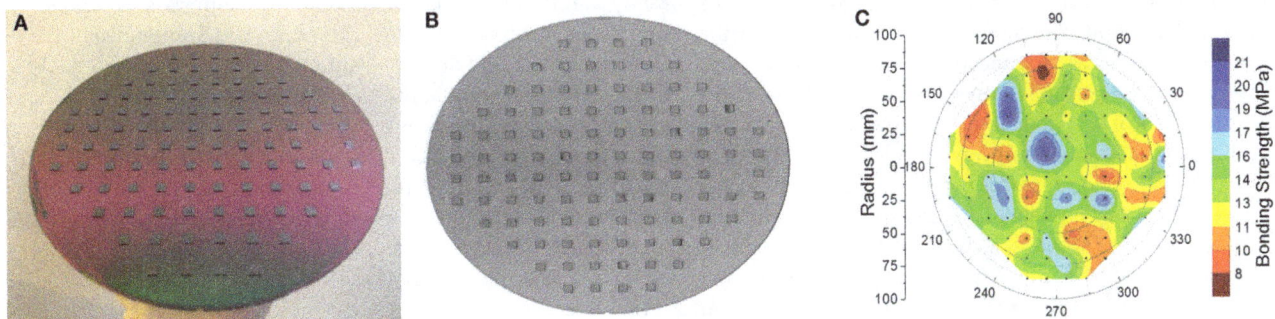

**FIGURE 5 | Demonstration of 104 III/V dies bonding to silicon wafer**. (A) Photo image of the bonded wafer, (B) CSAM results, (C) shear testing results.

0.182 nm, which is far below the required RMS of <1 nm for wafer direct bonding (Christiansen et al., 2006).

Prior the physical contact of the wafers for molecular bonding, the silicon wafer is performed with standard SPM clean for 10 min and SC1 clean with mega sonic for 5 min, while III/V die-attached handle wafer is first performed with sacrificial InGaAs cap layer etching in $H_3PO4$ solution for 1 min, followed with standard clean in $NH_4OH$ solution for 2 min. After that, $O_2$ plasma activation is applied to both silicon wafer and III/V dies in a RIE chamber for 1 min, followed with DI water rinsing and wafer drying. The III/V dies and the silicon wafer are then physically contacted with each other by notch alignment between two 8″ wafers, followed with pre-bonding in the 200 mm EVG bonder for 2 min with 1000 N mechanical force applied. The III/V dies are released from the handle wafer after pre-bonding, and all III/V dies are now transferred to the silicon device wafer. Finally, the bonded pairs are placed back to the EVG bonder for post-bonding annealing at 300°C for 12 h.

As a proof-of-concept demonstration, we show here the bonding of 104 InP dies to an 8″ silicon wafer. The silicon wafer is covered with 1 μm PECVD oxide after CMP. The InP dies are diced into 5 mm × 5 mm in size. **Figure 5A** shows the photo image of the bonded wafers with nearly all InP dies are successfully bonded to the silicon wafer. The only missing piece is peeled off during pick-and-place process. The CSAM shown in **Figure 5B** suggests a successful bonding. The dark areas, which suggest less strong bonding, come from the dies located in the InP wafer edge. **Figure 5C** shows the sheer testing results. The maximum bonding strength is larger than 20 MPa, with an averaged bonding strength of ~13 MPa, which is comparable with that of W2W bonding under the same process.

All in all, we believe that there are at least two significant implications of the proposed multiple D2W bonding technology:

(1) The significantly increased bonding efficiency owing to the simultaneous batch process. Through the batch process of the III/V dies (pre-clean, plasma activation, etc.), it is possible to bond unlimited number of dies. It also helps to avoid potential contamination by performing the pick-and-place before cleaning process, and eliminates the time link constraint of the bonding surface deactivation. This is the most significant processing advantage comparing to the conventional pick-and-place method.

(2) The scalability to whatever-sized silicon wafers. Such multiple D2W bonding technology can easily be adopted for even larger-sized silicon wafers, such as 300 mm wafer. This is the most critical step toward the potential manufacturability of H-OEIC.

## DESIGN OF III/V-on-Si LASERS

A hybrid III/V-on-silicon laser consists of a III/V epitaxial-layered structure and a silicon waveguide. It is a device that emits laser beams from silicon waveguides by electrical/optical injection to the III/V region. In this section, we will discuss the design of hybrid III/V-on-silicon lasers with regard to two fundamental laser elements, namely, optical gain medium and optical waveguide cavity.

## DESIGN OF III/V MQW STRUCTURES

There are two main material systems for the fabrication of long-wavelength lasers emitting at $1.55\,\mu m$, which are InGaAsP/InP and InGaAlAs/InP systems. Both kinds of materials can be used to fabricate hybrid lasers. InGaAlAs MQWs exhibit a larger conduction band discontinuity ($E_c = 0.72E_g$), and smaller valence band discontinuity compared with InGaAsP MQW. This leads to an improved electron confinement, which can improve the temperature characteristics of semiconductor laser diodes. Thus, InGaAlAs/InP material system is more suitable for high speed and uncooled operation of semiconductor laser diode. In this study, we select this material system for the hybrid silicon lasers demonstration. The MQW region includes eight $Al_{0.055}Ga_{0.292}In_{0.653}As$ quantum wells separated by nine $Al_{0.055}Ga_{0.292}In_{0.653}As$ barriers. The gain spectrum of the MQW is calculated and the wavelength of peak gain is designed at 1550 nm when the carrier injection density increases from $5 \times 10^{17}$ to $5 \times 10^{18}/cm^3$ as shown in **Figure 6A**. **Figure 6B** shows the measured photoluminescence (PL) spectrum for III/V epitaxial wafer at room temperature with the peak wavelength at about 1550 nm.

## DESIGN OF HYBRID LASER VERTICAL WAVEGUIDE STRUCTURE

As mentioned, the optical gain comes from overlying III/V stack layer, which needs to be structured to efficiently inject electrons or holes into the MQW regions. A high overlap between the optical mode and the MQW benefits to achieve a high optical gain,

which means that the optical mode needs to be well confined in the III/V waveguide. However, on the other hand, the light has to be confined sufficiently inside the silicon output waveguide for the efficient light extraction. In view of this, there are mainly two kinds of waveguide structures considering the optical power distribution for the hybrid laser with optical mode predominantly confined either in the silicon waveguide or in the III/V overlay. This leads to two different optical cavity designs. In the first design, the optical cavity comprises both III/V and silicon waveguides and the mode is mainly guided within the Si waveguide and evanescently coupled with the III/V waveguide. Such structure is also called as overlapped structure with hybrid mode (Fang et al., 2006, 2007a,b, 2008a,b,c, 2009). It has the advantage of making the coupling to a passive silicon waveguide straightforward and wavelength selective features can easily be defined in the silicon waveguide layer using CMOS fabrication techniques, which provides an accurate mechanism to control the emission wavelength of the laser. However, it requires a controllable thin bonding layer (<50 nm) for efficient optical coupling, which may increase the difficulty of bonding process. Furthermore, due to the weak interaction between the optical mode and gain material, it usually requires longer laser cavity, and thus resulting in high power consumption. In the second design, the mode in the hybrid section is mainly guided by the III/V waveguide, and the light is coupled from the III/V waveguide to the silicon waveguide through waveguide mode transformer, such as inverse tapers (Yariv and Sun, 2007; Sun et al., 2009a; Ben Bakir et al., 2011). In such design, the bonding interface can be relatively thick (typically from 30 to 150 nm) due to the released coupling constrain for the bonding interface. The advantage is that the optical mode experiences a high optical gain in the central region of the laser structure. However, the challenge of this structure is the fabrication of low-loss tapered waveguides. Hereafter, we name such design as adiabatic tapered coupling structure.

### Silicon waveguide thickness selection

The selection of the silicon waveguide thickness depends on the detailed device dimensions/structures and the fabrication process. For indium phosphide (InP)-based gain waveguides, the effective refractive index is typically larger than 3.2 if the waveguide width and height are larger than $1\,\mu m$. In order to achieve this index for silicon waveguides for effective coupling with the InP-based gain

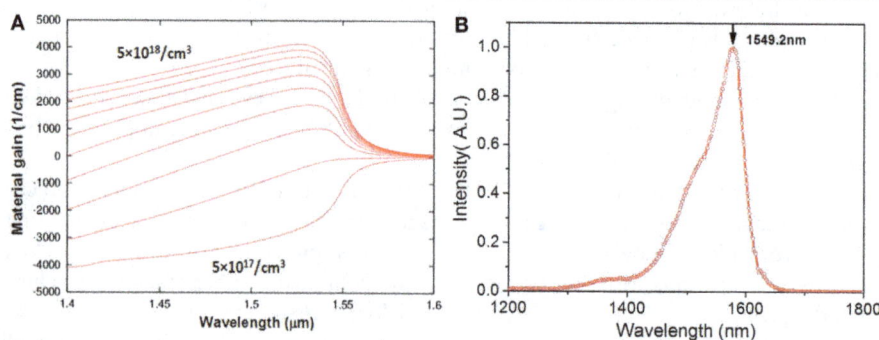

**FIGURE 6 | (A)** The calculated gain spectrum under different carrier injection concentration. **(B)** The measured photoluminescence (PL) spectrum for III/V epitaxial wafer.

region, the corresponding silicon waveguide thickness needs to be sufficiently large. **Figure 7** shows the calculated effective refractive index of the fundamental mode in silicon waveguide depending on the waveguide thickness. It indicates that the required silicon thickness needs to be larger than 450 nm to achieve an effective index of 3.2 for the waveguide with 2 $\mu$m in width. Such thick silicon layer does not match with the current mainstream silicon photonics. However, on the other hand, it is still possible to couple light from 220 nm silicon to InP waveguides by using very narrow InP waveguides (~200 nm) to push down the value of effective index, although the fabrication is difficult to form these narrow InP waveguides by conventional photolithography.

### Overlapped structure with hybrid mode

As mentioned above, there is a tradeoff between the optical mode confinement in the III/V and silicon regions for the overlapped

structure. The bonded III/V-Si structure forms the hybrid waveguide cavity. The effective refractive index of III/V active and Si regions are critical parameters for the hybrid waveguides, which, respectively, determine the light confinement factors in III/V and Si region. In our design, the confinement factors over the silicon and the quantum well regions are modified by altering the silicon waveguide thickness and the separate confinement heterostructure (SCH) thickness in order to ensure sufficiently low-threshold gain for lasing. While the thicker silicon waveguide pulls the optical mode into the silicon layer, the larger SCH thickness can drag back the optical mode into the III/V region.

**Figure 8A** shows the calculated optical confinement factors in MQW and Si depending on the Si waveguide width under different Si thickness, with assuming the III/V ridge width and the SCH thickness of 6 $\mu$m and 250 nm, respectively. It shows that when the waveguide widths of III/V and Si are fixed, the optical confinement in Si waveguide can be increased by using a thick Si layer. With the silicon waveguide thickness of 700 nm, large confinement of up to 70% in silicon waveguide is achieved. However, the device performance is very sensitive to the bonding interface quality due to the overlapping of the optical mode with the bonding interface between the III/V and the silicon. Based on the analysis, we adopt silicon thickness of 500 nm for the demonstration of hybrid Si lasers.

**Figure 8B** shows the simulated optical confinement factor in MQW and in silicon waveguide with different SCH thickness, with the fixed III/V and Si waveguide widths of 4 and 2 $\mu$m, and Si waveguide thickness of 500 nm. As the SCH thickness increases, the optical mode confinement in III/V region increase, which in turn significantly decreases the optical mode confinement in silicon waveguide. Inserts in **Figure 8B** show the simulated field distributions with SCH thicknesses of 0.1 and 0.5 $\mu$m. It shows obviously that for the small SCH thicknesses, the optical mode lies primarily in the silicon region, while the optical mode is dragged into III/V region with increased SCH thickness. The ability to control the optical mode with the SCH thickness is a key feature of this platform. For hybrid lasers, higher optical confinement is needed to achieve lower threshold current. Thus, we choose an optimized SCH thickness of 0.18 $\mu$m for the hybrid lasers.

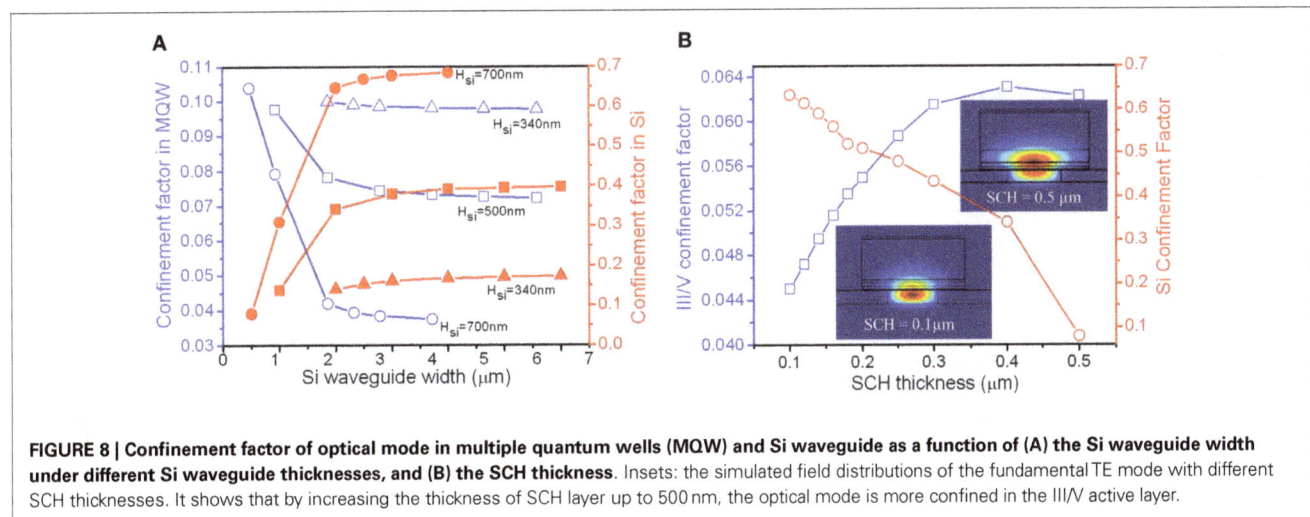

**FIGURE 7 | Effective refractive index of the silicon waveguide fundamental mode as the function of silicon waveguide thicknesses**. Both top and bottom claddings are oxide ($n = 1.45$) and silicon index is chosen as 3.48 at the wavelength of 1550 nm.

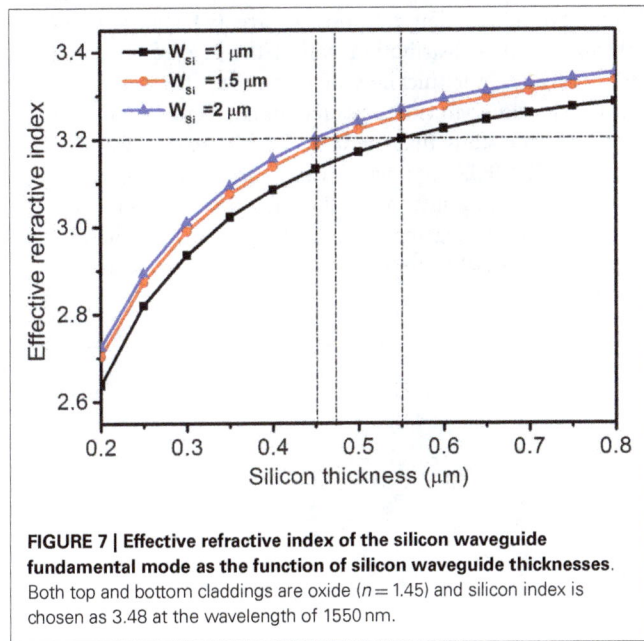

**FIGURE 8 | Confinement factor of optical mode in multiple quantum wells (MQW) and Si waveguide as a function of (A) the Si waveguide width under different Si waveguide thicknesses, and (B) the SCH thickness**. Insets: the simulated field distributions of the fundamental TE mode with different SCH thicknesses. It shows that by increasing the thickness of SCH layer up to 500 nm, the optical mode is more confined in the III/V active layer.

For such hybrid III/V-on-silicon lasers, another challenge arises from the control of the bonding layer thickness. Generally, a thin bonding layer (<50 nm) is needed for efficient optical coupling between III/V and silicon regions, while the thicker bonding layer benefits to the bonding quality of III/V layer and the bonding yield improvement. For direct bonding without oxide interlayer, it is easy to achieve such thin thickness, which is usually only the native oxide. However, this process is particularly sensitive to surface roughness and particles contamination, which would limit the bonding quality and bonding yield. DVS-BCB bonding can be used for the heterogeneous integration of III/V material on silicon to improve the yield. However, it is difficult to obtain a controllable thin bonding interlayer of <50 nm. In our case, we choose silicon oxide as interlayer between III/V and silicon, which is also compatible with the mainstream silicon phonics, in which all the devices are with oxide cladding.

**Figure 9A** shows the calculated optical confinement factor in MQW and silicon waveguide as the function of interlayer oxide thickness. In the simulation, we assume the fixed silicon thickness of 500 nm, the silicon waveguide width of 3 μm, and III/V ridge waveguide with of 6 μm. We observe from the results that the Si confinement factor largely decreases when the interlayer thickness increases from 10 to 100 nm. Only approximately 5% optical light is confined in the silicon waveguide when the interlayer thickness is 100 nm.

Additionally, the interlayer of oxide at the bonding interface also affects the characteristics temperature of hybrid III/V-on-silicon laser due to the poor thermal conductivity of as low as

1.3 W/m/K. The modal gain of laser is dependent on the temperature of the active region. As the temperature of active region increases, the modal gain decreases due to the increased carrier leakage out or not reaching the active region, and/or increased non-radiative recombination. The decrease of modal gain leads to high threshold current and low output optical power. In order to investigate the effect of interlayer on the thermal characteristics of the hybrid lasers, a two-dimensional model of the device structure is conducted using COMSOL by mapping out the heat dissipation of each layer. **Figure 9B** shows the simulation structure. In the simulation, the structure parameters are as follows: III/V ridge width, Si ridge width and thickness, and the laser cavity length are assumed to be 6 μm, 1 μm, 500 nm, and 1000 μm, respectively. Injection current is 500 mA and the corresponding voltage is 4 V. **Figure 9C** shows the calculated working temperatures in the III/V active region with different thicknesses of the oxide interlayer. The increase of temperature in the III/V active region versus interlayer thickness is about 0.02 K/nm. For illustration purpose, **Figure 9D** shows as an example the thermal distribution within the layered structure for the oxide interlayer thickness of 0 nm. The thermal distributions for the other thicknesses are similar. Actually, we can conclude from the study that the main hurdle for the heat dissipation is the SOI BOX layer, which can be seen from the results without oxide bonding interlayer (thickness = 0 nm). Thus, for enhanced thermal management, novel designs such as thermal shunt (Liang et al., 2012) are required to effectively remove the generated heat.

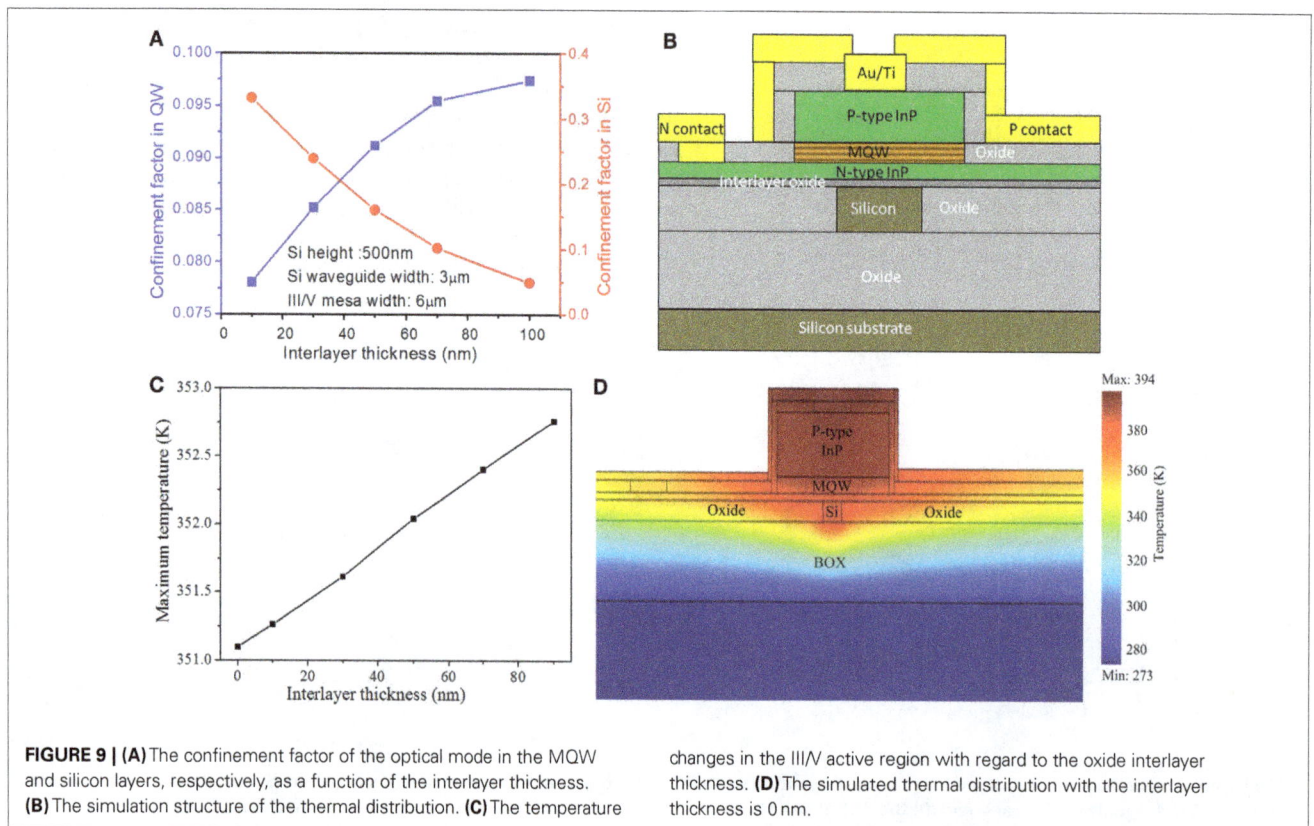

**FIGURE 9 | (A)** The confinement factor of the optical mode in the MQW and silicon layers, respectively, as a function of the interlayer thickness. **(B)** The simulation structure of the thermal distribution. **(C)** The temperature changes in the III/V active region with regard to the oxide interlayer thickness. **(D)** The simulated thermal distribution with the interlayer thickness is 0 nm.

### Adiabatic tapered coupling structure

For the adiabatic tapered coupling structure, the mode in the hybrid section is mainly guided by the III/V waveguide, and the light is coupled from the III/V waveguide to the silicon waveguide through a tapered waveguide. It shows that a tapering length ~100 μm is required for a sufficient light coupling with minimized optical loss. By using such tapered coupling, it eliminates the tricky tradeoff between the modal gain and vertical coupling efficiency, which is inherent in the overlapped waveguide structures. Therefore, hybrid lasers with a short cavity as pure III/V laser are possible. Up to now, the hybrid lasers with the high performances are achieved using such tapered coupling scheme (Levaufre et al., 2014; Zhang et al., 2014).

In order to efficiently couple the light between the Si-III/V hybrid waveguide and the silicon waveguide, the III/V waveguide and silicon waveguides are tapered simultaneously in the same direction. Here, we adopt a three-dimensional approximated model based on beam propagation method (BPM) in order to optimize the tapering structure of the silicon waveguide and III/V waveguide for an efficient coupling. **Figure 10A** schematically illustrates the design of such waveguide tapering structure. The coupling efficiency largely depends on the tapering design, especially the III/V waveguide taper width and taper length. Here, we simulate such dependency by varying the taper width and taper length, while fixing the III/V waveguide width of 5 μm, the silicon waveguide width of 1 μm, and the silicon taper length of 100 μm.

**Figures 10B,C**, respectively, show the simulated coupling efficiency from III/V-Si hybrid waveguide to silicon waveguide as functions of the III/V waveguide taper width and tape length. It suggests that the coupling efficiency from III/V-Si hybrid waveguide to Si waveguide can be as high as 85% by using an 80-μm-long III/V waveguide taper and a 100-μm-long silicon waveguide

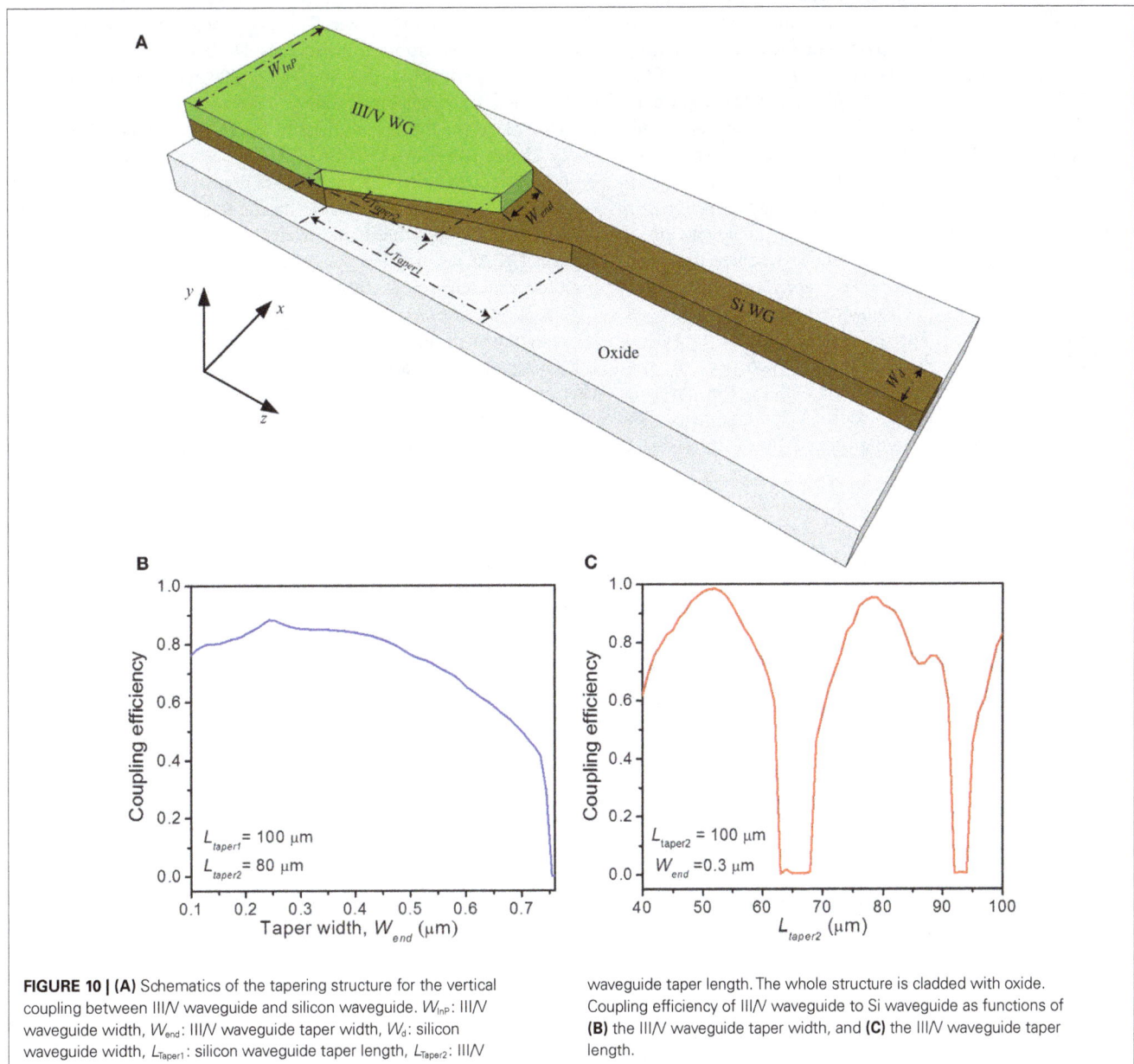

**FIGURE 10 | (A)** Schematics of the tapering structure for the vertical coupling between III/V waveguide and silicon waveguide. $W_{InP}$: III/V waveguide width, $W_{end}$: III/V waveguide taper width, $W_d$: silicon waveguide width, $L_{Taper1}$: silicon waveguide taper length, $L_{Taper2}$: III/V waveguide taper length. The whole structure is cladded with oxide. Coupling efficiency of III/V waveguide to Si waveguide as functions of **(B)** the III/V waveguide taper width, and **(C)** the III/V waveguide taper length.

tape. Through optimizing the III/V waveguide taper width and III/V waveguide taper length, the maximum coupling efficiency can be as high as 99%. However, due to the optoelectronic fabrication limitation, we are not able to demonstrate the hybrid laser using such adiabatic tapered coupling structure.

## DEMONSTRATION OF III/V-on-Si HYBRID LASERS

Using the proposed multiple D2W bonding technology, we have demonstrated various hybrid silicon lasers, including FP lasers, DBR lasers, sidewall-grating lasers, racetrack-shaped microring lasers, and MLL. In this section, we will first introduce the hybrid silicon laser fabrication process, leveraging on IME's CMOS-compatible silicon photonic fabrication facilities and NTU's expertise in optoelectronics fabrication capability, followed with showing some hybrid silicon laser demonstrations as the examples.

### III/V-on-Si HYBRID LASER FABRICATION PROCESS

The III/V-on-Si hybrid laser fabrication in our demonstrations includes two parts, namely silicon passive device fabrication using IME's CMOS line and multiple D2W bonding in IME's MEMS line, and III/V optoelectronics fabrication in NTU. **Figure 11** shows the fabrication process flow. We adopt commercially available SOI wafer with 340 nm silicon layer sitting on a 2 μm buried oxide (BOX) layer. The fabrication starts with the blanket silicon epitaxy to ~500 nm for refractive index matching between silicon waveguide device and InP gain medium. After the deposition of 70 nm oxide as the hard mask, the waveguide structures, including both grating coupler and inverse taper coupler are patterned by deep UV photolithography and transferred onto the silicon layer by using deep RIE etching. For the grating coupler, the silicon etching thickness is 377 nm. While for other waveguide devices, second silicon etch is applied down to the BOX layer by covering the surface grating coupler area with additional photo resist. Oxide cladding of 650 nm in thickness is deposited followed by a surface

planarization step, which includes oxide etch-back with 500 nm in depth and CMP process. Such planarization steps with CMP also help to smooth the bonding surface with very small surface roughness for molecular bonding. The interlayer oxide thickness can be well controlled by the CMP process, with only ~50 nm oxide left atop the silicon waveguide in our demonstration. For enhanced flatness and uniformity of the bonding surface, we only etch away the silicon surrounding the devices, remaining most of the silicon areas forming silicon plateaus.

The multiple D2W bonding is then performed after the preparation of III/V dies, followed with the process described in Section "III/V-to-Si Wafer-to-Wafer (W2W) Bonding Technology." As the designed devices are all within an area of 8 mm × 8 mm, the InP dies are all diced with 9 mm × 9 mm with the consideration of bonding misalignment of ±500 μm for the notch alignment, thus ensuring the full covering of all the silicon photonic devices within the III/V die area. For a 50 mm InP wafer, there are only 16 full dies with 9 mm × 9 mm in size. Thus, for the purpose of hybrid silicon laser demonstration using such D2W bonding technology, we only perform 16-dies bonding to 200 mm silicon wafers, with some of the silicon photonics device area being wasted. **Figure 12A** shows the photo image of the 16-dies bonded silicon wafer. **Figure 12B** shows the CSAM results. Except some particle-induced bonding defects, all the dies are bonded very well. However, we clearly observe that some of the die edge periphery regions are not bonded well due to the wafer dicing induced damage. **Figures 12C,D**, respectively, show the TEM and cross-SEM of the bonding structures, both suggesting very reliable bonding quality.

The III/V optoelectronic fabrication starts from the InP substrate removal using HCl solution. After photolithography, the InP mesa structures are formed by using $H_3PO_4$, $H_2O_2$, and HCl mixed solution to etch the InGaAs contact layer and p-InP cladding layer. The SCH layer and QW layer are also etched using $H_3PO_4{:}H_2O_2{:}H_2O$ solution, stopping on the n-InP cladding layer.

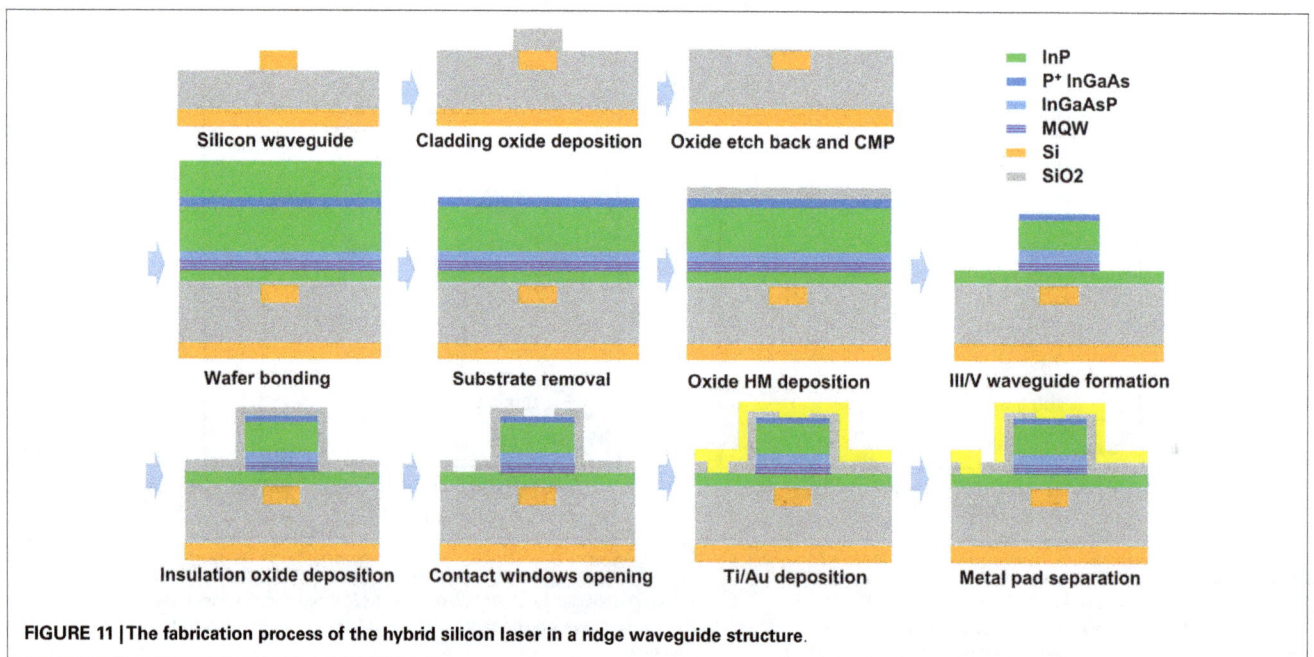

**FIGURE 11 | The fabrication process of the hybrid silicon laser in a ridge waveguide structure.**

Then, an $SiO_2$ insulator layer with the thickness of 300 nm is deposited, followed with the contact opening for p-type and n-type injection by one-time photolithography and the oxide is etched by HF solution. After that, Ti/Au metal contacts are formed by sputtering, with wet etching to form a Ti/Au slot between the n-type and p-type contacts using diluted HF and KI solution, respectively.

## FABRY–PEROT LASERS

The CW operation of the optical laser requires good thermal management to remove the generated heat. In the case of FP lasers, another way is to design narrow ridge waveguide to generate less heat. In our demonstration, we design a FP laser with ridge waveguide width of 6 μm. The laser facets are formed by lapping down the Si substrate to around 60 μm, followed with mechanical cleaving. The length of the FP laser is ~720 μm. The demonstrated FP laser is able to work at room temperature with CW operation. **Figure 13A** shows the measured $P–I$ curves under different temperatures. The threshold current at 264 K is ~45 mA,

and increases significantly to ~100 mA at room temperature. We attribute such fast increase of the threshold to the thick oxide interlayer, which prohibits the heat dissipation efficiently. The measured output power from a single facet without any reflection coating is more than 1 mW. This includes the coupling loss due to the un-optimized testing setup for light collection, which is estimated only with ~20% light collection efficiency. The thermal dissipation is very important for CW lasing. **Figure 13B** shows the measured lasing spectra under different temperatures. The wavelength shift with temperature is about 0.75 nm/K.

## LATERAL-COUPLED DISTRIBUTED-FEEDBACK LASERS WITH SIDE WALL GRATING

The FP laser is fabricated by lapping down the silicon substrate and mechanical cleaving to form the facets. From the optical communication and optical interconnect applications point of view, such FP lasers are not practical for photonic integration. Furthermore, how to achieve good facet is still a main challenge and a key limiting factor for high-performance hybrid lasers as reflection coating is always required in order to optimize the cavity transmission and reflection. In view of this, optical resonators, Bragg grating structures that form the cavities through fabrication are the good candidates for on-chip hybrid laser. We here show as an example of a hybrid laser using LC-DFB structure as the laser cavity.

**Figure 14A** schematically shows the perspective view of the LC-DFB hybrid laser with illustration of the key parameters, including the Bragg grating period $\Lambda$, the silicon ridge width $D$, and the grating teeth width, $W_1$. Considering the fabrication limitation, we design third-order later Bragg grating in order to achieve single-mode operation. With regard to the silicon thickness of 500 nm, the grating period $\Lambda$ is 670 nm with filling factor of 0.5. The ridge width $D$ and the teeth width $W1$ are, respectively, designed with 2 and 1 μm. The Bragg grating is centered beneath the III/V gain region, which is with the width of 12 μm. Both LC-DFB structure and III/V gain region are designed with the same lengths. In order to extract the output laser light for easy characterization, the vertical grating couplers are adopted. For the vertical grating coupler, the period is designed to be 640 nm with filling factor of 0.5, and the silicon etch depth is 377 nm. Such grating coupler design is purely based on theoretical calculation, without any process

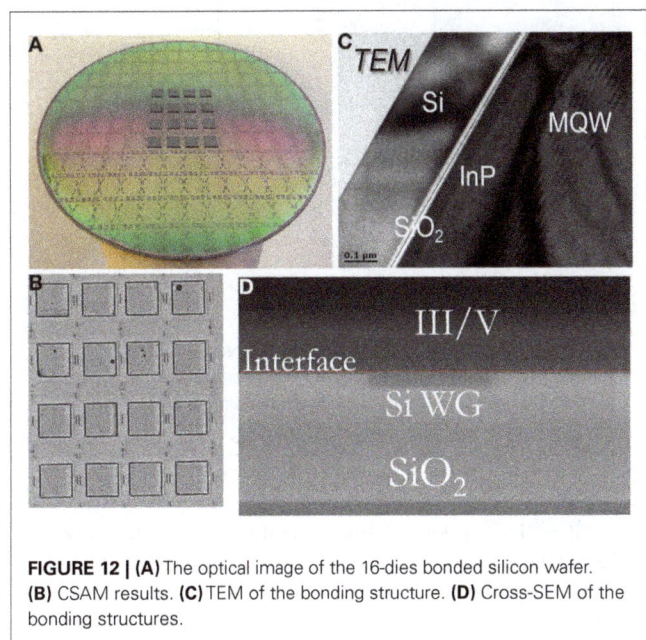

**FIGURE 12 | (A)** The optical image of the 16-dies bonded silicon wafer. **(B)** CSAM results. **(C)** TEM of the bonding structure. **(D)** Cross-SEM of the bonding structures.

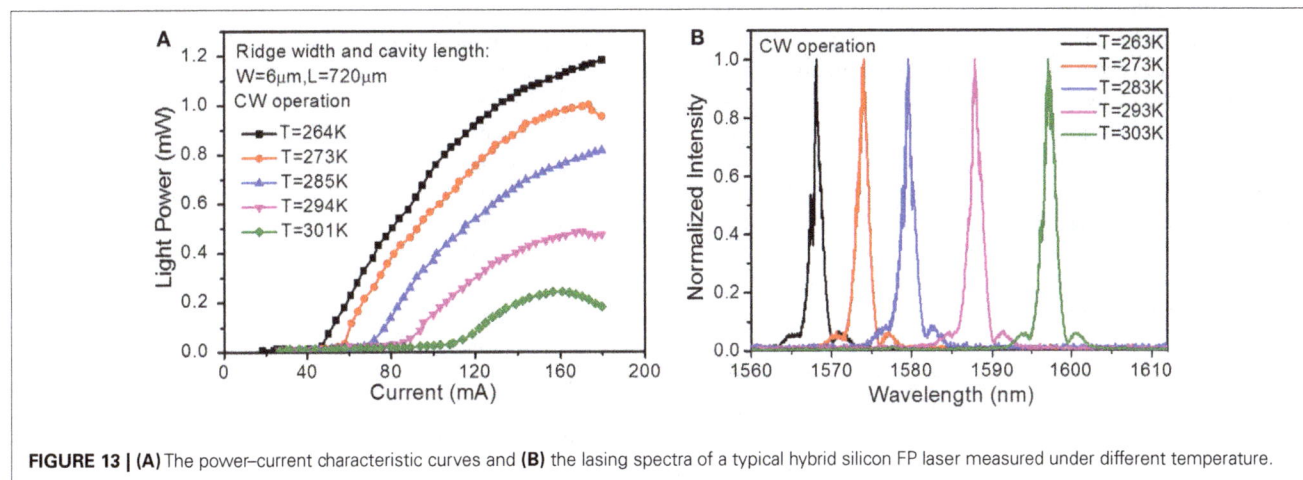

**FIGURE 13 | (A)** The power–current characteristic curves and **(B)** the lasing spectra of a typical hybrid silicon FP laser measured under different temperature.

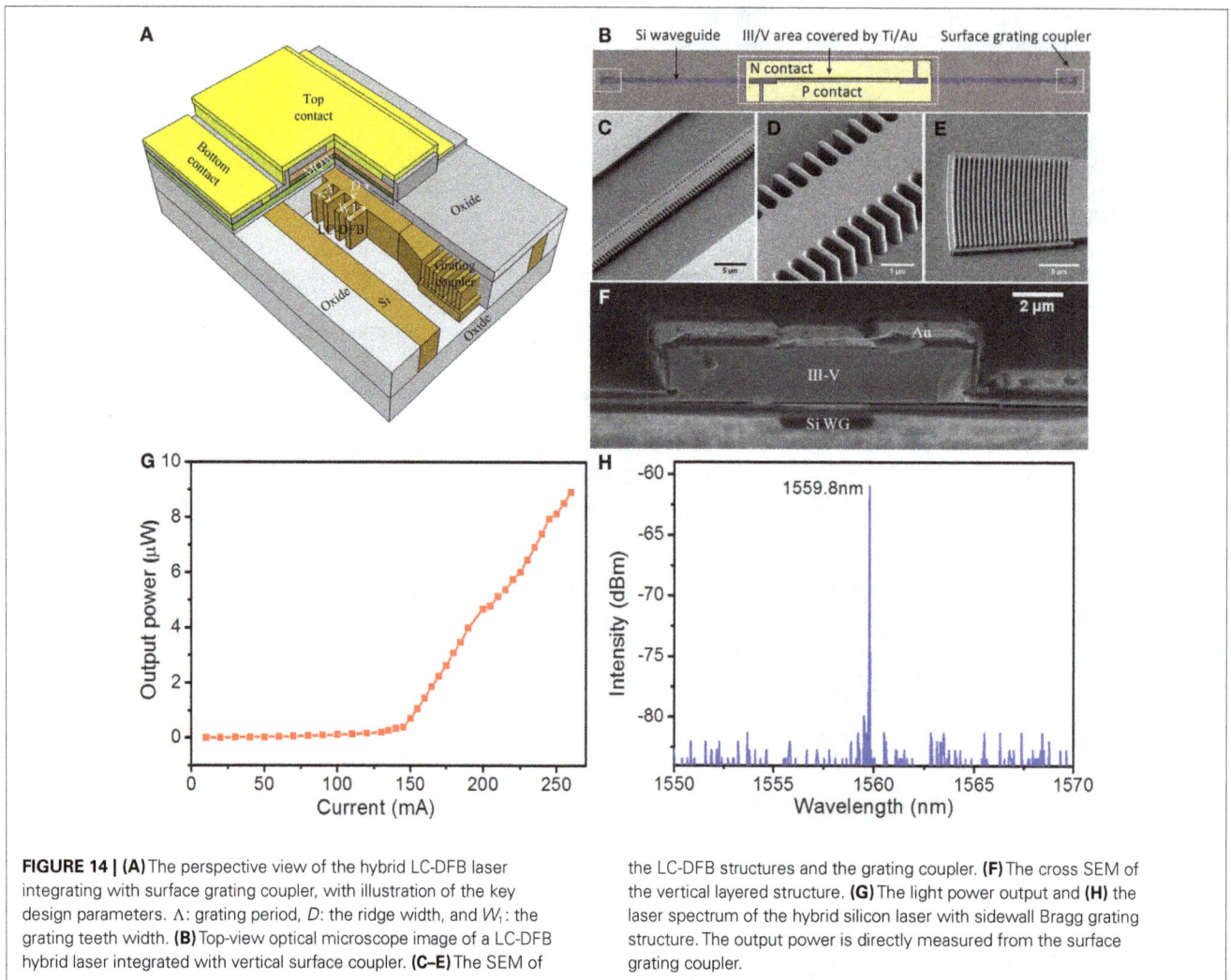

**FIGURE 14 | (A)** The perspective view of the hybrid LC-DFB laser integrating with surface grating coupler, with illustration of the key design parameters. $\Lambda$: grating period, $D$: the ridge width, and $W_t$: the grating teeth width. **(B)** Top-view optical microscope image of a LC-DFB hybrid laser integrated with vertical surface coupler. **(C–E)** The SEM of the LC-DFB structures and the grating coupler. **(F)** The cross SEM of the vertical layered structure. **(G)** The light power output and **(H)** the laser spectrum of the hybrid silicon laser with sidewall Bragg grating structure. The output power is directly measured from the surface grating coupler.

verification and optimization. For this demonstration, we did not design any mode transformer between III/V layer and silicon waveguide layer, thus expecting some transition loss. **Figure 14B** shows the optical microscope image of the fabricated hybrid LC-DFB laser with integrated vertical grating couplers. The LC-DFB structure and the III/V gain region are designed with same length of 700 $\mu$m, while the silicon device including two grating couplers is ~2750 $\mu$m. Due to the optoelectronic fabrication limitation, there is no designed taper between III/V waveguide and silicon waveguide, thus expecting relatively high transition loss. **Figures 14C–E** show the SEM images of the fabricated LC-DFB and vertical grating coupler, while **Figure 14F** shows the cross-sectional SEM of the vertical structures, illustrating the Si waveguide, the III/V layer, and the Ti/Au layer.

**Figures 14G,H** show the measured $P$–$I$ curve and the spectrum of the LC-DFB hybrid laser under pulse operation. The threshold current is ~120 mA, corresponding to a threshold current density of ~1.42 kA/cm$^2$. From the spectrum, we see clearly single-mode operation with the peak wavelength at 1559.8 nm and a side-mode-suppression ratio (SMSR) larger than 20 dB. This is expected from the LC-DFB design. However, the maximum output

power is only ~10 $\mu$W upon 250 mA current injection, which also can be observed from the spectrum measurement. We attribute the relatively low output power to the following two reasons, namely, the accumulated optical loss, and the inefficient vertical light coupling. First of all, the optical loss, which mainly includes the surface grating coupler coupling loss, the Bragg grating scattering loss, and the non-radiative recombination loss from the bonding interface, affects the light output significantly. From the reference measurement for the device only with passive silicon waveguide yet bonded III/V layer, the accumulated total loss is >40 dB, which is mainly due to the unoptimized surface grating coupler. Second, the oxide interlayer in our design, which might not be able to control precisely, will affect the light coupling efficiently from III/V layer to silicon waveguide. Furthermore, the polarization sensitivity of the surface grating coupler can also induce additional optical loss. Thus, the optimized grating coupler design for the light extraction from the silicon waveguide and the vertical light transition structure for light coupling from III/V layer to silicon waveguide can significantly increase the laser output power. Besides, by optimization of the Bragg grating period and silicon waveguide thickness, the SMSR can also be enhanced.

## PASSIVELY MODE-LOCKED LASERS

Semiconductor MLLs are excellent candidates for generating stable ultra-short optical pulses, which have a corresponding wide optical spectrum of phase correlated modes and high repetition rate. Optical frequency combs emitted by MLLs can have high extinction ratios, low jitter, and low chirp, which can be utilized in a variety of applications including AWG, optical clock generation and recovery, coherent communications systems, high-speed analog-to-digital conversion (ADC), and optical time-division multiplexing (OTDM), etc. Integration of MLLs on silicon is very promising as it combines the low-loss and low-dispersion characteristics of silicon material with high gain III/V material, thus ensuring improved performance. Furthermore, it will be possible for semiconductor MLLs to generate ultra-short optical pulses with low repetition rate on the silicon platform owing to the long cavity length. Here, we show our preliminary demonstration of a passive MLL using the developed heterogeneous integration platform.

The optical cavity of the MLL is defined by a 1250-$\mu$m-long gain section, a saturable absorber (SA) with the length of 30-100$\mu$m, and cleaved facet at the waveguide end. The gain section and SA are separated by a 20 $\mu$m electrical isolation region with isolation resistance >1 k$\Omega$. The SA is made up of the same active material as the gain section. The difference between the SA and gain section is that SA absorbs the light in the cavity upon applying a reverse bias, while the gain section amplifying light upon forward current injections.

The laser optical output is collected by a photodiode located in front of the cleaved facet. The typical threshold current with an unbiased 50-$\mu$m-long SA section is 88 mA. The device has a maximum single facet CW output power of 1 mW at room temperature when the injection current is 140 mA. The series resistance is about 8.5 $\Omega$, while the slope efficiency is about 0.02 mW/mA. **Figure 15A** shows the measured optical spectra at different injection currents. It shows that the widest optical emission is centered at about 1605 nm with a full-width at half-maximum (FWHM) of 5.4 nm at the injection current of 110 mA measured by an optical spectrum analyzer (OSA). Assuming the generated optical pulse is chirp-free and the shape of the pulse is with a Sech-function, the width of the optical pulse is calculated to 0.5 ps.

Passive mode locking of the device is obtained by forward biasing the gain section ($I_{gain}$) and reverse biasing ($V_{sa}$) or un-biasing the SA section. The mode locking behavior of the device is characterized by measuring the radio frequency (RF) spectrum using the spectrum analyzer (Agilent E4448A). **Figure 15B** shows the measured RF spectrum of the III/V-on-Si MLL at the injection current to the gain section ($I_{gain}$) of 110 mA and reverse biasing the SA section at −0.9 V. The resolution bandwidth (RBW) during measurement is 1 MHz. The repetition frequency is about 30.0 GHz with signal-to-noise ratio above 30 dB. By changing $I_{gain}$, it can be tuned to more than 30 GHz, giving clear evidence of passively mode locking of light signal. The measured RF linewidth of the injection locked laser is about 7 MHz by Lorentzian fitting the RF spectrum.

## SUMMARY AND FUTURE OUTLOOK
### KEY ACHIEVEMENTS

In summary, we reviewed in this paper the recent demonstrations of optical light source in silicon for the application of H-OEIC, with major focus on hybrid silicon lasers through wafer bonding technology. Furthermore, we proposed a proprietary high-throughput multiple dies-to-wafer (D2W) bonding technology by temporarily bonding III/V dies to a handle silicon wafer through pick-and-place process for subsequent simultaneous batch processing. Such high-throughput multiple D2W bonding technology features the merits of high bonding yield with unlimited III/V dies and scalability to whatever-size silicon wafers, thus is the key enabling technique toward potential manufacturability of large-scale H-OEIC. As proof-of-concept demonstration, we showed the III/V dies to silicon wafer bonding with up to 104 dies. Repeatable demonstrations of 16-III/V dies bonding to pre-patterned 200 mm silicon wafers are performed for the fabrication of hybrid silicon lasers with various laser cavities, including FP lasers, LC-DFB laser with side wall grating, and MLL.

### CHALLENGES AND FUTURE OUTLOOK

However, there are still many key issues need to be addressed before the hybrid silicon laser applied to optical interconnects system.

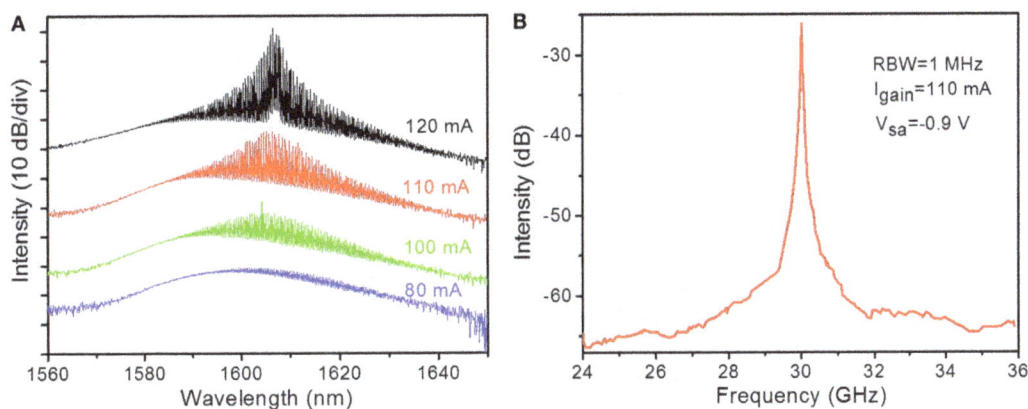

**FIGURE 15 | (A)** The measured optical spectra of the III/V-on-silicon mode-locked laser at different injection currents to the gain section while the SA section is floating. **(B)** Measured RF spectrum of the III/V-on-Si MLL at 110 mA injection current and saturation voltage of −0.9 V. The RBW during measurement is 1 MHz.

Here, we will only discuss three of the most important issues, including:

(1) the thermal management;
(2) the integration with other silicon photonic devices with full wafer processing capability; and
(3) the new platform beyond silicon for high-performance advanced hybrid lasers.

Thermal management is one of the major obstacles for achieving high-performance hybrid silicon lasers for practical applications (Sysak et al., 2011). Due to the poor thermal conductivity of the SOI BOX layer as well as the inter oxide layer between bonding surface, such layers would prevent the efficient heat dissipation to the silicon substrate, thus resulting in the poor laser performances, such as lower laser power. One of the simple ways to increase the thermal dissipation efficiency is to design the contact electrodes with thick and large area metal, serving as the top surface heat sink. Another efficient way for thermal dissipation is to remove the BOX layer in some areas and refill it with high thermal conductive materials such as polycrystalline silicon or metals, serving as thermal shunt (Liang et al., 2012). However, although such approach has been demonstrated with enhanced thermal management and increased laser performance, it still requires further development in order to further improve the performance.

The integration of such hybrid laser with the existing silicon photonics building blocks is another key issue before it is applied for H-OEIC. For most of the demonstrated hybrid silicon lasers, the silicon waveguide is normally with more than 500 nm thickness in order to ensure the optical index matching with III/V material for efficient light coupling to silicon waveguide. Such thick waveguide design is actually not compatible with the mainstream silicon photonics, with most of the key building blocks are demonstrated in 220 nm silicon wafers (Xu et al., 2014). Thus, novel designs taking care of both of these design considerations are required. Recently, Dong et al. (2014a,b) demonstrated novel integration scheme with associated transition structure via epitaxial growth of silicon in a pre-defined trench. Such epitaxial-grown silicon mesa also serves as the bonding interface with III/V gain material. Thus, the rest of the device area leaves with 220 nm silicon for other existing silicon photonic devices. Such novel demonstration sets a path toward the integration of hybrid silicon laser with existing silicon photonics building blocks. However, for practical integration with high-speed modulator and photodetector, which involves even complicated integration process with multiple oxide etch-back and CMP, it is still very challenging on how to ensure the flatness and smoothness of the bonding surface. More sophisticated design and further demonstration with integration of such are highly demanded.

The third issue is associated with current new demonstration trend that utilizes extremely low-loss SiN or SiON waveguide as the passive waveguide layer (Bovington et al., 2014; Luo et al., 2014). As we know, for some advanced type of hybrid lasers, such as the MLLs, extremely low optical loss is required for achieving high performances. The state-of-the-art demonstration of the silicon waveguide is still with propagation loss of ~2 dB/cm, which is higher comparing that of SiN waveguide of 0.1 dB/m (Bauters et al., 2011). Thus, III/V-SiN platform for hybrid lasers is another interest research area, which can address the loss issue. The integration between SiN waveguide and other SOI-based devices is also CMOS-compatible and ready for further application (Huang et al., 2014).

## ACKNOWLEDGMENTS

This work was supported by A*STAR SERC Future Data Center Technologies Thematic Strategic Research Programme under Grant No. 112 280 4038, and A*STAR – MINDEF Science and Technology Joint Funding Programme under Grant No.122 331 0076.

## REFERENCES

Alduino, A., and Paniccia, M. (2007). Interconnects: wiring electronics with light. *Nat. Photonics* 1, 153–155. doi:10.1038/nphoton.2007.17

Alexe, M., and Gösele, U. (2004). *Wafer Bonding: Applications and Technology*. Berlin: Springer-Verlag.

Bauters, J. F., Heck, M. J., John, D. D., Barton, J. S., Blumenthal, D. J., Bowers, J. E., et al. (2011). "Ultra-low-loss (< 0.1 dB/m) planar silica waveguide technology," in *Paper Presented at the IEEE Photonics Conference* (Arlington, VA: Marriott Crystal Gateway).

Beausoleil, R. G., Kuekes, P. J., Snider, G. S., Wang, S.-Y., and Williams, R. S. (2008). Nanoelectronic and nanophotonic interconnect. *Proc. IEEE* 96, 230–247. doi:10.1109/JPROC.2007.911057

Ben Bakir, B., Descos, A., Olivier, N., Bordel, D., Grosse, P., Augendre, E., et al. (2011). Electrically driven hybrid Si/III-V Fabry-Pérot lasers based on adiabatic mode transformers. *Opt. Express* 19, 10317–10325. doi:10.1364/OE.19.010317

Bogaerts, W., Baets, R., Dumon, P., Wiaux, V., Beckx, S., Taillaert, D., et al. (2005). Nanophotonic waveguides in silicon-on-insulator fabricated with CMOS technology. *J. Lightwave Technol.* 23, 401–412. doi:10.1109/JLT.2004.834471

Bovington, J., Heck, M., and Bowers, J. (2014). Heterogeneous lasers and coupling to Si 3 N 4 near 1060 nm. *Opt. Lett.* 39, 6017–6020. doi:10.1364/OL.39.006017

Boyraz, O., and Jalali, B. (2004). Demonstration of a silicon Raman laser. *Opt. Express* 12, 5269–5273. doi:10.1364/OPEX.12.005269

Boyraz, O., and Jalali, B. (2005). Demonstration of directly modulated silicon Raman laser. *Opt. Express* 13, 796–800. doi:10.1364/OPEX.13.000796

Camacho-Aguilera, R. E., Cai, Y., Patel, N., Bessette, J. T., Romagnoli, M., Kimerling, L. C., et al. (2012). An electrically pumped germanium laser. *Opt. Express* 20, 11316–11320. doi:10.1364/OE.20.011316

Cheng, S.-L., Lu, J., Shambat, G., Yu, H.-Y., Saraswat, K., Vuckovic, J., et al. (2009). Room temperature 1.6 μm electroluminescence from Ge light emitting diode on Si substrate. *Opt. Express* 17, 10019–10024. doi:10.1364/OE.17.010019

Christiansen, S. H., Singh, R., and Gosele, U. (2006). Wafer direct bonding: from advanced substrate engineering to future applications in micro/nanoelectronics. *Proc. IEEE* 94, 2060–2106. doi:10.1109/JPROC.2006.886026

Chu, T., Fujioka, N., and Ishizaka, M. (2009). Compact, lower-power-consumption wavelength tunable laser fabricated with silicon photonic-wire waveguide micro-ring resonators. *Opt. Express* 17, 14063–14068. doi:10.1364/OE.17.014063

Creazzo, T., Marchena, E., Krasulick, S. B., Yu, P. K., Van Orden, D., Spann, J. Y., et al. (2013). Integrated tunable CMOS laser. *Opt. Express* 21, 28048–28053. doi:10.1364/OE.21.028048

Dong, P., Hu, T.-C., Liow, T.-Y., Chen, Y.-K., Xie, C., Luo, X., et al. (2014a). Novel integration technique for silicon/III-V hybrid laser. *Opt. Express* 22, 26854–26861. doi:10.1364/OE.22.026854

Dong, P., Liu, X., Chandrasekhar, S., Buhl, L. L., Aroca, R., and Chen, Y.-K. (2014b). Monolithic silicon photonic integrated circuits for compact 100 Gb/s coherent optical receivers and transmitters. *IEEE J. Sel. Top. Quantum Electron.* 20, 1–8. doi:10.1109/JSTQE.2013.2295181

Dong, P., Hu, T.-C., Zhang, L., Dinu, M., Kopf, R., Tate, A., et al. (2013). 1.9 μm hybrid silicon/iii-v semiconductor laser. *Electron. Lett.* 49:664. doi:10.1049/el.2013.0674

Dong, P., Liao, S., Feng, D., Liang, H., Zheng, D., Shafiiha, R., et al. (2009). Low V pp, ultralow-energy, compact, high-speed silicon electro-optic modulator. *Opt. Express* 17, 22484–22490. doi:10.1364/OE.17.022484

Dumon, P., Bogaerts, W., Wiaux, V., Wouters, J., Beckx, S., Van Campenhout, J., et al. (2004). Low-loss SOI photonic wires and ring resonators fabricated with deep UV lithography. *IEEE Photonics Technol. Lett.* 16, 1328–1330. doi:10.1109/LPT. 2004.826025

Fang, A. W., Jones, R., Park, H., Cohen, O., Raday, O., Paniccia, M. J., et al. (2007a). Integrated AlGaInAs-silicon evanescent racetrack laser and photodetector. *Opt. Express* 15, 2315–2322. doi:10.1364/OE.15.002315

Fang, A. W., Park, H., Kuo, Y.-H., Jones, R., Cohen, O., Liang, D., et al. (2007b). Hybrid silicon evanescent devices. *Mater. Today* 10, 28–35. doi:10.1016/S1369-7021(07)70177-3

Fang, A. W., Koch, B. R., Gan, K.-G., Park, H., Jones, R., Cohen, O., et al. (2008a). A racetrack mode-locked silicon evanescent laser. *Opt. Express* 16, 1393–1398. doi:10.1364/OE.16.001393

Fang, A. W., Koch, B. R., Jones, R., Lively, E., Liang, D., Kuo, Y.-H., et al. (2008b). A distributed Bragg reflector silicon evanescent laser. *IEEE Photonics Technol. Lett* 20, 1667–1669. doi:10.1109/LPT.2008.2003382

Fang, A. W., Lively, E., Kuo, Y.-H., Liang, D., and Bowers, J. E. (2008c). A distributed feedback silicon evanescent laser. *Opt. Express* 16, 4413–4419. doi:10.1364/OE. 16.001393

Fang, A. W., Park, H., Cohen, O., Jones, R., Paniccia, M. J., and Bowers, J. E. (2006). Electrically pumped hybrid AlGaInAs-silicon evanescent laser. *Opt. Express* 14, 9203–9210. doi:10.1364/OE.14.009203

Fang, A. W., Sysak, M. N., Koch, B. R., Jones, R., Lively, E., Kuo, Y.-H., et al. (2009). Single-wavelength silicon evanescent lasers. *IEEE J. Sel. Top. Quantum Electron.* 15, 535. doi:10.1364/OE.22.005448

Fang, Q., Song, J., Luo, X., Jia, L., Yu, M., Lo, G., et al. (2012). High efficiency ring-resonator filter with NiSi heater. *IEEE Photonics Technol. Lett.* 24, 350–352. doi:10.1109/LPT.2011.2177816

Fedeli, J., Di Cioccio, L., Marris-Morini, D., Vivien, L., Orobtchouk, R., Rojo-Romeo, P., et al. (2008). Development of silicon photonics devices using microelectronic tools for the integration on top of a CMOS wafer. *Adv. Opt. Tech.* 2008. doi:10.1155/2008/412518

Feng, S., Lei, T., Chen, H., Cai, H., Luo, X., and Poon, A. W. (2012). Silicon photonics: from a microresonator perspective. *Laser Photonics Rev.* 6, 145–177. doi:10.1002/lpor.201100020

Fujioka, N., Chu, T., and Ishizaka, M. (2010). Compact and low power consumption hybrid integrated wavelength tunable laser module using silicon waveguide resonators. *J. Lightwave Technol.* 28, 3115–3120. doi:10.1109/JLT.2010.2073445

Goodman, J. W., Leonberger, F. J., Kung, S.-Y., and Athale, R. A. (1984). Optical interconnections for VLSI systems. *Proc. IEEE* 72, 850–866. doi:10.1109/PROC. 1984.12943

Grenouillet, L., Dupont, T., Philippe, P., Harduin, J., Olivier, N., Bordel, D., et al. (2012). Hybrid integration for silicon photonics applications. *Opt. Quant. Electron.* 44, 527–534. doi:10.1038/nnano.2014.215

Guha, B., Kyotoku, B. B., and Lipson, M. (2010). CMOS-compatible athermal silicon microring resonators. *Opt. Express* 18, 3487–3493. doi:10.1364/OE.18.003487

Guillot, G., and Pavesi, L. (2006). *Optical Interconnects: The Silicon Approach.* Berlin: Springer-Verlag.

Han, H.-S., Seo, S.-Y., and Shin, J. H. (2001). Optical gain at 1.54 μm in erbium-doped silicon nanocluster sensitized waveguide. *Appl. Phys. Lett.* 79, 4568–4570. doi:10.1063/1.1419035

Hattori, H. T., Seassal, C., Touraille, E., Rojo-Romeo, P., Letartre, X., Hollinger, G., et al. (2006). Heterogeneous integration of microdisk lasers on silicon strip waveguides for optical interconnects. *IEEE Photonics Technol. Lett.* 18, 223–225. doi:10.1109/LPT.2005.861542

Hong, T., Ran, G.-Z., Chen, T., Pan, J.-Q., Chen, W.-X., Wang, Y., et al. (2010). A selective-area metal bonding InGaAsP–Si laser. *IEEE Photonics Technol. Lett.* 22, 1141–1143. doi:10.1364/OE.22.005448

Huang, Y., Song, J., Luo, X., Liow, T.-Y., and Lo, G.-Q. (2014). CMOS compatible monolithic multi-layer Si3N4-on-SOI platform for low-loss high performance silicon photonics dense integration. *Opt. Express* 22, 21859–21865. doi:10.1364/OE.22.021859

Jalali, B., and Fathpour, S. (2006). Silicon photonics. *J. Lightwave Technol.* 24, 4600–4615. doi:10.1109/JLT.2006.885782

Justice, J., Bower, C., Meitl, M., Mooney, M. B., Gubbins, M. A., and Corbett, B. (2012). Wafer-scale integration of group III-V lasers on silicon using transfer printing of epitaxial layers. *Nat. Photonics* 6, 610–614. doi:10.1038/nphoton. 2012.204

Keyvaninia, S., Muneeb, M., Stanković, S., Roelkens, G., Van Thourhout, D., and Fedeli, J. (2012). "Multiple die-to-wafer adhesive bonding for heterogeneous integration," in *Paper Presented at the 16th European Conference on Integrated Optics (ECIO-2012).* Sitges, Barcelona, Spain.

Keyvaninia, S., Roelkens, G., Van Thourhout, D., Jany, C., Lamponi, M., Le Liepvre, A., et al. (2013a). Demonstration of a heterogeneously integrated III-V/SOI single wavelength tunable laser. *Opt. Express* 21, 3784–3792. doi:10.1364/OE.21. 003784

Keyvaninia, S., Verstuyft, S., Pathak, S., Lelarge, F., Duan, G.-H., Bordel, D., et al. (2013b). III-V-on-silicon multi-frequency lasers. *Opt. Express* 21, 13675–13683. doi:10.1364/OE.21.013675

Keyvaninia, S., Verstuyft, S., Van Landschoot, L., Lelarge, F., Duan, G.-H., Messaoudene, S., et al. (2013c). Heterogeneously integrated III-V/silicon distributed feedback lasers. *Opt. Lett.* 38, 5434–5437. doi:10.1364/OL.38.005434

Kissinger, G., and Kissinger, W. (1993). Void-free silicon-wafer-bond strengthening in the 200–400 C range. *Sens. Actuators A* 36, 149–156. doi:10.1016/0924-4247(93)85009-5

Kostrzewa, M., Di Cioccio, L., Zussy, M., Roussin, J., Fedeli, J., Kernevez, N., et al. (2006). InP dies transferred onto silicon substrate for optical interconnects application. *Sens. Actuators A* 125, 411–414. doi:10.1016/j.sna.2005.07.023

Kurczveil, G., Heck, M. J., Peters, J. D., Garcia, J. M., Spencer, D., and Bowers, J. E. (2011). An integrated hybrid silicon multiwavelength AWG laser. *IEEE J. Sel. Top. Quantum Electron.* 17, 1521–1527. doi:10.1109/JSTQE.2011.2112639

Lasky, J. (1986). Wafer bonding for silicon-on-insulator technologies. *Appl. Phys. Lett.* 48, 78–80. doi:10.1063/1.96768

Lee, A., Jiang, Q., Tang, M., Seeds, A., and Liu, H. (2012). Continuous-wave InAs/GaAs quantum-dot laser diodes monolithically grown on Si substrate with low threshold current densities. *Opt. Express* 20, 22181–22187. doi:10.1364/OE. 20.022181

Levaufre, G., Le Liepvre, A., Jany, C., Accard, A., Kaspar, P., Brenot, R., et al. (2014). "Hybrid III-V/silicon tunable laser directly modulated at 10Gbit/s for short reach/access networks," in *Paper Presented at the Optical Communication (ECOC), 2014 European Conference on,* Cannes France.

Liang, D., and Bowers, J. (2008). Highly efficient vertical outgassing channels for low-temperature InP-to-silicon direct wafer bonding on the silicon-on-insulator substrate. *J. Vac. Sci Technol. B* 26, 1560–1568. doi:10.1116/1.2943667

Liang, D., Fang, A. W., Chen, H.-W., Sysak, M. N., Koch, B. R., Lively, E., et al. (2009a). Hybrid silicon evanescent approach to optical interconnects. *Appl. Phys. A* 95, 1045–1057. doi:10.1007/s00339-009-5118-1

Liang, D., Fiorentino, M., Okumura, T., Chang, H.-H., Spencer, D. T., Kuo, Y.-H., et al. (2009b). Electrically-pumped compact hybrid silicon microring lasers for optical interconnects. *Opt. Express* 17, 20355–20364. doi:10.1364/OE.17. 020355

Liang, D., Fiorentino, M., Srinivasan, S., Bowers, J. E., and Beausoleil, R. G. (2011). Low threshold electrically-pumped hybrid silicon microring lasers. *IEEE J. Sel. Top. Quantum Electron.* 17, 1528–1533. doi:10.1109/JSTQE.2010.2103552

Liang, D., Roelkens, G., Baets, R., and Bowers, J. E. (2010). Hybrid integrated platforms for silicon photonics. *Materials* 3, 1782–1802. doi:10.1038/nnano.2014. 215

Liang, D., Srinivasan, S., Fiorentino, M., Kurczveil, G., Bowers, J., and Beausoleil, R. (2012). "A metal thermal shunt design for hybrid silicon microring laser," in *Paper Presented at the Optical Interconnects Conference, 2012* (Santa Fe, NM: IEEE).

Lim, A. E.-J., Song, J., Fang, Q., Li, C., Tu, X., Duan, N., et al. (2014). Review of silicon photonics foundry efforts. *IEEE J. Sel. Top. Quantum Electron.* 20, 8300112. doi:10.1109/JSTQE.2013.2293274

Liow, T.-Y., Song, J., Tu, X., Lim, A.-J., Fang, Q., Duan, N., et al. (2013). Silicon optical interconnect device technologies for 40 Gb/s and beyond. *IEEE J. Sel. Top. Quantum Electron.* 19, 8200312–8200312. doi:10.1109/JSTQE.2012. 2218580

Lipson, M. (2005). Guiding, modulating, and emitting light on silicon-challenges and opportunities. *J. Lightwave Technol.* 23, 4222. doi:10.1109/JLT.2005.858225

Liu, H., Wang, T., Jiang, Q., Hogg, R., Tutu, F., Pozzi, F., et al. (2011). Long-wavelength InAs/GaAs quantum-dot laser diode monolithically grown on Ge substrate. *Nat. Photonics* 5, 416–419. doi:10.1364/OE.20.022181

Liu, J., Sun, X., Camacho-Aguilera, R., Kimerling, L. C., and Michel, J. (2010). Ge-on-Si laser operating at room temperature. *Opt. Lett.* 35, 679–681. doi:10.1364/OL.35.000679

Liu, J., Sun, X., Kimerling, L. C., and Michel, J. (2009). Direct-gap optical gain of Ge on Si at room temperature. *Opt. Lett.* 34, 1738–1740. doi:10.1364/OL.34.001738

Liu, J., Sun, X., Pan, D., Wang, X., Kimerling, L. C., Koch, T. L., et al. (2007). Tensile-strained, n-type Ge as a gain medium for monolithic laser integration on Si. *Opt. Express* 15, 11272–11277. doi:10.1364/OE.15.011272

Luo, X., Song, J., Feng, S., Poon, A. W., Liow, T.-Y., Yu, M., et al. (2012). Silicon high-order coupled-microring-based electro-optical switches for on-chip optical interconnects. *IEEE Photonics Technol. Lett.* 24, 821–823. doi:10.1109/LPT.2012.2188829

Luo, X., Song, J., Zhou, H., Liow, T. Y., Yu, M., and Lo, P. G. Q. (2014). United States Patent No. US 2014/0153600 A1.

Maszara, W. (1991). Silicon-on-insulator by wafer bonding: a review. *J. Electrochem. Soc.* 138, 341–347. doi:10.1149/1.2085575

Mechet, P., Verstuyft, S., De Vries, T., Spuesens, T., Regreny, P., Van Thourhout, D., et al. (2013). Unidirectional III-V microdisk lasers heterogeneously integrated on SOI. *Opt. Express* 21, 19339–19352. doi:10.1364/OE.21.019339

Michel, J., Liu, J., and Kimerling, L. C. (2010). High-performance Ge-on-Si photodetectors. *Nat. Photonics* 4, 527–534. doi:10.1038/nphoton.2010.157

Miller, D. A. (2000). Optical interconnects to silicon. *IEEE J. Sel. Top. Quantum Electron.* 6, 1312–1317. doi:10.1109/2944.902184

Miller, D. A. (2009). Device requirements for optical interconnects to silicon chips. *Proc. IEEE* 97, 1166–1185. doi:10.1109/JPROC.2009.2014298

Ohashi, K., Nishi, K., Shimizu, T., Nakada, M., Fujikata, J., Ushida, J., et al. (2009). On-chip optical interconnect. *Proc. IEEE* 97, 1186–1198. doi:10.1109/JPROC.2009.2020331

Park, H., Fang, A., Kodama, S., and Bowers, J. (2005). Hybrid silicon evanescent laser fabricated with a silicon waveguide and III-V offset quantum wells. *Opt. Express* 13, 9460–9464. doi:10.1364/OPEX.13.009460

Pasquariello, D., and Hjort, K. (2002). Plasma-assisted InP-to-Si low temperature wafer bonding. *IEEE J. Sel. Top. Quantum Electron.* 8, 118–131. doi:10.1109/2944.991407

Pavesi, L., Dal Negro, L., Mazzoleni, C., Franzo, G., and Priolo, F. (2000). Optical gain in silicon nanocrystals. *Nature* 408, 440–444. doi:10.1038/35044012

Pavesi, L., and Lockwood, D. J. (2004). *Silicon Photonics*. Berlin: Springer-Verlag.

Poon, A. W., Luo, X., Xu, F., and Chen, H. (2009a). Cascaded microresonator-based matrix switch for silicon on-chip optical interconnection. *Proc. IEEE* 97, 1216–1238. doi:10.1109/JPROC.2009.2014884

Poon, A. W., Luo, X., Zhou, L., Li, C., Lee, J. Y., Xu, F., et al. (2009b). "Microresonator-based devices on a silicon chip: novel shaped cavities and resonance coherent interference," in *Practical Applications of Microresonators in Optics and Photonics*, ed. A. B. Matsko (Boca Raton, FL: CRC Press), 211–263.

Reed, G. T., and Knights, A. P. (2004). *Silicon Photonics: An Introduction*. England: John Wiley & Sons.

Reed, G. T., Mashanovich, G., Gardes, F., and Thomson, D. (2010). Silicon optical modulators. *Nat. Photonics* 4, 518–526. doi:10.1038/nphoton.2010.219

Roelkens, G., Van Thourhout, D., and Baets, R. (2005). Ultra-thin benzocyclobutene bonding of III–V dies onto SOI substrate. *Electron. Lett.* 41, 561–562. doi:10.1049/el:20050807

Roelkens, G., Van Thourhout, D., Baets, R., and Smit, M. (2006). Laser emission and photodetection in an InP/InGaAsP layer integrated on and coupled to a Silicon-on-Insulator waveguide circuit. *Opt. Express* 14, 8154–8159. doi:10.1364/OE.14.008154

Rong, H., Jones, R., Liu, A., Cohen, O., Hak, D., Fang, A., et al. (2005a). A continuous-wave Raman silicon laser. *Nature* 433, 725–728. doi:10.1038/nature03273

Rong, H., Liu, A., Jones, R., Cohen, O., Hak, D., Nicolaescu, R., et al. (2005b). An all-silicon Raman laser. *Nature* 433, 292–294. doi:10.1038/nature03273

Rong, H., Xu, S., Kuo, Y.-H., Sih, V., Cohen, O., Raday, O., et al. (2007). Low-threshold continuous-wave Raman silicon laser. *Nat. Photonics* 1, 232–237. doi:10.1038/nphoton.2007.29

Rotem, E., Shainline, J. M., and Xu, J. M. (2007a). Electroluminescence of nanopatterned silicon with carbon implantation and solid phase epitaxial regrowth. *Opt. Express* 15, 14099–14106. doi:10.1364/OE.15.014099

Rotem, E., Shainline, J. M., and Xu, J. M. (2007b). Enhanced photoluminescence from nanopatterned carbon-rich silicon grown by solid-phase epitaxy. *Appl. Phys. Lett.* 91, 051127. doi:10.1063/1.2766843

Seassal, C., Rojo-Romeo, P., Letartre, X., Viktorovitch, P., Hollinger, G., Jalaguier, E., et al. (2001). InP microdisk lasers on silicon wafer: CW room temperature operation at 1.6 μm. *Electron. Lett.* 37, 222–223. doi:10.1049/el:20010173

Shainline, J. M., and Xu, J. (2007). Silicon as an emissive optical medium. *Laser Photonics Rev.* 1, 334–348. doi:10.1002/lpor.200710021

Song, J., Luo, X., Tu, X., Jia, L., Fang, Q., Liow, T.-Y., et al. (2013). On-chip quasi-digital optical switch using silicon microring resonator-coupled Mach-Zehnder interferometer. *Opt. Express* 21, 12767–12775. doi:10.1364/OE.21.012767

Soref, R. (2006). The past, present, and future of silicon photonics. *IEEE J. Sel. Top. Quantum Electron.* 12, 1678–1687. doi:10.1021/ar900141y

Soref, R., and Lorenzo, J. (1985). Single-crystal silicon: a new material for 1.3 and 1.6 μm integrated-optical components. *Electron. Lett.* 21, 953–954. doi:10.1049/el:19850673

Sparacin, D. K., Spector, S. J., and Kimerling, L. C. (2005). Silicon waveguide sidewall smoothing by wet chemical oxidation. *J. Lightwave Technol.* 23, 2455. doi:10.1109/JLT.2005.851328

Stanković, S., Jones, R., Sysak, M. N., Heck, J. M., Roelkens, G., and Van Thourhout, D. (2011). 1310-nm hybrid III–V/Si Fabry–Pérot laser based on adhesive bonding. *IEEE Photonics Technol. Lett.* 23, 1781–1783. doi:10.1109/LPT.2011.2169397

Stanković, S., Jones, R., Sysak, M. N., Heck, J. M., Roelkens, G., and Van Thourhout, D. (2012). Hybrid III–V/Si distributed-feedback laser based on adhesive bonding. *IEEE Photonics Technol. Lett.* 24, 2155–2158. doi:10.1109/LPT.2012.2223666

Stanković, S., Van Thourhout, D., Roelkens, G., Jones, R., Heck, J., and Sysak, M. (2010). Die-to-die adhesive bonding for evanescently-coupled photonic devices. *ECS Trans.* 33, 411–420. doi:10.1149/1.3483531

Sun, X., Liu, H.-C., and Yariv, A. (2009a). Adiabaticity criterion and the shortest adiabatic mode transformer in a coupled-waveguide system. *Opt. Lett.* 34, 280–282. doi:10.1364/OL.34.000280

Sun, X., Liu, J., Kimerling, L. C., and Michel, J. (2009b). Direct gap photoluminescence of n-type tensile-strained Ge-on-Si. *Appl. Phys. Lett.* 95, 011911–011911–011913. doi:10.1364/OL.38.000652

Sun, X., Liu, J., Kimerling, L. C., and Michel, J. (2009c). Room-temperature direct bandgap electroluminesence from Ge-on-Si light-emitting diodes. *Opt. Lett.* 34, 1198–1200. doi:10.1364/OL.34.001198

Sysak, M. N., Liang, D., Jones, R., Kurczveil, G., Piels, M., Fiorentino, M., et al. (2011). Hybrid silicon laser technology: a thermal perspective. *IEEE J. Sel. Top. Quantum Electron.* 17, 1490–1498. doi:10.1109/JSTQE.2011.2109940

Tanabe, K., Guimard, D., Bordel, D., Iwamoto, S., and Arakawa, Y. (2010). Electrically pumped 1.3 μm room-temperature InAs/GaAs quantum dot lasers on Si substrates by metal-mediated wafer bonding and layer transfer. *Opt. Express* 18, 10604–10608. doi:10.1364/OE.18.010604

Tatsumi, T., Tanabe, K., Watanabe, K., Iwamoto, S., and Arakawa, Y. (2012). 1.3 μm InAs/GaAs quantum dot lasers on Si substrates by low-resistivity, Au-free metal-mediated wafer bonding. *J. Appl. Phys.* 112, 033107. doi:10.1063/1.4742198

Tong, Q.-Y., and Goesele, U. (1999). *Semiconductor Wafer Bonding: Science and Technology*. Wiley, 1998.

Tu, X., Chang, K.-F., Liow, T.-Y., Song, J., Luo, X., Jia, L., et al. (2014). Silicon optical modulator with shield coplanar waveguide electrodes. *Opt. Express* 22, 23724–23731. doi:10.1364/OE.22.023724

Tu, X., Liow, T.-Y., Song, J., Luo, X., Fang, Q., Yu, M., et al. (2013). 50-Gb/s silicon optical modulator with traveling-wave electrodes. *Opt. Express* 21, 12776–12782. doi:10.1364/OE.21.012776

Urino, Y., Shimizu, T., Okano, M., Hatori, N., Ishizaka, M., Yamamoto, T., et al. (2011). First demonstration of high density optical interconnects integrated with lasers, optical modulators, and photodetectors on single silicon substrate. *Opt. Express* 19, B159–B165. doi:10.1364/OE.19.00B159

Van Campenhout, J., Green, W. M., Assefa, S., and Vlasov, Y. A. (2009). Low-power 2x2 silicon electro-optic switch with 110-nm bandwidth for broadband reconfigurable optical networks. *Opt. Express* 17, 24020–24029. doi:10.1364/OE.17.024020

Van Campenhout, J., Liu, L., Romeo, P. R., Van Thourhout, D., Seassal, C., Regreny, P., et al. (2008). A compact SOI-integrated multiwavelength laser source based on cascaded InP microdisks. *IEEE Photonics Technol. Lett.* 20, 1345–1347. doi:10.1109/LPT.2008.926857

Van Campenhout, J., Rojo Romeo, P., Regreny, P., Seassal, C., Van Thourhout, D., Verstuyft, S., et al. (2007). Electrically pumped InP-based microdisk lasers integrated with a nanophotonic silicon-on-insulator waveguide circuit. *Opt. Express* 15, 6744–6749. doi:10.1364/OE.15.006744

Vivien, L., and Pavesi, L. (2013). *Handbook of Silicon Photonics*. Boca Raton, FL: CRC Press.

Xia, F., Sekaric, L., and Vlasov, Y. (2006). Ultracompact optical buffers on a silicon chip. *Nat. Photonics* 1, 65–71. doi:10.1038/nphoton.2006.42

Xiao, S., Khan, M. H., Shen, H., and Qi, M. (2007). Multiple-channel silicon micro-resonator based filters for WDM applications. *Opt. Express* 15, 7489–7498. doi:10.1364/OE.15.009386

Xu, D.-X., Schmid, J. H., Reed, G. T., Mashanovich, G. Z., Thomson, D. J., Nedeljkovic, M., et al. (2014). Silicon photonic integration platform-have we found the sweet spot? *IEEE J. Sel. Top. Quantum Electron.* 20, 189–205. doi:10.1109/JSTQE.2014.2299634

Yang, H., Zhao, D., Chuwongin, S., Seo, J.-H., Yang, W., Shuai, Y., et al. (2012). Transfer-printed stacked nanomembrane lasers on silicon. *Nat. Photonics* 6, 615–620. doi:10.1038/nphoton.2012.160

Yariv, A., and Sun, X. (2007). Supermode Si/III-V hybrid lasers, optical amplifiers and modulators: a proposal and analysis. *Opt. Express* 15, 9147–9151. doi:10.1364/OE.15.009147

Zhang, C., Srinivasan, S., Tang, Y., Heck, M. J., Davenport, M. L., and Bowers, J. E. (2014). Low threshold and high speed short cavity distributed feedback hybrid silicon lasers. *Opt. Express* 22, 10202–10209. doi:10.1364/OE.22.010202

Zhou, L., and Poon, A. W. (2007). Electrically reconfigurable silicon microring resonator-based filter with waveguide-coupled feedback. *Opt. Express* 15, 9194–9204. doi:10.1364/OE.15.009194

**Conflict of Interest Statement:** The authors declare that the research was conducted in the absence of any commercial or financial relationships that could be construed as a potential conflict of interest.

# 16

# Probing on growth and characterizations of $SnFe_2O_4$ epitaxial thin films on $MgAl_2O_4$ substrate

**Ram K. Gupta[1]\*, J. Candler[1], D. Kumar[2], Bipin K. Gupta[3] and Pawan K. Kahol[4]**

[1] Department of Chemistry, Pittsburg State University, Pittsburg, KS, USA
[2] Department of Mechanical Engineering, North Carolina A&T State University, Greensboro, NC, USA
[3] National Physical Laboratory (CSIR), New Delhi, India
[4] Department of Physics, Pittsburg State University, Pittsburg, KS, USA

**Edited by:**
Ibrahim Sayed Hussein, Ain Shams University, Egypt

**Reviewed by:**
Gregory Abadias, Institut Pprime, France
Ruiqin Tan, Ningbo University, China

**\*Correspondence:**
Ram K. Gupta, Department of Chemistry, Pittsburg State University, 1701 South Broadway, Pittsburg, KS-66762, USA
e-mail: ramguptamsu@gmail.com

Epitaxial tin ferrite ($SnFe_2O_4$) thin films were grown using KrF excimer (248 nm) pulsed laser deposition technique under different growth conditions. Highly epitaxial thin films were obtained at growth temperature of 650°C. The quality and epitaxial nature of the films were examined by X-ray diffraction technique. Furthermore, the phi-scans of the film and substrate exhibit fourfold symmetry, which indicates a cube-on-cube epitaxial growth of the film on $MgAl_2O_4$ substrate. Moreover, the magnetic force microscopy measurement shows domains with cluster-like structure, which is associated with ferromagnetic phase at room temperature. The coercive field and remnant magnetization of the films decrease with increase in temperature. These high quality ingenious magnetic films could be potentially used in data storage devices.

Keywords: epitaxial thin films, $SnFe_2O_4$, pulse laser deposition, ferromagnetic, bandgap

## INTRODUCTION

Recently, spinel ferrites with the general formula $MFe_2O_4$ (where M = Co, Mn, Mg, Sn, etc.) attract considerable research interest because of their wide applications in heterogeneous catalyst, sensors, transformers, magnetic recording, biomedical, etc. (Abdeen, 1998; Sedlár et al., 2000; Bao et al., 2007; Barcena et al., 2008; Xiang et al., 2010). The ferrites can be classified into different categories depending upon their cation distributions. Based on the cations distribution among the tetrahedral and octahedral sites of the coordinated oxygen, they can be either normal spinel $M^{2+}_{Tetrahedral}[Fe^{3+}Fe^{3+}]_{Octaheral}O_4$, or inverse spinel $Fe^{3+}_{Tetrahedral}[M^{2+}Fe^{3+}]_{Octaheral}O_4$ (Anantharaman et al., 1998).

Till date, most of the works on spinels have been reported on bulk in order to understand their magnetic behavior and correlate magnetic properties to their structural properties to improve their applications (Lüders et al., 2005). Epitaxial thin films of spinels have not drawn such a wide research attention despite of the fact that epitaxial films could modify the physical properties compared to the bulk material (Lüders et al., 2005). Epitaxial thin films of various ferrites have been grown using different techniques (Zimnol et al., 1997; Reisinger et al., 2003; Huang et al., 2007; Leung et al., 2008; Su et al., 2010). Among these techniques, pulsed laser deposition (PLD) is a very versatile and cost effective method which allows the stoichiometry transfer of multi-component materials from target to substrate (Green et al., 1995).

Pulsed laser deposition technique has been used for deposition of epitaxial thin films of magnesium ferrite on strontium titanate (Kim et al., 2010). The effect of post-annealing on the magnetic properties of epitaxial thin films of cobalt ferrite was studied (Axelsson et al., 2009). Nanostructured tin ferrites have been synthesized using different techniques (Liu et al., 2004; Liu and Li, 2005). It was also observed that the coercivity of the tin ferrite particles decreases with increase in the particle size (Liu et al.,

2004). Superparamagnetic behavior was observed for nanostructured tin ferrite (Liu and Li, 2005). PLD has been used to deposit (111) oriented epitaxial tin ferrite films on (0001) sapphire substrate (Gupta et al., 2011). In this communication, we report the epitaxial growth of tin ferrite films on (001) $MgAl_2O_4$ substrate using PLD technique. The quality and epitaxial nature of the films were evaluated by X-ray diffraction (XRD) diffraction technique. Magnetic domains with cluster-like structure were observed in the magnetic force microscopy (MFM) image of the film.

## EXPERIMENTS

$SnFe_2O_4$ target for PLD was made using solid state reaction method. $SnO_2$ (99.9%, Alfa Aesar, USA) and $Fe_2O_3$ (99.5%, Alfa Aesar, USA) were used as received. The well-ground mixture was heated at 1200°C for 10 h. The powder mixture was cold pressed at $6 \times 10^6$ N/m$^2$ load and sintered at 1400°C for 10 h. The films were deposited using KrF excimer PLD technique (Lambda Physik COMPex, $\lambda = 248$ nm and pulsed duration of 20 ns) at different substrate temperatures (550, 600, 650, and 690°C) under oxygen pressure of 0.1 mbar. The laser was operated at a pulse rate of 10 Hz, with energy of 300 mJ/pulse. The laser beam was focused onto a rotating target at a 45° angle of incidence. The target to substrate distance was 5 cm. Single crystal of (001) oriented $MgAl_2O_4$ was used as substrate. The substrate was ultrasonically cleaned in acetone and isopropanol for 10 min in each solvent.

The structural characterizations were performed using XRD. The XRD pattern of the films were recorded with Bruker AXS X-ray diffractometer using the $2\theta$–$\theta$ scan, rocking curve, and phi-scan with $CuK_{\alpha 1}$($\lambda = 1.5406$ Å) radiation which operated at 40 kV and 40 mA. The XRD measurements were performed using 0.1 mm aperture of the slits. The instrument broadening was corrected using $LaB_6$ as an instrumental broadening standard. MFM imaging was performed under ambient conditions using a Digital

Instruments (Veeco) Dimension-3100 unit with Nanoscope® III controller, operated in tapping mode. Magnetic measurements were performed on Quantum Design vibrating sample magnetometer (VSM). The optical transmittance measurements were made using UV–visible spectrophotometer (Ocean Optics HR4000).

## RESULTS AND DISCUSSION

The epitaxial nature of the films was investigated by XRD technique. The different scans such as $\theta$–$2\theta$, rocking ($\omega$) curve, and phi ($\phi$)-scans were used to study the quality and epitaxy of the films on (001) oriented $MgAl_2O_4$ substrate. Gupta and Yakuphanoglu (2011) and Gupta et al. (2011) have used sapphire and $SrTiO_3$ as substrate for epitaxial growth of $SnFe_2O_4$. In the present study, $MgAl_2O_4$ was chosen as substrate since both film and substrate have cubic crystal structure with small lattice mismatch (~3.8%). All the films grown at different temperatures showed preferred orientation along (002) direction. **Figure 1** shows the $\theta$–$2\theta$ and rocking curve for (002) peak for the film grown at 650°C. It is seen in the XRD pattern that only one peak oriented along (002) direction is observed, indicating the epitaxial nature of the film along (002) direction. The epitaxial nature of the film is due to the close lattice parameters of film and substrate as both $SnFe_2O_4$ (face-centered cubic, $a = 0.842$ nm) and $MgAl_2O_4$ (cubic, $a = 0.808$ nm) exhibit cubic symmetry (space-group Fd3m). The full width at half maximum (FWHM) of (002) peak was estimated using the rocking curve. The FWHM was calculated to be 0.42°, 0.42°, 0.39°, and 0.44° for the films grown at 550, 600, 650, and 690°C, respectively. The lowest FWHM was observed for film grown at 650°C, indicating highly quality of the film. The FWHM was for $SnFe_2O_4$ film grown on $SrTiO_3$ substrate was reported to be 0.96°, 0.94°, 0.56°, and 0.96° for the films grown at 550, 600, 650, and 690°C, respectively (Gupta and Yakuphanoglu, 2011). As observed, the FWHM for the $SnFe_2O_4$ films grown on $MgAl_2O_4$ are better than that on $SrTiO_3$, which is due to the close lattice match of $SnFe_2O_4$ and $MgAl_2O_4$. Although the lattice mismatch between the substrate and films is about 3.8%, the film shows strain of about 1.4%. We consider this high quality film for further characterizations. The phi ($\phi$)-scan of the film and substrate was recorded using (311) reflection plane ($2\theta = 34.28$ and $\psi = 25.24$) and is shown in **Figure 2**. The phi-scan of the film and substrate revealed fourfold symmetry for both. The phi-scan shows a cube-on-cube epitaxial growth of $SnFe_2O_4$ on $MgAl_2O_4$ substrate.

**Figure 3** shows the MFM image of the film recorded in the demagnetized state. The presence of magnetic domain due to grains of $SnFe_2O_4$ is quite evident in the MFM image. The grain size of the $SnFe_2O_4$ films was estimated to be 22 nm using (002) peak of XRD pattern (Gupta et al., 2011). The size of the magnetic domain was observed to be about 200 nm, indicating that about 10 grains make a domain. As seen in **Figure 3**, the magnetic image consists of domains with cluster-like structure where the magnetization is confined up and down with light and dark color, respectively.

The optical properties such as transparency and optical bandgap of the epitaxially grown tin ferrite were studied. **Figure 4** shows the optical transmittance spectra of the film. The optical

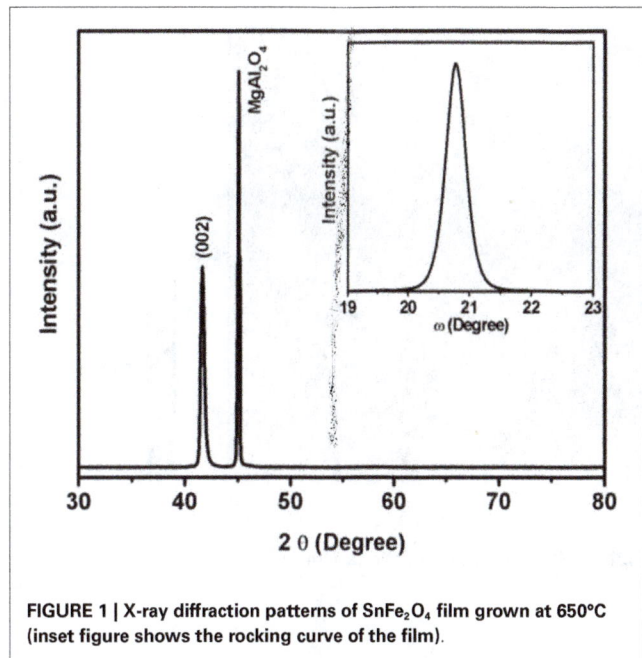

**FIGURE 1 | X-ray diffraction patterns of SnFe₂O₄ film grown at 650°C (inset figure shows the rocking curve of the film).**

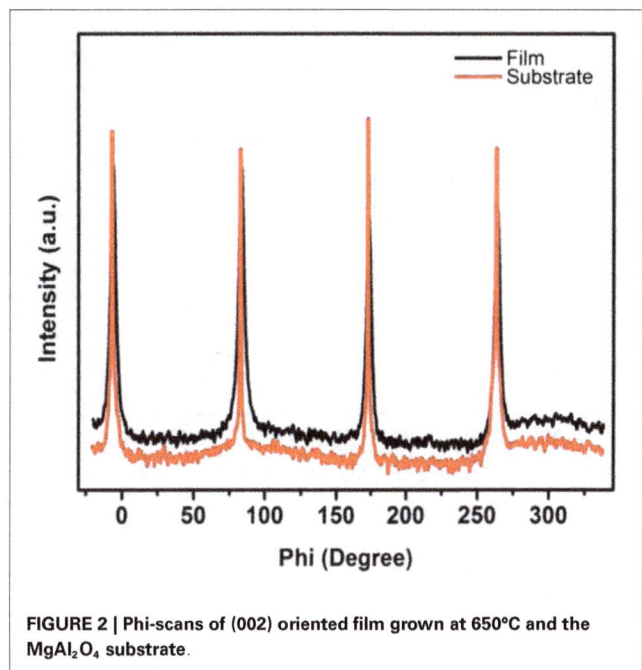

**FIGURE 2 | Phi-scans of (002) oriented film grown at 650°C and the MgAl₂O₄ substrate.**

bandgap of the film was calculated from absorption coefficient and photon energy. The absorption coefficient ($\alpha$) of the film was calculated using the following expression (Gupta et al., 2009)

$$\alpha = \ln\left(\frac{1}{T}\right) / d \qquad (1)$$

where $T$ is transmittance and $d$ is film thickness. The optical bandgap of the films was calculated using the following equation (Dolia et al., 2006)

$$(\alpha h \nu)^2 = A(h\nu - E_g) \qquad (2)$$

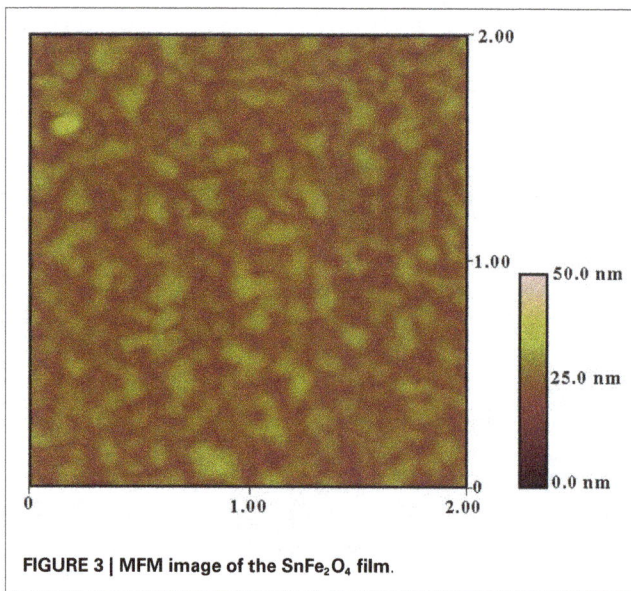

FIGURE 3 | MFM image of the SnFe₂O₄ film.

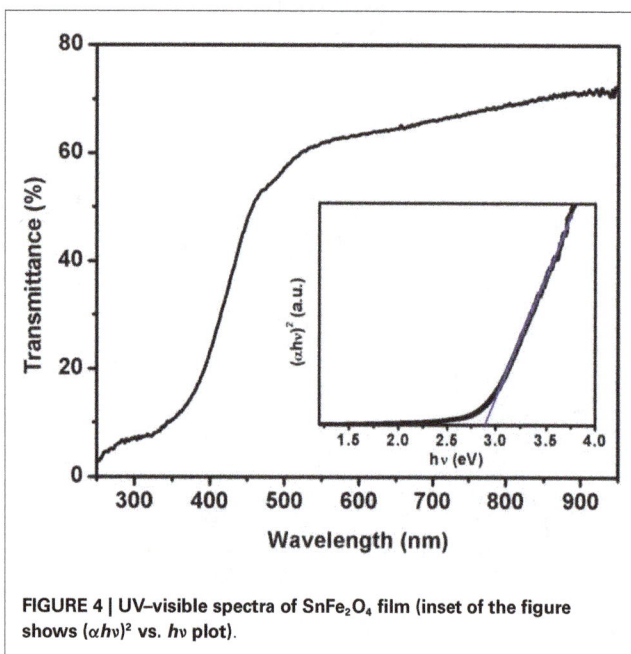

FIGURE 4 | UV–visible spectra of SnFe₂O₄ film (inset of the figure shows $(\alpha h\nu)^2$ vs. $h\nu$ plot).

where $A$ and $E_g$ are constant and optical bandgap, respectively. The $E_g$ can be determined by extrapolations of the linear regions of the plots to zero absorption. Inset of **Figure 3** shows $(\alpha h\nu)^2$ vs. $h\nu$ plot for the film. The bandgap of the film was calculated to be 2.8 eV. A bandgap of 2.7 eV is observed for tin ferrite film grown on sapphire substrate (Gupta et al., 2011). Dolia et al. (2006) have observed a bandgap of 2.5 eV for nickel ferrite film, whereas the bandgap of 2.7 eV was reported for zinc ferrite film (Wu et al., 2001).

The magnetic properties of the film were studied under different conditions. **Figure 5** shows the variation of magnetization with temperature ($M$ vs. $T$). As seen in **Figure 5**, the effect of temperature on magnetization was studied in zero-field-cooled (ZFC)

and field-cooled (FC) process under different applied magnetic fields. For the ZFC measurements, the film was cooled from high temperature to 10 K without applying any external magnetic field. After cooling to 10 K, an external magnetic field was applied and the magnetization of the film was recorded during the heating. For FC measurements, the magnetization is recorded while cooling the sample under an applied external magnetic field. As seen in **Figure 5**, during ZFC measurements the magnetization of the film increases with temperature up to ~275 K and then decreases with further increase in temperature. The nature of ZFC and FC curves is very similar but the magnitude of magnetization for FC curves is high. Similar results were observed for $M$ vs. $T$ process under high magnetic field. Furthermore, it should be noted that there is a distinct irreversibility between the ZFC and FC magnetization curves. This irreversibility persists up to high temperature of 375 K. Similar nature in $M$ vs. $T$ has been observed for the SnFe₂O₄ films grown on SrTiO₃ and sapphire substrate (Gupta and Yakuphanoglu, 2011; Gupta et al., 2011). The magnetization at 10 K in FC measurement was observed to be 12.9, 14.2 and 67.6 emu/cm³ for the SnFe₂O₄ film grown on MgAl₂O₄, SrTiO₃, and sapphire substrate, respectively. Although the maximum magnetization was observed on sapphire substrate, the difference in the values of magnetization measured during ZFC and FC at 10 K was almost constant (~10 emu/cm³) for the SnFe₂O₄ films on MgAl₂O₄, SrTiO₃, and sapphire substrate. The different values of magnetization for SnFe₂O₄ films on different substrates could be due to strain introduced by lattice mismatch of film and MgAl₂O₄, SrTiO₃, and sapphire substrates (Belenky et al., 2005). The lattice mismatch between SnFe₂O₄ and MgAl₂O₄, SrTiO₃, and sapphire was estimated to be 3.8, 7.3, and 8.4%, respectively. The strain introduced by lattice mismatch is an important parameter contributing to magnetic properties such as Curie temperature, coercivity, saturation magnetization, and anisotropy (Rao et al., 1998).

**Figure 6** shows the variation of magnetization with applied magnetic field ($M$ vs. $H$) at different temperatures. The $M$ vs. $H$ plots were measured at 10 and 300 K. The open hysteresis loop near origin at room temperature confirms the ferromagnetic nature of the film. It is observed that the coercive field and remnant magnetization of the film decrease with increase in the temperature. The coercive field of 4575 and 431 Oe is observed at 10 and 300 K, respectively. On the other hand, the value of remnant magnetization of 25.2 emu/cm³ and 8.3 emu/cm³ is observed at 10 and 300 K, respectively. The coercive field of 4861 and 1323 Oe was reported for SnFe₂O₄ film on sapphire substrate at 10 and 300 K, respectively (Gupta et al., 2011). On the other hand, the coercive field of 1853 and 801 Oe was observed for SnFe₂O₄ film on SrTiO₃ substrate at 10 and 300 K, respectively (Gupta and Yakuphanoglu, 2011). Again the difference in the remnant magnetization and coercive field for SnFe₂O₄ films on MgAl₂O₄, SrTiO₃, and sapphire substrates could be due to lattice mismatch between the film and substrates. The structural and magnetic characterizations of SnFe₂O₄ film on different substrates indicate that the properties of the film can be modified by using different substrates.

## CONCLUSION

We have successfully demonstrated the deposition of epitaxial tin ferrite thin films on MgAl₂O₄ substrate using PLD technique. XRD

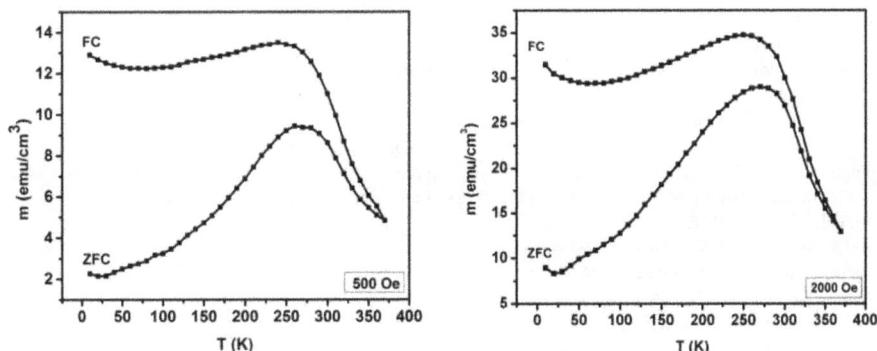

**FIGURE 5 |** *M* vs. *T* plots for $SnFe_2O_4$ film at different magnetic fields under FC and ZFC conditions.

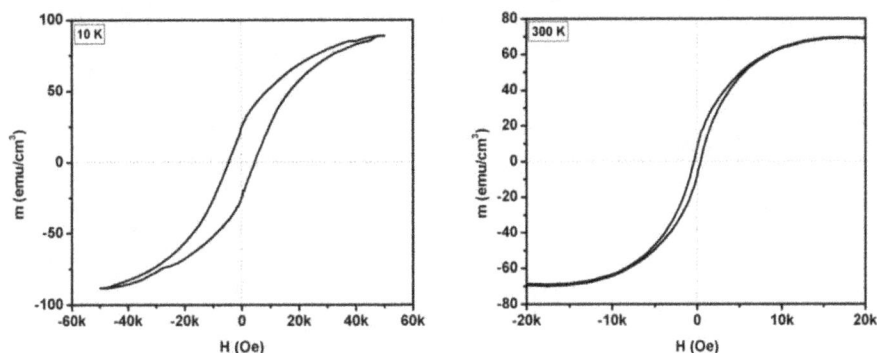

**FIGURE 6 |** *M* vs. *H* plot for $SnFe_2O_4$ film at different temperatures.

measurements confirm the epitaxial nature of the tin ferrite film. The phi-scan of the film and substrate shows fourfold symmetry, which evidenced the cube-on-cube epitaxial growth of tin ferrite on $MgAl_2O_4$ substrate. The optical bandgap of the film was observed to be 2.8 eV. Furthermore, the magnetic measurements exhibit the ferromagnetic nature of the film at room temperature. These epitaxial, transparent, and ferromagnetic films could be potentially used in the next generation data storage devices.

# REFERENCES

Abdeen, A. M. (1998). Electric conduction in Ni–Zn ferrites. *J. Magn. Magn. Mater.* 185, 199–206. doi:10.1016/S0304-8853(97)01144-X

Anantharaman, M. R., Jagatheesan, S., Malini, K. A., Sindhu, S., Narayanasamy, A., Chinnasamy, C. N., et al. (1998). On the magnetic properties of ultra-fine zinc ferrites. *J. Magn. Magn. Mater.* 189, 83–88. doi:10.1016/S0304-8853(98)00171-1

Axelsson, A.-K., Valant, M., Fenner, L., Wills, A. S., and Alford, N. M. (2009). Chemistry of post-annealing of epitaxial CoFe2O4 thin films. *Thin Solid Films* 517, 3742–3747. doi:10.1016/j.tsf.2009.01.142

Bao, N., Shen, L., Wang, Y., Padhan, P., and Gupta, A. (2007). A facile thermolysis route to monodisperse ferrite nanocrystals. *J. Am. Chem. Soc.* 129, 12374–12375. doi:10.1021/ja074458d

Barcena, C., Sra, A. K., Chaubey, G. S., Khemtong, C., Liu, J. P., and Gao, J. (2008). Zinc ferrite nanoparticles as MRI contrast agents. *Chem. Commun.* 2224–2226. doi:10.1039/B801041B

Belenky, L. J., Ke, X., Rzchowski, M., and Eom, C. B. (2005). Epitaxial La0.67Sr0.33MnO3/La0.67Ba0.33MnO3 superlattices. *J. Appl. Phys.* **97**, 10J107. doi:10.1063/1.1850384

Dolia, S. N., Sharma, R., Sharma, M. P., and Saxena, N. S. (2006). Synthesis, X-ray diffraction and optical band gap study of nanoparticles of NiFe2O4. *Indian J. Pure Appl. Phys.* 44, 774–776.

Green, S. M., Pique, A., and Harshavardhan, K. S. (1995). *Pulsed Laser Deposition of Thin Films.* New York: Wiley.

Gupta, R. K., Ghosh, K., and Kahol, P. K. (2011). Structural and magnetic properties of epitaxial SnFe2O4 thin films. *Mater. Letters* 65, 2149–2151. doi:10.1016/j.matlet.2011.04.059

Gupta, R. K., Ghosh, K., Patel, R., and Kahol, P. K. (2009). Highly conducting and transparent Ti-doped CdO films by pulsed laser deposition. *Appl. Surf. Sci.* 255, 6252–6255. doi:10.1016/j.apsusc.2009.01.091

Gupta, R. K., and Yakuphanoglu, F. (2011). Epitaxial growth of tin ferrite thin films using pulsed laser deposition technique. *J. Alloys Comp.* 509, 9523–9527. doi:10.1016/j.jallcom.2011.07.058

Huang, W., Zhou, L. X., Zeng, H. Z., Wei, X. H., Zhu, J., Zhang, Y., et al. (2007). Epitaxial growth of the CoFe2O4 film on SrTiO3 and its characterization. *J. Cryst. Growth* 300, 426–430. doi:10.1016/j.jcrysgro.2007.01.004

Kim, K. S., Muralidharan, P., Han, S. H., Kim, J. S., Kim, H. G., and Cheon, C. I. (2010). Influence of oxygen partial pressure on the epitaxial MgFe2O4 thin films deposited on SrTiO3 substrate. *J. Alloys Comp.* 503, 460–463. doi:10.1016/j.jallcom.2010.05.033

Leung, G. W., Vickers, M. E., Yu, R., and Blamire, M. G. (2008). Epitaxial growth of Fe3O4 on SrTiO3 substrates. *J. Cryst. Growth* 310, 5282–5286. doi:10.1016/j.jcrysgro.2008.07.126

Liu, F., Li, T., and Zheng, H. (2004). Structure and magnetic properties of SnFe2O4 nanoparticles. *Phys. Letters A* 323, 305–309. doi:10.1016/j.physleta.2004.01.077

Liu, F. X., and Li, T. Z. (2005). Synthesis and magnetic properties of SnFe2O4 nanoparticles. *Mater. Letters* 59, 194–196. doi:10.1016/j.matlet.2004.09.028

Lüders, U., Bibes, M., Bobo, J.-F., Cantoni, M., Bertacco, R., and Fontcuberta, J. (2005). Enhanced magnetic moment and conductive behavior in $NiFe_2O_4$ spinel ultrathin films. *Phys. Rev. B* 71, 134419. doi:10.1103/PhysRevB.71.134419

Rao, R. A., Lavric, D., Nath, T. K., Eom, C. B., Wu, L., and Tsui, F. (1998). Three-dimensional strain states and crystallographic domain structures of epitaxial colossal magnetoresistive $La_{0.8}Ca_{0.2}MnO_3$ thin films. *Appl. Phys. Lett.* 73, 3294–3296. doi:10.1063/1.122749

Reisinger, D., Schonecke, M., Brenninger, T., Opel, M., Erb, A., Alff, L., et al. (2003). Epitaxy of $Fe_3O_4$ on Si(001) by pulsed laser deposition using a TiN/MgO buffer layer. *J. Appl. Phys.* 94, 1857–1863. doi:10.1063/1.1587885

Sedlár, M., Matejec, V., Grygar, T., and Kadlecová, J. (2000). Sol–gel processing and magnetic properties of nickel zinc ferrite thick films. *Ceramics Int.* 26, 507–512. doi:10.1016/S0272-8842(99)00086-3

Su, H.-C., Dai, J.-Y., Liao, Y.-F., Wu, Y.-H., Huang, J. C. A., and Lee, C.-H. (2010). The preparation of Zn-ferrite epitaxial thin film from epitaxial $Fe_3O_4$:ZnO multilayers by ion beam sputtering deposition. *Thin Solid Films* 518, 7275–7278. doi:10.1016/j.tsf.2010.04.089

Wu, Z., Okuya, M., and Kaneko, S. (2001). Spray pyrolysis deposition of zinc ferrite films from metal nitrates solutions. *Thin Solid Films* 385, 109–114. doi:10.1016/S0040-6090(00)01906-4

Xiang, X., Fan, G., Fan, J., and Li, F. (2010). Porous and superparamagnetic magnesium ferrite film fabricated via a precursor route. *J. Alloys Comp.* 499, 30–34. doi:10.1016/j.jallcom.2010.03.125

Zimnol, M., Graff, A., Sieber, H., Senz, S., Schmidt, S., Mattheis, R., et al. (1997). Structure and morphology of $MgFe_2O_4$ epitaxial films formed by solid state reactions on MgO(100) surfaces. *Solid State Ionics.* 101–103, 667–672. doi:10.1016/S0167-2738(97)00321-4

**Conflict of Interest Statement:** The authors declare that the research was conducted in the absence of any commercial or financial relationships that could be construed as a potential conflict of interest.

# 17

# Multi-scale modeling of the impact response of a strain-rate sensitive high-manganese austenitic steel

*Orkun Onal, Cemre Ozmenci and Demircan Canadinc\**

*Advanced Materials Group (AMG), Department of Mechanical Engineering, Koç University, Istanbul, Turkey*

**Edited by:**
*Thomas Heine, Jacobs University Bremen gGmbH, Germany*

**Reviewed by:**
*Atilim Eser, RWTH Aachen, Germany*
*Reginald Felix Hamilton, The Pennsylvania State University, USA*

**\*Correspondence:**
*Demircan Canadinc, Rumeli Feneri Yolu, Sariyer, Istanbul, Turkey*
*e-mail: dcanadinc@ku.edu.tr*

A multi-scale modeling approach was applied to predict the impact response of a strain rate sensitive high-manganese austenitic steel. The roles of texture, geometry, and strain rate sensitivity were successfully taken into account all at once by coupling crystal plasticity and finite element (FE) analysis. Specifically, crystal plasticity was utilized to obtain the multi-axial flow rule at different strain rates based on the experimental deformation response under uniaxial tensile loading. The equivalent stress – equivalent strain response was then incorporated into the FE model for the sake of a more representative hardening rule under impact loading. The current results demonstrate that reliable predictions can be obtained by proper coupling of crystal plasticity and FE analysis even if the experimental flow rule of the material is acquired under uniaxial loading and at moderate strain rates that are significantly slower than those attained during impact loading. Furthermore, the current findings also demonstrate the need for an experiment-based multi-scale modeling approach for the sake of reliable predictions of the impact response.

Keywords: **high-manganese austenitic steel, crystal plasticity, finite element analysis, microstructure, strain rate sensitivity, impact**

## INTRODUCTION

Austenitic high-manganese (Mn) steels, a class of high strength steels, have received considerable attention since they offer a rare combination of exceptional work hardening capacity, high wear and abrasion resistance, high strength, and significant ductility (Owen and Grujicic, 1999; Bayraktar et al., 2004; Bouaziz et al., 2011). These extraordinary mechanical properties have been subject to several studies (Karaman et al., 2000, 2001; Bayraktar et al., 2004; Hutchinson and Ridley, 2006; Ueji et al., 2008; Niendorf et al., 2009, 2010), many of which revealed that the main mechanism underlying the observed mechanical behavior is the presence of twins in the microstructure accompanied by additional microstructural features, such as stacking faults, dynamic strain aging (DSA), and interaction of twins with dislocations (Karaman et al., 2000, 2001; Hutchinson and Ridley, 2006; Ueji et al., 2008). Simultaneous activity of all of the aforementioned micro-deformation mechanisms result in the improved strength and work hardening capacity of this class of steels, where all these mechanisms constitute obstacles against gliding dislocations (Karaman et al., 2000, 2001; Hutchinson and Ridley, 2006; Ueji et al., 2008). However, the complexity of the deformation behavior of these materials makes it difficult to clearly distinguish between the relative contributions of the hardening mechanisms.

A very good example to this complicated microstructure is that of Hadfield steel, high-Mn austenitic steel with a face-centered cubic (fcc) structure at room temperature (RT). Hadfield steel is well known for its deformation by both slip and twinning (Karaman et al., 2000; Toker et al., 2014), where the two mechanisms interact and further promote work hardening concomitant with increasing strain. To add further to the complexity of micro-deformation mechanisms, glide dislocations tend to form

high-density dislocation walls (HDDWs), which effectively hinder active glide dislocations from moving further, contributing to the unusual strain hardening exhibited by this material (Canadinc et al., 2005, 2007). However, dislocations are prevented from gliding not by twin boundaries or HDDWs only, but also by carbon (C), which can diffuse within the matrix throughout the deformation, as it becomes easily excited by the energy provided by the applied stresses (Canadinc et al., 2008a).

The excited C atoms diffuse within the matrix, however; when they meet dislocations at interstitial zones, they prevent dislocations from gliding further, which is referred to as pinning of the dislocation by the C atom (Owen and Grujicic, 1999; Canadinc et al., 2008a). This, indeed, gives way to DSA (Owen and Grujicic, 1999; Canadinc et al., 2008a), which further adds to the complexity of the microstructure by promoting strain rate sensitivity (SRS) and negative strain rate sensitivity (NSRS) (Canadinc et al., 2008a). Specifically, as steel is deformed at a faster rate, the rate of generation of forest dislocations increases, also leading to an increased level of stress at the same strain value, which is known as the SRS (Canadinc et al., 2008a). In Hadfield steel, however; the diffusion velocity of C also increases concomitant with strain rate. At lower strain rates, the C atoms stay longer at an interstitial site, potentially pinning a dislocation for a longer period, which is a positive contribution in terms of strain hardening. However, as the deformation rate increases, C atoms start to diffuse much faster, and they cannot pin dislocations for extended periods, which indeed takes away an important contribution to the overall hardening: the result is softening, and even though the strain rate increases, the material attains lower levels of stress at the same strains, which is known as the NSRS (Canadinc et al., 2008a). The NSRS, however; is eliminated at very high strain rates, such that the missing contribution to the overall hardening due to the pinning of dislocations becomes negligible as compared to the significant forest hardening, which constitutes the major mechanism of hardening at elevated rates of deformation (Canadinc et al., 2008a).

Despite their complicated micro-deformation mechanisms, other high-Mn austenitic steels with similar microstructures, such as the twinning-induced plasticity (TWIP) steels, continue to attract attention and find use in applications, mainly owing to their superior mechanical properties (Hutchinson and Ridley, 2006; Jeong et al., 2012; Xu et al., 2013; Wen et al., 2014). Automotive industry and other load bearing applications utilize high-Mn steels at an increasing rate, and ballistic applications constitute another potential area that may utilize this class of steels. For all these applications, however, resistance to impact loading and understanding of the deformation response under impact becomes of utmost importance (Toker et al., 2014). Considering the design process of any commercial product, on the other hand, one realizes that the capability to realistically predict the impact response of these alloys by numerical modeling is warranted, which constitutes the motivation of the current work.

The impact performance of a material is governed by its microstructure, and the corresponding parameters, such as grain boundaries, grain size, precipitates, and delaminations, and the material's ductile-to-brittle transition temperature (DBTT) altogether dictate the response to impact loading (Song et al., 2005; Kimura et al., 2008; Morris, 2008; Onal et al., 2014). Another important parameter affecting the impact response is the texture of the material, which defines the grain boundary–dislocation interactions, and thus, the crack propagation behavior under impact loading, as well as the degree of anisotropy, which dictates the relative slip activity in each grain (Onal et al., 2014), warranting incorporation of anisotropy effects into the design process.

The current study was undertaken with the motivation of addressing this issue, such that a combined experimental and numerical approach is proposed to predict the impact response of high-Mn austenitic steels in a realistic manner. For this purpose, uniaxial tensile deformation and impact responses of Hadfield steel were experimentally monitored, where the uniaxial deformation experiments featured four different strain rates ranging from moderate to high in order to also assess the role of NSRS exhibited by this material. Finite element (FE) simulations of the impact experiments were carried out, where the roles of texture, geometry, and SRS were successfully taken into account all at once by incorporating the proper multi-axial material flow rule obtained from crystal plasticity simulations into the FE analysis. Specifically, crystal plasticity was utilized to obtain the multi-axial flow rule at different strain rates based on the experimental deformation response under uniaxial tensile loading, and the equivalent stress – equivalent strain response was then incorporated into the FE model for the sake of a more representative hardening rule under impact loading. The current results demonstrate that reliable predictions can be obtained by proper coupling of crystal plasticity and FE analysis even if the experimental flow rule of the material is acquired under uniaxial loading and at moderate strain rates that are significantly slower than those attained during impact loading. Overall, the approach presented herein constitutes

an important guideline for the design process of impact bearing applications utilizing high-Mn austenitic steels.

## MATERIALS AND METHODS

The material investigated in this study is Hadfield steel, high-Mn austenitic steel with an fcc structure, and has a chemical composition of 12.44 wt% Mn, 1.10 wt% C, and balance iron. Small-scale dog-bone shaped tension samples were extracted from railroad frogs taken from service with the aid of electro-discharge machining to avoid any process-induced residual stresses and strains on the samples. The RT monotonic tensile deformation experiments were carried out on a servo-hydraulic test frame equipped with a digital controller and a miniature extensometer of 3 mm gage length. The results revealed a significant SRS prevalent in Hadfield steel polycrystals within the strain rate range of $1 \times 10^{-4}$ to $1 \times 10^{-1}$ 1/s (**Figure 1**). A serrated flow was exhibited by the material at all strain rates, however, the associated instability was much more prominent at $1 \times 10^{-3}$ 1/s, where NSRS was prevalent as evidenced by the lower stress levels attained despite the 10-fold increase in the strain rate from $1 \times 10^{-4}$ to $1 \times 10^{-3}$ 1/s.

The impact samples were extracted from both the same Hadfield railroad frog with dimensions of 2.8 mm $\times$ 25 mm $\times$ 4 mm, featuring a 60° notch with a radius of 0.1 mm and a depth of 1 mm (Onal et al., 2014). The specimens were mechanically polished down to a 4000 grit size in order to minimize the detrimental effects of machined surfaces on the impact response. The impact experiments were conducted at RT, and the specimens were subjected to deformation with an impact energy of 50 J and at a velocity of 3.8 m/s, where 8000 data points were collected during each experiment with a data acquisition frequency of 2 MHz. For both uniaxial deformation and impact experiments, three companion samples were tested in each case in order to ensure repeatability. The initial textures of the samples prior to deformation were determined by X-ray diffraction.

## RESULTS AND DISCUSSION
### FINITE ELEMENT SIMULATIONS OF EXPERIMENTAL IMPACT RESPONSE OF HADFIELD STEEL

It is well known that very large deformations taking place within very short time periods constitute the major difficulty of FE simulations of impact loading (Kormi et al., 1997; Raykhere et al., 2010; Onal et al., 2014). Specifically, the available time period is insufficient for the stabilization of the computed displacements in the case of impact loading, such that the material's behavior significantly deviates under this dynamic type of loading from that under normal conditions owing to the non-linear behavior under dynamic loading (Onal et al., 2014). The current FE simulations of the impact behavior were carried out utilizing the ANSYS® 15 commercial software and the FE analysis was based on an explicit dynamics system, where the LS-DYNA® solver was employed in computations for the sake of a reliable non-linear dynamic analysis.

The Lagrangian formulation and multilinear isotropic hardening were chosen to represent the non-linearity and the plastic flow, respectively. A fine "MultiZone" mesh was constructed with 13530 nodes and 11060 quadrilateral elements, where the geometry was automatically decomposed into a hex mesh, increasing both the

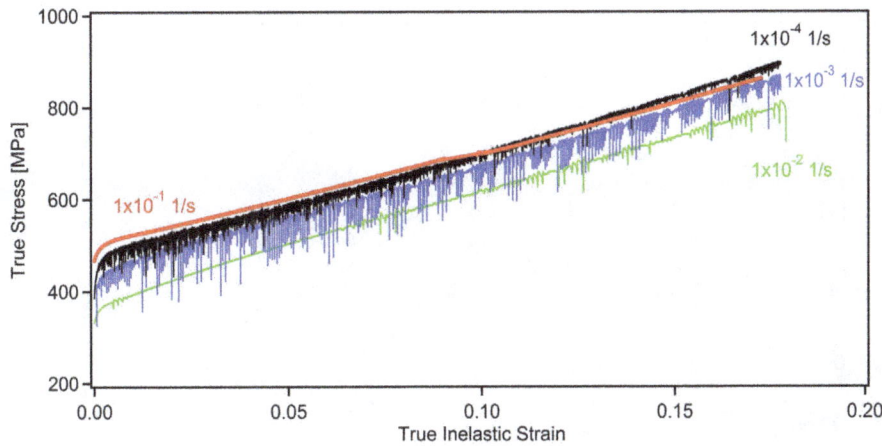

**FIGURE 1 | Room temperature uniaxial tensile deformation response of Hadfield steel obtained at different strain rates, demonstrating the NSRS.** Data were recompiled from Canadinc et al. (2008a).

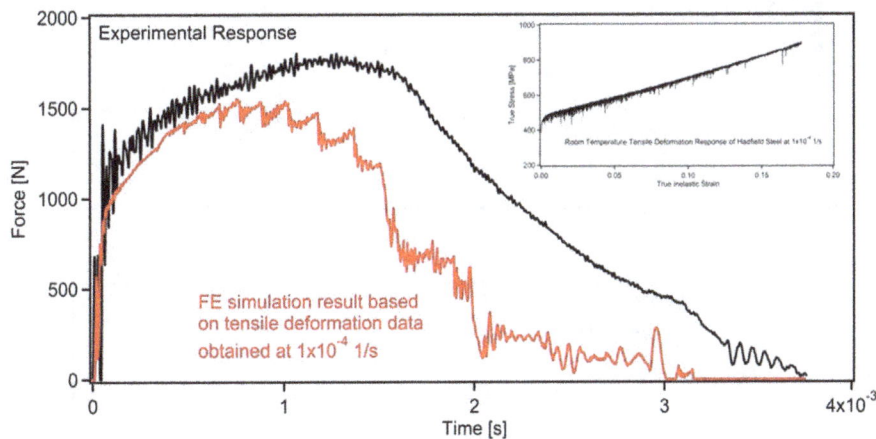

**FIGURE 2 | The RT experimental impact response of Hadfield steel and the corresponding FE simulation result, where the flow rule was defined based on the experimental uniaxial tensile deformation response obtained at a strain rate of $1 \times 10^{-4}$ 1/s.**

accuracy and computational efficiency of the simulations. The notch region was meshed with an element size of 0.1 mm, and the mesh had an orthogonal quality of 0.98 (out of 1.0). All contact regions of the impact sample were assumed to be frictional, and the corresponding static and dynamic friction coefficients were set to 0.2 and 0.09, respectively (Onal et al., 2014). As for the boundary conditions, a nodal displacement was imposed on the top notch surface and fixed supports were placed at the right/left sides and on the bottom surface. Beyond the fracture point of the specimens, i.e., where the sample, hammer contact terminates, the equivalent plastic strain (EPS) criterion was utilized to define failure of the material, namely an EPS of 0.68 was preset as the failure initiation point. The Cartesian coordinate system was considered while designating the dimensions and constraints.

The first set of simulations made use of the experimental RT uniaxial tensile deformation response of Hadfield steel obtained at a strain rate of $1 \times 10^{-4}$ 1/s to define the multilinear isotropic hardening rule (**Figure 2**). This strain rate is rather a moderate strain rate typically utilized in laboratory experiments employed to characterize the material's deformation response (Onal et al., 2014). It is evident that the corresponding simulation result significantly differs from the experimentally measured impact behavior, indicating that the flow rule of the material should be based on experiments featuring much higher strain rates as the impact deformation itself takes place very rapidly. Thus, in order to assess the role of strain rate, FE simulations considering higher rates of deformation, namely $1 \times 10^{-1}$, $1 \times 10^{-2}$, and $1 \times 10^{-3}$ 1/s, were carried out (**Figure 3**). As expected, a comparison of experimental and simulated impact responses (based on four different strain rates) expressed in terms of force–time data (**Figures 2** and **3**) indicates that the best predictions are made when the flow rule is based on the experimental deformation response obtained at

---

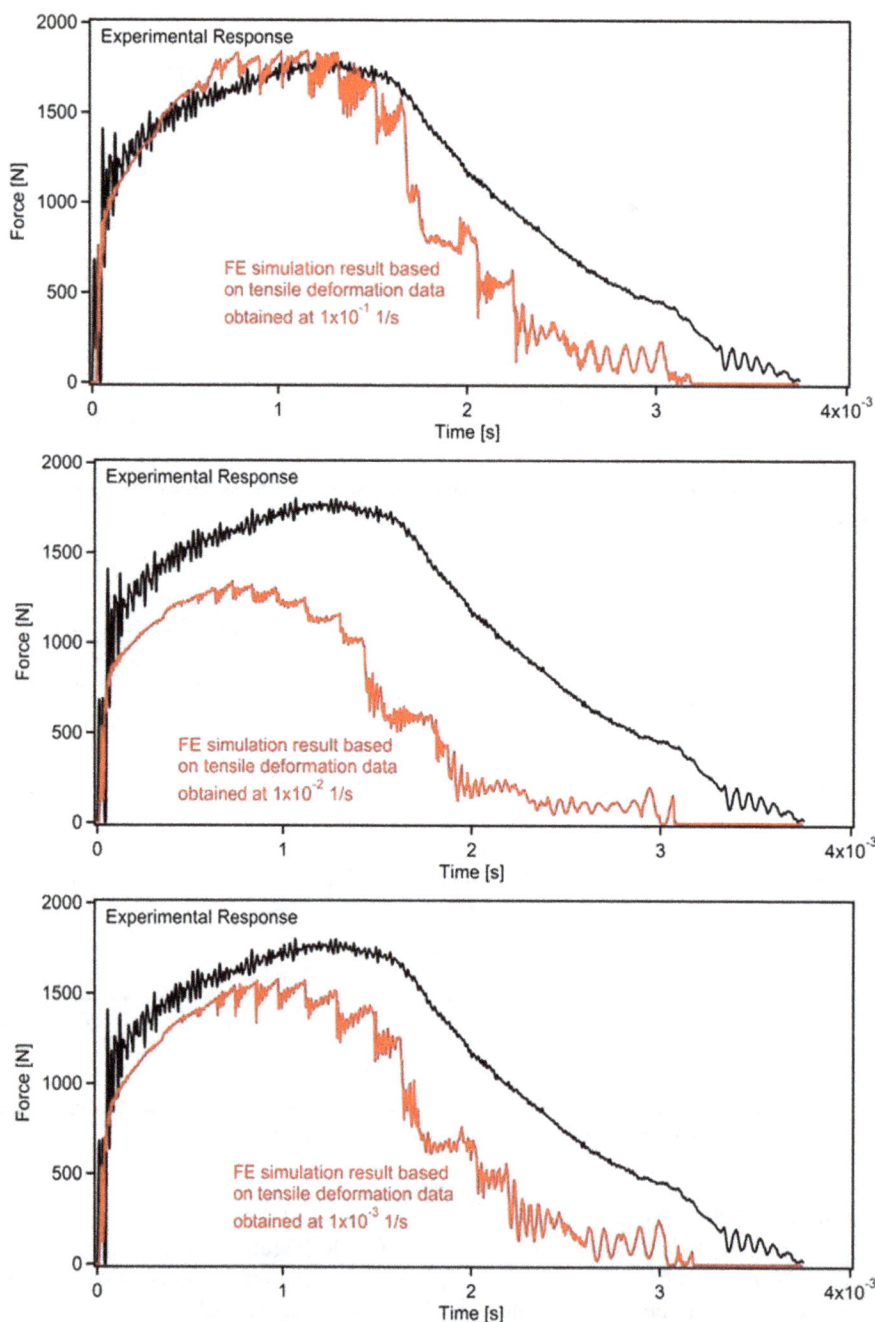

**FIGURE 3 | A comparison of the RT experimental impact response of Hadfield steel and the corresponding FE simulation results, where the flow rule was defined based on the experimental uniaxial tensile deformation responses obtained at strain rates of 1 × 10⁻¹, 1 × 10⁻², and 1 × 10⁻³ 1/s.**

the strain rate of $1 \times 10^{-3}$ 1/s. A further look at the results presented in **Figures 2** and **3** reveal a more interesting fact: the second best prediction is obtained when the flow rule is defined based on the uniaxial tensile deformation response obtained at a strain rate of $1 \times 10^{-4}$ 1/s. This implies that the experimental data obtained within the NSRS range should not be utilized to define the hardening rule when constructing a FE model to predict impact response, regardless of the strain rate.

Overall, all of the FE simulation results presented in **Figures 2** and **3** exhibited noticeable deviation from the experimental impact response, implying that the provided hardening response based on the experimental uniaxial tensile deformation data was not sufficiently representative of the material's mechanical behavior under impact loading at any of the considered strain rates. Further simulations featuring a finer mesh were also carried out in order to question the numerical procedure;

however, the same deviation from the experimental results persisted.

It has recently been demonstrated that a multi-axial stress–strain state is present in the critical region of the sample (Onal et al., 2014), and a comparison of normal and von Mises stress distributions reveals that employment of the uniaxial tensile deformation data as the input defining the material's flow rule under impact loading is not appropriate (**Figure 4**). Even though this approach might be appropriate for an isotropic material, it falls far from being realistic for a textured material exhibiting a significant degree of anisotropy (Onal et al., 2014). The numerical analyses carried out until this point (**Figures 2–4**) clearly demonstrate the need for a proper representation of microstructure under impact loading, in addition to the strain rate effects.

## INCORPORATION OF THE ROLE OF MICROSTRUCTURE INTO THE FINITE ELEMENT SIMULATIONS THROUGH CRYSTAL PLASTICITY

In order to utilize a proper flow rule to define hardening of Hadfield steel during impact and account for the role of texture and the corresponding anisotropy, in addition to the strain rate effects, in the current FE simulations, a crystal plasticity approach was adopted. Specifically, based on the aforementioned analysis of stress–strain distributions upon impact loading

(**Figure 4**), a flow rule based on equivalent stress–strain response was defined. In order to do so, a visco-plastic self-consistent (VPSC) algorithm was utilized to predict the von Mises stress–strain behavior of the material based on the experimental uniaxial tensile deformation data. Even though an alternative way of establishing the equivalent stress–strain response would have been carrying out multi-axial deformation experiments, the current methodology is much more efficient due to the impractical multi-axial experiments, especially in the case of high strain rates.

A successful crystal plasticity model should both predict the macroscopic deformation response and capture the deformation characteristics at the slip system level (Canadinc et al., 2011), such that the deformation response could be predicted under any type of loading (Canadinc et al., 2008b). With this motivation, the deformation of Hadfield steel was modeled at the microscopic level for each strain rate considered herein based on the corresponding experimentally obtained uniaxial deformation response (**Figures 5–8**). The VPSC model utilizes the initial texture of the material as an input, such that the loads on each grain, and thereby the slip activities in each grain, are dictated by the texture of the material (Lebensohn and Tomé, 1993; Kocks et al., 1998; Biyikli et al., 2010). Thereafter, the same micro level model established for

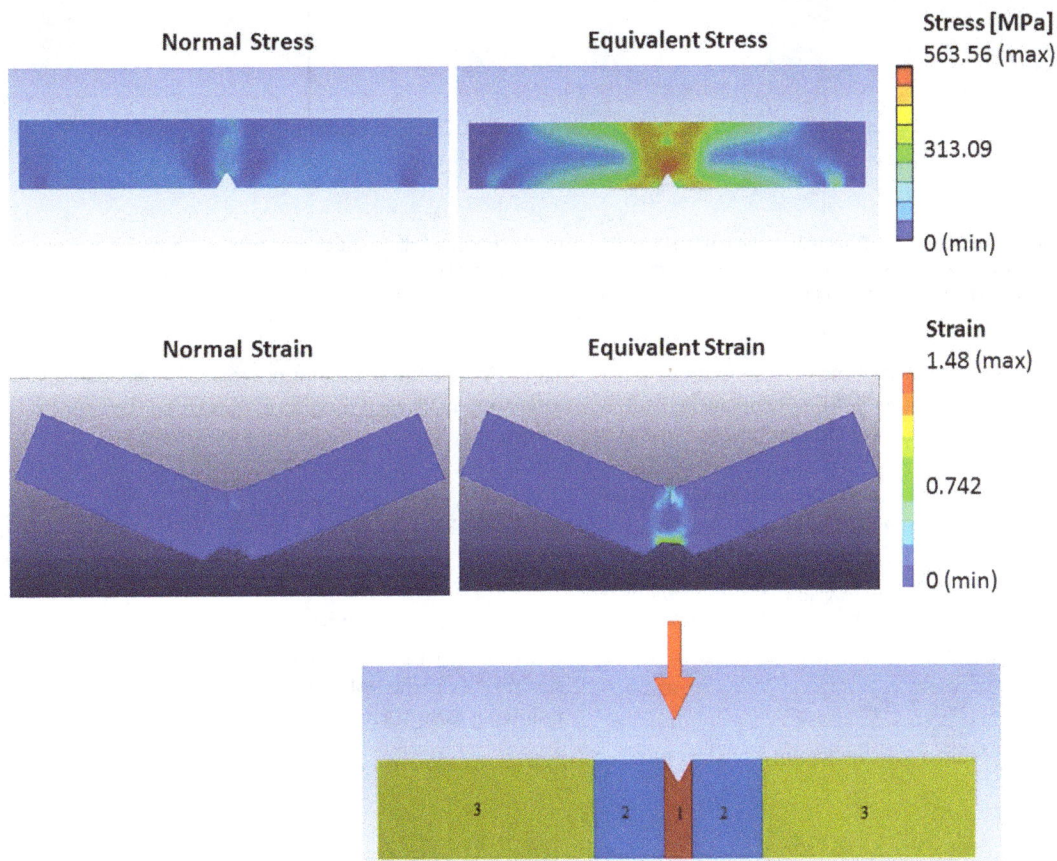

**FIGURE 4 | A material-independent FE simulation of impact loading demonstrating the typical distribution of normal and equivalent stress-strain fields (above the arrow), and the corresponding division of sample geometry based on stress intensities within the sample (Onal et al., 2014).**

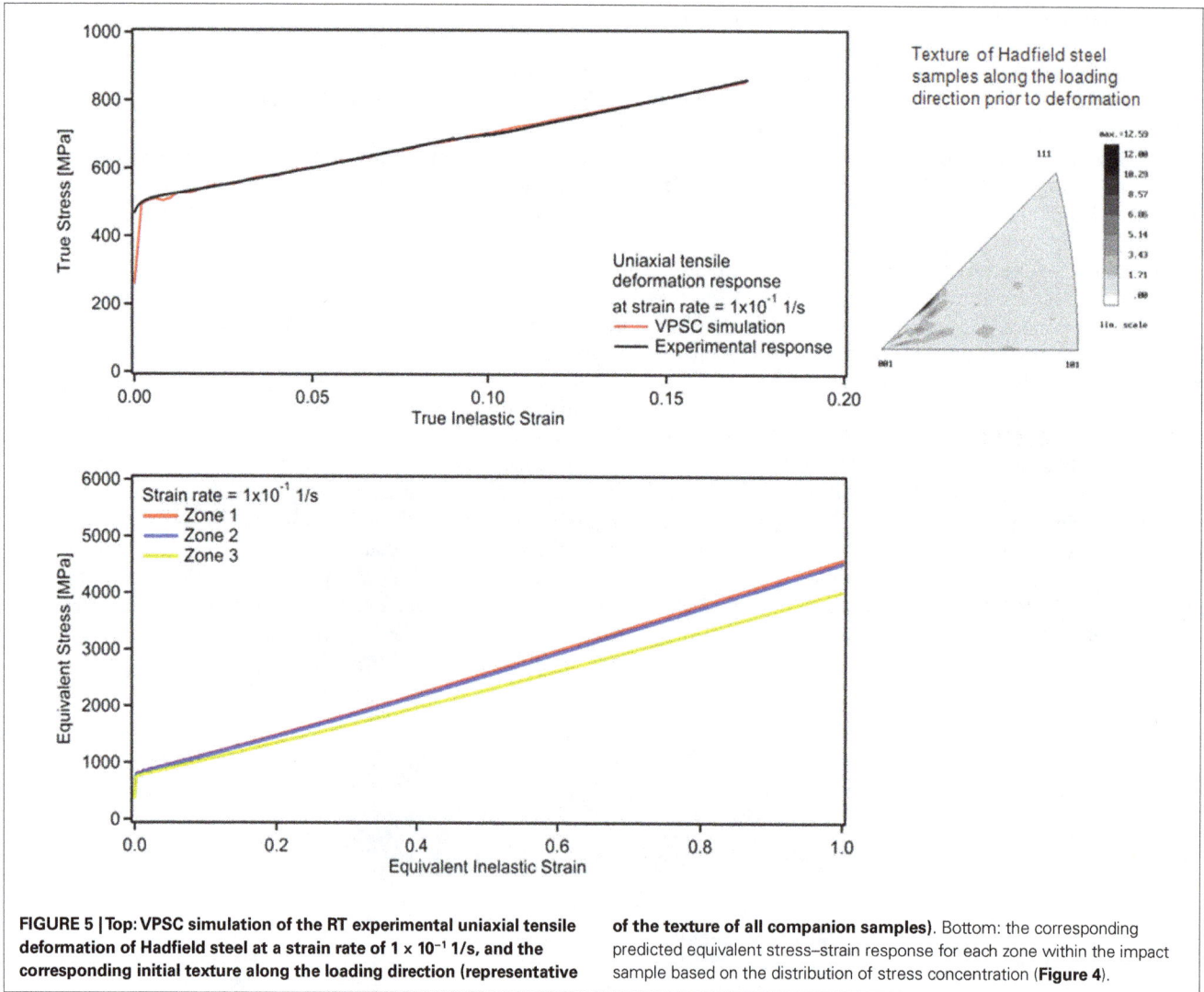

**FIGURE 5 | Top: VPSC simulation of the RT experimental uniaxial tensile deformation of Hadfield steel at a strain rate of 1 × 10⁻¹ 1/s, and the corresponding initial texture along the loading direction (representative** **of the texture of all companion samples).** Bottom: the corresponding predicted equivalent stress–strain response for each zone within the impact sample based on the distribution of stress concentration (**Figure 4**).

each strain rate was utilized to predict the corresponding equivalent stress–strain response (**Figures 5–8**), which can be utilized as a proper flow rule for the impact simulations, as discussed before.

The VPSC algorithm employed in the current study considers plastic deformation only, which takes place when one or more slip or twinning systems become active. For a slip system $S$, the corresponding resolved shear stress $\left(\tau_{RSS}^s\right)$ facilitating plastic deformation can be described in vector form based on the Schmid $\left(m_i^s\right)$ and the applied stress ($\sigma_i$) tensors:

$$\tau_{RSS}^s = m_i^s \sigma_i \qquad (1)$$

The non-linear shear strain rate in the system $S$ can be described as a function of $\tau_{RSS}^s$:

$$\dot{\gamma}^s = \dot{\gamma}_0 \left( \frac{\tau_{RSS}^s}{\tau_0^s} \right)^n = \dot{\gamma}_0 \left( \frac{m_i^s \sigma_i}{\tau_0^s} \right)^n \qquad (2)$$

where $\dot{\gamma}_0$ is a reference rate, $\tau_0^s$ is the threshold stress corresponding to this reference rate, and $n$ is the inverse of the rate sensitivity

index. When the contributions of all active systems in a single grain are superposed and pseudolinearized (Lebensohn and Tomé, 1993):

$$\dot{\varepsilon}_i = \left[ \dot{\gamma}_0 \sum_1^s \frac{m_i^s m_j^s}{\tau_0^s} \left( \frac{m_k^s \sigma_k}{\tau_0^s} \right)^{n-1} \right] \sigma_j = M_{ij}^{c(sec)} (\tilde{\sigma}) \, \sigma_j \qquad (3)$$

where $M_{ij}^{c(sec)}$ is the secant visco-plastic compliance of the crystal, which gives the instantaneous relation between stress and strain rate. At the polycrystal level, this relationship assumes the following form (Lebensohn and Tomé, 1993):

$$\dot{E}_i = M_{ij}^{(sec)} \left( \tilde{\Sigma} \right) \Sigma_j + \dot{\Sigma}^0 \qquad (4)$$

where $\dot{E}_i$ and $\Sigma$ represent the polycrystal strain rate and applied stress, respectively.

In a continuum comprising a matrix and inclusions, the deviations in strain rate and stress of the inclusions from those of the

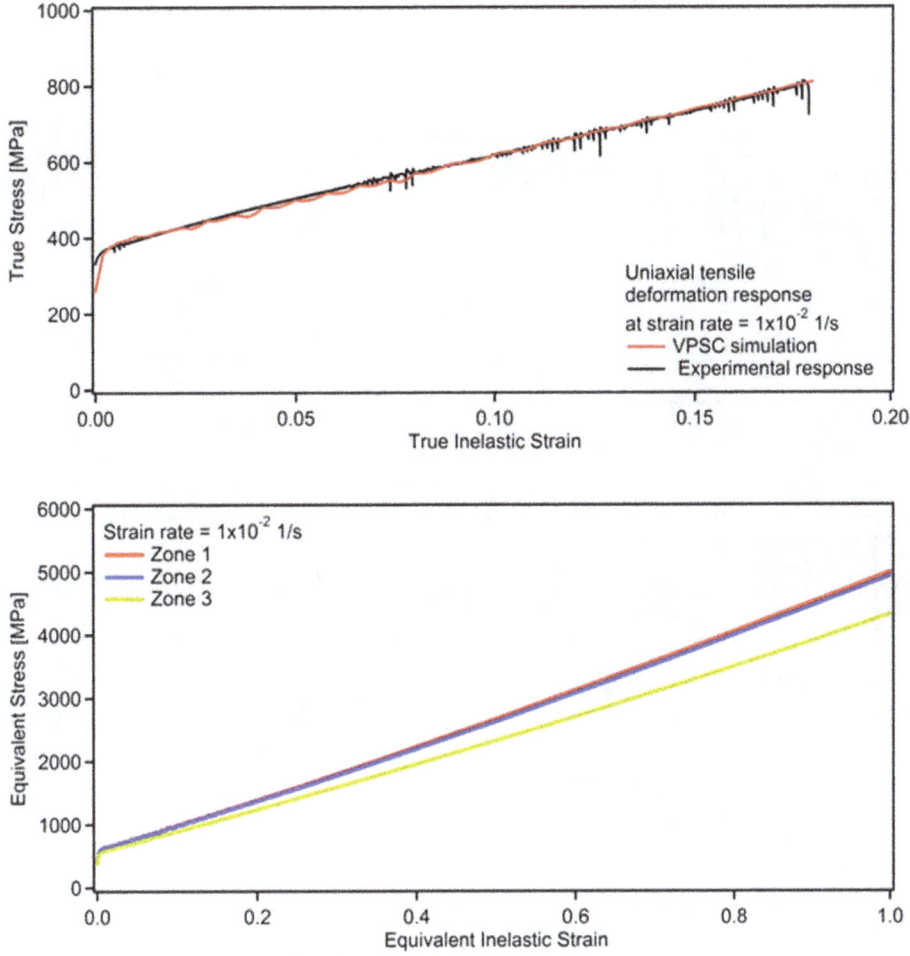

FIGURE 6 | Top: VPSC simulation of the RT experimental uniaxial tensile deformation of Hadfield steel at a strain rate of $1 \times 10^{-2}$ 1/s. Bottom: the corresponding predicted equivalent stress–strain response for each zone within the impact sample based on the distribution of stress concentration (Figure 4).

matrix can be defined as:

$$\dot{\tilde{\varepsilon}}_k = \dot{\varepsilon}_k - \dot{E}_k \quad (5)$$

$$\tilde{\sigma}_j = \sigma_j - \Sigma_j \quad (6)$$

where $\dot{\varepsilon}_k$ and $\sigma_j$ stand for the local (grain level) strain rate and stress. When the Eshelby inclusion formulation is employed to solve the stress equilibrium, one can obtain (Kocks et al., 1998):

$$\dot{\tilde{\varepsilon}} = -\tilde{M} : \tilde{\sigma} \quad (7)$$

The interaction tensor $\tilde{M}$ is defined as:

$$\tilde{M} = n'(I - S)^{-1} : S : M^{(sec)} \quad (8)$$

where $M^{(sec)}$ is the secant compliance tensor for the polycrystal aggregate and $S$ is the visco-plastic Eshelby tensor (Kocks et al., 1998). In Eq. 8, an effective value of $n' = 1$ was used, which ensures a rigid interaction (Lebensohn and Tomé, 1993).

Substitution of Eqs 3 and 4 into Eq. 7 yields the macroscopic secant compliance, $M^{(sec)}$, and the macroscopic strain rate is evaluated by taking the weighted average of crystal strain rates over all the grains as in Eq. 9:

$$M^{(sec)} = \left\langle M^{c(sec)} : \left( M^{c(sec)} + \tilde{M} \right)^{-1} : \left( M^{(sec)} + \tilde{M} \right) \right\rangle \quad (9)$$

Iterative solution of the Eqs 3, 7, and 9 gives the stress in each grain, the crystal's compliance tensor, and the polycrystal compliance consistent with the applied strain rate $\dot{E}_i$. In this work, the term $n$ in Eq. 2 was chosen as 20, which makes the formulation rate insensitive (Lebensohn and Tomé, 1993). This sounds contradictory to the overall aim of incorporating the NSRS of Hadield steel into the simulations, however; it should be noted that the uniaxial deformation response at each strain rate was modeled separately rather than utilizing a single micro-deformation model accounting for the SRS. Therefore, the term $n$ in Eq. 2 was assigned a value to ensure rate insensitivity within the same simulation, or in other

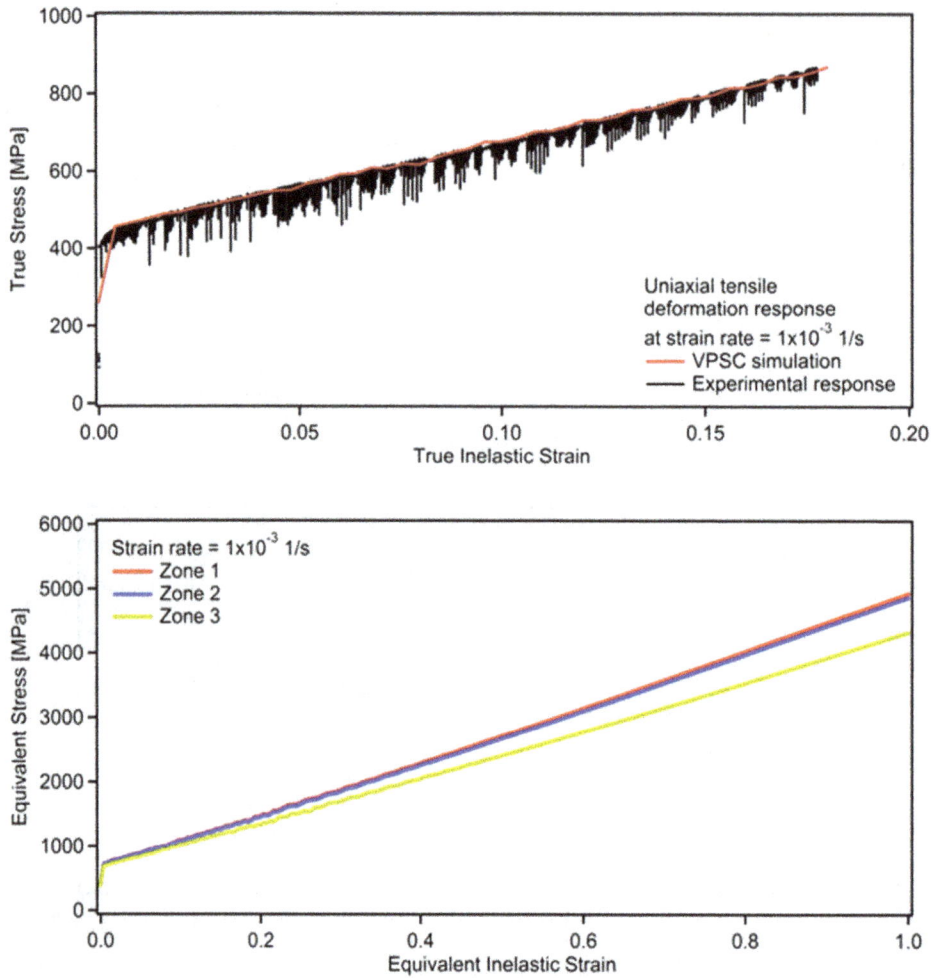

**FIGURE 7 | Top:** VPSC simulation of the RT experimental uniaxial tensile deformation of Hadfield steel at a strain rate of **1 × 10⁻³ 1/s**. Bottom: the corresponding predicted equivalent stress–strain response for each zone within the impact sample based on the distribution of stress concentration (**Figure 4**).

words, for each case. One reason for this is that, to the best of the authors' knowledge, a crystal plasticity model capable of predicting the NSRS exhibited by Hadfield steel has not been forwarded yet. Furthermore, the incorporation of NSRS requires coupling of crystal plasticity with atomistic simulations to properly account for the diffusivity of C and the corresponding consequences, which is beyond the scope of the current work.

The rate of overall dislocation density can be expressed as:

$$\dot{\rho} = \sum_n \left\{ k_1 \sqrt{\rho} - k_2 \rho \right\} \left| \dot{\gamma}^n \right| \qquad (10)$$

where $k_1$ and $k_2$ are geometric constants that define the athermal (statistical) storage of the moving dislocations (Kocks et al., 1998). The flow stress $\tau$ is defined in the traditional Taylor hardening format as:

$$\tau - \tau_0 = \alpha \mu b \sqrt{\rho} \qquad (11)$$

where $\alpha$ is the dislocation interaction parameter and $\tau_0$ is a reference strength, which is related to deformation at the grain level. The reference strength value for Hadfield steel was determined as 132 MPa in previous work (Canadinc et al., 2007), where the 0.2% offset yield strength value is normalized by the Taylor factor. The Taylor factor for the current materials was determined based on the experimentally measured texture (Canadinc et al., 2008b), and it is about 3.11, which represents a slightly textured material as compared to the fully random texture case that has a Taylor factor of 3.06. From Eq. 11, with $\tau_0$ constant, the rate of the flow stress is obtained by taking the time derivative as,

$$\dot{\tau} = \frac{\alpha \mu b \dot{\rho}}{2 \sqrt{\rho}} \qquad (12)$$

Substituting Eq. 10 into Eq. 12 results in:

$$\dot{\tau} = \sum_n \left\{ k_1 \frac{\alpha \mu b}{2} - k_2 \frac{\alpha \mu b}{2} \sqrt{\rho} \right\} \left| \dot{\gamma}^n \right| \qquad (13)$$

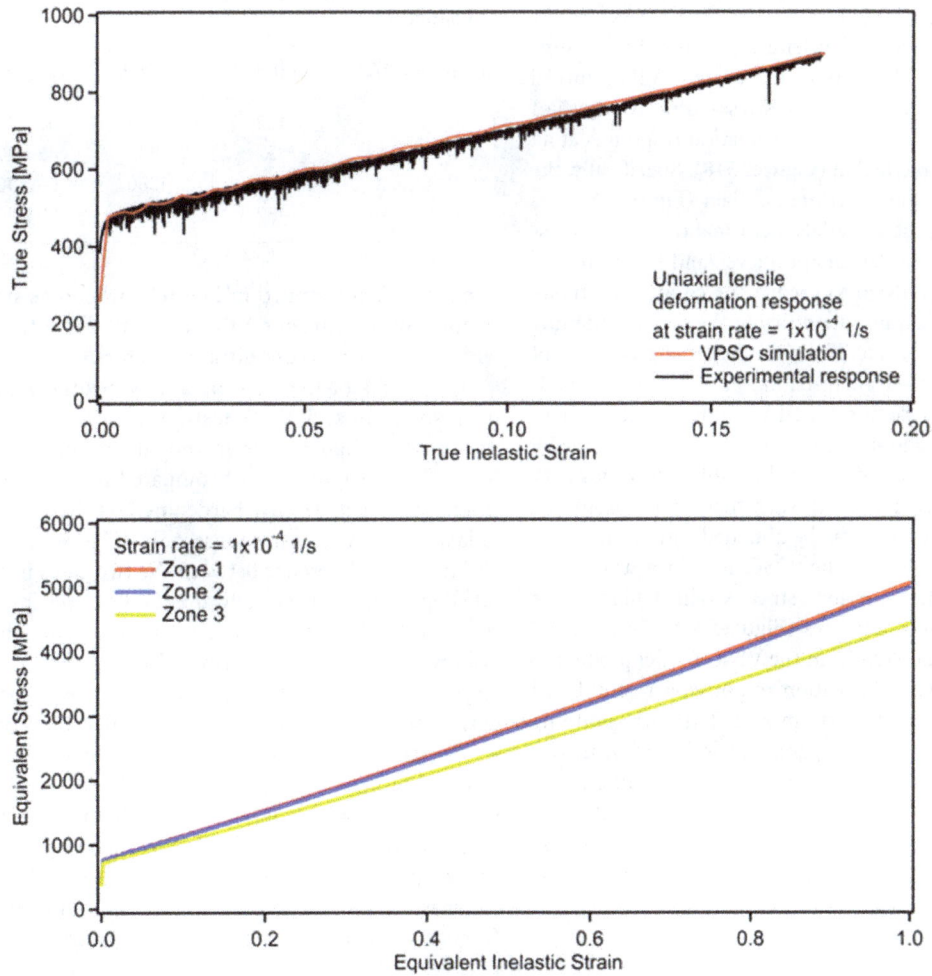

**FIGURE 8 | Top: VPSC simulation of the RT experimental uniaxial tensile deformation of Hadfield steel at a strain rate of 1 × 10⁻⁴ 1/s.** Bottom: the corresponding predicted equivalent stress–strain response for each zone within the impact sample based on the distribution of stress concentration (**Figure 4**).

From Eq. 11, the following identity is obtained for the square root of the density of dislocations:

$$\sqrt{\rho} = \frac{\tau - \tau_0}{\alpha \mu b} \tag{14}$$

Once Eq. 14 is substituted into Eq. 13, the rate of flow stress evolution is given by:

$$\dot{\tau} = \sum_n \left\{ k_1 \frac{\alpha \mu b}{2} - k_2 \frac{(\tau - \tau_0)}{2} \right\} |\dot{\gamma}^n| \tag{15}$$

One should note that the term $\left\{ \frac{\alpha \mu b}{2} k_1 - \frac{(\tau - \tau_0)}{2} k_2 \right\}$ in Eq. 15 is the linear Voce hardening term (Eq. 17). Having noted this, Eq. 15 can also be expressed as (Kocks et al., 1998):

$$\dot{\tau} = \sum_n \left\{ \theta_0 \left( \frac{\tau_s - \tau}{\tau_s - \tau_0} \right) \right\} |\dot{\gamma}^n| \tag{16}$$

where $\theta_0$ is the constant strain hardening rate, and $\tau_s$ represents the saturation stress in the absence of geometric effects, or the threshold stress. The hardening is defined by an extended Voce law (Kocks et al., 1998), which is characterized by the evolution of the threshold stress ($\tau^s$) with accumulated shear strain ($\Gamma$) in each grain of the form

$$\tau^s = \tau_0 + (\tau_1 + \theta_1 \Gamma) \left( 1 - \exp\left( -\frac{\theta_0 \Gamma}{\tau_1} \right) \right) \tag{17}$$

where $\tau_0$ is the reference strength, and $\tau_1$, $\theta_0$, and $\theta_1$ are the parameters that define the hardening behavior (Kocks et al., 1998). The hardening law defined by Eq. 17 characterizes the onset of plasticity and the saturation of threshold stress at larger strains.

The current VPSC model described by Eqs 1–17 was employed to solve for the stresses corresponding to the given strains throughout the deformation. The experimentally determined initial texture of Hadfield steel (inset of **Figure 5**) was utilized as input, and the macroscopic deformation responses were predicted as presented in **Figures 5–8** for all the strain rates considered in this work. The corresponding Voce hardening parameters for each strain rate are provided in **Table 1**.

In order to define a proper hardening rule for the FE simulations of the impact deformation, the current VPSC model was employed to predict the equivalent stress–strain response of Hadfield steel based on the uniaxial deformation responses at all four strain rates considered herein (**Figures 5–8**). Specifically, the successful prediction of the experimental data (**Figures 5–8**) is a strong indication that the materials' deformation was successfully modeled at the micro-deformation level, and therefore, the same VPSC model was utilized to predict the equivalent stress–strain response for each strain rate utilizing the same hardening parameters for each strain rate. Since the material-independent consideration of the stress–strain distribution under impact loading (**Figure 4**) had also demonstrated that the distribution of stresses and strains throughout the sample is heterogeneous, the FE mesh for each sample was divided into three different zones with three different flow rules (**Figure 4**), such that a more homogeneous stress–strain distribution can be obtained within each zone upon impact loading. Therefore, the VPSC model was utilized to predict the corresponding equivalent stress–strain state response for each zone at all four strain rates (**Figures 5–8**). Specifically, the same hardening parameters as in the VPSC model predicting the experimental uniaxial deformation response were employed in all three simulations for each strain rate. The corresponding deformation of the polycrystalline aggregate within each zone was defined to the VPSC algorithm through velocity gradient tensors, which were determined based on the material-independent strain distributions under impact loading (**Figure 4**). The corresponding velocity gradient tensors for each zone were computed as:

$$\dot{U}_1 = \begin{bmatrix} 4.2 & 1.0 & 0 \\ 1.0 & -2.1 & 0 \\ 0 & 0 & -2.1 \end{bmatrix},$$

$$\dot{U}_2 = \begin{bmatrix} 3.0 & 1.0 & 0 \\ 1.0 & -1.5 & 0 \\ 0 & 0 & -1.5 \end{bmatrix},$$

and

$$\dot{U}_3 = \begin{bmatrix} 0.3 & 1 & 0 \\ 1 & -0.15 & 0 \\ 0 & 0 & -0.15 \end{bmatrix},$$

for zones 1, 2, and 3, respectively (Onal et al., 2014).

The corresponding VPSC simulation results demonstrating the equivalent for each zone are presented in **Figures 5–8** for the strain rates of $1 \times 10^{-1}$, $1 \times 10^{-2}$, $1 \times 10^{-3}$, and $1 \times 10^{-4}$ 1/s, respectively. It should be noted that the strength levels attained by the equivalent deformation curves for each strain rate follows the same trend as that of the experimental uniaxial curves. This is not surprising since both the plastic deformation and the hindering of dislocations by diffusing C atoms are considered at the slip system level, where the latter leads to NSRS in Hadfield steel. Moreover, for each strain rate, higher stresses were obtained for zone 1 as compared to zones 2 and 3, and zone 3 exhibited the lowest stress levels (**Figures 5–8**), which stands in good agreement with the stress

**Table 1 | Voce hardening parameters utilized in the current VPSC simulations**.

| Strain rate (1/s) | $\tau_0$ (MPa) | $\tau_1$ (MPa) | $\theta_0$ (MPa) | $\theta_1$ (MPa) |
|---|---|---|---|---|
| $1 \times 10^{-1}$ | 132 | 925 | $53 \times 10^4$ | 350 |
| $1 \times 10^{-2}$ | 132 | 147 | $43 \times 10^3$ | 408 |
| $1 \times 10^{-3}$ | 132 | 1580 | $60 \times 10^4$ | 390 |
| $1 \times 10^{-4}$ | 132 | 918 | $55 \times 10^4$ | 398 |

intensities demonstrated in **Figure 4**, where the stresses decrease as one moves from zone 1 that contains the notch toward zone 3 with the least stress concentration factors.

The results of the FE simulations incorporating the roles of microstructure and NSRS texture through crystal plasticity are presented in **Figure 9**. Even though the predictions are much better for all strain rates as compared to those of the initial FE simulations that defined hardening based on the experimental uniaxial deformation response only (**Figures 2** and **3**), there is an important difference between the two cases in terms of strain rate dependence. Specifically, the results of the initial simulations revealed that the best predictions were obtained by defining the flow rule for Hadfield steel based on the uniaxial deformation response recorded at the highest strain rate ($1 \times 10^{-1}$ 1/s in the current work), in addition to the fact that the worst predictions were based on the flow rules defined within the NSRS range. Upon incorporation of microstructure into the FE model through crystal plasticity, however, it was evident that the best prediction was obtained by defining the flow rule based on the VPSC predictions of the equivalent stress–strain response at $1 \times 10^{-2}$ 1/s (**Figure 9**), which is the second highest strain rate and within the NSRS range (**Figure 1**). This contradictory result clearly demonstrates that reliable predictions can be obtained by proper coupling of crystal plasticity and FE analysis even if the experimental flow rule of the material is acquired under uniaxial loading and at strain rates that fall within the NSRS range. This is especially important in terms of utilizing standard laboratory experiments to characterize a material's fundamental properties, which are simple and may be restricted to strain rates within the NSRS owing to practical limitations, while predicting its deformation response under complicated loading scenarios, such as impact loading.

## CONCLUSION

The RT impact response Hadfield steel was studied with the aid of a multi-scale modeling approach coupling crystal plasticity and FE analysis. The roles of texture, geometry, and SRS were successfully taken into account all at once, where crystal plasticity was utilized to obtain the multi-axial flow rule at different strain rates based on the experimental deformation response under uniaxial tensile loading. The FE simulation results demonstrated that the utility of equivalent stress–equivalent strain response for defining the hardening rule under impact loading resulted in improved predictions. Interestingly, the simulation results indicated that a multi-axial definition of the material flow rule is the major parameter dictating the success of the predictions of impact loading deformation even in the presence of negative SRS, as in the case of Hadfield steel. Finally, the current set of results also demonstrated that reliable

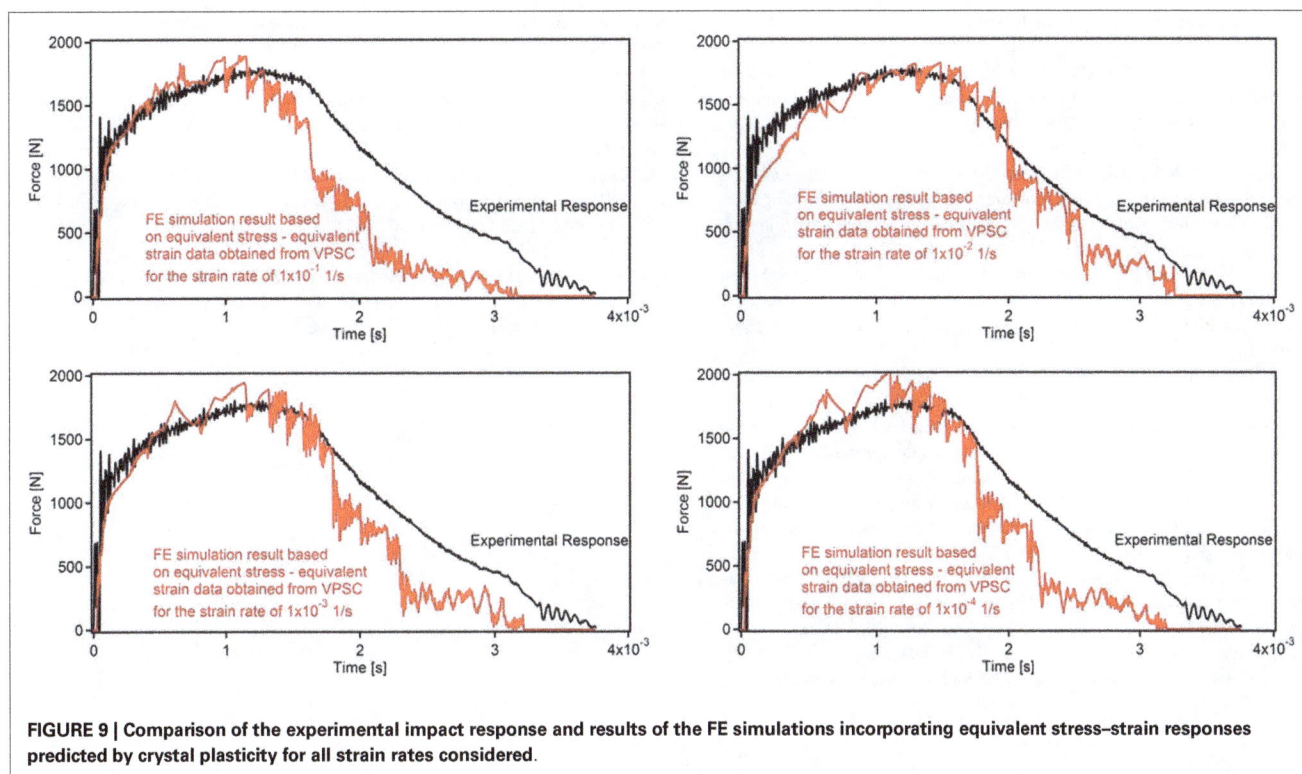

**FIGURE 9 | Comparison of the experimental impact response and results of the FE simulations incorporating equivalent stress–strain responses predicted by crystal plasticity for all strain rates considered.**

predictions can be obtained by proper coupling of crystal plasticity and FE analysis even if the experimental flow rule of the material is acquired under uniaxial loading and at moderate strain rates that are significantly slower than those attained during impact loading. This observation opens the venue for utilizing more practical and simpler laboratory experiments to characterize a material's fundamental properties while predicting its deformation response under complicated loading scenarios, such as impact loading.

## ACKNOWLEDGMENTS

This study was supported by the Scientific and Technological Research Council of Turkey (TÜBİTAK) under grant 112M806.

## REFERENCES

Bayraktar, E., Khalid, F. A., and Levaillant, C. (2004). Deformation and fracture behaviour of high manganese austenitic steel. *J. Mater. Process. Tech.* 147, 145–154. doi:10.1016/j.jmatprotec.2003.10.007

Biyikli, E., Canadinc, D., Maier, H. J., Niendorf, T., and Top, S. (2010). Three-dimensional modeling of the grain boundary misorientation angle distribution based on two-dimensional experimental texture measurements. *Mater. Sci. Eng. A* 527, 5604–5612. doi:10.1016/j.msea.2010.05.037

Bouaziz, O., Allain, S., Scott, C. P., Cugy, P., and Barbier, D. (2011). High manganese austenitic twinning induced plasticity steels: a review of the microstructure properties relationships. *Curr. Opin. Solid State Mater. Sci.* 15, 141–168. doi:10.1016/j.cossms.2011.04.002

Canadinc, D., Biyikli, E., Niendorf, T., and Maier, H. J. (2011). Experimental and numerical investigation of the role of grain boundary misorientation angle on the dislocation–grain boundary interactions. *Adv. Eng. Mater.* 13, 281–287. doi:10.1002/adem.201000229

Canadinc, D., Efstathiou, C., and Sehitoglu, H. (2008a). On the negative strain rate sensitivity of hadfield steel polycrystals. *Scripta Mater.* 59, 1103–1106. doi:10.1016/j.scriptamat.2008.07.027

Canadinc, D., Sehitoglu, H., Maier, H. J., and Kurath, P. (2008b). On the incorporation of length scales associated with pearlitic and bainitic microstructures into a visco-plastic self-consistent model. *Mater. Sci. Eng. A* 485, 258–271. doi:10.1016/j.msea.2007.08.049

Canadinc, D., Sehitoglu, H., and Maier, H. J. (2007). The role of dense dislocation walls on the deformation response of aluminum alloyed hadfield steel polycrystals. *Mater. Sci. Eng. A* 454–455, 662–666. doi:10.1016/j.msea.2006.11.122

Canadinc, D., Sehitoglu, H., Maier, H. J., and Chumlyakov, Y. I. (2005). Strain hardening behavior of aluminum alloyed hadfield steel single crystals. *Acta Mater.* 53, 1831–1842. doi:10.1016/j.actamat.2004.12.033

Hutchinson, B., and Ridley, N. (2006). On dislocation accumulation and work hardening in Hadfield steel. *Scripta Mater.* 55, 299–302. doi:10.1016/j.scriptamat.2006.05.002

Jeong, J. S., Woo, W., Oh, K. H., Kwon, S. K., and Koo, Y. M. (2012). In situ neutron diffraction study of the microstructure and tensile deformation behavior in Al-added high manganese austenitic steels. *Acta Mater.* 60, 2290–2299. doi:10.1016/j.actamat.2011.12.043

Karaman, I., Sehitoglu, H., Chumlyakov, Y. I., Maier, H. J., and Kireeva, I. V. (2001). Extrinsic stacking faults and twinning in hadfield manganese steel single crystals. *Scripta Mater.* 44, 337–343.

Karaman, I., Sehitoglu, H., Gall, K., Chumlyakov, Y. I., and Maier, H. J. (2000). Deformation of single crystal Hadfield steel by twinning and slip. *Acta Mater.* 48, 1345–1359. doi:10.1016/S1359-6454(99)00383-3

Kimura, Y., Inoue, T., Yin, F., and Tsuzaki, K. (2008). Inverse temperature dependence of toughness in an ultrafine grain-structure steel. *Science* 320, 1057–1060. doi:10.1126/science.1156084

Kocks, U. F., Tomé, C. N., and Wenk, H. R. (1998). *Texture and Anisotropy*. New York: Cambridge University Press.

Kormi, K., Webb, D. C., and Johnson, W. (1997). The application of the FEM to determine the response of a pretorsioned pipe cluster to static or dynamic axial impact loading. *Comp. Struct.* 62, 353–368. doi:10.1016/S0045-7949(96)00176-9

Lebensohn, R. A., and Tomé, C. N. (1993). A self-consistent anisotropic approach for the simulation of plastic deformation and texture development of polycrystals: application to zirconium alloys. *Acta Metall. Mater.* 41, 2611–2624.

Morris, J. W. (2008). Stronger, tougher steels. *Science* 320, 1022–1023. doi:10.1126/science.1158994

Niendorf, T., Lotze, C., Canadinc, D., Frehn, A., and Maier, H. J. (2009). The role of monotonic pre-deformation on the fatigue performance of a high-manganese austenitic TWIP steel. *Mater. Sci. Eng. A* 499, 518–524. doi:10.1016/j.msea.2008.09.033

Niendorf, T., Rubitschek, F., Maier, H. J., Niendorf, J., Richard, H. A., and Frehn, A. (2010). Fatigue crack growth–Microstructure relationships in a high-manganese austenitic TWIP steel. *Mater. Sci. Eng. A* 527, 2412–2417. doi:10.1016/j.msea.2009.12.012

Onal, O., Bal, B., Toker, S. M., Mirzajanzadeh, M., Canadinc, D., and Maier, H. J. (2014). Microstructure-based modeling of the impact response of a biomedical niobium-zirconium alloy. *J Mater. Res.* 29, 1123–1134. doi:10.1557/jmr.2014.105

Owen, W. S., and Grujicic, M. (1999). Strain aging of austenitic Hadfield manganese steel. *Acta Mater.* 47, 111–126. doi:10.1016/S1359-6454(98)00347-4

Raykhere, S. L., Kumar, P., Singh, R. K., and Parameswaran, V. (2010). Dynamic shear strength of adhesive joints made of metallic and composite adherents. *Mater. Des.* 31, 2102–2109. doi:10.1016/j.matdes.2009.10.043

Song, R., Ponge, D., and Raabe, D. (2005). Mechanical properties of an ultrafine grained C–Mn steelprocessed by warm deformation and annealing. *Acta Mater.* 53, 4881–4892. doi:10.1016/j.actamat.2005.07.009

Toker, S. M., Canadinc, D., Taube, A., Gerstein, G., and Maier, H. J. (2014). On the role of slip–twin interactions on the impact behavior of high-manganese austenitic steels. *Mater. Sci. Eng. A* 593, 120–126. doi:10.1016/j.msea.2013.11.033

Ueji, R., Tsuchida, N., Terada, D., Tsuji, N., Tanaka, Y., Takemura, A., et al. (2008). Tensile properties and twinning behavior of high manganese austenitic steel with fine-grained structure. *Scripta Mater.* 59, 963–966. doi:10.1016/j.scriptamat.2008.06.050

Wen, Y. H., Peng, H. B., Si, H. T., Xiong, R. L., and Raabe, D. (2014). A novel high manganese austenitic steel with higher work hardening capacity and much lower impact deformation than Hadfield manganese steel. *Mater. Design* 55, 798–804. doi:10.1016/j.matdes.2013.09.057

Xu, S., Ruan, D., Beynon, J. H., and Rong, Y. (2013). Dynamic tensile behaviour of TWIP steel under intermediate strain rate loading. *Mater. Sci. Eng. A* 573, 132–140. doi:10.1016/j.msea.2013.02.062

**Conflict of Interest Statement:** The authors declare that the research was conducted in the absence of any commercial or financial relationships that could be construed as a potential conflict of interest.

# 18

# Sedimentation upon different carrier liquid in giant electrorheological fluid and its application

*Yaying Hong and Weijia Wen\**

*Department of Physics, Hong Kong University of Science and Technology, Hong Kong, China*

**Edited by:**
*Weihua Li, University of Wollongong, Australia*

**Reviewed by:**
*Bo Hou, Soochow University, China*
*Xianzhou Zhang, G.H. Varley Engineering, Australia*

**\*Correspondence:**
*Weijia Wen, Department of Physics and Institute of Nano Science and Technology, The Hong Kong University of Science and Technology, Clear Water Bay, Kowloon, Hong Kong, China*
*e-mail: phwen@ust.hk*

When giant electrorheological (GER) fluid is settled after some time, particles can precipitate out of the oil in a multistep process that involves the formation of larger particles, the aggregation of colloids, and eventual sedimentation. Colloidal stability in GER fluid can influence the GER performance and the fluid flow steadiness. We investigated the sedimentation effect of the GER particles suspended in various carrier liquid. Different from the existing electrorheological (ER) fluids, GER particles consisting of oxalate core with urea coating are found oil synergistic. The sedimentation effect of the particles suspended in oils from the family of synthetic oil and mineral oil were checked by direct observation. The rheological behavior of the GER fluid upon electric field application was also investigated. These experiments showed that stable colloidal suspension and good GER effect can be achieved coherently by favorable particle–oil interaction. The resultant high yield stress and low sedimentation rate achieved due to the instrumental linking of hydrogen bond is showed in the hydrogenated silicone oil carrier liquid. With the anti-sedimentation characteristic upon the new carrier oil, hydrogenated silicone oil-GER fluid, we investigated their GER effect in a modified mono tube damper and the experimental result showed wide controllability range. Our investigations may broaden engineering applications.

**Keywords: giant electrorheological, sedimentation, GER damper**

## INTRODUCTION

Field responsive fluids, such as electrorheological (ER) fluid and magnetorheological (MR) fluid, are a trend to the new generation in design for product where power density, accuracy, and dynamic performance are the key features. These materials are different from the traditional smart materials, in that they are soft materials (typically dispersions or gels) rather than solids. For every system where it is desirable to control motion and vibration using a fluid with changing viscosity, a smart colloid may be an improvement in functionality and costs. Simplicity and more intelligence in the functionality are key features of this technology. Excellent features like fast response, simple interface between electrical power input and the mechanical power output, and controllability make smart material the next technology of choice for many applications (Stanway, 2004; Shen et al., 2006).

One significant discovery in ER field is the giant electrorheological (GER) effect made by Wen et al. (2003). The GER fluid consists of the nanoparticles of oxalate core with urea coating and silicone oil. Other than the breakthrough of the high yield stresses (100–200 kPa), there are other merits of such GER fluid including fast response time, reasonable sedimentation rate as well as high breakdown voltage, which open wide up for different industrial application such as car suspension and robotics.

Due to long inactivity duration of the particles in the fluid, undesired particle aggregation arises in concentrated GER fluid. As a result, the formation of sediments is facilitated. Hence, the poor re-dispersibility arises as a serious problem facing the technological application of the GER fluid. After years of intensive research, a fundamental correlation of the physic-chemical properties involved in the GER effect has not been adequately developed (Cho et al., 2004; Chen and Wei, 2006; Choi and Jhon, 2009). There have been reports on the ER enhancing sedimentation through the dispersing media, surfactants, agents, and none has systematically investigated the role of the dispersing liquid. This is largely due to the conventional wisdom that the dispersing liquid plays only a passive role in providing a large mismatch between the dielectric constants of solid particles and oil. It is known not to be the case for the recently discovered GER effect whose mechanism is based on the alignment of molecular dipoles through the hydrogen bonding network (Huang et al., 2006; Chen et al., 2010).

The aim of the present work is to investigate the role played by carrier liquid in the stability against particle aggregation and settling, the re-dispersibility of concentrated GER suspension and their GER effect. We systematically studied different types of oils including the synthetic oil and mineral oil influencing its colloidal stability by direct observation. The rheological behavior of the GER fluid upon electric field application was also investigated. The results will allow us to determine the relationship of the particle interaction with the carrier liquid. With the less sedimentation effect upon the tested carrier liquid, we investigated its GER performance in a modified mono tube damper for application usage.

## MATERIALS AND METHODS
### SYNTHESIS OF GER SAMPLES

Our GER particles, $BaTiO(C_2O_4)_2 + NH_2CONH_2$, were fabricated by the modification of Kudaka method in our previous work

(Gong et al., 2008). Barium chloride solution and oxalic acid solution were separately dissolved in distilled water at 65°C. Titanium tetrachloride solution was then slowly added to the prepared oxalic acid solution under ultrasonic. Both the solutions were then mixed immediately in an ultrasonic bath at 65°C to prevent hydrolysis. Nanometer sized core-shell particles were formed at this stage. Addition of urea, as an ER promoter, to the mixed solution led to the formation of a white colloid. The solution was drastically cooled to room temperature. The precipitate was washed, filtered, and then vacuum-dried at 100°C for 2 h to remove all traces water. All the chemicals aforementioned including two oil samples (white mineral oil and liquid paraffin) were supplied by Sigma Aldrich Chemical Company. Dimethyl terminated and hydrogenated silicone oils were supplied by Dow Corning. All the oils were dried at 120°C for 2 h before the experiment to avoid moisture. GER particles and each of the different oil samples were homogenized in a high-speed grinding mill for at least an hour. Consideration on longer mixing time is needed especially for the mineral oil samples. Concentration of the sample fluids can be denoted as the amount of oil, in units of milliliter, mixed with each gram of GER particles. Hence 0.5 means 10 g of GER particles mixed with 5 ml of oil.

## GER PARTICLE MEASUREMENT

JEOL-6700F scanning electron microscopy (SEM) with a target acceleration voltage of 5 kV was used to visualize the morphology of the GER particles. A SEM sample was prepared by dispersing 10 mg particle in 2 ml ethanol by ultrasonication. A drop of the suspension sample was transferred to a wafer for volatilization. The sample on the wafer was then gold-coated to enhance the electrical conductivity for the SEM. The size distribution of the batch samples was confirmed by PAnalytical X-Ray diffractometer (X'pert Pro).

## RHEOLOGICAL DATA COLLECTION

A circular-plate type viscometer (Haake RS1), with an 8 mm diameter rotating disk and a gap of 1 mm between the rotor and stator, was used to perform the rheological measurements. The step signals for driving the DC high-voltage source (SPELLMAN SL300) were generated by a functional generator (PM 5315, Philips). Software package RHEOWIN was used to collect experimental data. A 50 s square voltage pulse was applied to the sample with each measurement repeated at least three times due to their reproducibility and repeatability. Shear stress as a function of time was measured at a very low shear rate ($0.1 \, s^{-1}$). Yield point was reached when a stress-time curve changed its slope to be flat after an abrupt increase at the beginning of turning on the field. The yield stress at a given field was taken to be the maximum of the shear stress in the corresponding time span. The measurements were performed 24 h after dispersion at 24°C.

## SEDIMENTATION DATA COLLECTION

We have measured the sedimentation ratio of four GER fluid samples prepared with same GER particle concentrations of 0.5. Immediately after homogenization, the GER samples were put in a graduated square tube of constantly equal size under ambient gravity conditions. Over time, the settlement level becomes visible as an increasingly sharp interface between the concentrated phase

at the bottom and the diluted phase on the top. The homogeneity was determined as a percentage ratio of visible settlement level over the filling level with equation below.

$$\text{Sedimentation ratio} = \frac{\text{Opaque phase GER fluid}}{\text{Opaque phase} + \text{Clear oil phase}}$$

Re-dispersing measurements were carried out on the samples from the sedimentation tests (settling time of 12 weeks). The remixing was done with a planetary shaker, which homogenizes the sample by circular motions of the glass tube. The remixing behavior was determined by the time needed for full homogenization while the vortex-shaker was set to a constant value.

## GER DAMPER DATA COLLECTION

The modified GER damper, filled with hydrogenated silicone oil-GER fluid, was first compressed quasi-steadily with a loading speed of 200 mm/min on the UTM SINTECH 10/D. During the experiments, the UTM was first turned on to let the crosshead compress the piston rod of the GER damper, and then the electric field was applied to the GER duct when the displacement was about 10 mm. This procedure can avoid device damage as the static yield stress of the GER fluid is very large to cause jamming of the duct We collect the data right after the mixing of the GER fluid and the assembly of the damper. We then left the damper inactive for 3 months and collect data right after inactivity and re-dispersed phase.

## RESULTS
### GER PARTICLE CHARACTERIZATION

The dried GER particles were analyzed by SEM imaging (**Figure 1**). The particles fabricated are averagely spherical in shape with a regular diameter of 40 nm. X-ray diffractometer showed that the most frequent radius of the size distribution by volume is 15 nm and the average radius is 22 nm with a relative standard deviation of 51.97%. We then confirmed our fabricated GER particles match the description from Wen et al. (2004).

### RHEOLOGICAL DATA

The rheological effect of four types of oil samples from two different sources (**Table 1**) were measured by Haake RS1. Polysiloxanes from synthetic oil are semi-organic polymers and copolymers containing an inorganic backbone of repeating silicon-oxygen units and organic side chains substituted on the silicon atom along the polymer chain. Mineral oil consisted predominantly of carbon and hydrogen, which is non-polar.

With mixture concentration of 0.5, two extreme visual appearances such as sol-like and clay-like textures were obtained in the samples. When designing an ER application device, the zero-field viscosity of the mixture is always an important factor to be considered. The amplifying ratio is a constant (mixture viscosity divided by oil viscosity), which represented the GER particle interaction with oil. The highest amplifying ratio was credited to white mineral oil, which was meant to have the most unfavorable particle–oil interaction among the tested samples.

To give a simple physical picture of how the particle–oil interaction can play a crucial role in the GER effect, we measured the relevant yield stress (**Figure 2**) as a function of applied electric field.

## SEDIMENTATION AND RE-DISPERSING EFFECT

**Figure 3** showed the sedimentation ratio of the GER fluid samples within 3 months. It can be seen that the sedimentation rate for white mineral oil carrier liquid decreased initially, for several weeks, and then approached a stable asymptotic value after that. The rest of the tested samples were yet to be sedimented with significant optimum value in 3 months. The most relevant parameter for industrial application is the re-dispersing behavior, which could be improved significantly with hydrogenated silicone oil carrier liquid. As shown in the plotted result, the expended energy necessary for re-homogenization after 3 months of settlement is 15% less than the white mineral oil carrier liquid. Practically, the settling behavior of the hydrogenated silicone oil-GER fluid became less critical for industrial application where fast access to the full performance, i.e., damper, is necessary even after a long phase of inactivity.

## GER DAMPER TESTING RESULT

According to **Figures 2** and **3**, we can see that the best ER effect and sedimentation stability according to their measured characteristic was credited to hydrogenated silicon oil-GER fluid. An GER application test was carried out on a modified commercially available semi-active monotube damper in shear-mode. As shown in **Figure 4**, we measured the damping force performance of the prototype-test before, after inactivity and re-dispersing phase.

## DISCUSSION

In this study, we investigated the role played by carrier liquid in the stability against particle aggregation and settling, the re-dispersibility of concentrated GER suspension and their GER effect. Our results show that the hydrogen bondings in the carrier oil enhance the particle–liquid interaction that slowed down sedimentation, enhance easiness of re-dispersibility and maintained good GER performance.

With reference to the results presented in **Table 1**, some interesting observation found that although hydrogenated silicone oil-GER fluid were having similar mixture viscosity and amplifying ratio with the paraffin oil-GER fluid, the visual appearances are in both extreme. The mixing time needed for the latter samples were two times longer than hydrogenated silicone oil-GER fluid as it appeared that the GER particles in paraffin oil did not spread out evenly. There are many examples of such systems in which non-wetting are a known factor for leading to the same texture as observed in the mineral oil-ER fluid (Conrad and Sprecher, 1991; Shen et al., 2009). Different oil structure can play a significant role in the initial viscosity as well as in the visual appearances of the sample mixtures. The latter is suggestive of a strong interaction between the solid particles and oils.

In order to further connect the experimental results aforementioned, we found two motivating point. First, the results violated the norm for conventional ER fluid that high zero-field viscosity is usually accompanied by high yield stress (Halsey, 1992). Second, the plotted graphs contrary the conventional ER fluids that are only sensitive to the complex dielectric constant of the oil (Ma

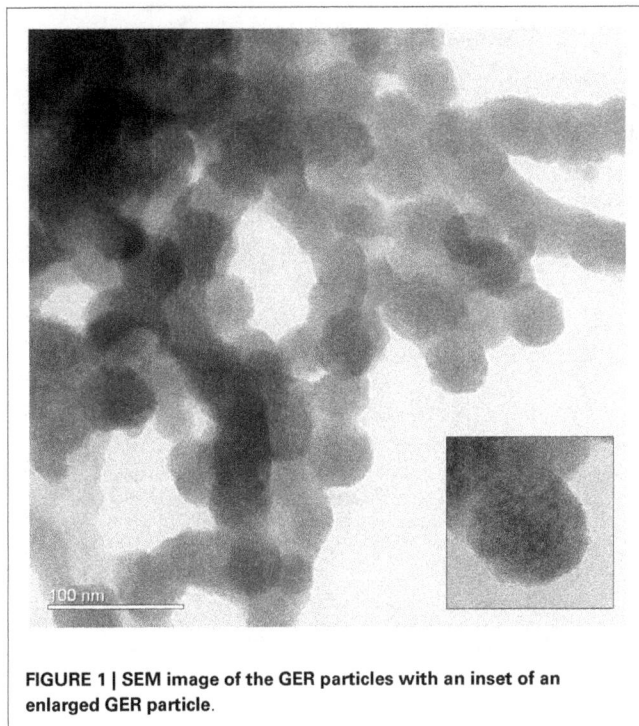

**FIGURE 1 | SEM image of the GER particles with an inset of an enlarged GER particle.**

Table 1 | Rheological data of various oil-GER fluids.

| Source | Oil type | Chemical formulae | Oil viscosity (m.Pas) | Mixture viscosity (m.Pas)[a] | Visual appearance | Amplifying ratio |
|---|---|---|---|---|---|---|
| Synthetic oil | Dimethyl terminated silicone oil | $H_3C\left[Si(CH_3)_2-O\right]_n Si$ | 12 | 169 | Liquid-like | 14.1 |
| | Hydrogenated silicone oil | $H_3C\left[Si(CH_3)(H)-O\right]_n Si$ | 20 | 400 | Liquid-like | 20 |
| Mineral oil | White mineral oil | $C_{25}H_{43}NO_3$ | 4 | 359 | Clay-like | 90 |
| | Liquid paraffin | $C_nH_{2n+2}\ n = 16 \sim 24$ | 20 | 477 | Clay-like | 23.85 |

[a] Mixture volume concentration = 0.5.

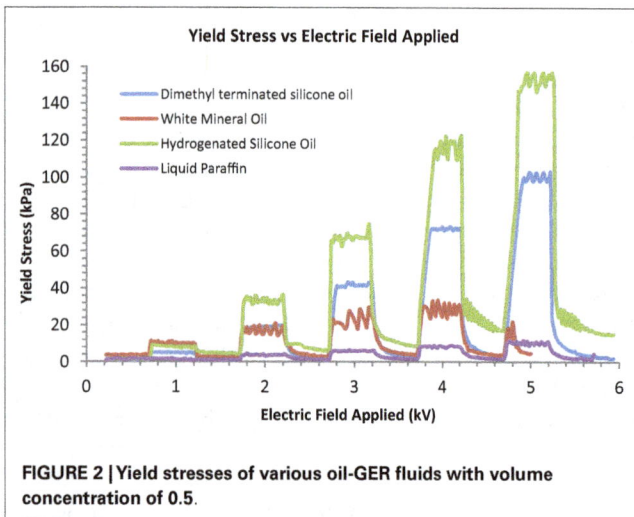

FIGURE 2 | Yield stresses of various oil-GER fluids with volume concentration of 0.5.

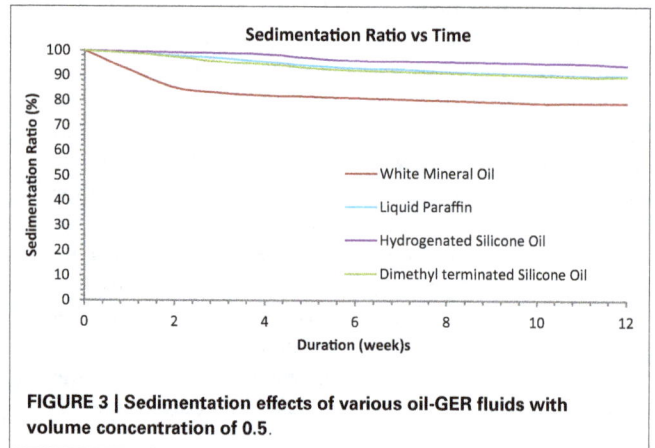

FIGURE 3 | Sedimentation effects of various oil-GER fluids with volume concentration of 0.5.

et al., 2003). **Figure 2** proofed that GER fluids are oil-sensitive upon their GER effect. For example, with hydrogenated silicone oil, one can obtain a very significant GER effect, but with the same particles dispersed in liquid paraffin, the GER effect is trivial.

Sedimentation occurs in colloidal systems due to the density mismatch between the solid and fluid phases, accentuated by particles aggregation through van der Walls interaction between the particles and the non-favorable particle–solvent interactions. **Figure 3** presented the sedimentation rate of the GER particles in various oils. Summarizing the plotted results, the particle–oil interaction contributes massive impact in the sedimentation effect, i.e., bad particle–oil interaction in white mineral oil and good particle–oil interaction in hydrogenated silicone oil. For the bad interaction case, GER particles are phase separated from the oil and the aggregation between two solid are large even with electric field applied. Hence, there can be no yield stress or even induce arcing since the solid aggregates and are always separated by oil (mineral oil family). Arcing is not allowed in the GER fluid system since they will cause irreversible destroy to the fluid. This phenomenon is shown (**Figure 2**) at the applied electric field of 5 kV/mm with white mineral oil-GER fluid.

The anti-sedimentation characteristic exhibited by the hydrogenated silicone oil-GER fluid is attributable to the fact that the oil has a very favorable particle–solvent interaction. **Figure 5A** illustrated the hydrogen bond effectively prevented direct contact between the GER particles, thus minimizing their aggregation and sedimentation. The hydrogen bond acted as the dispersant by suppressing colloidal aggregation in the GER fluid, and therefore also slowed down the separation from the bulk, as revealed through macroscopic sedimentation experiment.

On the other hand for the explanation of GER effect, the surface tension between the GER particles and hydrogenated silicone oil-ER fluid is greatly reduced due to the mediating effect of the hydrogen atom, thus allowing the particles to disperse and to move close together upon the application of an electric field. High GER effect is attributed to the presence of oxalate groups in the core nanoparticles and to the non-uniformity of the urea coating with the modified hydrogen atom in the oil (**Figure 5B**). The formation

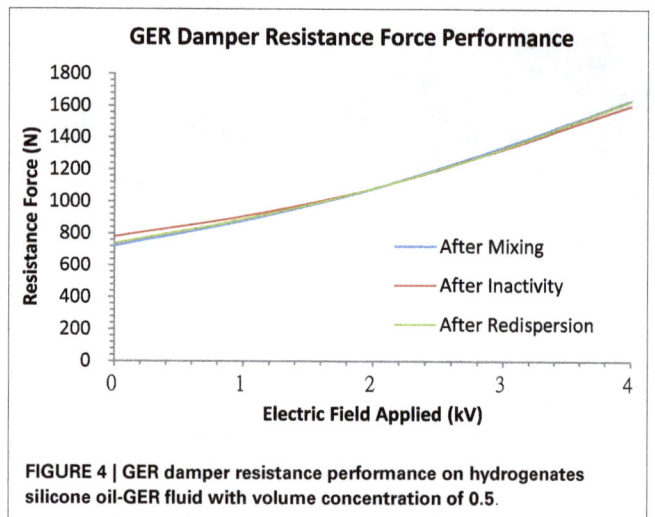

FIGURE 4 | GER damper resistance performance on hydrogenates silicone oil-GER fluid with volume concentration of 0.5.

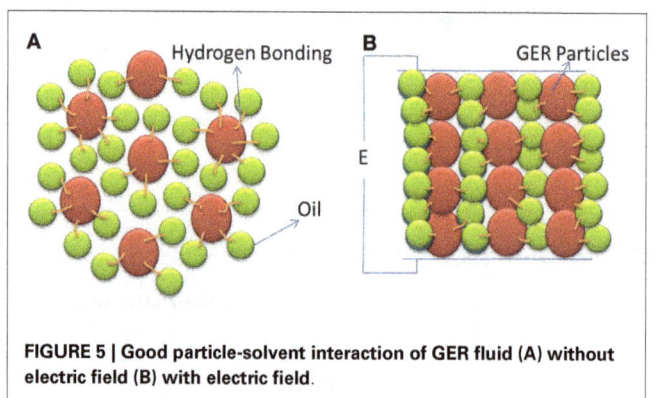

FIGURE 5 | Good particle-solvent interaction of GER fluid (A) without electric field (B) with electric field.

of the filaments, with an attendant lowering of the aligning field and a finite penetration length, can be attributed to the confinement effect exerted by oil chains. Urea molecules have a strong tendency to form hydrogen bonds with one another, and this influences their interactions with hydrophobic oil chains, which are incapable of forming hydrogen bonds.

**Figure 4** showed the quasi-steady loading curves of the GER damper under different electric field intensities. It can be seen that even when the electric field intensity is zero (E = 0 kV/mm), there is still a resistant force due to the gas chamber loaded with high pressure nitrogen gas. By increasing the electric field, the resistance force increased. However, the fluctuation of the force is very obvious, because the real flow inside the ER duct is not continuous. It is revealed in the result that sedimentation rate of the hydrogenated silicone oil-ER fluid works reasonably acceptable in GER damper with respect to their needed performance. Dynamic measurement was not performed as the damping force is still difficult to analyze. After some long duration test, the fluid and the sealing of the damper were affecting the output data due to the abrasiveness of the GER particles.

In summary, our study shows that the carrier liquid has synergistic effect on the GER particles. We demonstrated that good particle–oil interaction results in low sedimentation rate and high GER effect with good re-dispersibility after long inactivity duration. Our limitation of study is that our experiments have thus far been conducted only on quasi-steady loading GER damper performance. We did not construct well on the dynamic behavior of the GER fluid flow in the damper. Thus, we cannot discount the possibility of the colloidal stability enhancement during the dynamic fluid flow.

## ACKNOWLEDGMENTS

The authors acknowledge the financial support provided by the Hong Kong NSFC/RGC joint grant of N_HKUST601/11.

## REFERENCES

Chen, S., Huang, X., van der Vegt, N. F. A., Wen, W., and Sheng, P. (2010). Giant electrorheological effect: a microscopic mechanism. *Phys. Rev. Lett.* 105, 046001. doi:10.1103/PhysRevLett.105.046001

Chen, S., and Wei, C. (2006). Experimental study of the rheological behavior of electrorheological fluids. *Smart Mater. Struct.* 15, 377. doi:10.1088/0964-1726/15/2/018

Cho, M., Choi, H., and Ahn, W. (2004). Enhanced electrorheology of conducting polyaniline confined in MCM-41 channels. *Langmuir* 20, 202–207. doi:10.1021/la035051z

Choi, H., and Jhon, M. (2009). Electrorheology of polymers and nanocomposites. *Soft Matter* 5, 1562–1567. doi:10.1039/b818368f

Conrad, H., and Sprecher, A. (1991). Characteristic and mechanisms of electrorheological fluids. *J. Stat. Phys.* 64, 1073–1091. doi:10.1007/BF01048815

Gong, X., Wu, J., Huang, X., Wen, W., and Sheng, P. (2008). Influence of liquid phase on nanoparticle-based giant electrorheological fluid. *Nanotechnology* 19, 165602. doi:10.1088/0957-4484/19/16/165602

Halsey, T. (1992). Electrorheological Fluids. *Science* 258, 761–766. doi:10.1126/science.258.5083.761

Huang, X., Wen, W., Yang, S., and Sheng, P. (2006). Mechanisms of the giant electrorheological effect. *Solid State Commun.* 139, 581–588. doi:10.1016/j.ssc.2006.04.042

Ma, H., Wen, W., Tam, W. Y., and Sheng, P. (2003). Dielectric electrorheological fluids: theory and experiments. *Adv. Phys.* 343–383. doi:10.1080/0001873021000059987

Shen, C., Wen, W., Yang, S., and Sheng, P. (2006). Wetting-induced electrorheological effect. *J. Appl. Phys.* 99, 106104. doi:10.1063/1.2199749

Shen, R., Wang, X., Lu, Y., Wang, D., and Sun, G. (2009). Polar molecule dominated electrorheological fluids featuring high yield stresses. *Adv. Mat.* 22, 4631–4635. doi:10.1088/0953-8984/22/32/324105

Stanway, R. (2004). Smart fluid: current and future developments. *Mat. Sci. Technol.* 20, 931–939. doi:10.1179/026708304225019867

Wen, W., Huang, X., and Sheng, P. (2004). Particle size scaling of the giant electrorheological effect. *Appl. Phys. Lett.* 85, 299–301. doi:10.1063/1.1772859

Wen, W., Huang, X., Yang, S., Lu, K., & Sheng, P. (2003). The giant electrorheological effect in suspensions of nanoparticles. *Nat. Mater.* 2, 727–730. doi:10.1038/nmat993

**Conflict of Interest Statement:** The authors declare that the research was conducted in the absence of any commercial or financial relationships that could be construed as a potential conflict of interest.

# 19

# Atomistic model of metal nanocrystals with line defects: contribution to diffraction line profile

*Alberto Leonardi and Paolo Scardi\**

Department of Civil, Environmental and Mechanical Engineering, University of Trento, Trento, Italy

**Edited by:**
Simone Taioli, Bruno Kessler
Foundation, Italy

**Reviewed by:**
Andrea Piccolroaz, University of
Trento, Italy
Shangchao Lin, Florida State
University, USA

**\*Correspondence:**
Paolo Scardi, Department of Civil,
Environmental and Mechanical
Engineering, University of Trento, Via
Mesiano 77, Trento 38123, Italy
e-mail: paolo.scardi@unitn.it

Molecular Dynamics (MD) was used to simulate cylindrical Pd and Ir domains with ideal dislocations parallel to the axis. Results show significant discrepancies with respect to predictions of traditional continuum mechanics. When MD atomistic models are used to generate powder diffraction patterns, strong deviations are observed from the usual paradigm of a small crystal perturbed by the strain field of lattice defects. The Krivoglaz–Wilkens model for dislocation effects of diffraction line profiles seems correct for the screw dislocation case if most parameters are known or strongly constrained. Nevertheless the practical implementation of the model, i.e., a free refinement of all microstructural parameters, leads to instability. Possible effects of the experimental practice based on Line Profile Analysis are discussed.

**Keywords: x-ray powder diffraction, line profile analysis, nanocrystalline materials, dislocations, molecular dynamics simulation**

## INTRODUCTION

Line broadening has been used since the dawn of X-ray diffraction (XRD) to study the microstructure of crystalline phases. Besides the most basic information on the size and shape of crystalline domains, which in the simplest form is provided by the well-known Scherrer formula (Scherrer, 1918; Klug and Leroy, 1974), lattice defects have also been extensively studied [e.g., see (Warren, 1990; Krivoglaz, 1996; Snyder et al., 1999; Mittemeijer and Scardi, 2004)].

Analytical techniques to extract information from diffraction line profiles range from simple integral-breadth methods [see Scardi et al. (2004) for a recent review], including accurate studies of isolated peak tails (Wilson, 1955; Groma, 1998; Groma and Székely, 2000) to more complex Fourier analysis (Warren and Averbach, 1950; Warren, 1990), underlying the most recent Whole Powder Pattern Modeling (WPPM) approach (Scardi and Leoni, 2002; Scardi, 2008). All methods for studying dislocations rely on the early studies by Wilson, Krivoglaz, and Wilkens (Wilson, 1955; Wilkens, 1970a,b; Krivoglaz et al., 1983). The last author, in particular, provided an analytical expression for the Fourier Transform (FT) of the diffraction line profile of crystalline domains containing dislocations (Wilkens, 1970a,b)

$$A^D_{\{hkl\}}(L) = \exp\left[-2\pi^2 s^2 L^2 < \varepsilon^2_{\{hkl\}}(L) >\right]$$
$$= \exp\left[-\frac{\pi b^2}{2}\rho \overline{C}_{hkl}s^2 L^2 f^*\left(\frac{L}{R_e}\right)\right] \quad (1)$$

where $L$ is the Fourier length (distance between scattering centers) and $s$ the scattering vector ($s = 2\sin(\theta)/\lambda$). Main parameters are the average dislocation density ($\rho$) Burgers vector modulus ($b$) and effective outer cut-off radius ($R_e$), which is related to the extension of the effects of the dislocation strain field and, more generally, to dislocation interaction. The $f^*$ is a smooth function

of $L/R_e$ obtained by Wilkens in a heuristic way, to grant integrability of Eq. 1 (Wilkens, 1970b). The anisotropy of the elastic medium and of the dislocation strain field, which depends on the specific dislocation type (e.g., edge or screw) and slip system, is accounted for by the anisotropy or contrast factor $C_{hkl}$. For powder diffraction, an average is used for all equivalent components of a diffraction line profile, which depends on crystal symmetry, and more specifically on the Laue group; given the elastic tensor components, $C_{ij}$ and $S_{ij}$, the average contrast factor, $\overline{C}_{hkl}$, can be calculated for any desired crystalline phase and slip system $\langle uvz\rangle$ {hkl} [e.g., see Martinez-Garcia et al. (2009).

The Krivoglaz–Wilkens approach, using Eq. 1 or similar approximations, is useful and easily implemented in the experimental data analysis; it has been extensively used in materials science (e.g., see work by Klimanek and Kuzel (1988), Kuzel and Klimanek (1988), Kuzel and Klimanek (1989), Ungar et al. (1998), Ungar (2008), Scardi and Leoni (2002), Scardi et al. (2007)] even if, as a matter of fact, it has never been fully validated. A few studies (Kamminga and Delhez, 2000; Kaganer and Sabelfeld, 2011;Kaganer and Sabelfeld, 2014) have tested Eq. 1 against numerical simulations, but the latter were based on the same continuum mechanics expressions for the dislocation strain field underlying Eq. 1, and in any case referred to rather idealized microstructures. An experimental validation, e.g., by Transmission Electron Microscopy (TEM) is not straightforward: quantitative TEM evidence is hard to obtain when the dislocation density is in the range of interest to powder diffraction (typically, above $10^{14}\,\text{m}^{-2}$), especially after extensive plastic deformation and in small domains.

A useful and quite different point of view can be provided by Molecular Dynamics (MD). MD simulations can realistically describe nanocrystalline domains and clusters with the detail of atomistic models, which can be used to generate powder diffraction patterns from known, designed microstructures (Bulatov et al., 1998; Jacobsen and Schiotz, 2002; Yamakov et al., 2002; Yamakov et al., 2004; Li et al., 2010). The present work is a first step toward this direction, to shed some light on the validity of Eq. 1 and analytical methods relying on it, and the more general meaning of diffraction from defected polycrystalline materials.

In the present paper, the strain field in cylindrical nanocrystals containing screw or edge dislocations is first discussed, moving from the traditional continuous mechanics (stress–strain) description to atomistic simulations; strain effects on XRD patterns from simulated single-crystals and powders are then discussed in terms of broadening of the line profiles. In the last section, a state-of-the-art powder diffraction analysis is employed to investigate the simulated patterns, to assess validity of the Krivoglaz–Wilkens approach also considering the additional effects of surface relaxation.

## MATERIALS AND METHODS

Pd and Ir nanocrystals containing line defects were simulated by MD using the LAMMPS code [Large-scale Atomic/Molecular Massively Parallel Simulator – (Plimpton, 1995)], implementing geometrical conditions similar to those underlying the derivation of Eq. 1, i.e., straight dislocation lines in cylindrical regions. First step was the generation of isolated edge or screw dislocations in bulk microstructures, followed by a stabilization of defect-containing microstructures by the Embedded Atom Method (Daw and Baskes, 1984; Foiles et al., 1986; Sheng et al., 2011). **Figure 1** illustrates geometrical details of the generation process. The starting model of microstructure with screw dislocation was obtained by shifting the atomic coordinates of a perfect single crystal by $u_z = \frac{b\theta}{2\pi}$ along the [$hh0$] cylinder axis; to generate edge dislocations, instead, two (110) half-planes were removed along the cylinder axis. Periodic Boundary Conditions (PBCs) for the screw dislocation line were applied along the [$hh0$] axis (**Figure 1C**), whereas they were applied both along the $\left[\overline{hh}2h\right]$ dislocation axis and along $\left[\overline{h}h0\right]$ for the edge dislocation (**Figure 1A**). The initial microstructures were equilibrated for 1ns using the Langevin thermostat at 300 K combined with a constant Number of atom, Volume, and Energy (constant NVE) integration with 1fs time step. Next to the equilibration, a time trajectory was generated recording sequences of 100 microstructures at 1ps time intervals.

A time-average of the arrangement in space of the atomic positions was computed along the time trajectory, so to cancel the thermal effects out (Leonardi et al., 2011); this Time Averaged Microstructure (TAM) was then used to calculate powder diffraction patterns by the Debye scattering equation (DSE) (Debye, 1915; Gelisio et al., 2010), using the atomic coordinates in the cylindrical regions shown in **Figure 1** (black line wireframe). As in the Krivoglaz–Wilkens approach, dislocations were always straight lines running parallel to the cylinder axis. Effects related to the position of the dislocation line were also considered, randomly displacing the cylinder region of interest and then considering corresponding powder patterns and averages. For comparison, similar procedures were carried out for the same cylindrical regions without any line defect. To assess the role of domain size and surface, few other regions were also considered for different domain shapes (cube and sphere) and for smaller systems. In particular a $D = 20$ nm, $H = 28.7$ nm cylindrical domain was generated,

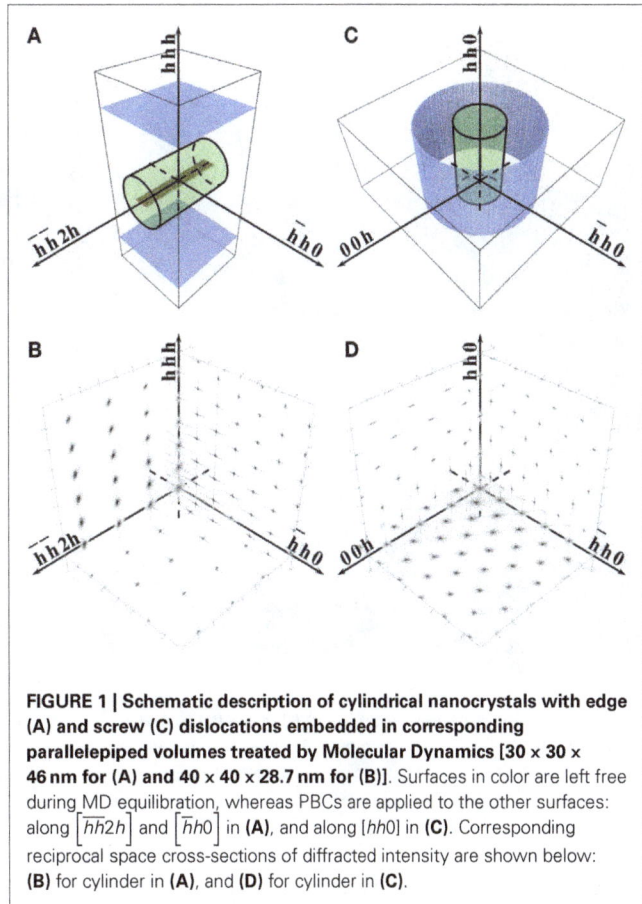

FIGURE 1 | Schematic description of cylindrical nanocrystals with edge (A) and screw (C) dislocations embedded in corresponding parallelepiped volumes treated by Molecular Dynamics [30 × 30 × 46 nm for (A) and 40 × 40 × 28.7 nm for (B)]. Surfaces in color are left free during MD equilibration, whereas PBCs are applied to the other surfaces: along $\left[\overline{hh}2h\right]$ and $\left[\overline{h}h0\right]$ in (A), and along [$hh0$] in (C). Corresponding reciprocal space cross-sections of diffracted intensity are shown below: (B) for cylinder in (A), and (D) for cylinder in (C).

both as a smaller version of that shown in **Figure 1C** ($D = 40$ nm, $H = 28.7$ nm), and carved out from it.

## RESULTS

**Figure 2** shows the isotropic (volumetric) and deviatoric strain fields for Pd bulk nanocrystals containing edge or screw dislocations. As expected, the edge case gives compressive/tensile (**Figure 2A**) and deviatoric (**Figure 2B**) strains, whereas the screw dislocation gives a predominantly deviatoric strain (**Figure 2D**). In both cases, the dislocation line splits in partials, according to the known reaction: $\frac{1}{2} \langle 110 \rangle \rightarrow \frac{1}{6} \langle 211 \rangle + \frac{1}{6} \langle 12\overline{1} \rangle$, a feature especially visible for the edge dislocation. The separation distance between edge partials (about 4 nm) is sensibly larger than predicted by simple considerations on surface energy of the faulted region between partials (~1 nm) (Hull and Bacon, 1965), but is well in agreement with other, more recent literature values (Hunter et al., 2011). Also the screw dislocation is not exactly as an ideal straight line, even though splitting and other deviations are less pronounced than in the edge case.

Extended dislocations have a complex effect on diffraction, well beyond a simple broadening of the diffraction line profiles. **Figure 3** shows the XRD intensity distribution on three orthogonal cross-sections of the reciprocal space (RS) for cylindrical nanocrystals containing edge (**Figure 3**, $D_e$, $E_e$, and $F_e$) or screw (**Figure 3**, $D_s$, $E_s$, and $F_s$) dislocations. It is also shown the intensity scattered by the corresponding perfect nanocrystal, i.e., the same

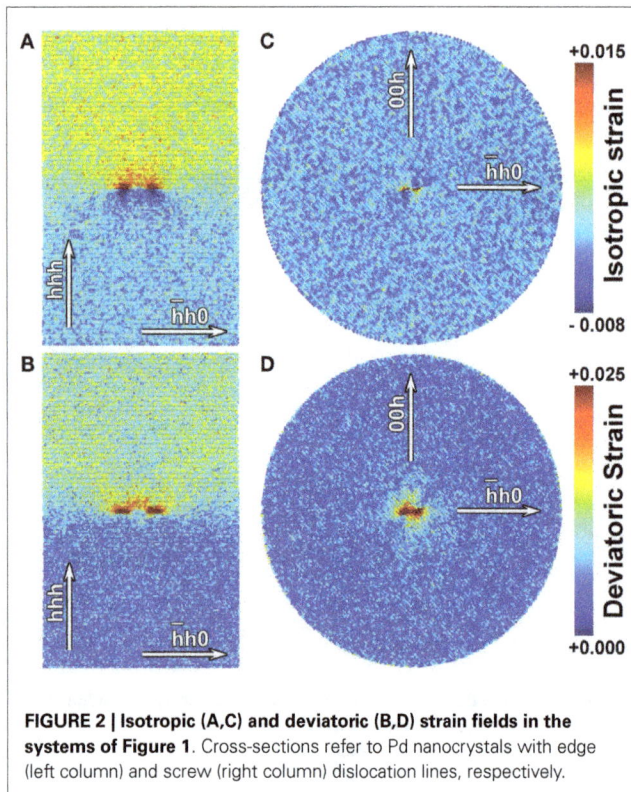

**FIGURE 2 | Isotropic (A,C) and deviatoric (B,D) strain fields in the systems of Figure 1**. Cross-sections refer to Pd nanocrystals with edge (left column) and screw (right column) dislocation lines, respectively.

cylinder with no line defect and atoms in perfect, ideal positions (**Figure 3**, $A_e$, $B_e$, $C_e$ and $A_s$, $B_s$, $C_s$ for edge and screw, respectively). As expected, the strain field gives a growing effect with the distance from the RS origin. Depending on the Miller indices the intensity distribution around points is differently affected, and in some cases splits in two distinct regions. This feature is clearly observed in the edge case, along the $\left[\overline{h}h0\right]$ direction, as an effect of the compressive and tensile strains, respectively above and below the dislocation slip plane (cf. **Figures 1** and **2**). Strain and faulted region in between the two partials affect in a rather complex way the distribution of intensity around all points in RS. Although mediated by the average over different orientations of the cylindrical domains, this complexity is expected to appear also in the corresponding powder patterns, as shown further below.

Another view-point on the atomic displacement effect of extended dislocations is provided by the Pair Distribution Function, shown in **Figure 4**. The same plot for perfect cylinders (i.e., domains with no line defects) gives an array of δ functions at all atom pair distances. These infinitely narrow bars, marked at the bottom of the plot of **Figure 4**, are broadened by the strain field of the dislocation lines, thus producing a sequence of distributions. Distribution widths for edge and screw cases are comparable, although the former gives additional peaks caused by the faulted region between the dislocation partials (**Figures 1** and **2**), which is responsible for a fraction of non-cubic sequence of atomic layer stacking. Effects on the diffraction pattern from a powder of dislocated cylindrical nanocrystals are therefore expected to be quite strong and different for the edge and screw dislocation cases.

**Figure 5** shows the powder patterns simulated by the DSE from the TAM of cylindrical Pd nanocrystals ($D = H = 16\,\text{nm}$) with edge or screw dislocations, and corresponding ideally perfect ("crystallographic") nanocrystals. Despite the different orientation of the cylindrical nanocrystals (cf. **Figure 1**), the nanocrystal shape with equal height and diameter make the crystallographic powder patterns quite similar. Visible differences are caused by the dislocations. The screw case gives a predominant effect of broadening, whereas, the edge case gives broadening and shape effects, the latter caused by the non-cubic atomic layer stacking in the faulted region between edge partials. Such effects are stronger for Pd than Ir (**Figure 6**), as the separation between the partials, and therefore the extension of the stacking fault ribbon in the cylindrical microstructure, is larger for Pd than for Ir.

**Figure 7** shows the effect of position of the dislocation line. To be compatible with assumptions underlying Eq. 1, all dislocations were straight lines parallel to the cylinder axis. The position of the dislocation line has negligible effects in the case of screw dislocations, as a consequence of the mainly deviatoric strain field introduced in the TAM. Differences are quite strong for the edge dislocation case. Features originating from the more complex strain field of the edge type, and the stacking fault region associated to the two partials are always visible. It can only be noted that a pattern obtained by averaging those for different random positions of the edge line is quite similar to the pattern with the edge dislocation along the cylinder axis. However, even such average pattern is clearly affected by strong deviations from the pattern expected for an fcc metal phase.

Molecular Dynamics simulations can easily be extended to different nanocrystal sizes and shapes. **Figure 8** shows the effect of changing the shape of the nanocrystal, following the same procedure described above for cylinders. Powder patterns from nanocrystals containing edge dislocations always show more complex details than the simple broadening provided by Eq. 1. The screw case seems qualitatively closer to the expected effects of Eq. 1, although some peaks present visible splitting [like the (642) line (Wilson, 1955) just above $Q = 12$ in **Figure 8**]. For the screw case in **Figure 9**, we also show the effect of changing the size of the cylindrical nanocrystal: largest effects are caused by changing the diameter, as can be easily explained considering that changing diameter acts on two dimensions, i.e., on the extension of the cylinder base.

## DISCUSSION

The DSE powder patterns described so far can be considered as "experimental" diffraction data and analyzed by a state-of-the-art method, like WPPM (Scardi and Leoni, 2002; Scardi, 2008). Besides using Eq. 1 for refining the dislocation parameters (like ρ, $R_e$, and $\overline{C}_{hkl}$), WPPM can model nanocrystalline domains of virtually any size and shape (Leonardi et al., 2012), also considering the presence of stacking faults (Scardi and Leoni, 2002; Scardi, 2008) and other microstructural features responsible for line broadening effects (Scardi, 2008). For example, the complex effect of relaxation of the nanocrystal surface can be described by an additional "strain" profile component, with a corresponding FT given by

$$A^{SR}_{\{hkl\}}(L) = \exp\left[-2\pi^2 s^2 L^2 < \varepsilon^2_{\{hkl\}}(L) >\right]$$
$$= \exp\left[-2\pi^2 s^2 L^2 \Gamma_{hkl}\left(aL + bL^2\right)\right] \quad (2)$$

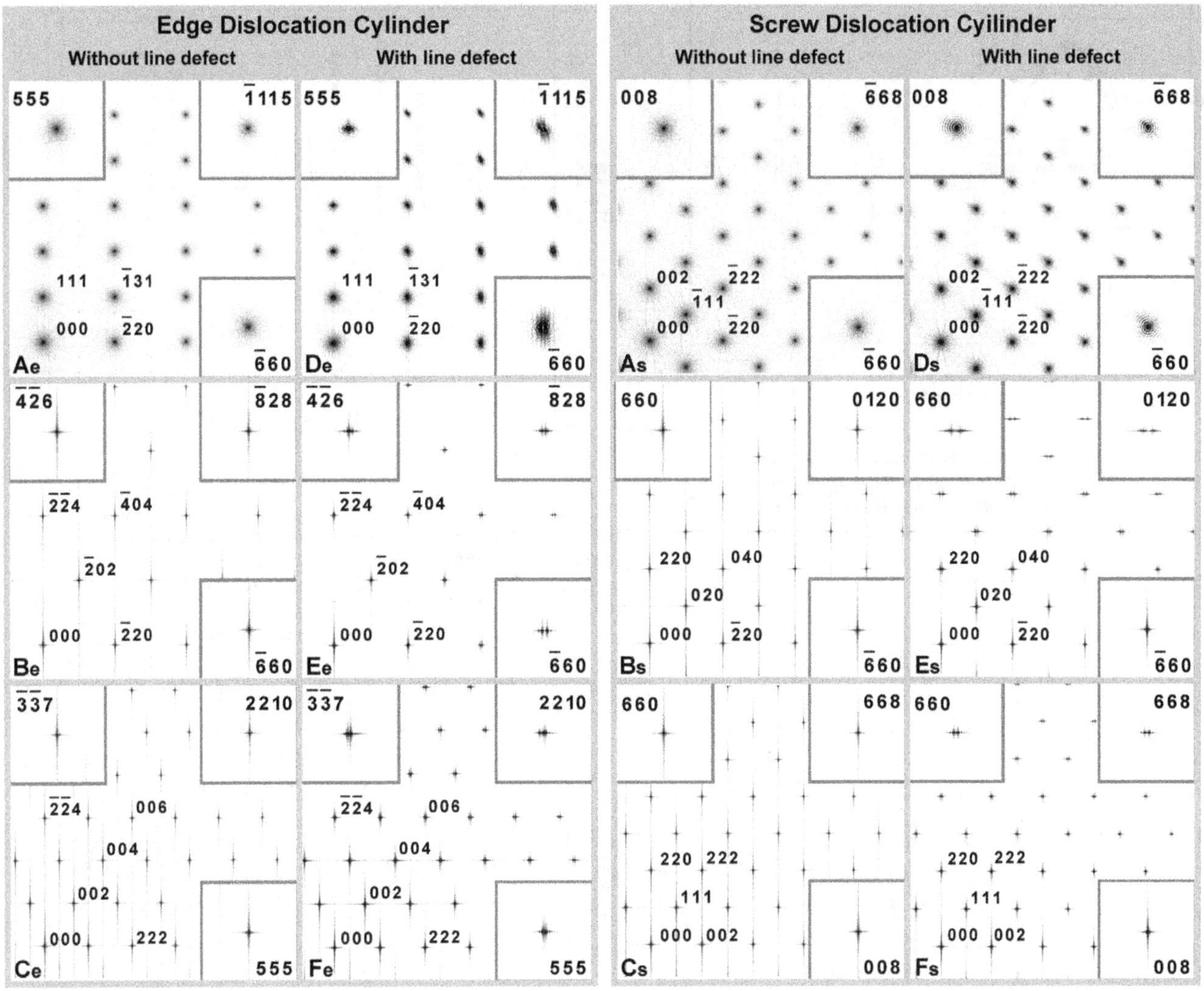

**FIGURE 3 | X-ray diffraction intensity distribution in reciprocal space**. Three different cross-sections are shown for cylindrical Pd nanocrystals ($D = H = 16$ nm) with edge dislocation ($D_e$, $E_e$, and $F_e$) and screw dislocation ($D_s$, $E_s$, and $F_s$). XRD intensity from corresponding cylinders without line defects are shown in ($A_e$, $B_e$, and $C_e$) and ($A_s$, $B_s$, and $C_s$). Refer to **Figure 1** for details on the cylindrical domains.

where $\Gamma_{hkl} = 1 + c(h^2k^2 + k^2l^2 + l^2h^2)/(h^2 + k^2 + l^2)^2 = 1 + c \cdot H^2$ accounts for strain anisotropy, referred to the Miller indices of the diffraction peak. Parameters $a$, $b$, $c$ can be adjusted to model the strain field and anisotropy.

The FT of the overall diffraction line profile, $A(L)$, can then be written as a product of all required FTs (Scardi, 2008):

$$A(L) = A_{\{hkl\}}^{D}(L) \cdot A_{\{hkl\}}^{S}(L) \cdot A_{\{hkl\}}^{SR}(L)\ldots \quad (3)$$

where $A_{\{hkl\}}^{S}(L)$ is the FT of the domain size effect, known in closed analytical form for a cylindrical shape (Langford and Louer, 1982). Therefore, in this specific case, besides refinable parameters of Eqs. 1 and 3 includes height ($H$) and base diameter ($D$) of the cylindrical domain, and adjustable parameters $a$, $b$, $c$ of Eq. 2. If necessary, the effect of stacking faults in reasonably low concentration can be described by a $A_{hkl}^{F}(L)$ component given by Warren's theory for faulted fcc metal structures (Warren, 1990; Scardi, 2008).

Results of **Figures 5–8** and relevant discussion point out the difficulty in modeling the pattern of a powder of cylindrical domains with edge dislocations. The traditional approach, underlying also Eq. 3, considers a perfect cylindrical fcc domain with defects; such a perturbation approach seems little convincing here, as it cannot account in any simple and accurate way for the faulted region between the two partials, with the corresponding hexagonal sequences, and the complex strain field, strongly dependent on split of the dislocation line in partials and their position inside the domain (cf. **Figure 7**). The screw dislocation case seems more "well-behaving," i.e., closer to the assumptions of Krivoglaz–Wilkens theory, mostly involving a line broadening effect. Therefore, in the following we focus on the screw case only, analyzing the corresponding powder patterns by the WPPM approach as described by Eq. 3.

**Figure 10** shows the modeling results for two different powders of cylindrical domains, respectively $D = 40/H = 28.7$ nm

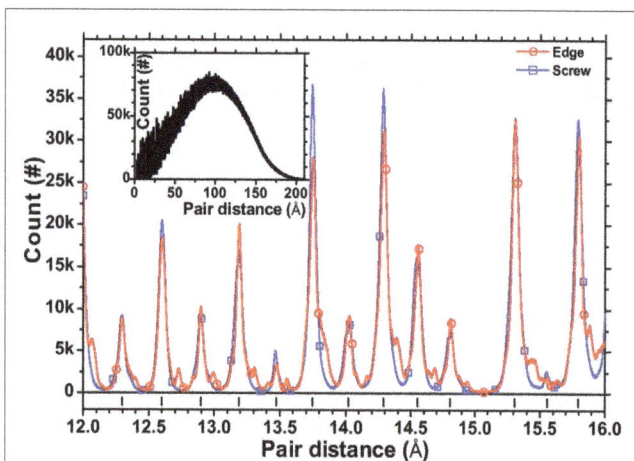

FIGURE 4 | A selected range of the distribution function of atom pair distances in the Time Averaged Microstructure of cylindrical Pd nanocrystals ($D = H = 16$ nm) with edge (red circular dot) and screw (blue square dot) dislocation lines. The inset shows the trend across the whole range of distances.

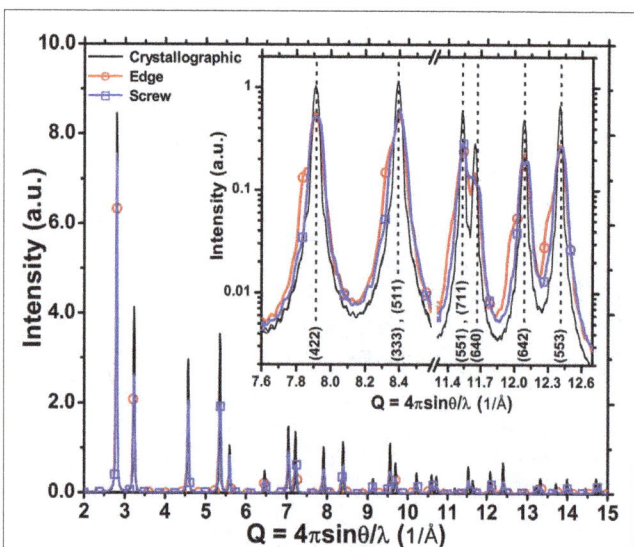

FIGURE 5 | X-ray powder diffraction pattern simulated by DSE from the time averaged microstructure of cylindrical Pd nanocrystals ($D = H = 16$ nm) with edge (red circular dot), screw (blue square dot) dislocations, and from the corresponding ideally perfect ("crystallographic") nanocrystals (black).

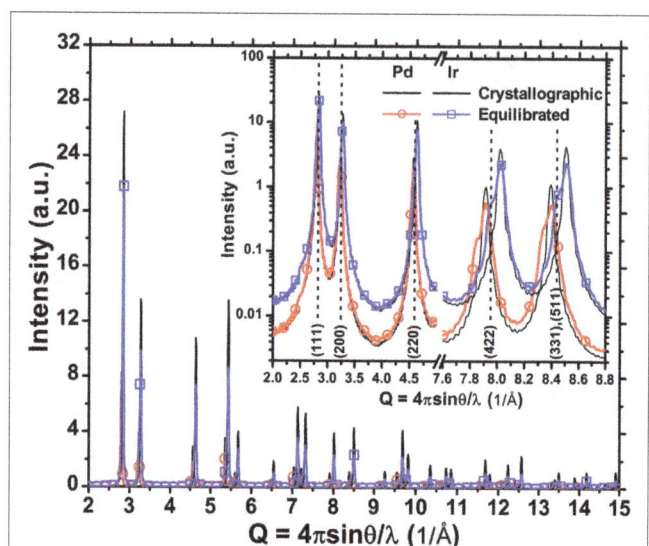

FIGURE 6 | X-ray powder diffraction pattern simulated by DSE from the ("equilibrated") TAM of cylindrical nanocrystals ($D = H = 16$ nm) with edge dislocations (line), and corresponding ideally perfect ("crystallographic") nanocrystals (dash). Results are shown for Pd (red circular dot) and Ir (blue square dot), having respectively larger and smaller separation distance between partial dislocations.

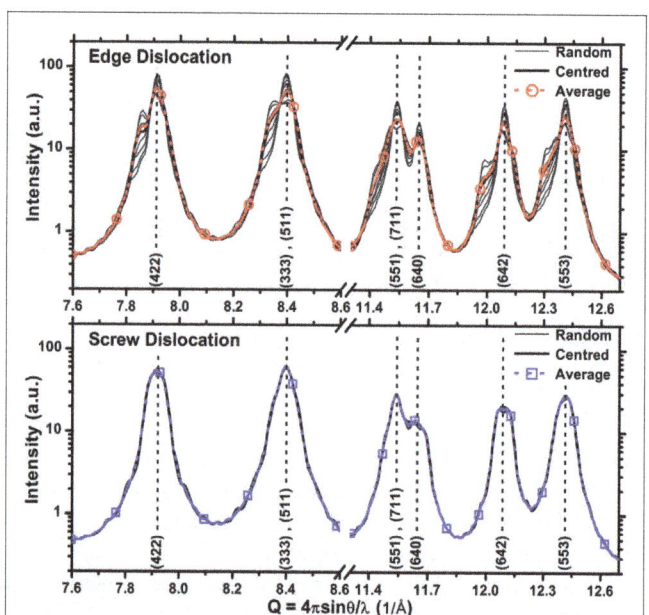

FIGURE 7 | X-ray powder diffraction pattern simulated by DSE from the ("equilibrated") TAM of cylindrical Pd nanocrystals ($D = H = 16$ nm) with edge (upper plot) and screw (lower plot) dislocation lines parallel to the cylinder axis, crossing the circular basis in different positions (center, random, and average).

(**Figures 10A–C**) and $D = 20/H = 28.7$ nm (**Figures 10D–F**). The result for ideal cylindrical domains (**Figures 10A,D**), i.e., with Pd atoms positioned according to ideal fcc structure, no dislocations, and no surface relaxation, is very good, as expected in case of domain size broadening effects only (Leonardi et al., 2012). Small deviations between DSE pattern and WPPM are expected, owing to the different hypotheses underlying DSE and WPMM, as the former is based on an intrinsically discrete, atomistic model, whereas, the last considers crystalline domains as ideal solid models, i.e.,

cylinders with a smooth surface [details can be found in Beyerlein et al. (2011)].

The modeling is still reasonably good when the atomistic model of the same cylindrical domain is equilibrated before, generating

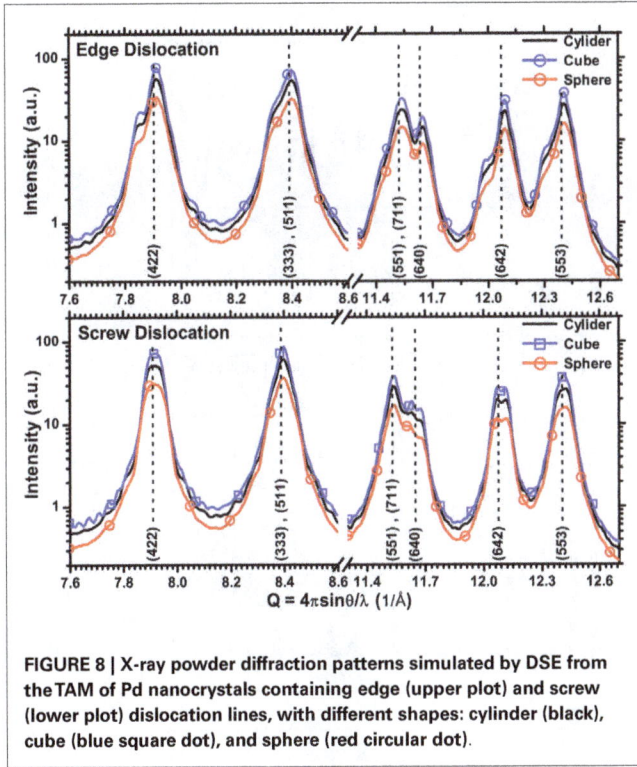

**FIGURE 8 | X-ray powder diffraction patterns simulated by DSE from the TAM of Pd nanocrystals containing edge (upper plot) and screw (lower plot) dislocation lines, with different shapes: cylinder (black), cube (blue square dot), and sphere (red circular dot).**

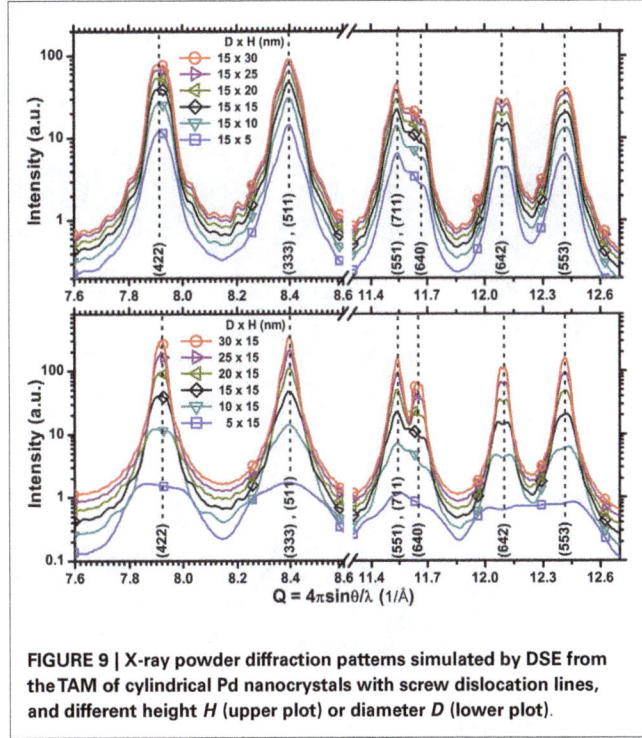

**FIGURE 9 | X-ray powder diffraction patterns simulated by DSE from the TAM of cylindrical Pd nanocrystals with screw dislocation lines, and different height $H$ (upper plot) or diameter $D$ (lower plot).**

the pattern by the DSE (**Figures 10B,E**). Energy minimization leads to a strain broadening effect caused by the relaxation of the free surface, which adds to the size broadening effect from the finite cylindrical domain. The effect is well represented by the model of Eq. 2. The WPPM result, instead, is much less satisfactory for the cylindrical domain containing a screw dislocation along the axis, showing marked and systematic deviations from the DSE pattern (**Figures 10C,F**): peak width is reasonably matched but details of the peak profile shape are definitely not reproduced. More in particular, the model of Eq. 1 is unstable: even if the contrast factor is fixed to the expected value for a screw dislocation (for the primary slip system of Pd, {111}⟨110⟩, $\overline{C}_{hkl} = A + B \cdot H^2 = 0.280476 - 0.64335 \cdot H^2$) when $\rho$ and $R_e$ are both allowed to vary, the last diverges (>1012 nm). In the case of the smaller cylinder ($D = 20/H = 28.7$ nm), even the cylinder height, when freely refined, tends to wrong (smaller) values. Such instability and drift of the refinement toward wrong values is partly due to the intrinsic correlation between parameters – quite clearly, between $\rho$ and $R_e$ – but is also a result of the complexity of the strain field, which is not fully captured by the model of Eq. 1.

The modeling improves if more parameters are fixed. Indeed, besides the contrast factor we can fix D and H to the model values (for the specific cylinder considered), and also set the dislocation density to the expected value, given by the ratio between dislocation length and cylinder volume: $\rho_t = H/[H \cdot \pi(D/2)^2] = [\pi(D/2)^2]^{-1} \text{m}^{-2}$. This gives values of $0.796 \times 10^{-15} \text{ m}^{-2}$ and $3.183 \times 10^{-15} \text{ m}^{-2}$, respectively for $D = 40/H = 28.7$ nm and $D = 20/H = 28.7$ nm cylinders.

Then the only microstructural parameter to be refined, besides unit cell parameters, is the effective outer cut-off radius. This is

how the refinements for the larger cylinder in **Figures 10C,F** were obtained. For a more robust convergence we refined together the patterns of both cylinders, without and with screw dislocation (**Figures 10B,C**, respectively), thus refining the same values of $a$, $b$, and $c$ in Eq. 3, together with $R_e$ for the cylinder with screw dislocation. The same procedure was repeated for the smaller cylinder (**Figures 10E,F**).

Best value of $R_e$ for the $D = 40/H = 28.7$ nm cylinder is 41.3(1) nm, quite close to twice the cylinder radius. This value is in close agreement with Wilkens model of Eq. 1, where line defects are assumed to be inside the so-called "restrictedly random dislocations regions," whose radius ($R_p$) is related to our definition of $R_e$ as $R_e \approx 2R_p = D$ [see Wilkens (1970b) for definitions, and Armstrong et al. (2006) for a more recent review on the validity of Eq. 1 and relation between $R_e$ and physical lengths of crystalline domains containing dislocations].

For the smaller cylinder, $D = 20/H = 28.7$ nm, $R_e$ is proportionally larger, 26.0(1) nm, but still not far from the expected value of $2R_p$. It is therefore verified the correctness of the hypotheses underlying the Krivoglaz–Wilkens model, although Eq. 1 can be unstable when trying to refine all parameters, especially those which correlate more strongly. The agreement with the model hypotheses increases with the domain size, and we can expect the model to be exact in the limit of very large diameters. Deviations in smaller domains are partly related to the non-ideality of dislocations, but also reflect the effect of the dislocation core region, which in the Krivoglaz–Wilkens model is excluded by an inner cut-off radius (Wilkens, 1970a,b).

Strain effects on the diffraction line profile analysis can be described by an r.m.s. strain (or microstrain) plot, originally proposed by Warren and Averbach (1950), Warren (1990).

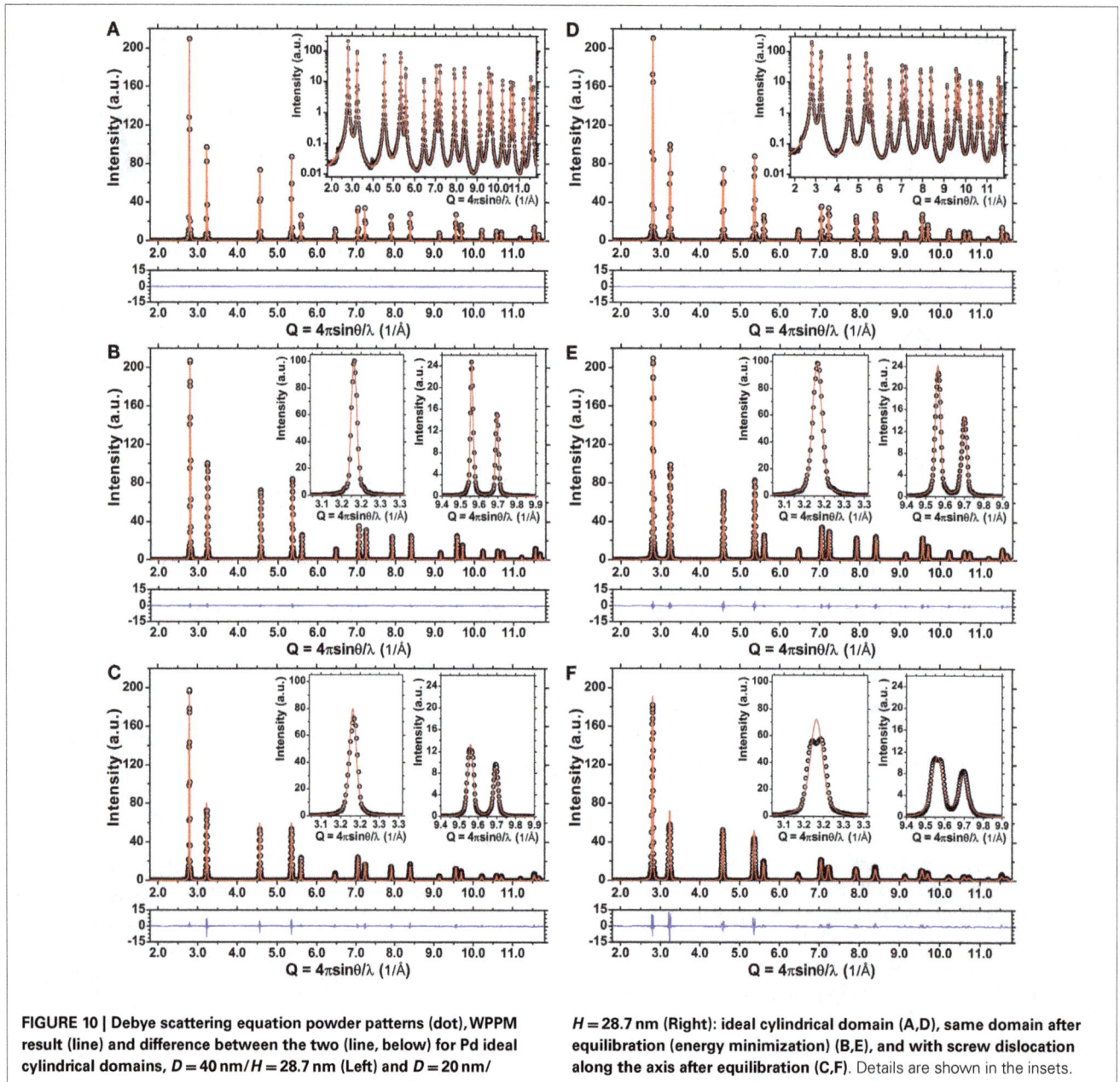

**FIGURE 10 | Debye scattering equation powder patterns (dot), WPPM result (line) and difference between the two (line, below) for Pd ideal cylindrical domains, $D = 40$ nm/$H = 28.7$ nm (Left) and $D = 20$ nm/** $H = 28.7$ nm (Right): ideal cylindrical domain (A,D), same domain after equilibration (energy minimization) (B,E), and with screw dislocation along the axis after equilibration (C,F). Details are shown in the insets.

This plot, shown in **Figure 11** for both models of cylindrical domains containing a screw dislocation, provides the r.m.s. strain $(\langle \varepsilon^2_{\{hkl\}} (L) \rangle^{1/2}$, the width of the strain distribution) over different distances $L$ inside the crystalline domain, taken along the scattering vector direction; as a consequence, the microstrain depends on the $\langle hkl \rangle$ crystallographic direction. **Figure 11** shows trends along $\langle 111 \rangle$, $\langle 200 \rangle$, and $\langle 220 \rangle$ for both strain components, respectively due to the dislocation (line) and to the grain boundary relaxation (dash). Strain values increase for smaller domain sizes, following the corresponding increase in dislocation density [from Eq. 1, $\langle \varepsilon^2_{\{hkl\}} (L) \rangle^{1/2} \propto \sqrt{\rho}$], whereas the effect of the grain boundary strain decreases with increasing diameter.

To better assess the effect of the cylindrical grain boundary, we carved a $D = 20$ nm/$H = 28.7$ nm cylinder from the larger one containing a screw dislocation along its axis, and then generated the powder pattern by the DSE. The WPPM analysis was made considering only the strain effect from the dislocation, with same contrast factor, fixed domain size and dislocation density $(3.183 \times 10^{-15}$ m$^{-2})$. The result gave $R_e = 26.4(1)$ nm, nearly the same value refined for the same cylinder with surface relaxation effect (**Figures 10E,F**), thus confirming that $R_e$ approaches the expected $2R_p$ value if all conditions of the Wilkens model are verified. Also in this case, however, the instability of Eq. (1) is confirmed, as a free refinement of all parameters gives diverging values of $R_e$, a dislocation density much smaller than expected, and a wrong height of the cylindrical domain.

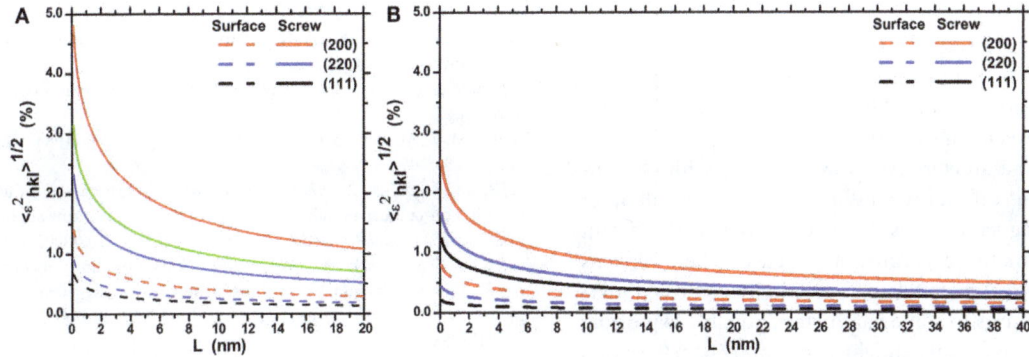

**FIGURE 11 | Plot of the r.m.s. strain as a function of the Fourier length, *L*, taken along the scattering vector direction inside the cylindrical domains: *D* = 20 nm/*H* = 28.7 nm (A) and**
*D* = 40 nm/*H* = 28.7 nm (B). Trends refer to the dislocation strain (screw, line) and surface relaxation strain (surface, dash) along different crystallographic directions.

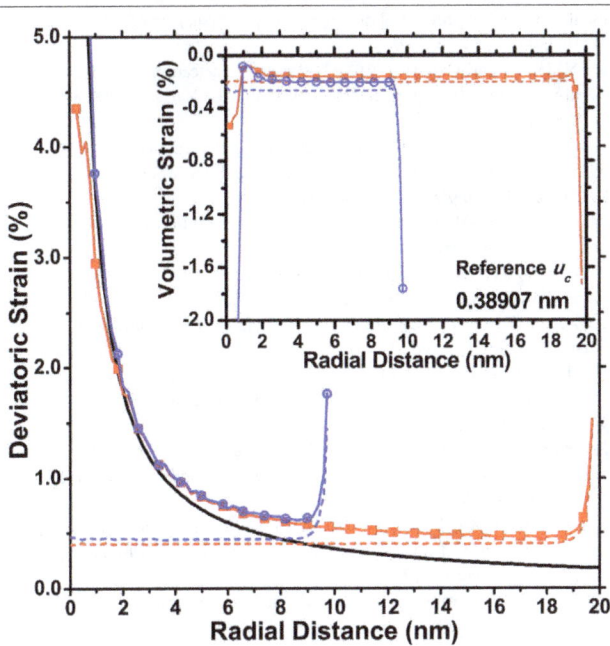

**FIGURE 12 | Deviatoric strain components as a function of the distance from the dislocation axis (cylinder axis), for an ideal screw dislocation (black line).** It is also shown the trend obtained by MD for the
*D* = 20 nm/*H* = 28.7 nm (blue circle) and *D* = 40 nm/*H* = 28.7 nm (red square) cylinders with dislocation and without dislocation, thus showing the effect of surface relaxation (dash lines). The inset shows the volumetric strain component.

It is interesting to consider the results shown so far in terms of the main strain component in the studied systems. **Figure 12** shows the deviatoric strain as a function of the radial distance from the cylinder axis, which is also the position of the screw dislocation line (cf. **Figure 1C**).

The strain for an ideal dislocation (Hull and Bacon, 1965) is shown together with the values obtained from MD, for both cylindrical domains ($D = 20$ nm/$H = 28.7$ nm and $D = 40$ nm/$H = 28.7$ nm) with and without screw dislocation.

Apart from the dislocation core region, where the continuum mechanics expression diverges while the MD values stay finite, it is quite evident that the MD trends for domains with screw dislocations result from a combination of the strain from the dislocation with that from the surface relaxation effect. The last steeply decays from the surface toward the inside the domain, and tends to decrease for increasing diameter of the cylindrical domain.

## CONCLUSION

Differences between real dislocations and the idealized models of continuum mechanics, become significant when dislocations are confined in nanocrystalline domains. MD simulations show that the traditional continuum mechanics expressions for the strain field only approximately agree with the actual strain field. This was demonstrated for the simple case of straight dislocations in cylindrical Pd nanocrystal, a condition closely matching the hypotheses underlying the Krivoglaz–Wilkens theory on dislocation effects in diffraction. Discrepancies between theory and MD simulations are especially evident for edge dislocations, as an effect of the split into partials and the corresponding stacking faults in between, responsible for a region of non-hexagonal layer stacking. Under these conditions, the traditional perturbation approach, based on a cubic (fcc) Pd phase with deformation caused by lattice defects, seems not appropriate to model XRD patterns obtained from MD simulations. Further discrepancy is observed if the edge dislocation line does not lay along the axis of the cylindrical domain.

The screw case seems more similar and compatible with the Krivoglaz–Wilkens theory. Diffraction patterns generated by MD seem little affected by the position of the dislocation line in the cylindrical domain, and the main effect is a broadening of the diffraction lines, as predicted by the theory. If all parameters of the system – cylindrical domain height and diameter, contrast factor of the screw dislocation in the primary slip system of Pd, dislocation density as given by line defect length divided by the cylinder volume – are fixed, then the dislocation outer cut-off radius is found in good agreement with the Wilkens model for restrictedly random dislocations: $R_e$ is about 10–20% larger than the diameter ($2R_p$)

of the restrictedly random regions, the discrepancy decreasing for increasing cylinder diameter.

Despite this positive result, if the powder diffraction pattern generated from the MD simulation is modeled according to the Wilkens theory, the non-ideality of the dislocations and the intrinsic correlations between parameters lead to a strong instability. If all microstructural parameters are allowed to vary without constraints, the outer cut-off radius tends to diverge, leading all other parameters to wrong values. It is therefore likely that significant errors may occur in the experimental practice, when Wilkens model is applied to real materials, e.g., plastically deformed metals; applying restrictions to some parameters, possibly exploiting evidence from other techniques, might significantly help to obtain reliable results from the analysis of the diffraction patterns. Effects are expected to be increasingly significant for smaller domain sizes. Further studies will be required to shed light on this important issue, in the effort to provide a realistic modeling of polycrystalline materials with lattice defects.

## REFERENCES

Armstrong, N., Leoni, M., and Scardi, P. (2006). Considerations concerning Wilkens' theory of dislocation line-broadening. *Z. Kristallogr. Suppl.* 23, 81–86. doi:10.1524/zksu.2006.suppl_23.81

Beyerlein, K. R., Snyder, R. L., and Scardi, P. (2011). Powder diffraction line profiles from the size and shape of nanocrystallites. *J. Appl. Crystallogr.* 44, 945–953. doi:10.1107/S0021889811030743

Bulatov, V., Farid, F. A., Ladislas, K., Benoit, D., and Sidney, Y. (1998). Connecting atomistic and mesoscale simulations of crystal plasticity. *Nature* 391, 669–672. doi:10.1038/35577

Daw, M. S., and Baskes, M. I. (1984). Embedded-atom method: derivation and application to impurities, surfaces, and other defects in metals. *Phys. Rev. B* 29, 6443–6453. doi:10.1103/PhysRevB.29.6443

Debye, P. (1915). Zerstreuung von Röntgenstrahlen. *Ann. Phys.* 351, 809–823. doi:10.1002/andp.19153510606

Foiles, S. M., Baskes, M. I., and Daw, M. S. (1986). Embedded-atom-method functions for the fcc metals Cu, Ag, Au, Ni, Pd, Pt, and their alloys. *Phy. Rev. B* 33, 7983–7991. doi:10.1103/PhysRevB.33.7983

Gelisio, L., Azanza Ricardo, C. L., Leoni, M., and Scardi, P. (2010). Real-space calculation of powder diffraction patterns on graphics processing units. *J. Appl. Crystallogr.* 43, 647–653. doi:10.1107/S0021889810005133

Groma, I. (1998). X-ray line broadening due to an inhomogeneous dislocation distribution. *Phy. Rev. B* 57, 7535–7542. doi:10.1103/PhysRevB.57.7535

Groma, I., and Székely, F. (2000). Analysis of the asymptotic properties of X-ray line broadening caused by dislocations. *J. Appl. Crystallogr.* 33, 1329–1334. doi:10.1107/S002188980001058X

Hull, D., and Bacon, D. J. (1965). *Introduction to Dislocations.* Oxford: Butterworth-Heinemann.

Hunter, A., Beyerlein, I. J., Germann, T. C., and Koslowski, M. (2011). Influence of the stacking fault energy surface on partial dislocations in fcc metals with a three-dimensional phase field dislocations dynamics model. *Phys. Rev. B* 84, 144108. doi:10.1103/PhysRevB.84.144108

Jacobsen, K. W., and Schiotz, J. (2002). Computational materials science: nanoscale plasticity. *Nat. Mater.* 1, 15–16. doi:10.1038/nmat718

Kaganer, V. M., and Sabelfeld, K. K. (2011). Short range correlations of misfit dislocations in the X-ray diffraction peaks. *Phys. Status Solidi A* 208, 2563–2566. doi:10.1002/pssa.201184255

Kaganer, V. M., and Sabelfeld, K. K. (2014). Strain distributions and diffraction peak profiles from crystals with dislocations. *Acta Cryst. A* 70, 457–471. doi:10.1107/S2053273314011139

Kamminga, J.-D., and Delhez, R. (2000). Calculation of diffraction line profiles from specimens with dislocations. A comparison of analytical models with computer simulations. *J. Appl. Crystallogr.* 33, 1122–1127. doi:10.1107/S0021889800006750

Klimanek, P., and Kuzel, R. (1988). X-ray diffraction line broadening due to dislocations in non-cubic materials. I. General considerations and the case of

elastic isotropy applied to hexagonal crystals. *J. Appl. Crystallogr.* 21, 59–66. doi:10.1107/S0021889887009580

Klug, H. P., and Leroy, E. A. (1974). *X-Ray Diffraction Procedures for Polycrystalline and Amorphous Materials.* New York, NY: Wiley.

Krivoglaz, M. A. (1996). *X-Ray and Neutron Diffraction in Nonideal Crystals.* Berlin: Springer.

Krivoglaz, M. A., Martynenko, O. V., and Ryaboshapka, K. P. (1983). Fiz. *Met. Metalloved.* 55, 5–17.

Kuzel, R., and Klimanek, P. (1988). X-ray diffraction line broadening due to dislocations in non-cubic materials. II. The case of elastic anisotropy applied to hexagonal crystals. *J. Appl. Crystallogr.* 21, 363–368. doi:10.1107/S002188988800336X

Kuzel, R., and Klimanek, P. (1989). X-ray diffraction line broadening due to dislocations in non-cubic crystalline materials. III. Experimental results for plastically deformed zirconium. *J. Appl. Crystallogr.* 22, 299–307. doi:10.1107/S0021889889001585

Langford, I. J., and Louer, D. (1982). Diffraction line profiles and scherrer constants for materials with cylindrical crystallites. *J. Appl. Crystallogr.* 15, 20–26. doi:10.1107/S0021889882011297

Leonardi, A., Beyerlein, K. R., Xu, T., Li, M., Leoni, M., and Scardi, P. (2011). Microstrain in nanocrystalline samples from atomistic simulation. *Z. Kristallogr. Proc.* 1, 37–42. doi:10.1524/zkpr.2011.0005

Leonardi, A., Leoni, M., and Scardi, P. (2012). Common volume functions and diffraction line profiles of polyhedral domains. *J. Appl. Crystallogr.* 45, 1162–1172. doi:10.1107/S0021889812039283

Li, X., Wei, Y., Lu, L., and Gao, H. (2010). Dislocation nucleation governed softening and maximum strength in nano-twinned metals. *Nature* 464, 877–880. doi:10.1038/nature08929

Martinez-Garcia, J., Leoni, M., and Scardi, P. (2009). A general approach for determining the diffraction contrast factor of straight-line dislocations. *Acta Cryst. A* 65, 109–119. doi:10.1107/S010876730804186X

Mittemeijer, E. J., and Scardi, P. (2004). *Diffraction Analysis of the Microstructure of Materials.* Berlin: Springer.

Plimpton, S. (1995). Fast parallel algorithms for short-range molecular dynamics. *J. Comput. Phys.* 117, 1–19. doi:10.1006/jcph.1995.1039

Scardi, P. (2008). "Microstructural properties: lattice defects and domain size effects," in *Powder Diffraction: Theory and Practice,* eds R. E. Dinnebier and S. J. L. Billinge (Oxford: Royal Society of Chemistry), 378–416.

Scardi, P., and Leoni, M. (2002). Whole powder pattern modelling. *Acta Cryst. A* 58, 190–200. doi:10.1107/S0108767301021298

Scardi, P., Leoni, M., and Delhez, R. (2004). Line broadening analysis using integral breadth methods: a critical review. *J. Appl. Crystallogr.* 37, 381–390. doi:10.1107/S0021889804004583

Scardi, P., Leoni, M., and D'Incau, M. (2007). Whole powder pattern modelling of cubic metal powders deformed by high energy milling. *Z. Kristallogr.* 222, 129–135. doi:10.1524/zkri.2007.222.3-4.129

Scherrer, P. (1918). Bestimmung der Grösse und der inneren Struktur von Kolloidteilchen mittels Röntgensrahlen. *Nachr. Ges. Wiss. Gottingen Math. Phys. Kl.* 98.

Sheng, H. W., Kramer, M. J., Cadien, A., Fujita, T., and Chen, M. W. (2011). Highly optimized embedded-atom-method potentials for fourteen fcc metals. *Phy. Rev. B* 83, 134118. doi:10.1103/PhysRevB.83.134118

Snyder, R. L., Fiala, J., and Bunge, H.-J. (1999). *Defect and Microstructure Analysis by Diffraction.* New York, NY: Oxford University Press.

Ungar, T. (2008). Dislocation model of strain anisotropy. *Powder Diffr.* 23, 125–132. doi:10.1154/1.2918549

Ungar, T., Ott, S., Sanders, P. G., Borbely, A., and Weertman, J. R. (1998). Dislocations, grain size and planar faults in nanostructured copper determined by high resolution X-ray diffraction and a new procedure of peak profile analysis. *Acta Mater.* 46, 3693–3699. doi:10.1016/S1359-6454(98)00001-9

Warren, B. E. (1990). *X-Ray Diffraction.* New York, NY: Dover.

Warren, B. E., and Averbach, B. L. (1950). The effect of cold-work distortion on X-ray patterns. *J. Appl. Phys.* 21, 595–599. doi:10.1063/1.1699713

Wilkens, M. (1970a). The determination of density and distribution of dislocations in deformed single crystals from broadened X-ray diffraction profiles. *Phys. Status Solidi A* 2, 359–370. doi:10.1002/pssa.19700020224

Wilkens, M. (1970b). "Theoretical aspects of kinematical X-ray diffraction profiles from crystals containing dislocation distributions," in *Fundamental Aspects of Dislocation Theory,* eds J. A. Simmons, R. de Wit, and R. Bullough (Washington, DC: NBS Spec. Publ.), 317.

Wilson, A. J. C. (1955). The effects of dislocations on X-ray diffraction. *Il Nuovo Cimento* 1, 277–283. doi:10.1007/BF02900634

Yamakov, V., Wolf, D., Phillpot, S. R., Mukherjee, A. K., and Gleiter, H. (2002). Aluminium reveals some surprising behaviour of nanocrystalline aluminium by molecular-dynamics simulation. *Nat. Mater.* 1, 1–4. doi:10.1038/nmat700

Yamakov, V., Wolf, D., Phillpot, S. R., Mukherjee, A. K., and Gleiter, H. (2004). Deformation-mechanism map for nanocrystalline metals by molecular-dynamics simulation. *Nat. Mater.* 3, 43–47. doi:10.1038/nmat1035

**Conflict of Interest Statement:** The authors declare that the research was conducted in the absence of any commercial or financial relationships that could be construed as a potential conflict of interest.

# 20

# Impact of atomic layer deposition to nanophotonic structures and devices

*Muhammad Rizwan Saleem[1,2] \*, Rizwan Ali[1], Mohammad Bilal Khan[2], Seppo Honkanen[1] and Jari Turunen[1]*

[1] Institute of Photonics, University of Eastern Finland, Joensuu, Finland
[2] Center for Energy Systems (CES), USAID Center for Advance Studies, National University of Sciences and Technology (NUST), Islamabad, Pakistan

**Edited by:**
Mohammed Es-Souni, Kiel University of Applied Sciences, Germany

**Reviewed by:**
Venu Gopal Achanta, Tata Institute of Fundamental Research, India
Bouchta Sahraoui, University of Angers, France

**\*Correspondence:**
Muhammad Rizwan Saleem, Center for Energy Systems (CES), USAID Center for Advance Studies, National University of Sciences and Technology (NUST), Sector H-12, Islamabad 44000, Pakistan
e-mail: rizwan@casen.nust.edu.pk; rizwan.saleem@uef.fi

We review the significance of optical thin films by Atomic Layer Deposition (ALD) method to fabricate nanophotonic devices and structures. ALD is a versatile technique to deposit functional coatings on reactive surfaces with conformal growth of compound materials, precise thickness control capable of angstrom resolution, and coverage of high aspect ratio nano-structures using wide range of materials. ALD has explored great potential in the emerging fields of photonics, plasmonics, nano-biotechnology, and microelectronics. ALD technique uses sequential reactive chemical reactions to saturate a surface with a monolayer by pulsing of a first precursor (metal alkoxides or covalent halides), followed by reaction with second precursor molecules such as water to form the desired compound coatings. The targeted thickness of the desired compound material is controlled by the number of ALD-cycles of precursor molecules that ensures the self-limiting nature of reactions. The conformal growth and filling of $TiO_2$ and $Al_2O_3$ optical materials on nano-structures and their resulting optical properties have been described. The low temperature ALD-growth on various replicated sub-wavelength polymeric gratings is discussed.

**Keywords: atomic layer deposition, optical materials, nanophotonics, nano-optical devices, plasmonics**

## 1. INTRODUCTION

Thin film deposition techniques have been significantly stimulated by the advancement in the high-tech applications in optical systems (Pedrotti, 1993; Macleod, 2001). Dielectric transparent optical thin films in conventional optical filters select a spectral range of transmitted or reflected light such as anti reflecting coatings, pass band filters, athermal optical filters, edge filters, lenses etc., for several precision instruments (Spiller, 1984; Dobrowolski et al., 1996; Szeghalmi et al., 2009; Huber et al., 2014). Analogously, the rapidly developing advanced applications include narrow band filters for dense wavelength division multiplexing, low laser damage devices, optical elements for deep ultraviolet, and extreme ultraviolet lithography (Weber et al., 2012b), optical waveguides (slot waveguides, resonant waveguide structures, plasmonic structures, etc.) for optical signal processing in optical communications, optical quantum computations (Jia et al., 2011; Weber et al., 2012b; Jalaluddin and Magnusson, 2013).

All such advanced applications require optical coatings of precise thickness control and of high quality (high packing density, good surface uniformity, low defect density, good adhesion with underlying substrate, etc.), which depends on choice of deposited film material and method being employed (Martinu et al., 2014). The optical films must exhibit good mechanical properties such as good adhesion, acceptable scratch, abrasion resistance, low stress, low crack density, and thermally and environmentally stable are suitable for various applications in nanophotonics. Traditionally, optical coatings have been fabricated by Physical Vapor Deposition (PVD) techniques from solid material sources through evaporation (resistive heating and electron beam evaporation), sputtering (Chung et al., 2009; Jalaluddin and Magnusson, 2013),

laser assisted evaporation, and ion assisted deposition with typically high deposition rates. In order to confine waveguide modes in dielectric films over tight tolerances, the control of films homogeneity, composition, thickness uniformity, adhesion with underlying substrate, and better control of microstructure (high packing density) are highly desirable and can be achieved by employing Atomic Layer Deposition (ALD) (Riihelä et al., 1996; Saleem et al., 2012b, 2013c). ALD operates in cycles consisting of four essential steps: (1) exposure of first precursor material, (2) evacuation or purging of additional precursor material and reaction byproducts from the chamber, (3) exposure of second precursor material, typically oxidants or reagents, and (4) evacuation or purging of the reaction byproduct molecules from the chamber as shown schematically in **Figure 1**.

Atomic layer deposition is a modified form of Chemical Vapor Deposition (CVD) technique, which possesses angstrom level resolution, layer-by-layer growth of ultra-thin compound films on planar, and high aspect ratio micro- and nano-structures employed in various potential applications (Kim et al., 2009). It distinguishes from CVD in terms that the precursor material pulses are introduced in the reactor on to the substrate alternately, once at a time (Ponraj et al., 2013). The reactor is continuously purged with a commonly used inert gas such as nitrogen after each precursor pulse, which extracts all the reaction byproducts except those which are chemisorbed on the substrate. As a result, film growth proceeds through sequential surface reactions and enable self-limiting and self-controlled growth (Im et al., 2012). Owing to the self-controlled film growth, the extremely accurate film thickness in angstroms depends on the number of ALD-cycles (Riihelä et al., 1996). Another important consequence of self-controlled

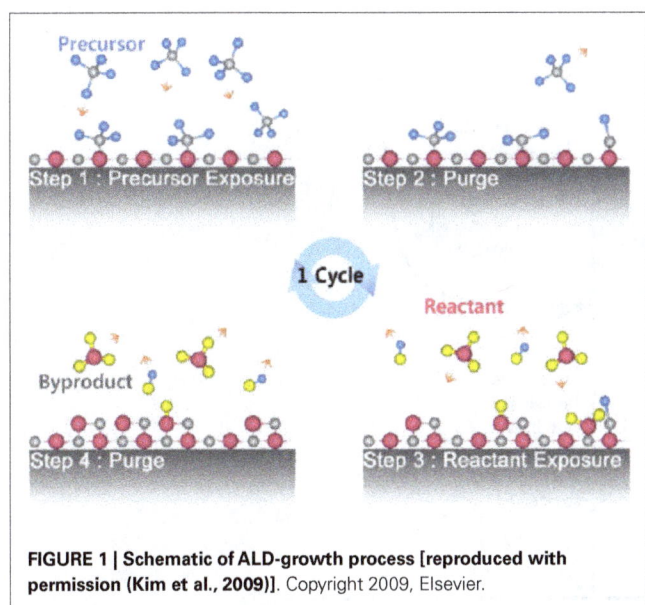

**FIGURE 1 | Schematic of ALD-growth process [reproduced with permission (Kim et al., 2009)].** Copyright 2009, Elsevier.

ALD-growth is that complex shaped high aspect ratio structures are uniformly coated over large-area substrates (George, 2010; Wang et al., 2014). ALD-Al$_2$O$_3$ is employed to uniformly coat silica spheres in synthetic opals (Sechrist et al., 2006). The wavelength of photonic crystals was investigated to shift progressively to longer wavelengths as a function of filled opal volume fraction by Al$_2$O$_3$ with pinhole defect free characteristics (Sechrist et al., 2006). This is worth emphasizing that films grown by ALD exhibit high density (Saleem et al., 2012b, 2013c, 2014d) e.g., the measured hydrogen contents in ALD-grown TiO$_2$ (Ritala and Leskelä, 1993; Ritala et al., 1993) films are of lowest compared to those prepared by other techniques (Bennett et al., 1989). The high density consequently verifies, relative low concentrations of penetrated H$_2$O molecules into ALD-grown films (Saleem et al., 2012b, 2013c, 2014d).

Typically, ALD-growth proceeds in slow fashion and produces sub-monolayer per cycle, which is a drawback of ALD technology. On the other hand, slow growth rates facilitate several high-tech processes as advantage and are paramount to ensure uniform high quality and pinhole defect free films over large areas. Since the origin of ALD technology is to deposit epitaxial layers on semiconductor substrates and were known as Atomic Layer Epitaxy (ALE) (Pimbley and Lu, 1985; Goodman and Pessa, 1986; Ide et al., 1988; Gong et al., 1990; Yu, 1993; Bedair, 1994), which demands thin dielectric gate oxide films (barrier layers) to control leakage current at minimum film thickness (Puurunen, 2005). Furthermore, in photonics high quality and dense thin films are of significant importance to confine propagating modes via lithographically fabricated nano-structured topographies (Riyanto et al., 2012). ALD-growth on such nano-structures proceeds as an independent process in order to mitigate any dimensional inaccuracies (Saleem et al., 2011a). Such nano-structural inaccuracies are possibly be adjusted by the thickness of ALD-growth as a function of number of ALD controlled cycles to deposit films of the order of Å.

## 2. CHARACTERISTICS OF ALD FILMS FOR THE FABRICATION OF NANOPHOTONIC STRUCTURES

Nanophotonic structures consist of periodic alternation of refractive index as high-low-high-low contrast that can change the state of polarization, amplitude, or phase of an incident light (Knop, 1978). The waveguide layer confines, propagates, and emerges the light as narrow spectral peaks by resonance of light modes within the structure. ALD-coated thin films play a significant role in the fabrication of nanophotonic structures to provide the following:

i.   Excellent conformality of high aspect ratio nano-structures.
ii.  Excellent step coverage uniformity and thickness control over large scale smooth surfaces.
iii. Low temperature growth, in particular on polymer materials below their glass transition temperatures as well as to deposit optical materials in amorphous phase.
iv.  Reasonably good precursor material properties with minimum damage to substrate materials.
v.   Self-saturative growth when the surface binding sites are filled and/or deposited film thickness is highly reproducible after each ALD-cycle.

### 2.1. LOW TEMPERATURE ALD-GROWTH

One of the important feature of ALD is to grow high quality thin films at low growth temperatures 50–250°C (Triani et al., 2009; Im et al., 2012). Precursor molecules are thermally activated to react on the substrate surface and chemisorbed. Low temperature ALD-growth is extremely important for heat-sensitive materials such as polymers, biomaterials, and heat-sensitive structures: self-assembly of molecules, self-rolled microtubes with ultra-thin wall thickness (Purniawan et al., 2010; Im et al., 2012). TiO$_2$ films at 80°C are deposited on plasma-treated polycarbonate substrate, which shows enhanced adhesion to ALD-coated films (Latella et al., 2007). Uniformly smooth thin amorphous ALD-TiO$_2$ films possessing high refractive index are coated on replicated nano-structures on various polymer substrates such as polycarbonate, OrmoComp, Cyclic-Olefin-copolymer (COC) at low deposition temperature of 120°C and have been employed as Resonant Waveguide Gratings (RWGs) or Guided Mode Resonance Filters (GMRFs) (Saleem et al., 2011a, 2012a,d, 2013b, 2014a,c). ALD-TiO$_2$ growth temperature has an effect on the film's surface roughness, density, and refractive index owing to crystallization of TiO$_2$ material to anatase phase. Increase in ALD-TiO$_2$ growth temperature from 175 to 225°C led to increase surface roughness due to crystallization, consequently the fabricated films are porous with relatively low density and refractive index compared to amorphous films prepared at 125–150°C (Aarik et al., 2000), for further details see sections 3.1 and 3.2. A schematic representation of ALD-growth rate with growth temperature is shown in **Figure 2** (Kääriäinen et al., 2013).

In order to explore the best operating regime of ALD one needs to adjust temperatures and growth rates. At high growth temperatures, the second precursor material decompose on the surface before reaction with first precursor and consequently deposit on the substrate by increasing the growth rate. Alternatively, the first stable precursor may desorb before reaction with second precursor, thereby decreasing the growth rate. On the other hand, if the

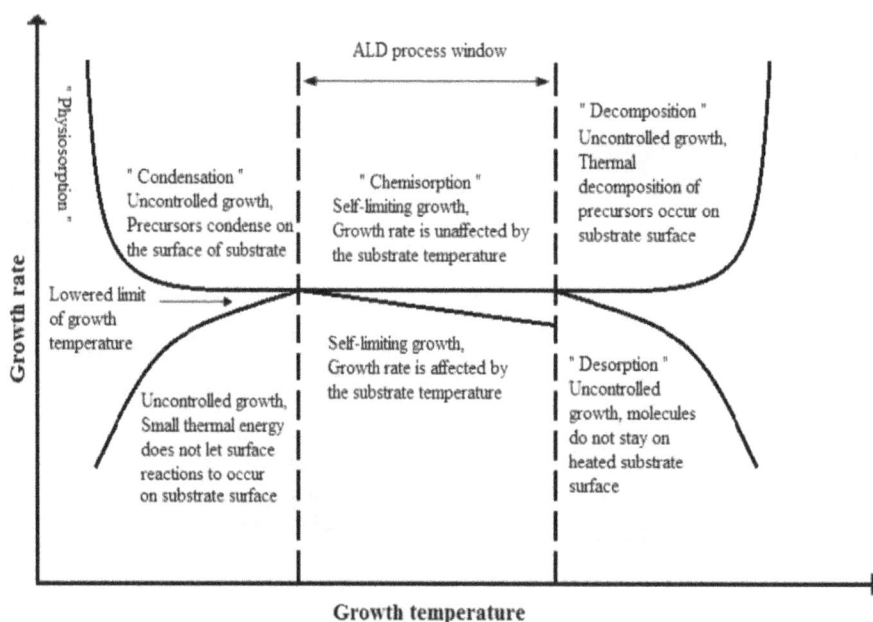

**FIGURE 2 | Schematic of ALD-growth process with temperature (Reproduced with permission © Picosun).**

temperature is too low, the precursor materials may desorb without a complete reaction or even condense a liquid or solid on the surface and results to increase in growth rate. Alternatively, precursor materials do not possess sufficient thermal energy to activate appropriate reactions and thus results in to lower the deposition rate than expected. Keeping all these parameters in consideration, a complete or partial monolayer of molecules can be deposited by selection of suitable temperature zone through ALD temperature window (Kääriäinen et al., 2013). Temperature variations during ALD-growth have significant influence on the surface roughness e.g., in infiltration of silica opals to fabricate $TiO_2$ inverse opals with spherical size templates ranging from 200 to 440 nm in diameter show different roughness (King et al., 2005a). Atomic Force Microscopy (AFM) studies shows that $TiO_2$ films yielded a root-mean-square (RMS) roughness of 0.2 nm at 100°C, 2.1 nm at 300°C, and 9.6 nm at 600°C, which shows that low temperature deposition is of much more significance for even filling and preparing extremely smooth interfaces (King et al., 2005a).

## 2.2. SELF-CONTROLLED FILM THICKNESS AND MATERIAL COMPOSITION

Atomic layer deposition technique is capable to deposit thin atomic films with precisely controlled thickness and free from occurrence of light losses in optical micro- and nano-devices resulted by thickness roughness (Vahala, 2003; Wang et al., 2012). Scanning electron microscopy (SEM) images in **Figure 3A** show the conformal growth of ALD-$TiO_2$ films of various thicknesses of (0–120 nm) on Si triangular lattice of photonic crystals (PCs) (Graugnard et al., 2006). Transmission Electron Microscopy (TEM) images of four-bilayer nanolaminates of W/$Al_2O_3$ grown by ALD at temperature of 177°C are shown in **Figure 3B** (Sechrist et al., 2005). Each nanolaminate in the structure employing 111 ALD-cycles

of $Al_2O_3$ and 32 to that of W, which are clearly shown in **Figure 3B**. Furthermore, it reveals that interfacial roughness is more prominent where $Al_2O_3$ nucleates on W and possesses less roughness at interfaces. Tungsten W grows on $Al_2O_3$ sites, which may results due to polycrystallinity of W nanolayer on $Al_2O_3$. More often, the ALD film growth is a fraction of one monolayer and has higher film density to those of films deposited by other techniques without pinhole defects (Carcia et al., 2009; Saleem et al., 2012b, 2013c).

## 2.3. CONFORMAL GROWTH FOR 2D AND 3D STRUCTURES

Atomic layer deposition is a unique technique relies on the sequential self-terminating surface saturating reactions between gas-phase precursor molecules and a solid surface. Owing to the self-limiting reactions, it gives better control on thickness, conformality, film quality and uniformity, and large-area coverage of high aspect ratio nano-structures, 3D complex structures such as PC, opals, nanopores, nanowires, and nanotubes (King et al., 2005b; Knez et al., 2007). **Figure 4A** shows the cross-sectional views of SEM images of ZnS:Mn and $TiO_2$ infiltrated opals and inverted structures into silica opal templates. **Figures 4B,C** show the $TiO_2/SiO_2$ opals and $TiO_2$ inverse opals where the growth in small central void space is precisely controlled by self-limiting nature of ALD, which enables accurately controlled deposition on complex geometries (King et al., 2005b). **Figures 4D,E** show a multi-layered PC structure using $TiO_2$/ZnS:Mn/$TiO_2$ in order to combine the luminescent properties of ZnS:Mn to that of higher refractive index of $TiO_2$. The structure is fabricated with 10 nm of ZnS:Mn in a sintered opal followed by $TiO_2$ deposition in remaining volume to fill. An Ion-mill is used to expose $SiO_2$ opal and etch by HF to form an inverse opal, which is subsequently infiltrated by 10 nm of high refractive index ALD-$TiO_2$ layer (King et al., 2005b).

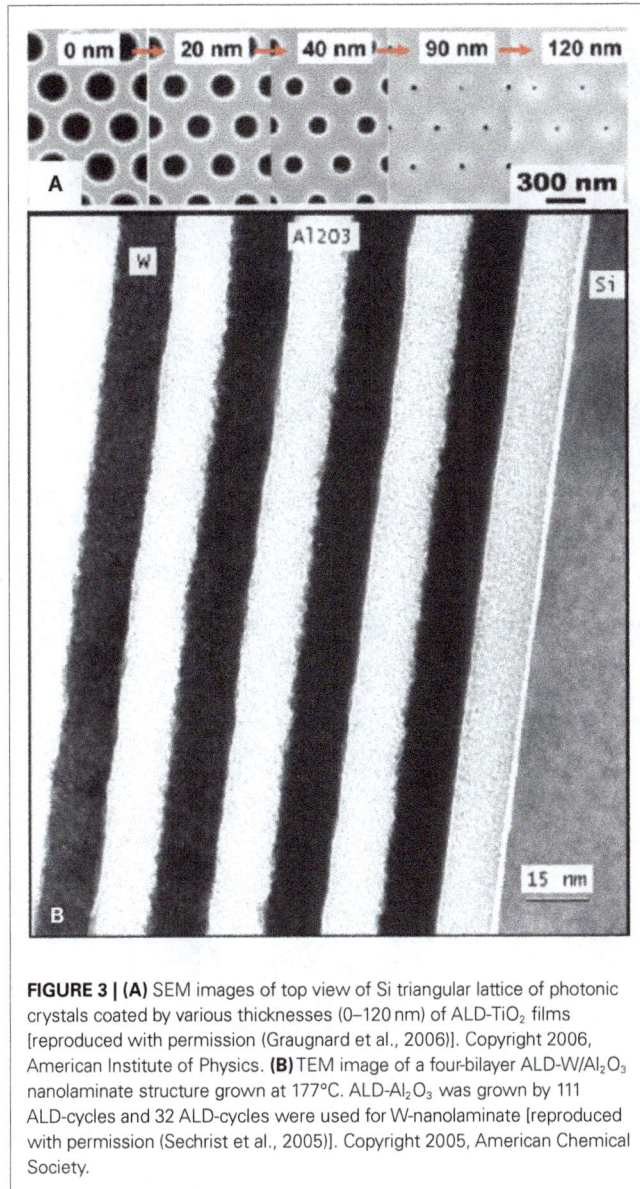

**FIGURE 3 | (A)** SEM images of top view of Si triangular lattice of photonic crystals coated by various thicknesses (0–120 nm) of ALD-TiO$_2$ films [reproduced with permission (Graugnard et al., 2006)]. Copyright 2006, American Institute of Physics. **(B)** TEM image of a four-bilayer ALD-W/Al$_2$O$_3$ nanolaminate structure grown at 177°C. ALD-Al$_2$O$_3$ was grown by 111 ALD-cycles and 32 ALD-cycles were used for W-nanolaminate [reproduced with permission (Sechrist et al., 2005)]. Copyright 2005, American Chemical Society.

Uniform arrays of dense and aligned TiO$_2$ nanotubes are fabricated by employing ALD on nanoporous Al$_2$O$_3$ on Si substrate as shown in **Figure 4F** (Sander et al., 2004). The catalytic properties of such TiO$_2$ nanomembranes have been demonstrated by (Kemell et al., 2007). Wang et al. (2004) patterned self-assembled monolayer of hexagonally arranged polystyrene and SiO$_2$ nanospheres. ALD-TiO$_2$ and Al$_2$O$_3$ films are deposited on the substrate. TiO$_2$ surface layer was removed by ion beam followed by removal of polystyrene beads to form a micrometer-scaled surface with TiO$_2$ nanobowls as shown in **Figures 4G–J**. The structure of ALD-TiO$_2$ nanobowls holding removable polystyrene nanospheres of size 450 nm are shown in **Figure 4K**. Such uniformly fabricated structures demonstrated the major advantages of ALD compared to other deposition methods such as CVD or PLD in terms of conformal growth on complex geometries. Highly complex morphology of butterfly wing's structures are replicated after ALD

coating (Huang et al., 2006) shown in **Figures 4L,M**. Modifications of such structures by ALD can improve their potential use as optical elements e.g., in holography. ALD-TiO$_2$ was employed to tune the static photonic band of a 2D Si triangular lattice PC slab waveguide by nanoscale control of the dielectric contrast, propagation, and dispersion (Graugnard et al., 2006). This research group also reported the fabrication of 3D nano-structures through synthetic silica opals and holographically patterned polymeric templates to control PC band gap properties (Graugnard et al., 2006).

## 3. ALD THIN FILMS OPTICAL PROPERTIES

The primary parameter to access the optical quality of a film is the complex refractive index with real part $n$ and imaginary part $k$ (extinction coefficient) and is related to the complex relative permittivity, $\varepsilon_r = \varepsilon_{r,r} - i\varepsilon_{r,i}$

$$N(\lambda) = n(\lambda) - ik(\lambda) = \sqrt{\varepsilon_r} = \sqrt{\varepsilon_{r,r} - i\varepsilon_{r,i}} \qquad (1)$$

where $\varepsilon_{r,r}$ and $\varepsilon_{r,i}$ represent the real and imaginary parts of $\varepsilon_r$, respectively. Equation (1) relates the dielectric properties of the materials with optical properties and depends on wavelength called dispersion. When light passes through a lossy material that absorbs or scatters light and loses its energy called attenuated light. Attenuation may occur due to several mechanisms such as generation of phonons (lattice waves), photogeneration, free carrier absorption, and scattering (Kasap and Capper, 2006). Equation (1) gives:

$$n^2 - k^2 = \varepsilon_{r,r} \qquad (2)$$

and

$$2nk = \varepsilon_{r,i} \qquad (3)$$

where the optical constants $n$ and $k$ can be determined from the reflectance of thin films for a specified polarization and angle of incidence of light. For normal incidence, the reflection coefficient $r$ is given as:

$$r = \frac{1-N}{1+N} = \frac{1-n+ik}{1+n-ik}. \qquad (4)$$

Reflectance $R$ is given as:

$$R = |r|^2 = \left| \frac{1-n+ik}{1+n+ik} \right|^2 = \frac{(1-n)^2 + k^2}{(1-n)^2 - k^2}. \qquad (5)$$

Larger values of $k$ results in strong absorption and leads to reflectance almost unity. As a result light is reflected and any light in the medium is highly attenuated. In general, optical filter demands $k(\lambda)$ below $10^{-4}$ with corresponding optical losses well below 1 dB/cm. In most optical thin films, the material selection requires amorphous (to avoid losses due to grain boundaries), isotropic, without birefringence and scattering losses below $10^{-4}$ (Martinu et al., 2014).

**Figures 5A,B** show the transmittance of ALD-TiO$_2$ as deposited amorphous films of thickness ~200 nm on fused silica substrate

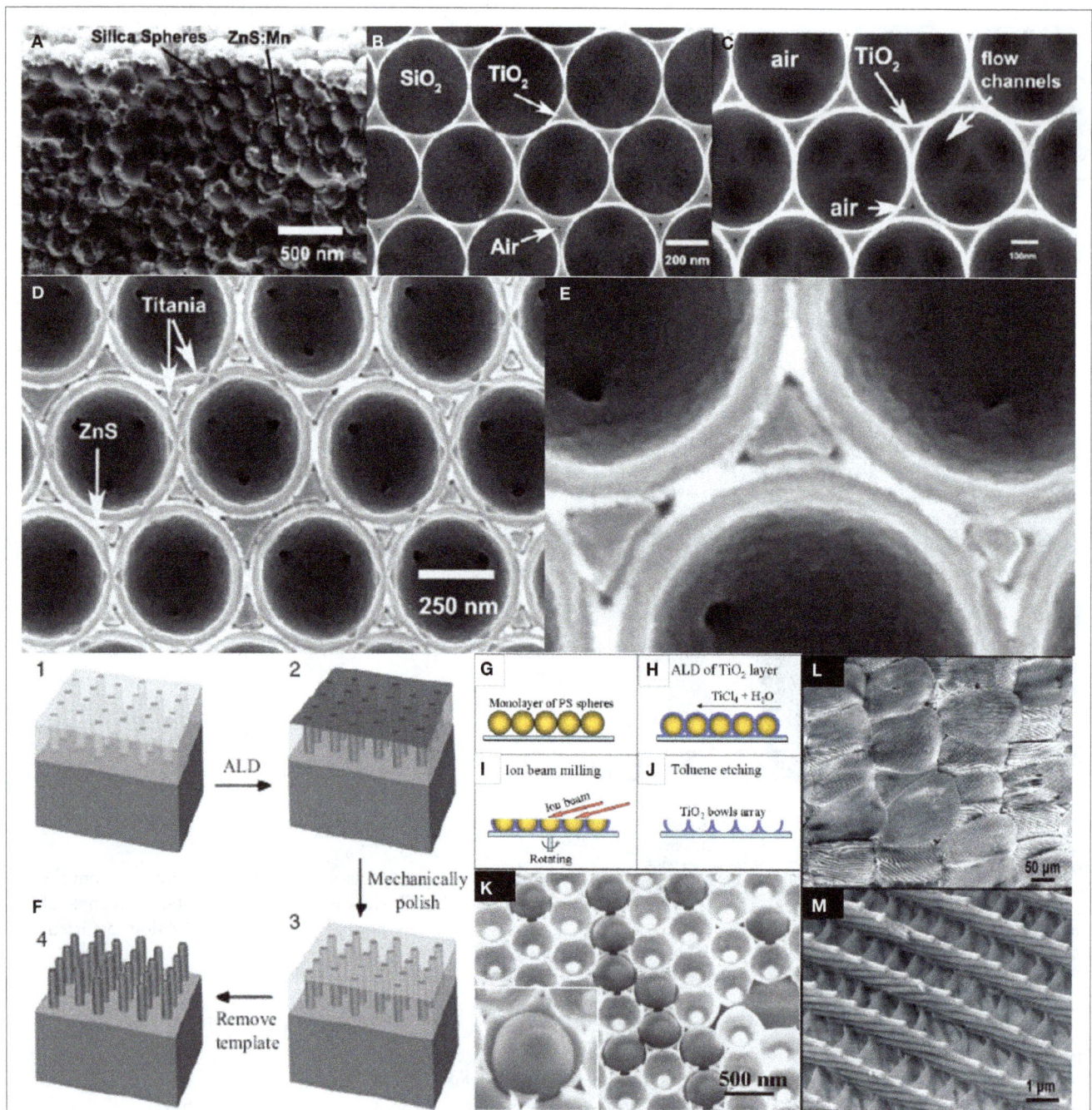

**FIGURE 4 | (A)** SEM images of cross-sectional view of ZnS:Mn and TiO₂ infiltrated opals and inverse opals into SiO₂ opal templates. **(B,C)** SEM images of TiO₂/SiO₂ and TiO₂ inverse opal. **(D,E)** Multi-layered PC structure of TiO₂/ZnS:Mn/TiO₂ into SiO₂ opal template [reproduced with permission (King et al., 2005b)]. Copyright 2004, Elsevier. **(F)** Schematic fabrication of TiO₂ nanotube arrays on Si substrate; (i) preparation of nanoporous Al-template by anodization of Al film, (ii) ALD-TiO₂ deposited on Al-template, (iii) surface layer removal of TiO₂, and (iv) chemical etching of Al-template to reveal dense, aligned, and uniform array of TiO₂ nanotubes [reproduced with permission (Sander et al., 2004)]. Copyright 2004, Wiley. **(G–J)** Schematic of TiO₂ nanobowls preparation. **(K)** Polystyrene (PS) spheres staying on a TiO₂ nanobowls. The inset shows a PS sphere of 450 nm in size inside a TiO₂ nanobowl [reproduced with permission (Wang et al., 2004)]. Copyright 2004, American Chemical Society. **(L,M)** SEM images of alumina replicas of butterfly wing scale on Si substrate after removal of butterfly template [reproduced with permission (Huang et al., 2006)]. Copyright 2006, American Chemical Society.

with SEM image and heat treated ALD-TiO₂ films in crystalline (anatase phase) with corresponding SEM image, respectively (Saleem et al., 2013a). Kumar et al. (2009) demonstrated the

optical properties of thin Al₂O₃ films grown by ALD in the spectral range of 400–1800 nm and retrieved optical constants from reflection spectra using Sellmeier's formula with mean square

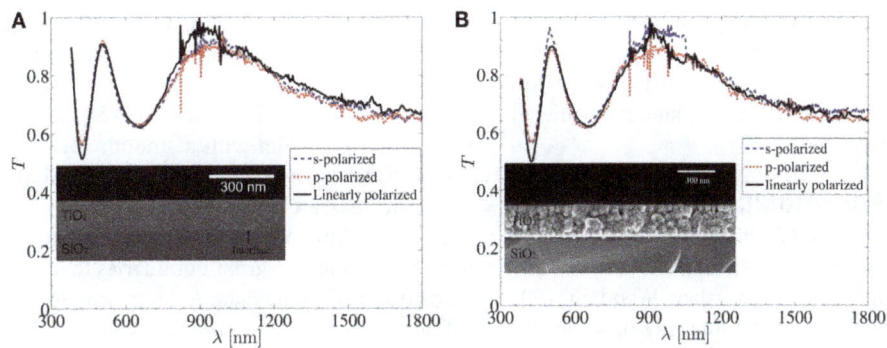

**FIGURE 5 | (A)** Transmittance (T) of ALD-TiO₂ as deposited film of thickness ~200 nm on fused silica substrate with inset of SEM image. **(B)** Transmittance (T) of ALD-TiO₂ heat treated film (crystalline phase-anatase) of thickness ~200 nm on fused silica substrate with inset of SEM image [reproduced with permission (Saleem et al., 2013a)]. Copyright 2013, Optical Society of America.

error values of 0.0006 and 0.0070 on silicon and soda lime glass substrates, respectively.

### 3.1. EFFECT OF ALD-GROWTH TEMPERATURE ON CRYSTAL STRUCTURE AND OPTICAL SCATTERING LOSSES

ALD-TiO₂ films grow at 100–140°C exhibit amorphous phase while containing chlorine contaminations, which decreases significantly with increase in growth temperature. The composition of TiO₂ films in terms of Ti/O ratio has a weak dependence on the growth temperature. The films grow at higher temperatures at 165–350°C possess polycrystalline anatase phase where preferential orientation of crystallites occur while the films at 400°C contained rutile phase. The scattering losses due to crystalline phases was attributed and characterized quantitatively by the half width of {220} and {440} reflections of Reflection High Energy Electron Diffraction (RHEED) patterns (Aarik et al., 1997a). A significantly perfect preferential orientation of crystallites results in a relatively smaller RHEED half widths, which predominantly occurred around 300°C for anatase phase while no preferential orientation at 400°C (rutile phase) appears (Aarik et al., 1997a). Furthermore, films with non-preferential orientation characteristics at growth temperature 400°C have significant absorption in transmission at 400–900 nm wavelength ranges and dominate scattering losses owing to low packing density and consequent surface roughness. The polycrystalline films absorption may also be attributing to the crystallites' surfaces and/or grain interfaces.

### 3.2. EFFECT OF ALD-GROWTH TEMPERATURE ON DENSITY AND REFRACTIVE INDEX

The refractive index of the films grown at 300°C is higher than that the films grown at 100 and 400°C. This is attributed to a decrease in the scattering losses due to preferential orientation of crystallites (Aarik et al., 1997a). Unlike the anatase films, no preferential orientation occurs in rutile films at 400°C with substantial void contents between the crystallites, which result in decrease of packing density (Aarik et al., 1997a). This fact described a linear increase in refractive index with increase in film's packing density (Ottermann and Bange, 1996) and a decrease in orientation scattering and surface roughness (Aarik et al., 1995). Similarly,

Groner et al. (2004) demonstrated that the density of ALD-Al₂O₃ films decreases with decreasing growth temperature and estimated average densities ranged from 3.0 g/cm³ at 177°C to 2.5 g/cm³. ALD-Al₂O₃ film densities have been calculated by (Ott et al., 1997) using the Lorentz–Lorenz relationship based on refractive index. Groner et al. (2004) demonstrated that the refractive index of Al₂O₃ films decreases slightly with decreasing growth temperature with a reported refractive index $n \approx 1.67$ at growth temperature 300–500°C. Other studies investigated that decrease in refractive index is caused by a density decrease with a rise in impurity levels at low growth temperatures (Kukli et al., 1997).

### 3.3. EFFECT OF ALD-GROWTH CONDITIONS ON PHYSICAL PROPERTIES OF FILMS

The polycrystalline phases (rutile/TiO₂-II) observed at a film thickness as minimum as 3 nm at temperature 375–550°C using TiCl₄ and H₂O as precursor materials (Aarik et al., 1997b). The crystal structure is controlled by the pressure of H₂O precursor. Films grown at low H₂O pressure possesses TiO₂-II phase while high H₂O pressures are favorable and result in rutile phase. Such phase formation is influenced by the H₂O dose on the hydroxyl surface groups or on the abundance of H₂O in gas-phase. In addition, the structure depends on the film thickness. A pure TiO₂-II phase appears below a critical thickness $t_c$ while rutile phase appeared at thickness above $t_c$. Both structures co-exist at a thickness range of 150–180 nm and $t_c$ depends on the precursor pressures and subsequent purge times. Critical thickness $t_c$ reduces abruptly with short purge times while growth rate increase with decreasing purge times. This is attributed to overlapping of the reactant pulses and increase gas-phase reaction on the surface. The growth rate varies slowly with precursor doses while $t_c$ depends on both precursor doses (Aarik et al., 1997b).

ALD-Al₂O₃ growth has been studied more extensively than any other ALD system (George et al., 1996). Typical Trimethylaluminum (TMA) and H₂O are used for the growth of ALD-Al₂O₃ as precursor materials. Groner et al. (2004) demonstrated maximum achievable growth rate ~1.33 Å/cycle of Al₂O₃ at 100 and 125°C, which is explained by essentially required thermal activation energy at lower temperatures and decreasing Al–OH and

Al–CH$_3$ surface coverage at higher temperatures (Ott et al., 1997). Growth rate depends on necessary surface species and subsequent reaction kinetics. At lower temperatures, the surface coverage becomes high, which dictates slow reaction kinetics while low surface coverage results due to rapid reaction kinetics at higher temperatures (Ott et al., 1997).

## 4. ALD-GROWTH IN NANOPHOTONIC STRUCTURES

### 4.1. ALD-GROWTH IN RESONANT WAVEGUIDE GRATINGS

#### 4.1.1. Resonance anomalies and origination of RWG

Briefly describing the history of RWGs, when Wood in 1902 observed sharp spectral variations in the intensity of a metallic grating for TM-polarized light (electric field vector is perpendicular to grating lines), which he named anomalies (Wood, 1902). A theoretical explanation of these anomalies was given by Rayleigh in 1907 as one spectral order appearing at grazing incidence and occurred at a particular wavelength called *Rayleigh wavelength* and found close to Wood anomalies (Hessel and Oliner, 1965). Thereby, anomalies were categorized as Rayleigh and Resonance, which both were used for filtering applications. The employment of such anomalies of grating structures in dielectric materials referred to the term: RWGs or GMRFs (Saarinen et al., 2005). A diffraction grating consists of a periodic modulation of refractive index and guided mode resonance phenomena occurs when diffracted light from a diffraction grating couples with a leaky waveguide mode and satisfying phase-matching conditions, which results in a narrow linewidth peak at a particular wavelength that depends on a selectable range of optical parameters (Avrutskii et al., 1985). In this section, we briefly discuss the role of ALD-grown thin films in the development of RWGs.

#### 4.1.2. Structure of a RWG

The structure of a RWG is shown in **Figure 6A**, which consists of a substrate with refractive index $n_3$, an integrated diffraction grating with refractive index distribution $n_2(x)$ along $x$-direction and a medium from where light is incident (usually air) with refractive index $n_1$. **Figure 6A** shows the direction of various propagating diffraction orders and can be simply calculated from the Eq. (6) (Saleem, 2012)

$$n_2 \sin\theta_m = n_1 \sin\theta_1 + m\lambda/d, \qquad (6)$$

where $d$ is grating period, $\lambda$ is wavelength of incident light, $\theta_1$ is incident angle, $m = 0, \pm 1, \pm 2, \pm 3, \ldots$ is the index of diffraction order, $n_1$ and $n_2$ are the refractive indices before and after the air-dielectric interface.

#### 4.1.3. ALD-growth in corrugated structures

Resonant waveguide gratings have been extensively studied since last couple of decades and several researchers fabricated RWG structures on various substrates employing ALD films to investigate their waveguiding properties, propagation losses, and fine tune the optical spectra (Alasaarela et al., 2010, 2011b; Saleem et al., 2011a, 2012a,d, 2013a,b, 2014a,c,d; Saleem, 2012). The fabrication of silicon photonic nano-structures is established about a decade using state-of-the-art 198 or 248 nm deep UV-lithographic techniques, which limit the structure size around 100 or 160 nm,

respectively. For Si-slot waveguides, the achieved feature size was below 100 nm by partial filling with ALD-grown materials as well as to reduce the propagation losses (Alasaarela et al., 2009, 2011a; Säynätjoki et al., 2011; Karvonen et al., 2014). **Figure 6B** shows cross-section view of SEM image of a Si-slot waveguide coated by amorphous TiO$_2$ film with a smooth surface showing peak-to-peak surface roughness <0.4 nm over an AFM measured area of $1 \times 1\,\mu$m (Alasaarela et al., 2010).

Polycrystalline ALD-TiO$_2$ films grown at 350°C exhibit propagation losses due to grain boundaries that can be minimized by depositing an intermediate ALD-grown layer of Al$_2$O$_3$ as shown in **Figures 6B,C**. As an example of conformal ALD-growth, RWG structure of a binary grating profile in fused silica substrate is coated by amorphous film of TiO$_2$ as shown in **Figure 6D**. **Figures 6E,F** show SEM images of TiO$_2$ films on a silicon binary grating and a sinusoidal grating profile fabricated on a hydrogen bonded polymer (Azobenzene Complex) that was spin coated on a glass substrate, respectively. The grating pattern was written using circularly polarized light of 488 nm Ar$^+$ laser source of 300 mW/cm$^2$ intensity to expose polymer. The fabricated polymer grating profile was coated with conformal TiO$_2$ thin film by ALD at 80°C to avoid thermal degradation (Alasaarela et al., 2011b). Likewise, **Figures 6G,H** show the ALD conformal growth of Al$_2$O$_3$ on high aspect ratio corrugated structures (Ritala et al., 1999; Weber et al., 2012a). In **Figures 6I–K** silicon slot waveguide patterns to be filled by ALD-grown materials and the subsequent filling by ALD-Al$_2$O$_3$ and ALD-TiO$_2$ in nano slots are shown, respectively. Moreover, the slots are filled as to demonstrate extremely uniform nanolaminate structure of five layers with 10 nm of each ALD-Al$_2$O$_3$ and ALD-TiO$_2$ with reduced propagation losses compared to crystalline ALD-TiO$_2$ films (Alasaarela et al., 2011a).

#### 4.1.4. Propagation loss reduction with ALD-coated titanium dioxide thin films

The optical field in the waveguide structures must be confined and guided through with minimum field loss owing to controlled sidewall surface roughness without leakage into substrate that has been accompanied through uniform surface coatings of high index material(s). Alasaarela et al. (2010, 2011a) have shown that such propagation losses due to surface roughness could be reduced by conformal coatings of ALD-TiO$_2$ films for which a further reduction in losses was measured by increasing film thickness. This attributes to increase in an effective index of waveguide to strongly confine the propagating waveguide modes and a significant reduction in surface roughness. **Table 1** illustrates the values of waveguide losses (in decibels per centimeter) with an increase in the thickness of TiO$_2$ film. We recently studied, theoretically, the optical dispersion engineering properties of silicon-strip waveguides using ALD-TiO$_2$ films as overlayer on nano-structures (Erdmanis et al., 2012).

#### 4.1.5. ALD-growth in nanoreplicated corrugated profiles

Low temperature ALD-growth is of much more importance for polymeric substrate waveguides in applications for various biomolecular sensors and RWGs (Triani et al., 2010; Magnusson et al., 2011). Nanoimprint technology was proposed by (Chou et al., 1997) in 1990s to fabricate replicated nano-structures.

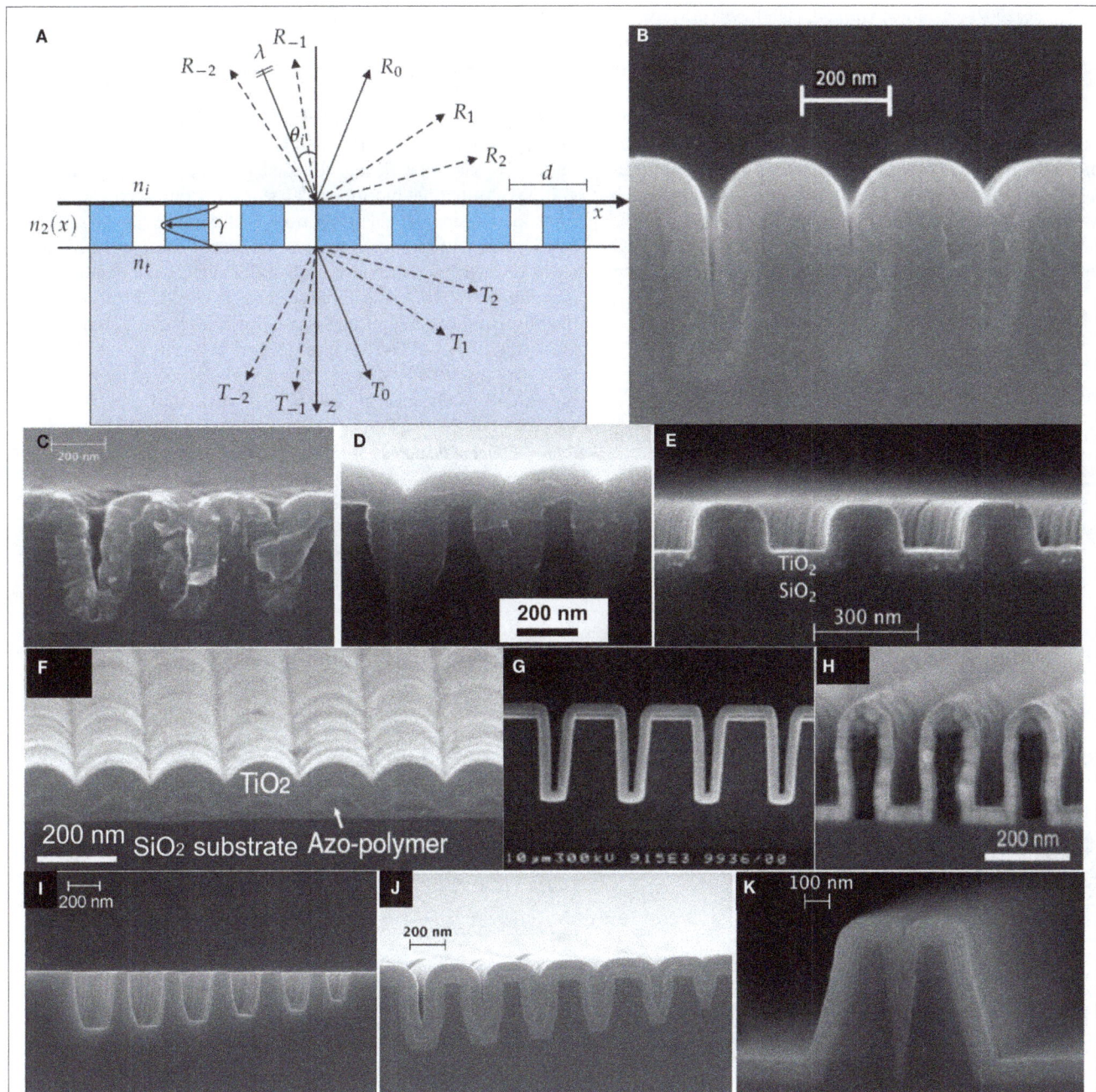

**FIGURE 6 | (A)** A schematic illustration of resonant waveguide grating with forward and backward diffraction [reproduced with permission (Saleem, 2012)]. Copyright 2012, University of Eastern Finland. **(B)** SEM image of silicon slot structure coated by ALD-TiO$_2$ amorphous layer at 120°C. ALD-grown amorphous layer of TiO$_2$ on silicon slot structure [reproduced with permission (Alasaarela et al., 2010)]. Copyright 2010, Optical Society of America. **(C)** Without and **(D)** with intermediate Al$_2$O$_3$ layer [reproduced with permission (Alasaarela et al., 2010)]. Copyright 2010, Optical Society of America. **(E)** SEM image of TiO$_2$ film on a silicon binary grating [reproduced with permission (Alasaarela et al., 2010)]. Copyright 2010, Optical Society of America. **(F)** SEM image of TiO$_2$-coated sinusoidal profile of Azo-polymer on a glass substrate [reproduced with permission (Alasaarela et al., 2011b)]. Copyright 2011, Optical Society of America. **(G)** Cross-sectional view of SEM image of 300 nm ALD-grown Al$_2$O$_3$ film on Si wafer with trench structure [reproduced with permission (Ritala et al., 1999)]. Copyright 1999, WILEY-VCH Verlag GmbH. **(H)** SEM image of an overcoated polymer grating [reproduced with permission (Weber et al., 2012b)]. Copyright 2012, Elsevier. **(I–K)** Si-slot waveguide structures, filling by ALD-TiO$_2$/Al$_2$O$_3$ film and filling by nanolamintes of ALD-TiO$_2$/Al$_2$O$_3$, respectively [reproduced with permission (Säynätjoki et al., 2011)]. Copyright 2011, Optical Society of America.

Transparent optical polymers have emerged as potential candidates in nanophotonic functional devices through replication in a wide variety of thermo-plastics (Herzig, 1997; Jaszewski et al., 1998; Mönkkönen et al., 2002; Liou and Chen, 2006) using replication tools of high replication fidelity and resolution to fabricate high aspect ratio structures (Pietarinen et al., 2007;

Siitonen et al., 2007; Cui, 2008; Worgull, 2009). Recently, we have demonstrated conformal growth of ALD technique to fill various sub-wavelength RWG structures. A silicon master stamp was fabricated using hydrogen silsesquioxane (HSQ) resist without Reactive Ion Etching (RIE) process, which employed as to replicate sub-wavelength grating structures with different periods in thermoplastic polymers as shown in **Figures 7A,D** (Saleem et al., 2012c). The replicated grating structures in polycarbonate substrate with periods $d = 425$ and 368 nm are shown in **Figures 7B,E** and subsequently coated by thin amorphous films of ALD-TiO$_2$ as shown in **Figures 7C,H**. Likewise, the replication of corrugated

profiles in other polymers such as COC and OrmoComp and their overlayer ALD-TiO$_2$ coatings are shown in **Figures 7F,G** and **7I,J**, respectively. **Figure 7K** shows a replicated grating profile with period $d = 540$ nm, coated by 60 nm of ALD-TiO$_2$ in applications of non-polarizing gratings (Saleem et al., 2012e). A high aspect ratio replicated structure followed by ALD-growth is depicted in **Figure 7L**. Saleem et al. (2011b) used such replicated polymeric gratings as athermal bio-molecular sensors within accuracy of a fraction of a nanometer. Furthermore, such sub-wavelength replicated grating structures in various polymers were compared in terms of their resonance peak stability, efficiency, full width half maximum (FWHM), ease of fabrication, and residual stresses for their use in potential applications (Saleem et al., 2013d). The interfacial adhesion of ALD-TiO$_2$ films on polycarbonate substrate has been demonstrated by theoretical models based on experimental measurements (Triani et al., 2005; Latella et al., 2007).

**Table 1 | Propagation loss values of silicon slot waveguides as a function of ALD-TiO$_2$ film thickness.**

| ALD-TiO$_2$ thickness (nm) | Propagation loss (dB/cm) |
|---|---|
| 0 | ~65 |
| 20 | ~14 |
| 30 | ~11 |
| 50 | ~07 |

### 4.1.6. Effect of thickness of TiO$_2$ films on replicated structures

A filtering design is recently proposed by Saleem et al. (2014b) with its real experimental demonstration to present the influence of ALD-TiO$_2$ thickness layer on replicated nanostrcutures of RWGs. Polymeric rectangular grating profile with a ridge height

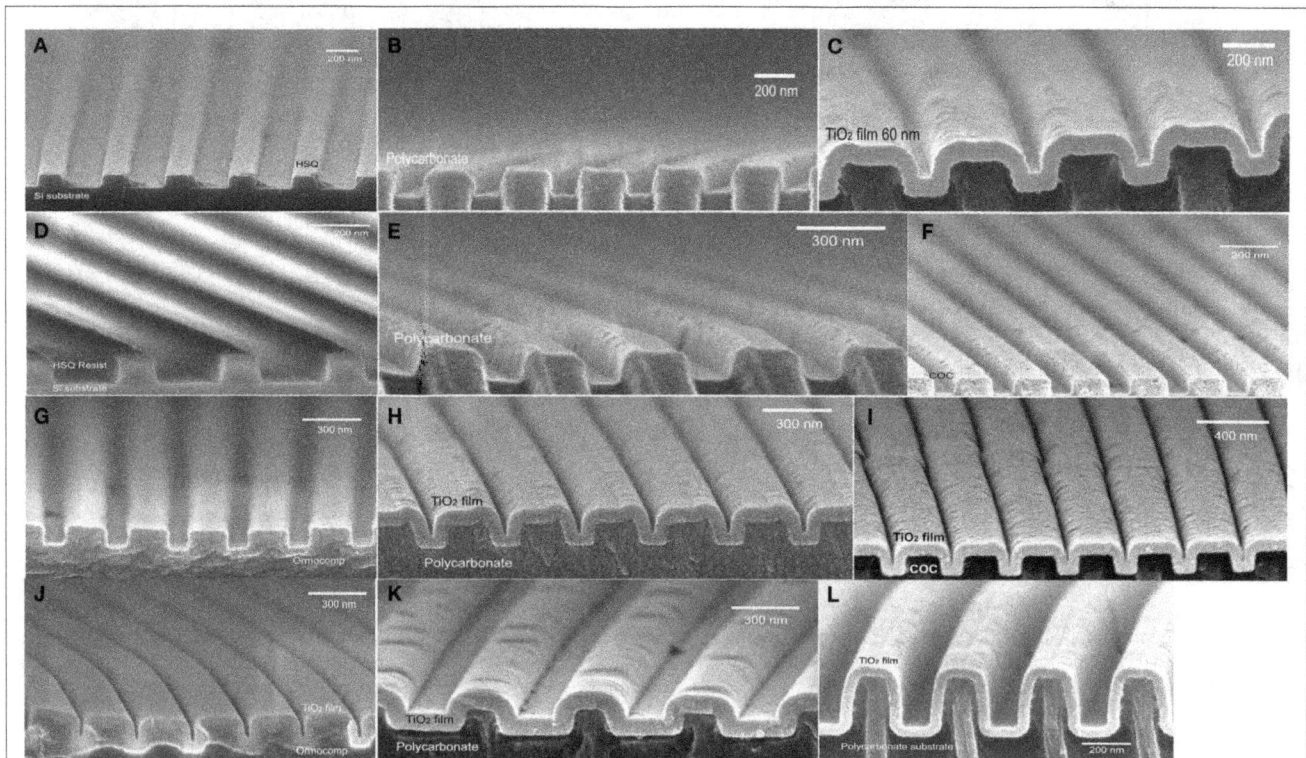

**FIGURE 7 | (A)** Si-stamp of RWG with period $d = 425$ nm by HSQ resist without reactive ion etching. **(B)** Nanoimprinted profile of RWG in polycarbonate substrate by Si-Stamp of $d = 425$ nm. **(C)** Replicated grating profile coated by thin amorphous layer of ALD-TiO$_2$.
**(D)** Si-stamp of sub-wavelength RWG with period $d = 325$ nm by HSQ resist. **(E)** Replicated grating profile in polycarbonate. **(F)** Replicated grating profile in COC. **(G)** Replicated grating profile in OrmoComp. **(H)** Replicated grating profile in polycarbonate ($d = 368$ nm) coated by

thin amorphous layer of ALD-TiO$_2$. **(I)** Replicated grating profile in COC ($d = 325$ nm) coated by thin amorphous layer of ALD-TiO$_2$.
**(J)** Replicated grating profile in OrmoComp ($d = 325$ nm) coated by thin amorphous layer of ALD-TiO$_2$. **(K)** Replicated grating profile in polycarbonate ($d = 540$ nm) coated by thin amorphous layer of ALD-TiO$_2$ for non-polarizing properties. **(L)** Replicated grating profile in polycarbonate coated by thin amorphous layer of ALD-TiO$_2$ showing high aspect ratio structure.

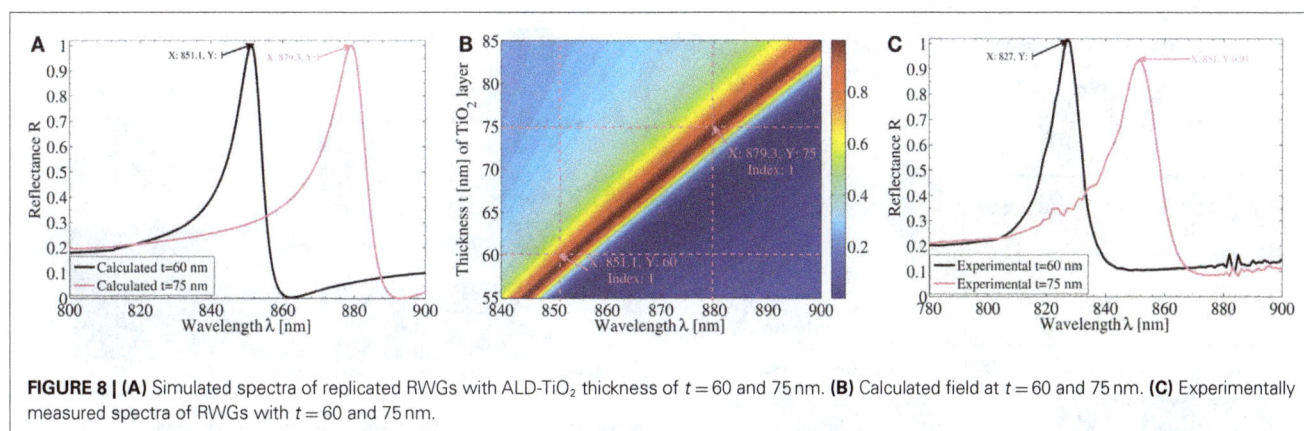

**FIGURE 8 | (A)** Simulated spectra of replicated RWGs with ALD-TiO$_2$ thickness of $t = 60$ and 75 nm. **(B)** Calculated field at $t = 60$ and 75 nm. **(C)** Experimentally measured spectra of RWGs with $t = 60$ and 75 nm.

$h = 120$ nm, period $d = 425$ nm, ridge width $w = 268$ nm, and duty cycle $ff = w/d = 0.63$ was covered by a thin dielectric TiO$_2$ cover layer of thicknesses $t = 60$ and 75 nm, grown by ALD. The device was illuminated by a linearly polarized plane wave (TE polarization) from air at an angle of incidence $\theta_i = 20°$ and the specularly reflection was obtained at $\theta_o = 20°$. **Figure 8A** shows the calculated field at TiO$_2$ thicknesses of 60 and 75 nm and **Figure 8B** shows the simulated results by Fourier Modal Method (FMM) to predict spectral shift at (100%) maximum reflectance as a function of variation in waveguide thickness (TiO$_2$ layer). A spectral shift of ~28 nm was calculated theoretically with an increase in ALD-TiO$_2$ thickness by ~15 nm. Experimental measurements show specular reflectance resonance peaks at wavelengths 827 and 851 nm for $t$ values of 60 and 75 nm, respectively as shown in **Figure 8C**. Such spectral shifts are attributed to increase in effective refractive index of waveguide, which result in an interaction of grating structure at longer wavelengths to originate resonance. Furthermore, simulated and experimentally measured results show slightly different spectral locations, which may arise due to surface roughness, true refractive indices of materials, structure linewidths, and slight variations in the periodicity after a number of fabrication steps (Saleem et al., 2011b).

## 4.2. ALD-GROWTH IN NANOPLASMONICS

Metallic nanoparticles have a number of valuable optical properties derived from their ability to collectively support the light-induced electronic oscillations to harness electromagnetic surface waves called, as surface plasmon polaritons (SPPs). SPPs are density fluctuations of conduction electrons in metals and propagate along metal surface with exponentially decaying evanescent field perpendicular to the metal-dielectric interface (Ritchie, 1957) and extend ~100–300 nm for visible light (Wei and Xu, 2014). SPPs have attractive features with metallic nanogaps, nanoholes, nanotips, nanoapertures, nanowires, or nanoparticles and concentrate optical field to nanoscale volumes and allow fundamental study of light–matter interactions at length scales, which were not accessible otherwise (Brongersma and Shalaev, 2010). Moreover, owing to sharp corners of engineered nano-structures, another important class of localized plasmon wave, which confines optical energy in a more tightly fashion at a relatively shorter decay length of 10–50 nm called, localized surface plasmons (LSPs) as shown in

**Figure 9A** (Im et al., 2012) and likewise to SPPs confine maximum optical energy at metallic surfaces.

Surface plasmon polaritons arise due to coupling between photons and conduction electrons, thereby, carrying higher momentum than free-space light. The momentum of light can be enhanced by illuminating a thin metallic film through a prism or strong grating coupling of patterned metallic surfaces (Raether, 1986).

### 4.2.1. ALD-Passive layer on metallic nanoplasmonic structures

ALD-grown nano-overlayers on metallic nanoplasmonic structures not only protect surface but also enhance the optical transmission. In most plasmonic devices operating at visible or infrared regimes, Ag or Au metals are used. At wavelengths below 600 nm, Ag has low ohmic losses, higher propagation lengths (low SPPs damping), higher sensitivity for biosensors, and are used in applications where a direct interface with biological molecules is not required, such as electrodes for plasmonic solar cells (Atwater and Polman, 2010). Silver Ag is used due to cost-effective solutions, better optical properties and relatively strong interfacial adhesion with glass substrates than Au (Ferry et al., 2008; Lindquist et al., 2008). Despite these advantages, Ag has lower chemical stability and readily oxidized in the air and shows a reduction in optical intensity in nanoplasmonic structures without passive layer (Im et al., 2010). While the Ag surfaces can be protected and improved in functionality by an overlayer of dielectric coating (~10–20 nm), which must be less than mean decay length of SPPs in order to annul from being absorption of SP field in thicker layers.

These ultra-thin barrier films must be dense, pinhole free, encapsulate the patterned structure entirely (Saleem et al., 2013c) and grown at a low deposition temperature to avoid degrading or oxidizing of underlying Ag film. Similarly, the use of silver nanoparticles coated by ~1 nm Al$_2$O$_3$ layer was investigated to preserve plasmon resonance peak after annealing at 500°C and have shown to degrade resonance peaks without Al$_2$O$_3$ layer even at annealing temperature of 200°C (Whitney et al., 2007). Likewise, Sung et al. (2008) showed enhanced thermal stability of Al$_2$O$_3$ coated silver nanoparticles against high power femtosecond laser. Self-assembly based plasmonic arrays tuned by minimum thickness of ALD fabricated (<20 nm) Au dots achieved enhanced visible light absorption per volume-equivalent thickness

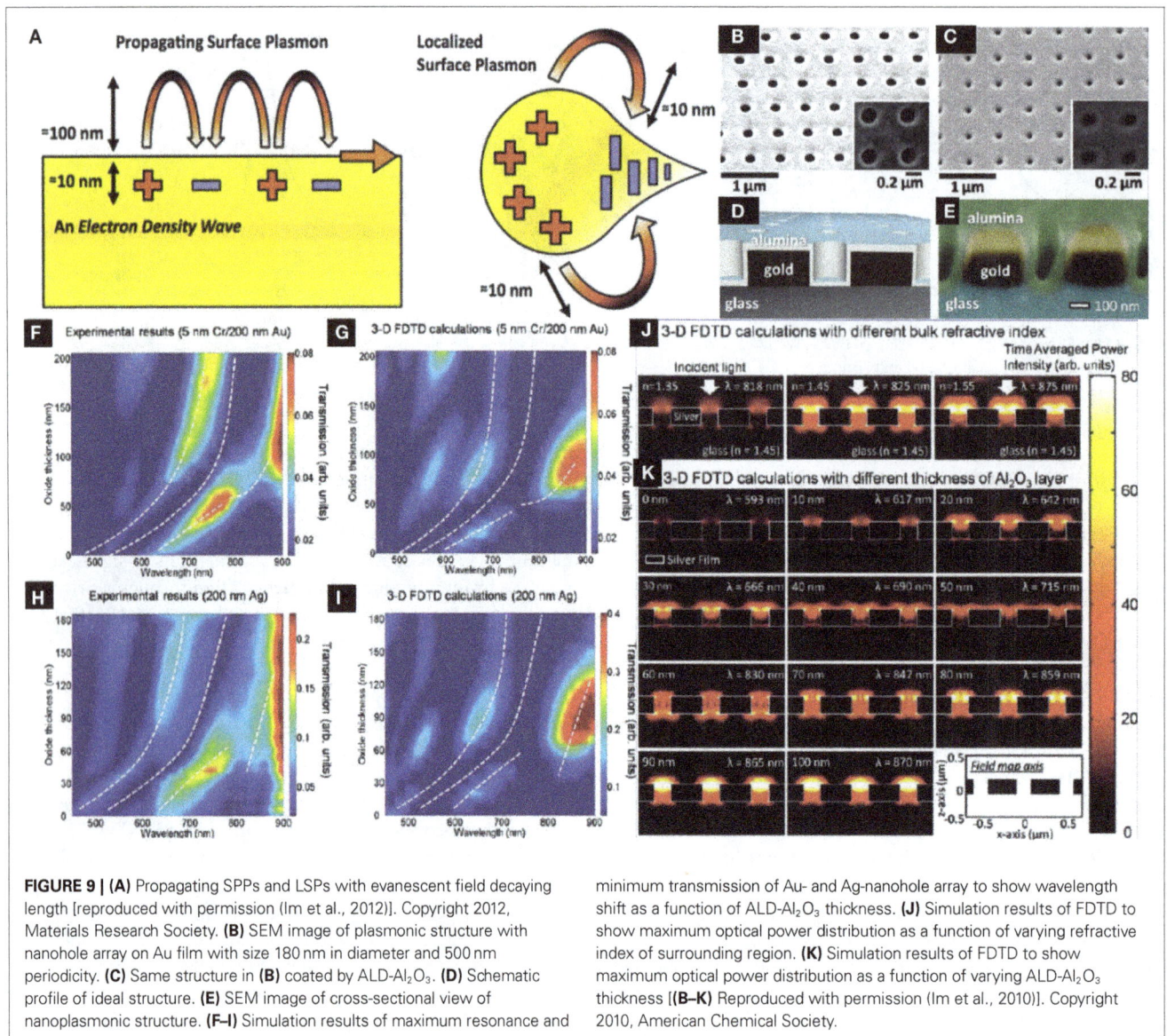

**FIGURE 9 | (A)** Propagating SPPs and LSPs with evanescent field decaying length [reproduced with permission (Im et al., 2012)]. Copyright 2012, Materials Research Society. **(B)** SEM image of plasmonic structure with nanohole array on Au film with size 180 nm in diameter and 500 nm periodicity. **(C)** Same structure in **(B)** coated by ALD-Al$_2$O$_3$. **(D)** Schematic profile of ideal structure. **(E)** SEM image of cross-sectional view of nanoplasmonic structure. **(F–I)** Simulation results of maximum resonance and minimum transmission of Au- and Ag-nanohole array to show wavelength shift as a function of ALD-Al$_2$O$_3$ thickness. **(J)** Simulation results of FDTD to show maximum optical power distribution as a function of varying refractive index of surrounding region. **(K)** Simulation results of FDTD to show maximum optical power distribution as a function of varying ALD-Al$_2$O$_3$ thickness [**(B–K)** Reproduced with permission (Im et al., 2010)]. Copyright 2010, American Chemical Society.

in applications including conversion of solar energy into electrical power (Hägglund et al., 2013). One of the possible reason of enhancing optical transmission is conformal and precisely uniform deposition of dielectric thin films on nano-structures is its predominance over rough topographies.

### 4.2.2. Tuning local refractive index of nanoplasmonic structures

The plasmonic resonance depends on the geometric parameters, incident angle, or wavelength, surrounding environmental conditions (dielectric constants), molecular layer bindings or adsorption, surface roughness, and the thickness of an overlayer on the metallic nano-structures. A change in any one of these parameters modify the resonance properties and ultimately shifts the resonance peak. **Figure 9B** shows an SEM image of nanohole array structure fabricated on Au (with 5 nm Cr adhesive layer) metallic film of 200 nm thickness by Focused Ion Beam with a period of 500 nm and hole diameter of 180 nm. A schematic of a nanoplasmonic structure coated by ALD-Al$_2$O$_3$ is shown in **Figure 9D** while top and cross-sectional views of SEM images of ALD-Al$_2$O$_3$ overlayer on metallic nanohole array are shown in **Figures 9B,C** and **Figure 9E**, respectively. Since, the ALD-Growth fills the nanoholes conformably and gives rise resonance spectral shifts as a result of tuning effective index seen by SPs, which is approximated by a simple model (Jung et al., 1998):

$$n_{eff} = n_{bulk} + (n_{film} - n_{bulk})\left[1 - e^{(-2t/l)}\right], \qquad (7)$$

where $n_{bulk}$ and $n_{film}$ are the refractive indices of the surrounding bulk materials and thin deposited film, respectively, $t$ is the thickness of the film and $l$ is the decay length of SPs evanescent field perpendicular to the surface. It can be seen from Eq. (7) that as the thickness $t$ of Al$_2$O$_3$ film increases, effective index increases from $n_{bulk}$ to $n_{film}$, which results in a red-shift: here $n_{bulk}$ is air index and $n_{film}$ is $Al_2O_3$ index. Thereby, by varying $t$ one can tune the resonance frequency of SPs precisely, since ALD grows films up to angstrom scale resolution. Furthermore, an increase in the $t$ results in higher intensity that might be attributed to low SPs damping at relatively longer wavelengths as well as an increase in the periodicity of the structure reported by Przybilla et al. (2006). Furthermore, filling of nanoholes by ALD-$Al_2O_3$ not only results in higher effective index and reduction in waveguide-like attenuation of sub-wavelength structures for optical transmission enhancement but also encompasses index matching on both sides of the film. Thus, making an asymmetric peak to symmetric one with a transformation of (air–metal–glass) interfaces to (air–alumina–metal–glass) that boost optical transmission at a certain $Al_2O_3$ thickness where SPs wave sees almost a similar index above and below metallic structures (Krishnan et al., 2001).

### 4.2.3. Effect of $Al_2O_3$ thickness on spectral shifts

The effects of $Al_2O_3$ thickness on the spectral shifts have been investigated by theoretical calculations using 3D Finite Difference Time Domain (FDTD) simulations and compared with experimental results, not shown here [see Im et al. (2010)]. **Figures 9F,G** show a comparison of calculated results with ALD-$Al_2O_3$ deposition with a step of 6 and 10 nm on gold sample, respectively and white dashed lines depict resonance maxima and transmission minima. An increase in ALD-$Al_2O_3$ thickness not only results in red-shift but also increases transmission intensity by a factor of about 5 at certain thickness value. Likewise, Ag-nanohole arrays show the calculated results and their consistency with experimentally measured ones, as seen in **Figures 9H,I**.

3D FDTD simulations were carried out to validate the optical field distribution in each nanohole as a function of index matching on top and bottom sides of metallic films. In these simulations the bulk refractive index over nanohole arrays is varied from 1.35 to 1.55. **Figure 9J** shows that the maximum transmitted field intensity and output power occurs when the $n_{bulk}$ above and inside of each nanohole matches to that of the glass substrate with index 1.45 and Krishnan et al. (2001) demonstrated that under such symmetric conditions optical transmission enhances by a factor of 10 compared to asymmetric interfaces. **Figure 9K** shows FDTD calculated results of filed distributions in nanohole arrays for different values of $Al_2O_3$ overlayer by considering the refractive indices of glass and alumina 1.45 and 1.65, respectively. At symmetric condition one expects that the effective index seen by the SP waves approximately matches to that of substrate (glass) and maximum electromagnetic power distribution occurs at certain thickness value of $t = 60$–$70$ nm. Hence, ALD overlayer can tune and shift transmission spectrum in a precisely and controlled manner, which is of significant importance in biochemical sensing, surface plasmon resonance imaging (Chinowsky et al., 2004), biomimetic sensing, and membrane protein research.

## 5. CONCLUSION

Due to continuous device miniaturization, precise control of thin film growth is essential at the atomic level to fabricate nano-optical and semiconductor devices. To target such requirements, ALD methods have been developed for the growth of ultra-thin and highly conformal and uniform films. ALD is a vapor phase technique based on sequential and self-limiting growth on functional surfaces as well as on substrates with high aspect ratios after chemical reactions. The use of ALD deposited optical materials such as $TiO_2$ and $Al_2O_3$ in nanophotonic devices are discussed.

Atomic layer deposition is capable of smooth and uniform deposition on a wide variety of organic and inorganic materials in a number of high-Tech applications including bio-molecular sensors, biomimetic sensing, membrane proteins, nanophotonic and microelectronic devices. ALD coatings can conformably fill the PC of metallic nanohole arrays that increases the extraordinary optical transmission of plasmonic structures and to reduce the feature size of the silicon waveguides and propagation losses. The increase in ALD-$TiO_2$ thickness results in a further reduction in propagation losses in Si-slot waveguides and increase in transmission through plasmonic structures due to increase in effective index of the structures. The low deposition rate of ALD is extremely important to deposit high density, high quality, and uniform thin films in microelectronics industry as high-k gate oxide thin barrier layers as well as to propagate evanescent modes with relatively longer decay lengths in photonics. The low temperature growth enables and emerges ALD as a unique technique to deposit coatings on polymer substrates, temperature sensitive materials, and bio-molecular materials to avoid their degradation. ALD-growth on several substrates also promises to protect the underneath surfaces/structures from environmental effects due to extremely low penetration through high density and pinhole free films.

The future prospects of ALD are very promising. ALD should also play an integral role in new paradigms for optical materials in nanophotonics. The number of applications for ALD also continues to grow outside of the semiconductor arena. The future should see ALD continue to expand into new areas and find additional applications and challenges in demanding protective coatings, microelectromechanical systems, nanoelectronics, solar cells etc., that benefit from its precise thickness control and conformality.

## ACKNOWLEDGMENTS

Financial support from Strategic funding Initiative TAILOR of the University of Eastern Finland, Academy of Finland, Tekes (Finland), Higher Education Commission (HEC), USAID Center for Advance Studies, CES, NUST, (Pakistan) are greatly appreciated.

## REFERENCES

Aarik, J., Aidla, A., Kiisler, A.-A., Uustare, T., and Sammelselg, V. (1997a). Effect of crystal structure on optical properties of $TiO_2$ films grown by atomic layer deposition. *Thin Solid Films* 305, 270–273. doi:10.1016/S0040-6090(97)00135-1

Aarik, J., Aidla, A., Sammelselg, V., and Uustare, T. (1997b). Effect of growth conditions on formation of $TiO_2$-ii thin films in atomic layer deposition process. *J. Cryst. Growth* 181, 259–264. doi:10.1016/S0022-0248(97)00279-0

Aarik, J., Aidla, A., Mändar, H., and Sammelselg, V. (2000). Anomalous effect of temperature on atomic layer deposition. *J. Cryst. Growth* 220, 531–537. doi:10.1016/S0022-0248(00)00897-6

Aarik, J., Aidla, A., Uustare, T., and Sammelselg, V. (1995). Morphology and structure of $TiO_2$ thin films grown by atomic layer deposition. *J. Cryst. Growth* 148, 268–275. doi:10.3762/bjnano.5.7

Alasaarela, T., Korn, D., Alloatti, L., Säynätjoki, A., Tervonen, A., Palmer, R., et al. (2011a). Reduced propagation loss in silicon strip and slot waveguides coated by atomic layer deposition. *Opt. Express* 19, 11529–11538. doi:10.1364/OE.19.011529

Alasaarela, T., Zheng, D., Huang, L., Priimagi, A., Bai, B., Tervonen, A., et al. (2011b). Single-layer one-dimensional nonpolarizing guided-mode resonance filters under normal incidence. *Opt. Lett.* 36, 2411–2413. doi:10.1364/OL.36.002411

Alasaarela, T., Saastamoinen, T., Hiltunen, J., Aäynätjoki, A., Tervonen, A., Stenberg, P., et al. (2010). Atomic layer deposited titanium dioxide and its application in resonant waveguide grating. *Appl. Opt.* 49, 4321–4325. doi:10.1364/AO.49.004321

Alasaarela, T., Säynätjoki, A., Hakkarainen, T., and Honkanen, S. (2009). Feature size reduction of silicon slot waveguide by partial filling using atomic layer deposition. *Opt. Eng.* 48, 080502. doi:10.1117/1.3206731

Atwater, H. A., and Polman, A. (2010). Plasmonics for improved photovoltaic devices. *Nat. Mater.* 9, 205–213. doi:10.1038/nmat2629

Avrutskii, D., Golubenko, G., Sychugov, V., and Tishchenko, A. (1985). Light reflection from the surface of a corrugated waveguide. *Sov. Tech. Phys. Lett.* 11, 401–402.

Bedair, S. (1994). Atomic layer epitaxy deposition processes. *J. Vac. Sci. Technol.* B12, 179. doi:10.1116/1.587179

Bennett, J. M., Pelletier, E., Albrand, G., Borgogno, J., Lazarides, B., Carniglia, C. K., et al. (1989). Comparison of the properties of titanium dioxide films prepared by various techniques. *Appl. Opt.* 28, 3303–3317. doi:10.1364/AO.28.003303

Brongersma, M. L., and Shalaev, V. M. (2010). The case for plasmonics. *Appl. Phys.* 328, 440–441. doi:10.1126/science.1186905

Carcia, P., McLean, R., Groner, M., Dameron, A., and George, S. (2009). Gas diffusion ultrabarriers on polymer substrates using Al$_2$O$_3$ atomic layer deposition and SiN plasma-enhanced chemical vapor deposition. *J. Appl. Phys.* 106, 23533. doi:10.1063/1.3159639

Chinowsky, T. M., Mactutis, T., Fu, E., and Yager, P. (2004). Optical and electronic design for a high-performance surface plasmon resonance imager. *SPIE Proc.* 5261, 173–182. doi:10.1117/12.538536

Chou, S., Krauss, P., Zhang, W., Guo, L., and Zhuang, L. (1997). Sub-10 nm imprint lithography and application. *J. Vac. Sci. Technol.* B 15, 2897–2904. doi:10.1116/1.589752

Chung, C., Liao, M., and Lai, C. (2009). Effect of oxygen flow ratios and annealing temperatures on Raman and photoluminescence of titanium oxide thin films deposited by reactive magnetron sputtering. *Thin Solid Films* 518, 1415–1418. doi:10.1016/j.tsf.2009.09.076

Cui, Z. (ed.) (2008). *Nanofabrication.* New York, NY: Springer.

Dobrowolski, J. A., Tikhonravov, A. V., Trubetskov, M. K., Sullivan, B. T., and Verly, P. G. (1996). Optimal single-band normal incidence antireflection coatings. *Appl. Opt.* 35, 644–658. doi:10.1364/AO.35.000644

Erdmanis, M., Karvonen, L., Saleem, M. R., Ruoho, M., Pale, V., Tervonen, A., et al. (2012). Ald-assisted multiorder dispersion engineering of nanophotonic strip waveguides. *J. Lightwave Technol.* 30, 2488–2493. doi:10.1109/JLT.2012.2200235

Ferry, V. E., Sweatlock, L. A., Pacifici, D., and Atwater, H. A. (2008). Plasmonic nanostructure for efficient light coupling into solar cells. *Nano Lett.* 8, 4391–4397. doi:10.1021/nl8022548

George, S., Ott, A., and Klaus, J. (1996). Surface chemistry for atomic layer growth. *J. Phys. Chem.* 100, 13121–13131. doi:10.1021/jp9536763

George, S. M. (2010). Atomic layer deposition: an overview. *Chem. Rev.* 110, 111–131. doi:10.1021/cr900056b

Gong, J., Jung, D., El-Masry, N., and Bedair, S. (1990). Atomic layer epitaxy of AlGaAs. *J. Appl. Phys.* 57, 400.

Goodman, C. H., and Pessa, M. V. (1986). Atomic layer epitaxy. *J. Appl. Phys.* 60, R65. doi:10.1063/1.337344

Graugnard, E., Gaillot, D. P., Dunham, S. N., Neff, C. W., Yamashita, T., and Summers, C. J. (2006). Photonic band tuning in two-dimensional photonic crystal slab waveguides by atomic layer deposition. *Appl. Phys. Lett.* 89, 181108. doi:10.1063/1.2360236

Groner, M., Fabreguette, F., Elam, J., and George, S. (2004). Low-temperature Al$_2$O$_3$ atomic layer growth. *J. Chem. Mater.* 16, 639–645. doi:10.1021/cm0304546

Hägglund, C., Zeltzer, G., Ruiz, R., Thomann, I., Lee, H.-B.-R., Brongersma, M. L., et al. (2013). Self-assembly based plasmonic arrays tuned by atomic layer deposition for extreme visible light absorption. *Nano Lett.* 13, 3352–3357. doi:10.1021/nl401641v

Herzig, H. (ed.) (1997). *Micro-Optics: Elements, Systems and Applications.* London: Taylor & Francis.

Hessel, A., and Oliner, A. (1965). A new theory of wood's anomalies on optical coatings. *Appl. Opt.* 4, 1275–1297. doi:10.1364/AO.4.001275

Huang, J., Wang, X., and Wang, Z. L. (2006). Controlled replication of butterfly wings for achieving tunable photonic properties. *Nano Lett.* 6, 2325–2331. doi:10.1021/nl061851t

Huber, S., der Kruijs, R., Yakshin, A., and Zoethout, E. (2014). Subwavelength single layer absorption resonance antireflection coatings. *Opt. Express* 22, 490–497. doi:10.1364/OE.22.000490

Ide, Y., McDermott, B., Hashemi, M., Bedair, S., and Goodhue, W. (1988). Sidewall growth by atomic layer epitaxy. *J. Appl. Phys.* 53, 2314.

Im, H., Lindquist, N. C., Lesuffleur, A., and Oh, S.-H. (2010). Atomic layer deposition of dielectric overlayers for enhancing the optical properties and chemical stability of plasmonic nanoholes. *ACS Nano* 4, 947–954. doi:10.1021/nn901842r

Im, H., Wittenberg, N. J., Lindquist, N. C., and Oh, S.-H. (2012). Atomic layer deposition: a versatile technique for plasmonics and nanobiotechnology. *J. Mater. Res.* 27, 663–671. doi:10.1557/jmr.2011.434

Jalaluddin, M., and Magnusson, R. (2013). Guided-mode resonant thermo-optic tunable filters. *IEEE Photonics Technol. Lett.* 25, 1412–1415. doi:10.1109/LPT.2013.2266272

Jaszewski, R., Schift, H., Gobrecht, J., and Smith, P. (1998). Hot embossing in polymers as a direct way to pattern resist. *Microelectron. Eng.* 4, 575–578. doi:10.1016/S0167-9317(98)00135-X

Jia, K., Zhang, D., and Ma, J. (2011). Sensitivity of guided mode resonance filter-based biosensor in visible and near infrared ranges. *Sens. Actuators B Chem.* 156, 194–197. doi:10.1016/j.snb.2011.04.013

Jung, L. S., Campbell, C. T., Chinowsky, T. M., Mar, M. N., and Yee, S. S. (1998). Quantitative interpretation of the response of surface plasmon resonance sensors to adsorbed films. *Langmuir* 14, 5636–5648. doi:10.1021/la971228b

Kääriäinen, T., Cameron, D., Kääriäinen, M.-L., and Sherman, A. (2013). *Atomic Layer Deposition: Principles, Characteristics, and Nanotechnology Applications.* Beverly, MA: Scrivener publishing, Wiley.

Karvonen, L., Säynätjoki, A., Roussey, M., Kuittinen, M., and Honkanen, S. (2014). Application of atomic layer deposition in nanophotonics. *SPIE Proc.* 8988, 89880Z–89881Z. doi:10.1364/OE.21.032417

Kasap, S., and Capper, P. (eds) (2006). *Handbook of Electronic and Photonic Materials.* New York, NY: Springer.

Kemell, M., Pore, V., Tupala, J., Ritala, M., and Leskelä, M. (2007). Atomic layer deposition of nanostructured TiO$_2$ photocatalysts via template approach. *Chem. Mater.* 19, 1816–1820. doi:10.1021/cm062576e

Kim, H., Lee, H.-B., and Maeng, W.-J. (2009). Applications of atomic layer deposition to nanofabrication and emerging nanodevices. *Thin Solid Films* 517, 2563–2580. doi:10.1016/j.tsf.2008.09.007

King, J. S., Graugnard, E., and Summers, C. J. (2005a). TiO$_2$ inverse opals fabricated using low-temperature atomic layer deposition. *Adv. Mater.* 17, 1010–1013. doi:10.1002/adma.200400648

King, J., Heineman, D., Graugnard, E., and Summers, C. (2005b). Atomic layer deposition in porous structures: 3D photonic crystals. *Appl. Surf. Sci.* 244, 511–516. doi:10.1016/j.apsusc.2004.10.110

Knez, M., Nielsch, K., and Niinistö, L. (2007). Synthesis and surface engineering of complex nanostructures by atomic layer deposition. *Adv. Mater.* 19, 3425–3438. doi:10.1002/adma.200700079

Knop, K. (1978). Rigorous diffraction theory for transmission phase gratings with deep rectangular grooves. *J. Opt. Soc. Am.* 68, 1206–1210. doi:10.1364/AO.17.003598

Krishnan, A., Thio, T., Kim, T., Lezec, H., Ebbesen, T., Wolff, P., et al. (2001). Evanescently coupled resonance in surface plasmon enhanced transmission. *Opt. Commun.* 200, 1–7. doi:10.1016/S0030-4018(01)01558-9

Kukli, K., Ritala, M., Leskelä, M., and Jokinen, J. (1997). Atomic layer epitaxy growth of aluminum oxide thin films from a novel Al(CH$_3$)2Cl precursor and H$_2$O. *J. Vac. Sci. Technol.* A 15, 2214–2218. doi:10.1116/1.580536

Kumar, P., Wiedmann, M. K., Winter, C. H., and Avrutsky, I. (2009). Optical properties of Al$_2$O$_3$ thin films grown by atomic layer deposition. *Appl. Opt.* 48, 5407–5412. doi:10.1364/AO.48.005407

Latella, B., Triani, G., Zhang, Z., Short, K., Bartlett, J., and Ignat, M. (2007). Enhanced adhesion of atomic layer deposited titania on polycarbonate substrates. *Thin Solid Films* 515, 3138–3145. doi:10.1016/j.tsf.2006.08.022

Lindquist, N. C., Luhman, W. A., Oh, S.-H., and Holmes, R. J. (2008). Plasmonic nanocavity arrays for enhanced efficiency in organic photovoltaic cells. *Appl. Phys. Lett.* 93, 123308. doi:10.1063/1.2988287

Liou, A., and Chen, R. (2006). Injection molding of polymer micro- and sub-micron structures with high-aspect ratios. *Int. J. Adv. Manuf. Technol.* 28, 1097–1103. doi:10.1007/s00170-004-2455-2

Macleod, H. A. (2001). *Thin-Film Optical Filters.* Bristol: Institute of Physics.

Magnusson, R., Wowro, D., Zimmerman, S., and Ding, Y. (2011). Resonant photonic biosensors with polarization-based multiparametric discrimination in each channel. *Sensors* 11, 1476–1488. doi:10.3390/s110201476

Martinu, L., Hichwa, B., and Klemberg-Sapieha, J. E. (2014). *Advances in Optical Coatings Stimulated by the Development of Deposition Techniques and the Control of Ion Bombardment.* Berlin: Springer, 36–45.

Mönkkönen, K., Hietala, J., Pääkkönen, P., Kaikuranta, T., Pakkanen, T., and Jääskeläinen, T. (2002). Replication of sub-micron features using amorphous thermoplastics. *Polym. Eng. Sci.* 42, 1600–1608. doi:10.1002/pen.11055

Ott, A., Klaus, J., George, S. M., and Johnson, J. M. (1997). Al$_2$O$_3$ thin film growth on si (100) using binary reaction sequence chemistry. *Thin Solid Films* 292, 135–144. doi:10.1016/S0040-6090(96)08934-1

Ottermann, C., and Bange, K. (1996). Correlation between the density of TiO$_2$ films and their properties. *Thin Solid Films* 286, 32–34. doi:10.1016/S0040-6090(96)08848-7

Pedrotti, F. (1993). *Introduction to Optics.* New York, NJ: Prentice-Hall, Inc.

Pietarinen, J., Siitonen, S., Immonen, J., Suvanto, M., Kuittinen, M., Mönkkönen, K., et al. (2007). Transparent thermoplastics: replication of diffractive optical elements using micro-injection molding. *Opt. Mater.* 30, 285–291. doi:10.1016/j.optmat.2006.11.046

Pimbley, J., and Lu, T.-M. (1985). Two-dimensional atomic correlations of epitaxial layers. *J. Appl. Phys.* 57, 4583. doi:10.1063/1.335364

Ponraj, J. S., Attolini, G., and Bosi, M. (2013). Review on atomic layer deposition and applications of oxide thin films. *Crit. Rev. Solid State Mater. Sci.* 38, 203–233. doi:10.1080/10408436.2012.736886

Przybilla, F., Degiron, A., Laluet, J.-Y., Genet, C., and Ebbesen, T. (2006). Optical transmission in perforated noble and transition metal films. *J. Opt.* 8, 458–456. doi:10.1088/1464-4258/8/5/015

Purniawan, A., French, P., Pandraud, G., and Sarro, P. (2010). TiO$_2$ ald nanolayer as evanescent waveguide for biomedical sensor applications. *Procedia Eng.* 5, 1131–1135. doi:10.1016/j.proeng.2010.09.310

Puurunen, R. L. (2005). Surface chemistry of atomic layer deposition: a case study for the trimethylaluminum/water process. *J. Appl. Phys.* 97, 121301. doi:10.1063/1.1940727

Raether, H. (1986). *Surface Plasmons on Smooth and Rough Surfaces and on Gratings.* Berlin: Springer-Verlag.

Riihelä, D., Ritala, M., Matero, R., and Leskelä, M. (1996). Introducing atomic layer epitaxy for the deposition of optical thin films. *Thin Solid Films* 289, 250–255. doi:10.1016/S0040-6090(96)08890-6

Ritala, M., and Leskelä, M. (1993). Growth of titanium dioxide thin films by atomic layer epitaxy. *Thin Solid Films* 225, 288–295. doi:10.1016/0040-6090(93)90172-L

Ritala, M., Leskelä, M., Dekker, J.-P., Mutsaers, C., Soininen, P. J., and Skarp, J. (1999). Perfectly conformal TiN and Al$_2$O$_3$ films deposited by atomic layer deposition. *Chem. Vap. Deposition* 5, 7–9. doi:10.1002/(SICI)1521-3862(199901)5:1<7::AID-CVDE7>3.0.CO;2-J

Ritala, M., Leskelä, M., Niinistö, L., and Haussalo, P. (1993). Titanium isopropoxide as a precursor in atomic layer epitaxy of titanium dioxide thin films. *Chem. Mater.* 5, 1174–1181. doi:10.1021/cm00032a023

Ritchie, R. (1957). Plasma losses by fast electron in thin films. *Phys. Rev.* 106, 874–881. doi:10.1103/PhysRev.106.874

Riyanto, E., Rijanto, E., and Prawara, B. (2012). A review of atomic layer deposition for nanoscale devices. *Mechatron. Electrical Power Vehicular Technol* 3, 65–72. doi:10.14203/j.mev.2012.v3.65-72

Saarinen, J., Noponen, E., and Turunen, J. (2005). Guided mode resonance filters of finite aperture. *J. Opt. Eng.* 34, 2560–2566. doi:10.1117/12.208079

Saleem, M., Khan, M., Khan, Z., Stenberg, P., Alasaarela, T., Honkanen, S., et al. (2011a). Thermal behavior of waveguide gratings. *SPIE Proc.* 8069, 80690A–80691A. doi:10.1117/12.885708

Saleem, M., Stenberg, P., Alasaarela, T., Silfsten, P., Khan, M., Honkanen, S., et al. (2011b). Towards athermal organic-inorganic guided mode resonance filters. *Opt. Express* 19, 24241–24251. doi:10.1364/OE.19.024241

Saleem, M. R. (2012). *Resonant Waveguide Gratings by Replication and Atomic Layer Deposition.* Ph.D., thesis, Department of Physics and Mathematics, University of Eastern Finland, Joensuu.

Saleem, M. R., Ali, R., Honkanen, S., and Turunen, J. (2014a). Determination of thermo-optic properties of atomic layer deposited thin TiO$_2$ films for athermal resonant waveguide gratings by spectroscopic ellipsometry. *SPIE Proc.* 9130, 9130A–9131A. doi:10.1117/12.2052299

Saleem, M. R., Ali, R., Honkanen, S., and Turunen, J. (2014b). Effect of waveguide thickness layer on spectral resonance of a guided mode resonance filter. *IEEE (IBCAST) Proc.* 14197105, 39–43. doi:10.1109/IBCAST.2014.6778117

Saleem, M. R., Honkanen, S., and Turunen, J. (2014c). Mode-splitting of a non-polarizing guided mode resonance filter by substrate overetching effect. *SPIE Proc.* 8974, 897417–897411. doi:10.1117/12.2038116

Saleem, M. R., Honkanen, S., and Turunen, J. (2014d). Thermal properties of TiO$_2$ films fabricated by atomic layer deposition. *IOP Conf. Ser. Mater. Sci. Eng.* 60, 012008. doi:10.1088/1757-899X/60/1/012008

Saleem, M. R., Honkanen, S., and Turunen, J. (2012a). Partially athermalized waveguide gratings. *SPIE Proc.* 8428, 842817–842811. doi:10.1117/12.922070

Saleem, M., Honkanen, S., and Turunen, J. (2012b). Thermal properties of TiO$_2$ films grown by atomic layer deposition. *Thin Solid Films* 520, 5442–5446. doi:10.1016/j.tsf.2012.04.008

Saleem, M., Stenberg, P., Khan, M., Khan, Z., Honkanen, S., and Turunen, J. (2012c). Hydrogen silsesquioxane resist stamp for replication of nanophotonic components in polymers. *J. Micro Nanolithogr. MEMS MOEMS* 11, 013007. doi:10.1117/1.JMM.11.1.013007

Saleem, M., Stenberg, P., Khan, M., Khan, Z., Honkanen, S., and Turunen, J. (2012d). HSQ resist for replication stamp in polymers. *SPIE Proc.* 8249, 82490G. doi:10.1117/12.907862

Saleem, M., Zheng, D., Bai, B., Kuittinen, P. S. M., Honkanen, S., and Turunen, J. (2012e). Replicable one-dimensional non-polarizing guided mode resonance gratings under normal incidence. *Opt. Express* 20, 16974–16980. doi:10.1364/OE.21.000345

Saleem, M. R., Honkanen, S., and Turunen, J. (2013a). Effect of substrate overetching and heat treatment of titanium oxide waveguide gratings and thin films on their optical properties. *Appl. Opt.* 52, 422–432. doi:10.1364/AO.52.000422

Saleem, M. R., Honkanen, S., and Turunen, J. (2013b). Non-polarizing single layer inorganic and double layer organic-inorganic one-dimensional guided mode resonance filters. *SPIE Proc.* 8613, 86130C–86131C. doi:10.1117/12.2001692

Saleem, M., Ali, R., Honkanen, S., and Turunen, J. (2013c). Thermal properties of thin Al$_2$O$_3$ films and their barrier layer effect on thermo-optic properties of TiO$_2$ films grown by atomic layer deposition. *Thin Solid Films* 542, 257–262. doi:10.1016/j.tsf.2013.06.030

Saleem, M., Honkanen, S., and Turunen, J. (2013d). Thermo-optic coefficient of ormocomp and comparison of polymer materials in athermal replicated sub-wavelength resonant waveguide gratings. *Opt. Commun.* 288, 56–65. doi:10.1016/j.optcom.2012.09.061

Sander, M., Côté, M., Gu, W., Kile, B., and Tripp, C. (2004). Teplate-assisted fabrication of dense, aligned arrays of titania nanotubes with well-controlled dimensions on substrates. *Adv. Mater.* 16, 2052–2057. doi:10.1002/adma.200400446

Säynätjoki, A., Alasaarela, T., Khanna, A., Karvonen, L., Stenberg, P., Kuittinen, M., et al. (2011). Angled sidewalls in silicon slot waveguides: conformal filling and mode properties. *Opt. Express* 17, 21066–21076. doi:10.1364/OE.17.021066

Sechrist, Z., Fabreguette, F., Heintz, O., Phung, T., Johnson, D., and George, S. (2005). Optimization and structural characterization of W/Al$_2$O$_3$ nanolaminates grown using atomic layer deposition techniques. *Chem. Mater.* 17, 3475–3485. doi:10.1021/cm050470y

Sechrist, Z., Schwartz, B., Lee, J., McCormick, J., Piestun, R., Park, W., et al. (2006). Modification of opal photonic crystals using Al$_2$O$_3$ atomic layer deposition. *J. Chem. Mater.* 18, 3562–3570. doi:10.1021/cm060263d

Siitonen, S., Pietarinen, J., Laakkonen, P., Jefimovs, K., and Kuittinen, M. (2007). Replicated polymer light guide interconnector with depth modified surface relief grating couplers. *Opt. Rev.* 14, 304–309. doi:10.1007/s10043-007-0304-x

Spiller, E. (1984). Totally reflecting thin-film phase retarders. *Appl. Opt.* 23, 3544–3549. doi:10.1364/AO.23.003544

Sung, J., Kosuda, K., Zhao, J., Elam, J., Spears, K., and Duyne, R. V. (2008). Stability of silver nanoparticles fabricated by nanosphere lithography and atomic layer deposition to femtosecond laser excitation. *J. Phys. Chem. C* 112, 5707–5714. doi:10.1021/jp0774140

Szeghalmi, A., Helgert, M., Brunner, R., Gösele, F. H. U., and Knez, M. (2009). Atomic layer deposition of Al$_2$O$_3$ and TiO$_2$ multilayers for applications as bandpass filters and antireflection coatings. *Appl. Opt.* 48, 1727–1732. doi:10.1364/AO.48. 001727

Triani, G., Campbell, J., Evans, P., Davis, J., Latella, B., and Burford, R. (2009). Low temperature atomic layer deposition of titania thin films. *Thin Solid Films* 518, 3182–3189. doi:10.1021/nn201167j

Triani, G., Campbell, J., Evans, P., Davis, J., Latella, B., and Burford, R. (2010). Low temperature atomic layer deposition of titania thin films. *Thin Solid Films* 518, 3182–3189. doi:10.1016/j.tsf.2009.09.010

Triani, G., Evans, P. J., David, R. G., Mitchell, D. J. A., Finnie, K. S., James, M., et al. (2005). Atomic layer deposition of $TiO_2/Al_2O_3$ films for optical applications. *SPIE Proc.* 58870, 587009. doi:10.1117/12.638039

Vahala, K. J. (2003). Optical microcavities. *Nature* 424, 839–846. doi:10.1038/nature01939

Wang, B. J., Huang, G., and Mei, Y. (2014). Modification and resonance tuning of optical microcavities by atomic layer deposition. *Chem. Vap. Deposition* 20, 103–111. doi:10.1002/cvde.201300054

Wang, J., Zhan, T., Huang, G., Cui, X., Hu, X., and Mei, Y. (2012). Tubular oxide microcavity with high-index-contrast walls: mie scattering theory and 3D confinement of resonant modes. *Opt. Express* 20, 18555–18567. doi:10.1364/OE.20. 018555

Wang, X. D., Graugnard, E., King, J. S., Wang, Z. L., and Summers, C. J. (2004). Large-scale fabrication of ordered nanobowl arrays. *Nano Lett.* 4, 2223–2226. doi:10.1021/nl048589d

Weber, T., Käsebier, T., Helgert, M., Kley, E.-B., and Tünnermann, A. (2012a). Tungsten wire grid polarizer for applications in the DUV spectral range. *Appl. Opt.* 51, 3224–3227. doi:10.1364/AO.51.003224

Weber, T., Kasebier, T., Szeghalmi, A., Knez, M., Kley, E.-B., and Tunnermann, A. (2012b). High aspect ratio deep UV wire grid polarizer fabricated by double patterning. *Microelectron. Eng.* 98, 433–435. doi:10.1016/j.mee.2012.07.044

Wei, H., and Xu, H. (2014). Plasmonics in composite nanostructures. *Mater. Today.* doi:10.1016/j.mattod.2014.05.012

Whitney, A., Elam, J., Stair, P., and Duyne, R. V. (2007). Toward a thermally robust operando surface-enhanced Raman spectroscopy substrate. *J. Phys. Chem. C* 111, 16827–16832. doi:10.1021/jp074462b

Wood, R. (1902). On a remarkable case of uneven distribution of light in a diffraction grating spectrum. *Philos. Mag.* 4, 396–402. doi:10.1080/14786440209462857

Worgull, M. (2009). *Hot Embossing: Theory and Technology of Microreplication.* Oxford: Elsevier.

Yu, M. L. (1993). Mechanism of atomic layer epitaxy of GaAs. *J. Appl. Phys.* 73, 716. doi:10.1063/1.353328

**Conflict of Interest Statement:** The authors declare that the research was conducted in the absence of any commercial or financial relationships that could be construed as a potential conflict of interest.

# 21

# Ultrasonic assisted consolidation of commingled thermoplastic/glass fiber rovings

*Francesca Lionetto, Riccardo Dell'Anna, Francesco Montagna and Alfonso Maffezzoli\**

Materials Science and Technology Group, Department of Engineering for Innovation, University of Salento, Lecce, Italy

**Edited by:**
Patricia Krawczak, École Nationale
Supérieure des Mines de Douai,
France

**Reviewed by:**
Chung Hae Park, École Nationale
Supérieure des Mines de Douai,
France
Véronique Michaud, École
Polytechnique Fédérale de Lausanne,
Switzerland
Ralf Schledjewski, Montanuniversität
Leoben, Austria

**\*Correspondence:**
Alfonso Maffezzoli, Materials Science
and Technology Group, Department of
Engineering for Innovation, University
of Salento, via per Monteroni, Lecce
73100, Italy
e-mail: alfonso.maffezzoli@
unisalento.it

Thermoplastic matrix composites are finding new applications in different industrial area, thanks to their intrinsic advantages related to environmental compatibility and processability. The approach presented in this work consists in the development of a technology for the simultaneous deposition and consolidation of commingled thermoplastic rovings through to the application of high energy ultrasound. An experimental equipment, integrating both fiber impregnation and ply consolidation in a single process, has been designed and tested. It is made of an ultrasonic welder, whose titanium sonotrode is integrated on a filament winding machine. During winding, the commingled roving is at the same time in contact with the mandrel and the horn. The intermolecular friction generated by ultrasound is able to melt the thermoplastic matrix and impregnate the reinforcement fibers. The heat transfer phenomena occurring during the *in situ* consolidation have been simulated solving by finite element (FE) analysis, an energy balance accounting for the heat generated by ultrasonic waves and the melting characteristics of the matrix. To this aim, a calorimetric characterization of the thermoplastic matrix has been carried out to obtain the input parameters for the model. The FE analysis has enabled to predict the temperature distribution in the composite during heating and cooling. The simulation results have been validated by the measurement of the temperature evolution during ultrasonic consolidation. The reliability of the developed consolidation equipment has been proved by producing hoop wound cylinder prototypes using commingled continuous E-glass rovings and polypropylene filaments. The consolidated composite cylinders are characterized by high mechanical properties, with values comparable with the theoretical ones predicted by the micromechanical analysis.

**Keywords: thermoplastic composites, ultrasonic, consolidation, viscoelastic heating, filament winding, finite element method, process modeling, polypropylene**

## INTRODUCTION

Continuous fiber reinforced thermoplastic matrix composites are showing a great potential for many different applications, thanks to their easy processing without requiring long curing times, their long shelf life, low level of moisture uptake, easy welding ability, and higher repairability potential (Ahmed et al., 2006; Sinmazcelik, 2006; Ning et al., 2007). Filament winding, widely used for the production of continuous fiber reinforced composites with thermosetting matrix, has been recently adopted for manufacturing continuous fiber reinforced thermoplastic materials (Henninger and Friedrich, 2002; Dobrzanski et al., 2007). The need of fast and efficient technologies has led to the development of simultaneous deposition and *in situ* consolidation of the commingled yarns or tapes with the application of heat in order to melt the thermoplastic matrix at the deposition interface. Several automated tape deposition techniques are available under the name of automated tape laying (ATL) or automated fiber placement (AFP) (Ye et al., 1995; Henninger et al., 2002; Gennaro et al., 2011; Mondo et al., 2012).

The heat sources currently used for on-line impregnation during filament winding are hot air jets or flames, which are characterized by an acceptable cost but also by a limited ability to control the temperature in narrow ranges, or laser sources, much easier to control but more expensive. *In situ* consolidation presents several critical aspects related to the temperature control, considering the high temperature needed for matrix melting. Additional issues occur when the deposition must take place on surfaces with single or double curvature with the need to orient the fibers in any arbitrary direction, obtaining at the same time a fully consolidated component (Tierney and Gillespie, 2006).

High energy ultrasound has a great potential for *in situ* composite consolidation. It can be easily automated for the deposition of thermoplastic matrix semi-pregs or prepregs, both on the flat and curved surfaces. More recently, it has been applied also to the joining and assembly of thermoplastic matrix composites (Liu et al., 2001; Yousefpour et al., 2004; Amancio-Filho and dos Santos, 2009; Fernandez Villegas and Bersee, 2010; Fernandez Villegas et al., 2013). Ultrasonic welding is also used during lay-up of thermoplastic tapes reinforced with carbon fibers to keep in place the different plies before full consolidation in autoclave.

In ultrasonic assisted consolidation, a sonotrode (or horn) transfers low amplitude (typically 5–100 μm) vibratory energy at high frequency (typically 20–40 kHz) to the joining parts. Vibrations are responsible of surface and intermolecular friction, which melt the thermoplastic matrix (Ávila-Orta et al., 2013). As observed by Tolunay et al. (1983), the heating occurs over the whole volume under the sonotrode tip. Ultrasonic consolidation of thermoplastic matrix composites does not require the use of any foreign material in the joint, such as the resistive inserts used for resistance welding (Ageorges and Ye, 2001; Ageorges et al., 2001).

In this work, high power ultrasound has been applied to the simultaneous deposition, impregnation, and consolidation of commingled thermoplastic rovings, made of thermoplastic filaments and glass fibers. An experimental set-up, integrating a laboratory filament winding machine with the sonotrode of an ultrasonic welding head, has been developed. The propagation of ultrasonic waves has been used to achieve melting of the thermoplastic matrix, impregnation of the reinforcement fibers and, finally, consolidation of the different plies of fiber reinforced thermoplastic composite materials. The system has the potential to consolidate layer by layer both flat and cylindrical samples with hoop windings. The heat transfer phenomena, which occurs during consolidation of thermoplastic rovings, have been simulated solving by finite element (FE) analysis, an energy balance accounting for the heat generated by ultrasonic waves and the melting characteristics of the matrix. A characterization of the physical and mechanical properties of samples obtained with this equipment has been performed.

## EXPERIMENTAL
### MATERIALS

The material used in this study is a dry pseudo-prepreg, made by commingling continuous E-glass rovings and polypropylene (PP) filaments. It is supplied by Fiber Glass Industries Inc. under the trade name Twintex R PP 60 1870 N. The content of glass fibers is 60% by weight. A cross-section of the commingled roving is shown in **Figure 1A**. The ultrasonic horn, applying at the same time, ultrasonic waves and pressure, leads to the matrix melting and consolidation, as schematically sketched in **Figure 1B**.

### ULTRASONIC CONSOLIDATION EQUIPMENT

The ultrasonic equipment developed in this study for the impregnation and *in situ* consolidation of commingled rovings is schematically represented in **Figure 2**. It integrates both fiber impregnation and ply consolidation in a single process. The equipment consists of an ultrasonic welder of thermoplastic polymers (Sonic Italia, s.r.l.), characterized by a maximum power of 2000 W and a frequency of 20 kHz, whose titanium sonotrode is mounted on a filament winding machine (VEM S.p.a.). This latter is a 2 degree of freedom machine equipped with a mandrel of 150 mm diameter.

The commingled roving, after being tensioned by a tension controller, is wound on a rotating mandrel with a defined rotational speed, where it is at the same time in contact with an ultrasonic horn, which applies ultrasonic waves and pressure. The intermolecular friction generated by ultrasound is able to melt the thermoplastic matrix and impregnate the reinforcement fibers. It should be underlined that the developed system does not use a pre-heater unit before winding.

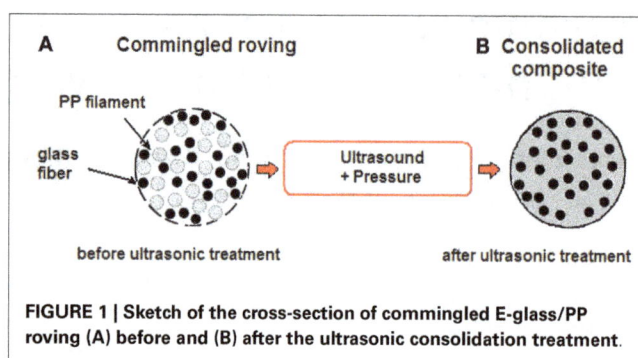

FIGURE 1 | Sketch of the cross-section of commingled E-glass/PP roving (A) before and (B) after the ultrasonic consolidation treatment.

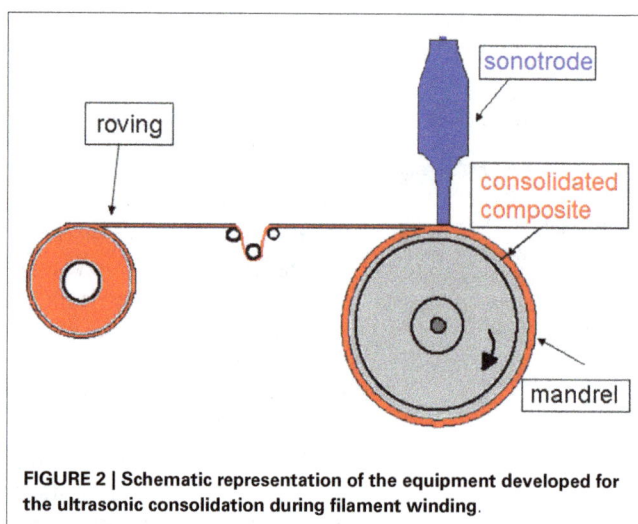

FIGURE 2 | Schematic representation of the equipment developed for the ultrasonic consolidation during filament winding.

## PRODUCTION AND CHARACTERIZATION OF CONSOLIDATED COMPOSITES

The developed consolidation equipment has been used to produce cylinder prototypes by hoop winding, i.e., with the fibers lied at almost 90° with respect to the mandrel axis direction. Tubular components with an inner diameter of 150 mm and a length of 200 mm have been produced by winding two layers of commingled roving on a mandrel with a rotating speed of 0.7 rad/s.

The efficiency of fiber impregnation during processing can be evaluated by calculating the void fraction $\Phi_{void}$ in the consolidated composite according to the following equation:

$$\Phi_{void} = \frac{\rho_t - \rho_a}{\rho_t} \tag{1}$$

where $\rho_t$ is theoretical density of the composite (i.e., the density of a completely consolidated part) and $\rho_a$ is the actual density of the composite. The actual density has been experimentally determined according to the ASTM D 792 standard (ASTM D 792, 2008) on ten samples with dimensions 40 mm × 10 mm cut from the composite. The theoretical density has been calculated by the simple rule of mixtures.

The effectiveness of the ultrasonic consolidation has been also assessed by measuring the shear modulus by dynamic mechanical analysis. Five rectangular specimens (40 mm × 10 mm × 1 mm) have been tested in torsion mode on an ARES rheometer (TA

Instruments). A frequency sweep test between 0.08 and 15 Hz with an amplitude of deformation of 0.016% has been carried out.

For comparison purposes, composite specimens with the same dimensions have been preheated at 200°C for 2 min and consolidated by compression molding at 200°C and 10 bar for 30 s.

The morphology of the samples has been analyzed by a Zeiss EVO 40 scanning electron microscope at variable pressure operating with an accelerating voltage of 20 KV. Some selected samples have been embedded in an epoxy resin and polished with a polishing machine using 500, 1000, and, finally, 2400 grit size silicon carbide grinding paper.

## EXPERIMENTAL DETERMINATION OF THE INPUT PARAMETERS FOR THE MODEL

The FEM model developed in this work needs some input parameters related to thermal behavior of the material used in this study. To this aim, a complete thermal characterization of the commingled E-glass/PP roving has been carried out by differential scanning calorimetry (DSC) using a Mettler DSC 822e calorimeter. The melting behavior of the PP matrix of commingled roving has been analyzed by dynamic DSC scans from 25 to 250°C at 10°C/min.

The temperature changes during ultrasonic consolidation have been monitored in real time using a high speed data acquisition system including NiCr/NiAl (type K) thermocouples and a Pico TC-08 thermocouple data logger (Pico Technology Ltd.).

## PROCESS MODELING
### GEOMETRY AND MATERIAL PROPERTIES
The analyzed 2D Cartesian geometry, schematically sketched in **Figure 3**, neglects the mandrel curvature. The commingled thermoplastic roving, assumed as a single domain with properties calculated from those of glass fibers and PP, is considered always in contact with the titanium sonotrode and the steel mandrel. The sonotrode operates at 20 kHz, a frequency widely used in the ultrasonic welding technology.

### GOVERNING EQUATIONS
The governing equations used to model the non-isothermal problem are the conservation equation for energy coupled with the equations accounting for the heat generated by ultrasonic waves and the heat absorbed by the melting of the matrix:

$$\rho Cp \frac{\partial T}{\partial t} = k\frac{\partial^2 T}{\partial x^2} + k\frac{\partial^2 T}{\partial y^2} + Q - \rho \dot{H}_m \qquad (2)$$

where $\rho$ is the density, Cp the specific heat capacity, and $k$ the thermal conductivity. $Q$ represents the heat generation produced by ultrasonic heating and $\dot{H}_m$ the melting heat of the PP matrix of the commingled thermoplastic roving.

The heat generation term $Q$ in Eq. 2 is obtained when a viscoelastic material undergoing a sinusoidal deformation at high frequency dissipates a fraction of energy as a heat due to intermolecular friction. $Q$ depends on the applied frequency, the square of strain amplitude ($\varepsilon_0$) of the ultrasonic vibration, and the loss modulus $E''$ of the material (Benatar and Cheng, 1989;

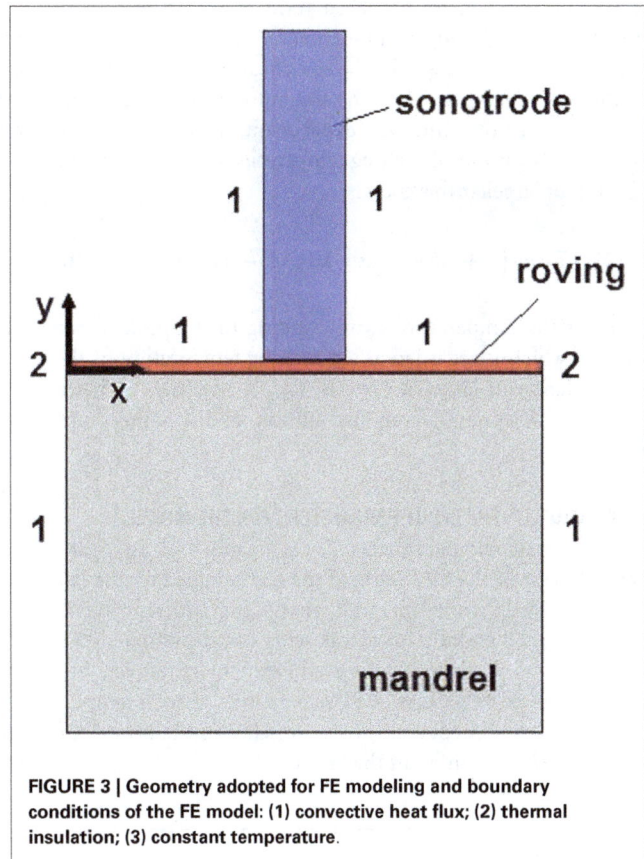

**FIGURE 3 | Geometry adopted for FE modeling and boundary conditions of the FE model: (1) convective heat flux; (2) thermal insulation; (3) constant temperature.**

Benatar and Gutowski, 1989; Suresh et al., 2007; Roopa Rani and Rudramoorthy, 2013), according to the following equation:

$$Q = \frac{\omega \times \varepsilon_0^2 \times E''}{2} \qquad (3)$$

where $\omega = 2\pi f$, with $f$ the ultrasonic frequency (20 kHz). The strain amplitude $\varepsilon_0$ is obtained as the ratio between the maximum displacement amplitude of the ultrasonic sonotrode and the thickness of the commingled roving under consolidation. The loss modulus $E''$ for a viscoelastic material is the out of phase modulus and it is a measure of the energy dissipated through intermolecular friction (Ferry, 1980). The values of $E''$ at 20 kHz for PP is 0.32 GPa, as experimentally found by Benatar et al. (1989) at room temperature. This constant value is assumed in the model.

The term $\dot{H}_m$ in Eq. 2, representing the heat necessary to promote the polymer melting, is a function of the degree of melting $X_m$:

$$\dot{H}_m = H_T \times \frac{dX_m}{dt} \qquad (4)$$

where $H_T$ is a reference value, which is assumed to be the total heat absorbed in the melting process, and the degree of melting $X_m$ is defined as:

$$X_m(T) = \frac{H(T)}{H_T} \qquad (5)$$

where $H(T)$ is the enthalpy absorbed at the temperature $T$. With this assumptions, $X_m$ ranges in the interval $(0, 1)$. The degree of melting $X_m$ can be expressed by the statistical approach of Greco and Maffezzoli (2003), based on the assumption that the melting peak, obtained in a dynamic DSC experiment, can be regarded as a statistical distribution of melting temperatures resulting from a distribution of lamellar thicknesses:

$$X_m(T) = \left\{ 1 + (d - 1) \exp\left[ k_{mb}(T - TC) \right] \right\}^{\frac{1}{(1-d)}} \quad (6)$$

where $T_C$ is the temperature corresponding to the peak of the DSC signal, which is regarded as the melting temperature of the larger population of lamellar crystals; $k_{mb}$ is an intensity factor related to the sharpness of the distribution, and $d$ is the shape factor.

## DETERMINATION OF THE INPUT PARAMETERS FOR THE MODEL

In order to obtain the parameters $T_C$, $k_{mb}$, and $d$ of Eq. 6, the melting behavior of the PP matrix of the commingled roving has been analyzed by dynamic DSC scans at 10°C/min. As reported in **Figure 4A**, a broad endothermic peak with a maximum at 438 K is observed when PP melts. As observed for many semicrystalline polymers, the melting process of PP occurs over a broad temperature interval as a consequence of the presence in the crystalline regions of lamellae of different thickness.

The degree of melting can be determined by the non-linear regression of DSC dynamic experiments, assuming that the heat evolved during melting is proportional to the extent of melting. The melting enthalpy absorbed at the temperature $T$ can be obtained as a function of the temperature:

$$H(T) = \int_{T_0}^{T} dH \times dT \quad (7)$$

where $T_0$ is the starting temperature. The degree of melting ($X_m$) of the PP matrix, defined as the integral curve of the DSC melting peak, is reported in **Figure 4B**. The non-linear regression of this curve according to Eq. 5 has enabled to obtain the parameters for melting modeling, reported in **Table 1**, which have been given as an input to the simulation.

The thermal conductivity $k$ and heat capacity Cp of the unidirectional composite are anisotropic properties with different values in the longitudinal direction and transversal direction. Moreover, $k$ and Cp significantly vary with temperature. At room temperature, i.e., in the dry state, the composite is formed of three phases (polymer, glass, and air), while, during melting and consolidation, i.e., in the wet state, only two phases (polymer and glass) are present. In order to account for this, the values in the dry state, $k_{dry}$ and $Cp_{dry}$ have been introduced

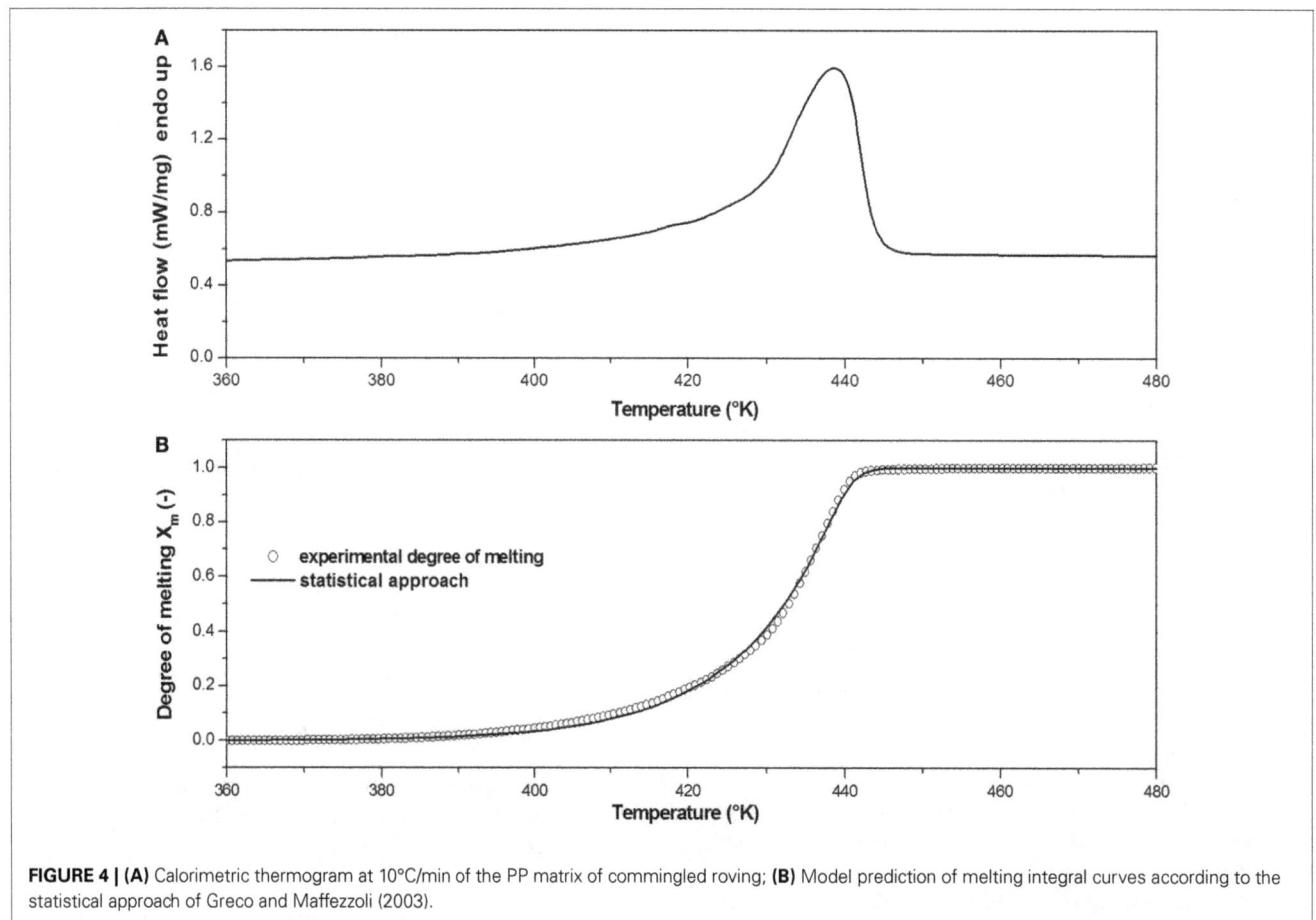

**FIGURE 4 | (A)** Calorimetric thermogram at 10°C/min of the PP matrix of commingled roving; **(B)** Model prediction of melting integral curves according to the statistical approach of Greco and Maffezzoli (2003).

**Table 1 | Parameters for melting model obtained from the non-linear regression of the experimental data with the statistical approach.**

| d (-) | $k_{mb}$ (1/K) | $T_c$ (K) |
|-------|----------------|-----------|
| 11.9  | 0.9            | 438.0     |

as follows:

$$k_{dry} = k_{pp} \times \Phi_{PP} + k_{glass} \times \Phi_{glass} + k_{air} \times \Phi_{air} \qquad (8)$$

$$Cp_{dry} = Cp_{pp} \times \Phi_{PP} + Cp_{glass} \times \Phi_{glass} + Cp_{air} \times \Phi_{air} \qquad (9)$$

where $\Phi_{PP}$, $\Phi_{glass}$, and $\Phi_{air}$ represents the volume fraction of PP, glass fiber, and air, respectively. In order to calculate $k_{dry}$ and $Cp_{dry}$ along the x and y direction, the volume fraction of air $\Phi_{air}$ has been experimentally determined. Its value is 0.33 for the dry roving under a pressure value of 17 bar, which is the pressure applied by the sonotrode.

In a similar way, the values in the wet state, $k_{wet}$ and $Cp_{wet}$ have been determined as follows:

$$k_{wet} = k_{pp} \times \Phi_{PP} + k_{glass} \times \Phi_{glass} \qquad (10)$$

$$Cp_{wet} = Cp_{pp} \times \Phi_{PP} + Cp_{glass} \times \Phi_{glass}. \qquad (11)$$

Moreover, in order to account for the variation of $k$ and Cp with the temperature, the following relationships have been implemented in the model:

$$k = k_{dry} \times (1 - X_m) + k_{wet} \times X_m \qquad (12)$$

$$Cp = Cp_{dry} \times (1 - X_m) + Cp_{wet} \times X_m \qquad (13)$$

where $X_m$ is the degree of melting expressed by Eq. 6.

The values of thermal conductivity and heat capacity along the longitudinal axis, $k_{x\text{-}wet}$ and $Cp_{x\text{-}wet}$, have been determined by the rule of mixtures, while those along the transversal axis, $k_{y\text{-}wet}$ and $Cp_{y\text{-}wet}$, have been determined by the inverse rule of mixtures commonly used also for the determination of mechanical properties of composite materials (Lionetto et al., 2014). The material properties used in the FE model are reported in **Table 2**.

## BOUNDARY CONDITIONS

The model defined by Eqs 2, 3, and 6 has been solved with the FE method using the commercial software Comsol Multiphysics 4.4. As boundary conditions, convective heat exchange has been adopted at the lateral walls of sonotrode and mandrel and on the upper and lower surfaces of the consolidated roving undergoing cooling, as schematically shown in **Figure 3**. The convective heat transfer coefficient of 5 W/m$^2$ × K, typical of a natural convection, is used. A condition of constant temperature (298 K) has been kept on the external part of mandrel and sonotrode, far from the welding interface. Adiabatic conditions are assumed at the other boundaries.

A mesh with triangular elements with the maximum dimensions smaller than one tenth of the smallest thickness has been used.

A time dependent study has been used with the Heat Transfer module in Comsol Multiphysics 4.4 (Comsol Inc.). The time

**Table 2 | Material properties for the FE model.**

|  | Titanium sonotrode | Steel mandrel | PP/glass fiber | | | |
|--|--------------------|---------------|-------|-------|-------|-------|
| $\rho$ (kg/cm$^3$) | 4507 | 7860 | 1446 | | | |
|  |  |  | dry_x | wet_x | dry_y | wet_y |
| k (W/m × K) | 21.9 | 51.9 | 0.38 | 0.56 | 0.07 | 0.30 |
| Cp (J/kg × K) | 520 | 472 | 1360 | 1541 | 1179 | 1298 |

interval chosen for the simulation is between 0 and 3 s with time steps of 0.001 s.

The movement of the roving due to the mandrel rotation has been accounted in the model by setting the time range in which the sonotrode is active equal to the contact time of the roving and the sonotrode during a dynamic consolidation test. If $\omega$ and $R$ are the rotating speed and the radius of the mandrel, respectively, then the velocity of the roving is:

$$v = \omega \times R = \frac{L}{t_c} \qquad (14)$$

If $L$ is the length of the sonotrode in contact with the roving, then the contact time $t_c$ of the roving under the sonotrode is:

$$t_c = \frac{L}{\omega \times R} \qquad (15)$$

In the studied case, considering four different values of rotating speed of the mandrel (0.2, 0.3, 0.7 and 1.3 rad/s), a mandrel radius of 0.075 m and a contact length of 5 mm, a point of the commingled roving is in contact with the sonotrode for a time $t_c$ of about 0.3, 0.2, 0.1, and 0.05 s, respectively. For this reason, in order to simulate the movement of the roving, the sonotrode has been considered active only in the first 0.05, 0.1, 0.2, and 0.3 s of the simulation time interval.

## RESULTS AND DISCUSSION

### MODEL RESULTS

The FE analysis provides the temperature distribution in the composite during the ultrasonic impregnation and consolidation. The prediction of the temperature distribution in the composite during processing is very important since it strongly affects the quality of the thermoplastic composites. The temperature map after 0.1 s in the modeled geometry interested by the thermal gradient for a contact time $t_c = 0.1$ s is reported in **Figure 5**. Only the roving in contact with the sonotrode is significantly heated reaching a temperature of about 212°C, which is well above the melting temperature of PP (about 165°C).

The temperature distribution at different points of the commingled roving in contact with the sonotrode is shown in **Figure 6** for a contact time $t_c = 0.1$ s. After 0.1 s, a temperature of 212°C, much higher than the melting temperature of the PP matrix, is reached in the central area below the sonotrode and in the mid-plane of the roving. Then, when the sonotrode is switched off, the composite is cooled until the end of simulation (3 s).

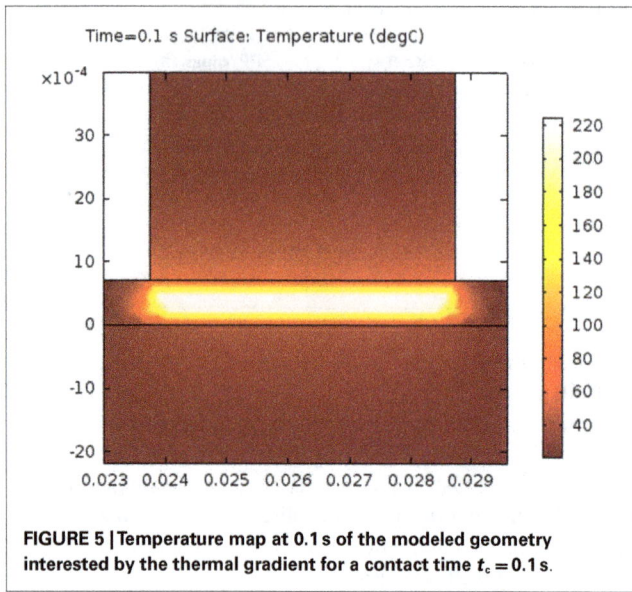

FIGURE 5 | Temperature map at 0.1 s of the modeled geometry interested by the thermal gradient for a contact time $t_c = 0.1\,\text{s}$.

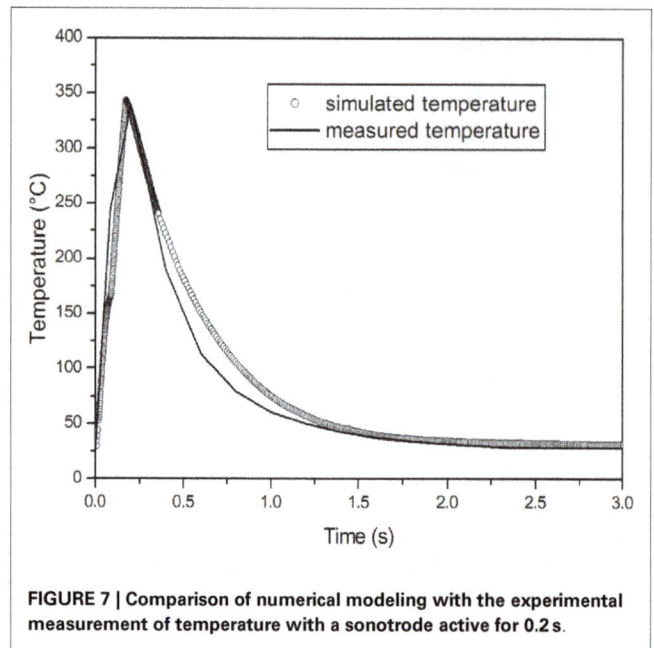

FIGURE 6 | Temperature distribution in different points of the commingled roving in contact with the sonotrode for a contact time $t_c = 0.1\,\text{s}$.

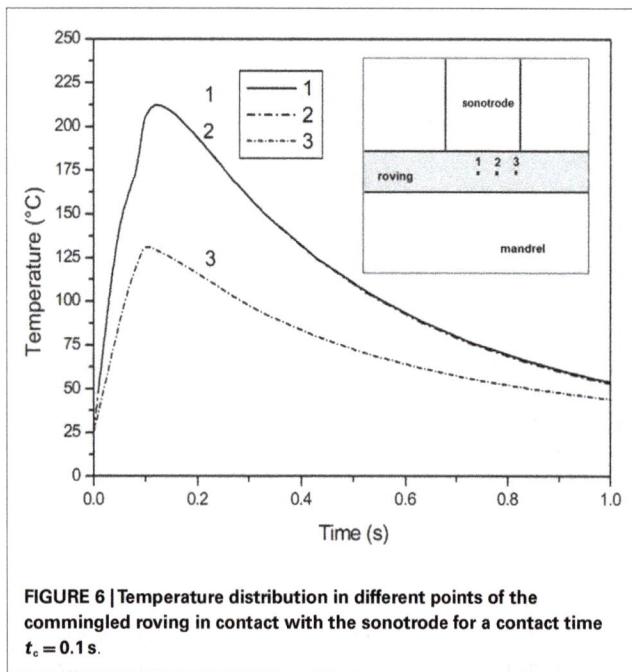

FIGURE 7 | Comparison of numerical modeling with the experimental measurement of temperature with a sonotrode active for 0.2 s.

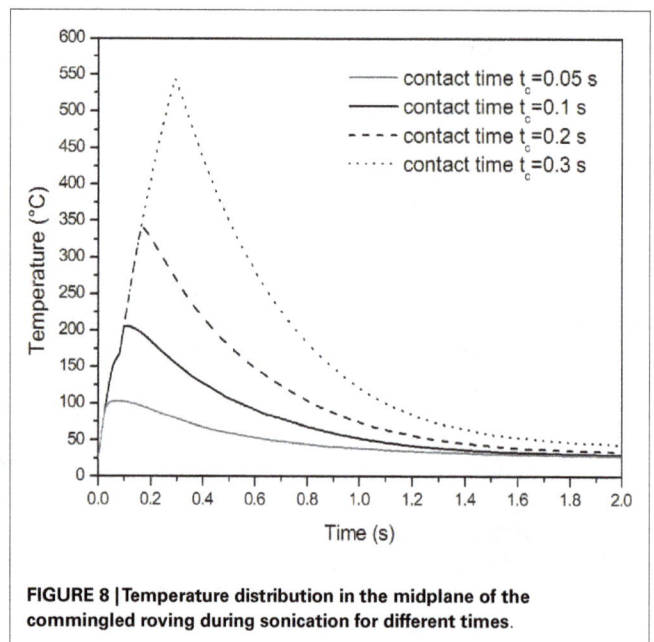

FIGURE 8 | Temperature distribution in the midplane of the commingled roving during sonication for different times.

The FE results are validated by the experimental measurement of the temperature obtained exposing the composite to ultrasonic waves for 0.2 s, as reported in **Figure 7**. A K-type thermocouple has been inserted between two roving plies. When the ultrasonic device is turned on, a very rapid heating is observed with the temperature reaching a value over the melting temperature of the thermoplastic matrix. A peak value of 339°C has been measured, which is in good agreement with the simulation, where the peak is reached at 343°C.

**Figure 8** shows the temperature distribution in the midplane of the commingled roving in contact with the sonotrode for different exposure times. It is clearly observable how the exposure time affects the maximum temperature reached in the roving.

Long exposure times, corresponding in a dynamic consolidation experiment to a slower mandrel rotation speed, can cause the degradation of the polymer matrix, while with a very short exposure time, e.g., 0.05 s in **Figure 8**, the PP matrix does not melt. The optimum value of contact time between the sonotrode and the roving, determined using the static FE model, is 0.1 s.

**Figure 9A** shows the calculated temperature evolution along the $y$ axis in correspondence of the midplane of sonotrode for a process time of 0.1 s. The mandrel and the sonotrode act as heat sinks reaching 35 and 44°C, respectively, at the interface with the roving midplane between the mandrel and the sonotrode.

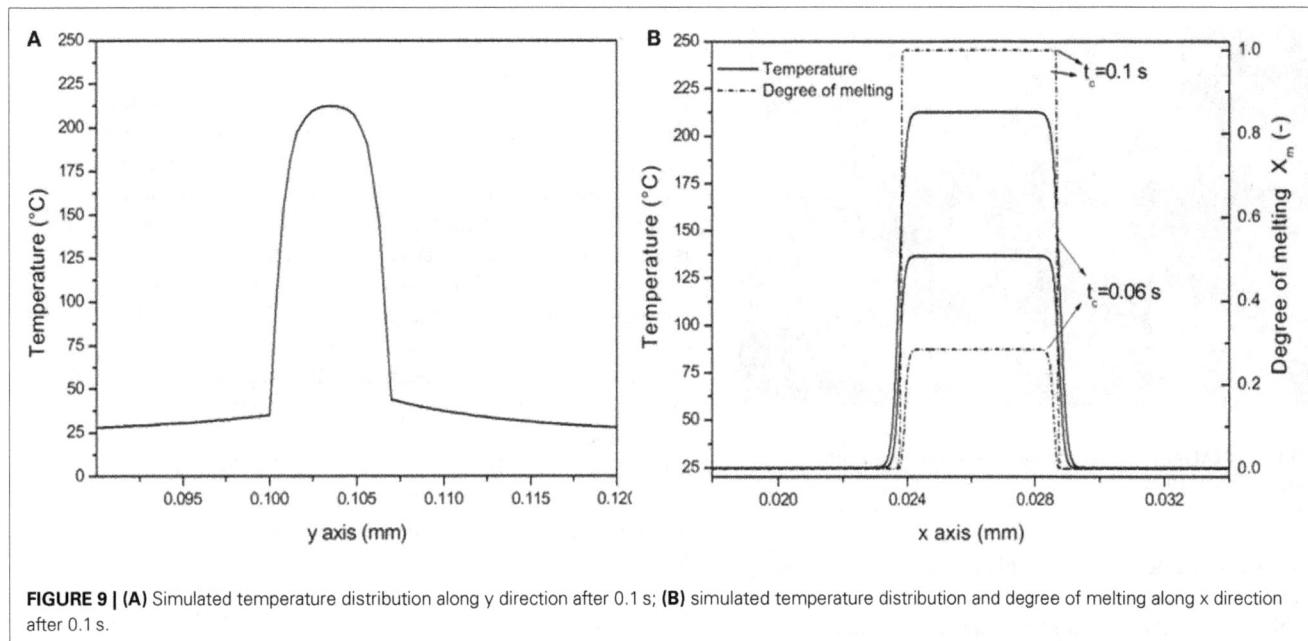

**FIGURE 9 | (A)** Simulated temperature distribution along y direction after 0.1 s; **(B)** simulated temperature distribution and degree of melting along x direction after 0.1 s.

**Figure 9B** shows the calculated temperature evolution along the $x$ axis, the direction along which the roving moves under the sonotrode during dynamic consolidation experiments. The temperature of the commingled roving is high where it is in contact with the sonotrode, sharply increasing up to 212°C. Then, it decreases reaching the room temperature value in the area not in contact with the sonotrode.

The degree of melting as a function of the space coordinates as shown in **Figure 9B** is also calculated, confirming the full melting of the PP matrix.

It should be noted that the temperature distribution along y direction is different from that along x direction, without a zone at constant temperature. This behavior depends on the simulation of a static process with mandrel and sonotrode acting as heat sinks. It is expected that, when simulating a dynamic process with a mandrel and a sonotrode which can heat up, the heat removed by roving at these boundaries is lower, enabling thus the roving to reach a constant temperature across the $y$-axis.

## SAMPLE MANUFACTURING AND CHARACTERIZATION

A two stage approach has been adopted:

1. dry hoop winding of two layers of the commingled roving has been performed on a mandrel with a diameter equal to 150 mm;
2. the sonotrode is moved along the mandrel axis with a contact pressure of 17 bar while the mandrel is kept under rotation at 0.7 rad/s in order to perform matrix melting, fiber impregnation, and ply consolidation. The mandrel speed has been set in order to keep the contact time between the sonotrode and the roving at 0.1 s, the optimum value determined using the static FE model.

The pressure applied by the sonotrode, when it is contact with the roving, has been estimated by dividing the force of the

**Table 3 | Physical and mechanical properties of the prototype cylinders (two layers) obtained by filament winding and consolidated by ultrasound or by compression molding.**

| Consolidation method | Density (g/cm$^3$) | Void content (%) | $G'_{12}$ modulus_ 90°(MPa) |
|---|---|---|---|
| Ultrasonic | 1.38 ± 0.02 | 4.6 ± 0.5 | 928 ± 24 |
| Compression molding | 1.42 ± 0.02 | 1.8 ± 0.2 | 995 ± 130 |

sonotrode (85 N) by the contact area. Since the contact area is not perfectly parallel to the sonotrode, the obtained value of 17 bar is the minimum value. This value is however in the range of molding pressures used for thermoplastic matrix composites, which is between 10 and 60 bars (Wakeman et al., 1998; Santulli et al., 2002a,b).

The density, void content, and storage shear modulus, $G_{12}'$, of samples taken from the consolidated cylinders have been measured. DMA tests in torsion on specimens cut parallel to prototype axis (90° to fiber direction) have been carried out, being the shear modulus $G_{12}'$ a property dominated by the matrix and by fiber impregnation. For comparison purposes, samples with the same number of layers have been consolidated by compression molding at 200°C and 10 bar for 30 s. In **Table 3** the physical and mechanical properties of the composite cylinders (two layers), obtained by filament winding and consolidated by ultrasound or by compression molding, are reported. The void content of ultrasonically consolidated samples is within the typical range found in literature for composites processed by filament winding (Henninger et al., 2002; Gennaro et al., 2011; Stefanovska et al., 2014) and compression molding (Long et al., 2001; Santulli et al., 2002a,b; Greco et al., 2007), even if with this technology the consolidation times and the compaction rates are quite different from those characteristics of the ultrasonic assisted filament winding.

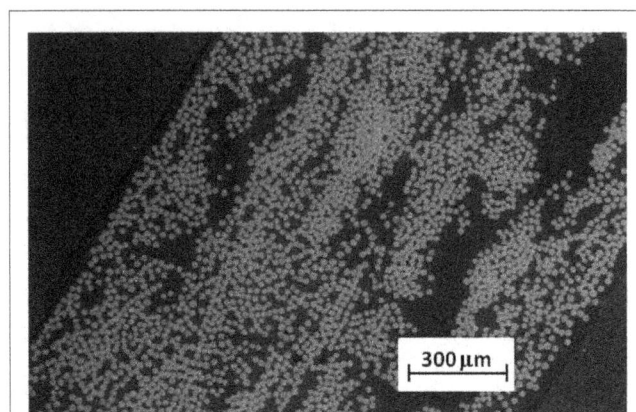

**FIGURE 10 | SEM micrographs of a polished cross-section of a consolidated composite.**

As shown in **Table 3**, the consolidation obtained by ultrasonic exposure is satisfactory. During processing, the ultrasonic irradiation is able to melt the thermoplastic matrix and the contact time and pressure are able to impregnate the fiber and consolidate the composite. These results suggest that the sonotrode is able to act at the same time as a heater and as a consolidating device.

The experimental values of $G'_{12}$ are close to the theoretical one, 1010 MPa, which has been obtained using the Halpin–Tsai equations (Nielsen and Landel, 1994):

$$\frac{M_c}{M_m} = \frac{1 + \xi \eta V_f}{1 - \eta V_f} \qquad (16)$$

where

$$\eta = \frac{(M_f / M_m) - 1}{(M_f / M_m) + \xi} \qquad (17)$$

$\xi$ is a measurement of the shape ratio of the reinforcement and depends on the fiber geometry, packing, and load conditions. For the calculation of $G_{12}$' according to Eqs 15 and 16, $\xi = 1$ (Nielsen and Landel, 1994), a fiber volume fraction $V_f = 0.34$, and a shear modulus of glass fibers and PP equal to 30 and 0.51 GPa, respectively, have been adopted (Mark, 1999).

The morphology of the consolidated composite samples analyzed by scanning electron microscopy (SEM) is reported in **Figure 10**. As can be seen, fibers are well impregnated but are distributed non-uniformly. Most fibers are very well impregnated while the porosity is mainly in the form of macro voids located in polymer rich regions.

## CONCLUSION

In this work, high power ultrasound has been applied for the simultaneous deposition, impregnation and, consolidation of commingled thermoplastic rovings, made of thermoplastic filaments and glass fibers. An experimental set-up, integrating a laboratory filament winding machine with the sonotrode of an ultrasonic welding head, has been developed. During winding, the commingled roving is at the same time in contact with the mandrel and the sonotrode. The sonotrode is able to melt the matrix and to apply a pressure on the consolidating material.

With the developed experimental set-up, several two-plies prototypes of composite cylinders have been produced starting from commingled rovings made of E-glass fibers and PP filaments. The values of shear modulus obtained by mechanical characterization are very close to those predicted by the micromechanical theory of unidirectional continuous fiber composites. The void content is comparable with the values reported in literature.

The obtained physical, mechanical, and microstructural results confirm the reliability of the proposed technology for the *in situ* consolidation of commingled thermoplastic rovings during filament winding. Further research is in progress, focused on the optimization of the winding parameters and the implementation of a compaction roller able to increase the compaction time. In our opinion, the optimization of the proposed technique will further reduce the void content and increase the mechanical properties, thus increasing the competitive advantage of the ultrasonic assisted filament winding.

Finally, a FE model has been proposed in order to compute the temperature distribution in the composite during exposure to ultrasound in static experiments, using an energy balance accounting for the heat generated by ultrasound and for polymer melting. The simulation results have been validated by the measurement of the temperature evolution during static ultrasonic consolidation.

## REFERENCES

Ageorges, C., and Ye, L. (2001). Resistance welding of thermosetting composite/thermoplastic composite joints. *Compos. Part A Appl. Sci. Manuf.* 32, 1603–1612. doi:10.1016/S1359-835X(00)00183-4

Ageorges, C., Ye, L., and Hou, M. (2001). Advances in fusion bonding techniques for joining thermoplastic matrix composites: a review. *Compos. Part A Appl. Sci. Manuf.* 32, 839–857. doi:10.1016/S1359-835X(00)00166-4

Ahmed, T. J., Stavrov, D., Bersee, H. E. N., and Beukers, A. (2006). Induction welding of thermoplastic composites-an overview. *Compos. Part A Appl. Sci. Manuf.* 37, 1638–1651. doi:10.1016/j.compositesa.2005.10.009

Amancio-Filho, S. T., and dos Santos, J. F. (2009). Joining of polymers and polymer–metal hybrid structures: recent developments and trends. *Polym. Eng. Sci.* 49, 1461–1476. doi:10.1002/pen.21424

ASTM D 792. (2008). *Standard Test Method for Density and Specific Gravity (Relative Density) of Plastics by Displacement.* West Conshohocken, PA: ASTM International.

Ávila-Orta, C., Espinoza-González, C., Martínez-Colunga, G., Bueno-Baqués, D., Maffezzoli, A., and Lionetto, F. (2013). An overview of progress and current challenges in ultrasonic treatment of polymer melts. *Adv. Polym. Technol.* 32, E582–E602. doi:10.1002/adv.21303

Benatar, A., and Cheng, Z. (1989). Ultrasonic welding of thermoplastics in the far field. *Polym. Eng. Sci.* 29, 1699–1704. doi:10.1002/pen.760292312

Benatar, A., Eswaran, R. M., and Nayar, S. K. (1989). Ultrasonic welding of thermoplastics in the near-field. *Polym. Eng. Sci.* 29, 1689–1698. doi:10.1002/pen.760292311

Benatar, A., and Gutowski, T. G. (1989). Ultrasonic welding of PEEK graphite APC-2 composites. *Polym. Eng. Sci.* 29, 1705–1721. doi:10.1002/pen.760292313

Dobrzanski, L. A., Domagala, J., and Silva, J. F. (2007). Application of Taguchi method in the optimisation of filament winding of thermoplastic composites. *Archives Mater. Sci. Eng.* 28, 133–140.

Fernandez Villegas, I., and Bersee, H. E. N. (2010). Ultrasonic welding of advanced thermoplastic composites: an investigation on energy-directing surfaces. *Adv. Polym. Technol.* 29, 112–121. doi:10.1002/adv.20178

Fernandez Villegas, I., Moser, L., Yousefpour, A., Mitschang, P., and Bersee, H. E. N. (2013). Process and performance evaluation of ultrasonic, induction and resistance welding of advanced thermoplastic composites. *J. Thermoplast. Compos. Mater.* 26, 1007–1024. doi:10.1177/0892705712456031

Ferry, J. D. (1980). *Viscoelastic Properties of Polymers*. New York, NY: Wiley.

Gennaro, R., Montagna, F., Maffezzoli, A., Fracasso, F., and Fracasso, S. (2011). On-line consolidation of commingled polypropylene/glass roving during filament winding. *J. Thermoplast. Compos. Mater.* 24, 789–804. doi:10.1177/0892705711401849

Greco, A., and Maffezzoli, A. (2003). Statistical and kinetic approaches for linear low-density polyethylene melting modeling. *J. Appl. Polym. Sci.* 89, 289–295. doi:10.1002/app.12079

Greco, A., Musardo, C., and Maffezzoli, A. (2007). Flexural creep behaviour of PP matrix woven composite. *Compos. Sci. Technol.* 67, 1148–1158. doi:10.1016/j.compscitech.2006.05.015

Henninger, F., and Friedrich, K. (2002). Thermoplastic filament winding with online-impregnation. Part A: process technology and operating efficiency. *Compos. Part A Appl. Sci. Manuf.* 33, 1479–1486. doi:10.1016/S1359-835X(02)00135-5

Henninger, F., Hoffmann, J., and Friedrich, K. (2002). Thermoplastic filament winding with online-impregnation. Part B. Experimental study of processing parameters. *Compos. Part A Appl. Sci. Manuf.* 33, 1677–1688. doi:10.1016/S1359-835X(02)00135-5

Lionetto, F., Calò, E., Di Benedetto, F., Pisignano, D., and Maffezzoli, A. (2014). A methodology to orient carbon nanotubes in a thermosetting matrix. *Compos. Sci. Technol.* 96, 47–55. doi:10.1016/j.compscitech.2014.02.016

Liu, S. J., Chang, T., and Hung, S. W. (2001). Factors affecting the joint strenght of ultrasonically welded polypropylene composite. *Polym. Compos.* 22, 132–141. doi:10.1002/pc.10525

Long, A. C., Wilks, C. E., and Rudd, C. D. (2001). Experimental characterisation of the consolidation of a commingled glass/polypropylene composite. *Compos. Sci. Technol.* 61, 1591–1603. doi:10.1016/S0266-3538(01)00059-8

Mark, J. E. (1999). *Polymer Data Handbook*. New York, NY: Oxford University Press.

Mondo, J., Wijskamp, S., and Lenferink, R. (2012). "Overview of thermoplastic composite ATL and AFP technologies," in *2nd Internat. Confer. Exhibit. Thermoplastic Composites ITHEC 2012*, (Bremen: WFB Wirtschaftsforderung Bremen GmbH), 74–78.

Nielsen, L. E., and Landel, R. F. (1994). *Mechanical Properties of Polymers and Composites*. New York, NY: Marcel Dekker.

Ning, H., Vaidya, U., Janowski, G. M., and Husman, G. (2007). Design, manufacture and analysis of a thermoplastic composite frame structure for mass transit. *Compos. Struct.* 80, 105–116. doi:10.1016/j.compstruct.2006.04.090

Roopa Rani, M., and Rudramoorthy, R. (2013). Computational modeling and experimental studies of the dynamic performance of ultrasonic horn profiles used in plastic welding. *Ultrasonics* 53, 763–772. doi:10.1016/j.ultras.2012.11.003

Santulli, C., Gil, R. G., Long, A. C., and Clifford, M. J. (2002a). Void content measurements in commingled E-glass/polypropylene composites using image analysis from optical micrographs. *Sci. Eng. Compos. Mater.* 10, 77–90. doi:10.1515/SECM.2002.10.2.77

Santulli, C., Brooks, R., Rudd, C. D., and Long, A. C. (2002b). Influence of microstructural voids on the mechanical and impact properties in commingled E-glass/polypropylene thermoplastic composites. *Proc. IME J. Mater. Des. Appl.* 216, 85–100.

Sinmazcelik, T. (2006). Natural weathering effects on the mechanical and surface properties of polyphenylene sulphide (PPS) composites. *Mater. Des.* 27, 270–277. doi:10.1016/j.matdes.2004.10.022

Stefanovska, M., Samakoski, B., Risteska, S., and Maneski, G. (2014). Influence of some technological parameters on the content of voids in composite during on-line consolidation with filament winding technology. *Int. J. Chem. Nucl. Mater. Metall. Eng.* 8, 398–402.

Suresh, K. S., Roopa Rani, M., Prakasan, K., and Rudramoorthy, R. (2007). Modeling of temperature distribution in ultrasonic welding of thermoplastics for various joint designs. *J. Mater. Process. Technol.* 186, 138–146. doi:10.1016/j.jmatprotec.2006.12.028

Tierney, J., and Gillespie, J. W. (2006). Modeling of in-situ strength development for the thermoplastic composite tow placement process. *J. Compos. Mater.* 40, 1487–1506. doi:10.1177/0021998306060162

Tolunay, M. N., Dawson, P. R., and Wang, K. K. (1983). Heating and bonding mechanisms in ultrasonic welding of thermoplastics. *Polym. Eng. Sci.* 23, 726–733. doi:10.1002/pen.760231307

Wakeman, M. D., Cain, T. A., Rudd, C. D., Brooks, R., and Long, A. C. (1998). Compression moulding of glass and polypropylene composites for optimized macro- and micro-mechanical properties. Part I: commingled glass and polypropylene. *Compos. Sci. Technol.* 58, 1879–1898. doi:10.1016/S0266-3538(98)00011-6

Ye, L., Friedrich, K., Kastel, J., and May, Y. W. (1995). Consolidation of unidirectional CF/PEEK composites from commingled yarn prepreg. *Compos. Sci. Technol.* 54, 349–358. doi:10.1016/0266-3538(95)00061-5

Yousefpour, A., Hojjati, M., and Immarigeon, J. P. (2004). Fusion bonding/welding of thermoplastic composites. *J. Thermoplast. Compos. Mater.* 17, 303–339. doi:10.1177/0892705704045187

**Conflict of Interest Statement:** The authors declare that the research was conducted in the absence of any commercial or financial relationships that could be construed as a potential conflict of interest.

# 22

# Dentistry on the bridge to nanoscience and nanotechnology

*Marco Salerno\* and Alberto Diaspro*

*Department of Nanophysics, Istituto Italiano di Tecnologia, Genova, Italy*

**Edited by:**
*Partha Pratim Mondal, Indian Institute of Science, India*

**Reviewed by:**
*Jose L. Toca-Herrera, University of Natural Resources and Life Sciences, Austria*
*Shivani Sharma, University of California Los Angeles, USA*

**\*Correspondence:**
*Marco Salerno, Department of Nanophysics, Istituto Italiano di Tecnologia, via Morego 30, Genova I-16163, Italy*
*e-mail: marco.salerno@iit.it*

Dentistry is the area of medical sciences that is most resistant to the introduction of the novel methods arisen from the development of nanoscience and nanotechnology in the last 20 years. Without moving on to science-fiction-like views pointing to times far ahead in the future, we show that the available nanoscale devices and processes of current science and technology, partly inherited from the areas of microscopy and microelectronics, have already proven to be useful for research and development in different fields of dental research. To this goal, we review some results obtained in the last few years at our Institute in the area of dental materials and their characterization, which showed successful application of our background in microscopy and nanoengineering.

Keywords: **dental materials, microscopy, surface roughness, nanofabrication, resin composites, implants**

## THE BROAD SCOPE OF DENTISTRY

Dentistry is considered to be a branch of medicine, such as orthopedics or physiology or neurology. In fact, while based on common foundations of medicine, dentistry is a large stand-alone area, which has its own grasps to the different fields mentioned above (American Dental Association, 2014). For example, topics of periodontal ligament health and dental implant osteointegration clearly relate to a peculiar "dental orthopedics"; the fact that teeth are not a mineralized appendix but rather a living organ in a living environment (gingiva, oral liquids) involves a specific "dental physiology"; the sensitivity of the living tooth, in connection with the pulp and related nerves, can be associated to a possible "dental neurology." Indeed, the oral environment, which is the place where dental functions are operated, is one of the most complex ones in the human body.

As a result, within the dental area several specialized disciplines exist (see **Figure 1**). *Conservative dentistry* aims to maintain the original denture and largely overlaps with *restorative dentistry*, the latter encompassing prosthodontics, periodontics, and endodontics. *Prosthodontics* (or prosthetic dentistry) is required in case of massive irreversible tooth damage that cannot be fixed with simple resin composite filling, and requires the use of prosthesis for teeth, namely crowns, bridges, and even whole dentures. *Periodontics* (or periodontology) addresses the diseases of the periodontum, i.e., the teeth ligaments. *Endodontics* focuses on therapy of the root canals and connected pulp diseases. *Oral surgery* relates to the extraction of teeth that cannot be saved anymore and should be replaced by implants, and is usually combined with *maxillo-facial surgery*, extending the treated area around the denture. Prosthodontics, periodontics, and surgery correlate to *implantology*, which deals with the installation of dental implants. *Orthodontics* concerns the alignment and straightening of teeth, as well as fixing midface and mandibular growth issues. *Pedodontics* (or pediatric dentistry) relates to the specific dental issues of children's

temporary dentures. Of course, oral and maxillofacial surgery is accompanied by related *pathology and radiology* specialties, as well as dental *anesthesiology*. Similarly, *oral biology* exists as a field on its own, also extending to craniofacial biology. In addition to the obvious extension to animals in *veterinary dentistry*, more niche specialties exist such as *forensic odontology*, using dental evidence in law to document people identity, and *geriodontics* (or geriatric dentistry), associating the delivery of dental care in old adults with aging issues. In US, *Dental public health* is yet another specialty, involved with social and political issues of dental therapy and follow-up. Sometimes the term *odontostomatology* is also found, from Greek "odontos" for teeth and "stomatos" for mouth, which is not a dental specialty but a general term including dental and other non-teeth related diseases such as oral cancer. In Italy, the term *gnatology* (or stomatognatology) is also common, as the discipline focused on occlusion of maxillary bones and their geometrical–functional relations to teeth, muscles, and nerves. *Dental hygiene* has in the recent years become an independent specialty as well.

Such a broad range of topics clearly makes use of a large spectrum of materials. In parallel, advanced technological solutions are required for the application of the underlying scientific principles. Actually, in dentistry, many special processes and practice protocols have been developed over the past century. As a result, dentistry is separated from the other medical areas, and the recent integration of nanotechnology and nanoscience into dentistry is progressing more slowly than for general medicine. The goal of this review is to report about the current application of nanotechnology and nanoscience in dentistry. This work does not pretend to be comprehensive, and will address sparse examples mainly from the fields of restorative materials and implantology, where we envisage the highest chances of successful application of nanoscience and nanotechnology. Additionally, the chance of having pharmaceutical treatment and diagnostics will be mentioned shortly. After reading this work, it should be clear that a more

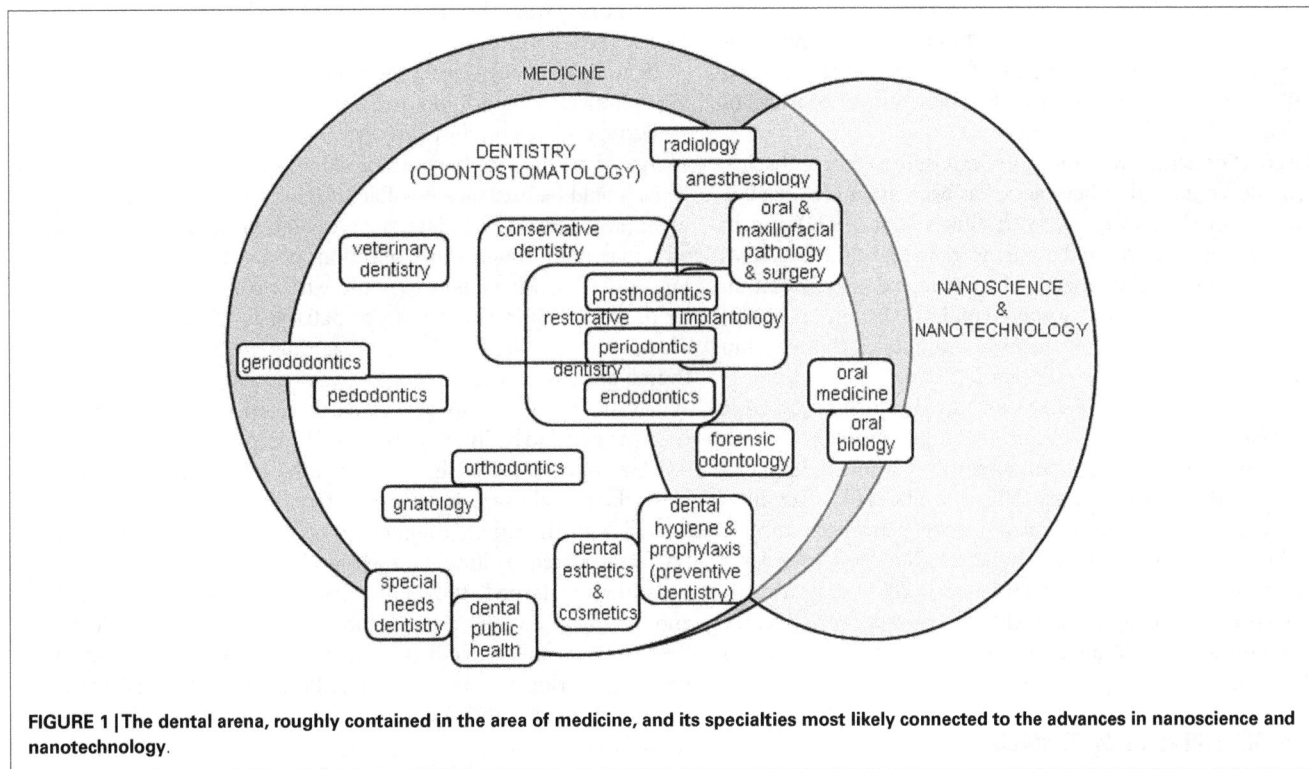

FIGURE 1 | The dental arena, roughly contained in the area of medicine, and its specialties most likely connected to the advances in nanoscience and nanotechnology.

thorough application of concepts and tools emerged in the last two decades in nanoscience and nanotechnology could make dentistry benefit significantly and progress faster in the close future.

## NANOTECHNOLOGY AND NANOSCIENCE IN MEDICINE: NOT JUST SCIENCE FICTION

The Drexler's view (Freitas, 2000; Kumar and Vijayalakshmi, 2006), with billions of colloidal nanorobots injected in sick organ tissue or its environment (the mouth here) and working independently, is not under discussion in this work. The limitations of this utopia (such as the problem of producing and programing these nanounits efficiently, and having them move in a realm where adhesion forces overcome macroscale ones such as gravity) are well-known and not yet successfully addressed. Since we want to avoid approaches still too far from reality, we are not considering what could eventually be done in 20 years. This review describes common topics in dentistry addressed recently by nanotechnological tools with at least partial success. In particular, we will focus on the probably most important nanotechnological tool, used for both imaging and nanomanipulation, which is the atomic force microscope (AFM) (Gerber and Lang, 2006). Therefore, first we will shortly review the applications of AFM in dentistry carried out by other groups in the past 20 years. Then, according to different topics of dentistry, we will review our own work in the area.

## AFM IN DENTISTRY

Since its birth in the late 80s, the AFM has progressively expanded its use from physical systems toward biological and biomedical applications (Kasas and Thomson, 1997). The plus of AFM is that samples require little preparation: no conductive coating is

necessary, different from the scanning tunneling microscope and the scanning electron microscope (SEM), and no thin slices have to be cut such as in the transmission electron microscope (TEM). Instead, real 3D surfaces can be measured directly, even on bulky objects for stand-alone AFM heads. Furthermore, the environment can be ambient air, different from SEM/TEM, which usually work in high vacuum, or even liquid, which is the type of medium preferred by biological matter[1].

For these reasons, and particularly for the non-destructive imaging, researchers have started to use AFM also in the field of dentistry since the mid-90s. The earliest works aimed to inspect the native dental tissue, especially dentin, after acid treatment (Marshall et al., 1993). In Buzalaf et al. (2014), the profile of mice teeth enamel crystals as responding to fluorosis was investigated, instead. An interesting comparison of AFM vs. SEM imaging of dentin was made in Kubinek et al. (2007), who pointed out the presence of artifacts due to dehydration in SEM. In particular, dehydration makes the collagen fibers in the tubules to collapse, which may be critical for the adhesion to dentin of the restorative composites. A difference in local mechanical properties surrounding the tubules has been observed by AFM in Kinney et al. (1996),

---

[1]In fact, this promise for easy operation often makes non-expert end-users think that practically every sample can be put under the AFM, whereas still basic requirements exist: (1) the sample must be a stable solid, i.e., not too gelly or sticky and properly bonded to the substrate; (2) the features must be sparse to be spatially resolved, and with the substrate appearing in between as the reference height, since AFM measures step-heights; (3) AFM is intrinsically slow and low in amount of contents, different from, e.g., SEM: indeed, even recent fast-scanning AFMs require acquisition times of minutes, and one AFM image allows for limited zoom-in since it typically contains ~0.3 vs. ~1.2 megapixels of SEM and ≥3 of digital photography.

who found fourfold hardness (~2.3 GPa) for peritubular vs. intertubular dentin. The mechanical properties of dental materials as measured by AFM have been the focus of a work by Pustan and Belcin (2009), who demonstrated also the measurement of frictions coefficient.

The effect of demineralization after exposure to different acids, for example, contained in beverages, has been often investigated, both in dentin (Eliades et al., 1999; Silikas et al., 1999) and in enamel (Pyne et al., 2009), in the latter work by real-time fast AFM. Conversely, the protective effect against enamel demineralization offered by a paste of casein phosphopeptides (Poggio et al., 2009) as well as different recent formulation toothpastes (Lombardini et al., 2014) has also been investigated.

Despite the limitation of AFM on samples formed by large and curved objects (see discussion in Section "Implant Surface Morphology"), it has been used successfully also on implants and orthodontic brackets wires (Silikas et al., 2001). Ceramics are still other dental materials on which AFM has been applied successfully, to characterize their surface modifications after laser (Folwaczny et al., 1998) or etching (Luo et al., 2001) or heat treatment (Gatin et al., 2013). Finally, AFM has proven useful also for fractographic analysis of failed surfaces in restorative composites (Jandt, 1998).

## DENTAL RESTORATIVE MATERIALS

One major field of application of microscopy and nanotechnology to dentistry is in the characterization and fabrication of dental restorative composites, both in surface morphology and elastic properties. In recent years, several composites claimed to be "nano," such as Filtek Supreme (FS) by 3M-ESPE (3M-ESPE, 2014) and Venus diamond (VD) by Kulzer (2014). The latter in particular is an advanced formulation of hybrid composite, using filler particles with multiple size populations across both the micro- and the nano-scale (Ferracane, 2011), and was often chosen in our works as a reference material.

### SURFACE CHARACTERIZATION OF RESTORATIVE COMPOSITES

The most advanced tool of surface characterization in nanotechnology is the AFM (Vahabi et al., 2013) along with the many other scanning probe microscopes derived from it (Cricenti et al., 2011). This instrument relies on physical contact of a very sharp tip with the specimen. As such, it can provide a direct measurement of 3D surfaces in the real space.

In Salerno et al. (2010a,b), we investigated the effect of air-polishing (AP) on VD restorative composite as a reference. The outcome of this analysis was the confirmation, as suggested by the counseling dentists that AP in itself does not only remove plaque but also damages the surface of composite-based dental restorations. This was assessed by measuring the change in the most common parameter of surface roughness, namely the root mean square (RMS) of heights, $S_q$, after AP. The combinations of AP conditions among two abrading powders (bicarbonate and glycine), two jet distances (2 and 7 mm), and three times (5, 10, and 30 s) were investigated. Even if the conditions for best AP treatment (i.e., least damage) were identified, still some level of damage is unavoidable (see **Figures 2A,B**). This can have a consequence on long-term success of the restorations, due to recurring secondary

caries, arising from the increased bacterial adhesion occurring on AP roughened surfaces.

In addition to measuring the dental surface damage and finding optimized AP conditions to limit it, in Salerno et al. (2010a,b), we also speculated on the possibility that an advanced AFM image analysis can correlate the surface damage to fractal dimensions, which could be used as a possible tool for screening of the treatment (see **Figure 2C**). In fact, surfaces both untreated and treated with maximum roughening conditions presented a fractal behavior, exhibiting self-affinity over the whole scale range accessible by the AFM, approximately from 300 nm to 100 μm scan size. This is roughly three orders of magnitude, which is a rule-of-thumb minimum requirement for a surface to have fractal nature. When the fractal dimensions were calculated, the above surfaces (non-treated and badly treated) showed comparable values of 2.4–2.7. We concluded that the random non-optimized AP does not affect the fractal character of the surface. Differently, the surfaces treated with minimum roughening conditions lost their native fractal character, exhibiting no self-affinity.

In the mentioned study, both the experimental tool of AFM and the image analysis using fractal concepts came to the dental field from nanotechnology and advanced analytical practices of nanoengineering and nanoscience. Additional multifractal analysis carried out recently has confirmed the results of the basic fractal study (Tălu et al., 2015).

In another work (Salerno et al., 2012a,b), we studied the $S_q$ roughness of dental restorative materials after another process similar to AP, which is polishing for dental restoration finishing. This polishing is applied to the dental surfaces not to remove the plaque and refresh their surfaces, but rather soon after their restoration, to decrease the surface roughness of the excess material used by the dentist to fill the dental cavity. As compared to AP, more massive and harder matter is milled during polishing, which requires solid abrading tools such as rotating grommets and burs. In this case, the starting surfaces are rougher with respect to the treated ones (see **Figure 3C**). The smoothest surface obtained is the best, same as for AP, for both medical (decrease of bacterial adhesion) and esthetical reasons (improve optical properties as reflectance and translucency).

In this study, not only the roughness was assessed but also the surface effects on the elastic properties of the materials were also investigated (**Figures 3A,B**. The null hypothesis was that the polishing would not cause degradation to the mechanical properties of the surface. To test this, we used a nanoindenter (Hay, 2009). As compared to old-fashioned microindenter, where only the final conditions of tip imprint size and maximum load reached are recorded, in nanoindentation [or instrumented indentation (Cripps and Anthony, 2004)] the datapoints at all intermediate instants of different load and penetration depth are measured during the experiment. This makes it possible to collect a whole load-indentation dataset, which is converted into a stress–strain engineering plot (**Figure 3A**). By fitting this plot to physical models, mainly Oliver–Pharr (Oliver and Pharr, 1992), both the surface hardness of the specimen and the elastic modulus of the material can be retrieved. In Salerno et al. (2012a,b), five load-indentation cycles were carried out on the specimens at increasing maximum (final) load between 40 and 200 mN (**Figure 3A**). From each cycle,

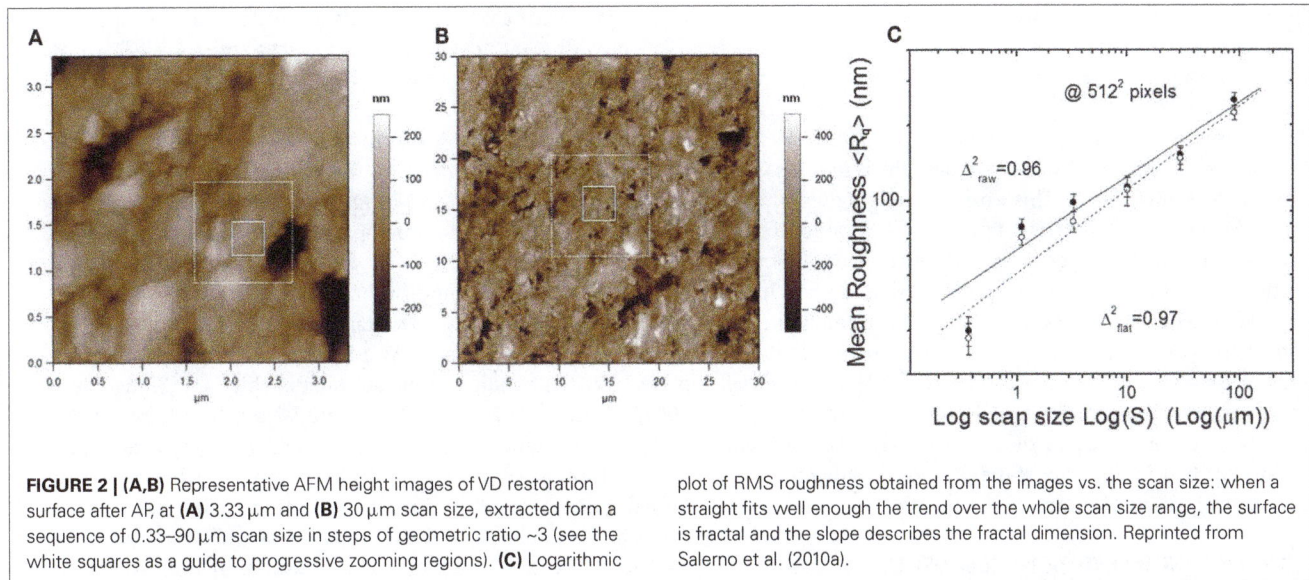

FIGURE 2 | (A,B) Representative AFM height images of VD restoration surface after AP, at (A) 3.33 μm and (B) 30 μm scan size, extracted form a sequence of 0.33–90 μm scan size in steps of geometric ratio ~3 (see the white squares as a guide to progressive zooming regions). (C) Logarithmic plot of RMS roughness obtained from the images vs. the scan size: when a straight fits well enough the trend over the whole scan size range, the surface is fractal and the slope describes the fractal dimension. Reprinted from Salerno et al. (2010a).

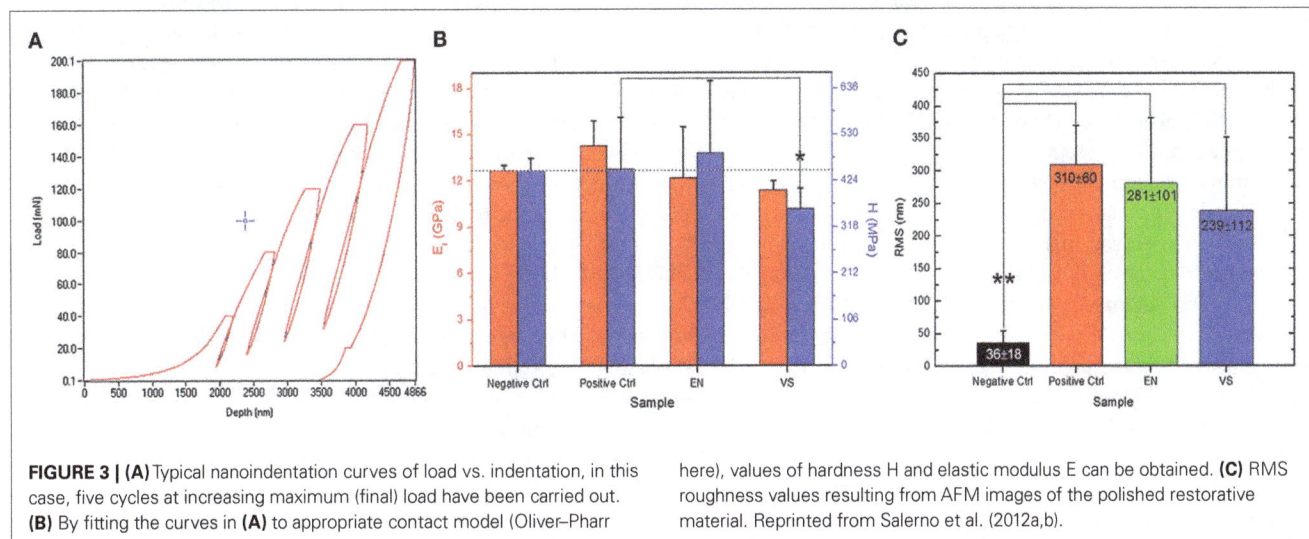

FIGURE 3 | (A) Typical nanoindentation curves of load vs. indentation, in this case, five cycles at increasing maximum (final) load have been carried out. (B) By fitting the curves in (A) to appropriate contact model (Oliver–Pharr here), values of hardness H and elastic modulus E can be obtained. (C) RMS roughness values resulting from AFM images of the polished restorative material. Reprinted from Salerno et al. (2012a,b).

the apparent elastic modulus was calculated, resulting in a profile of elasticity at different depths in the polished material. At the less deep indentations of 1–2 μm (**Figure 3B**), a lower modulus was observed than that found at 5 μm depth. In the latter case, depth was in excess of the surface roughness and the smear layer of polishing debris left on the surface, so the true elastic modulus was measured. Finally, the null hypothesis was confirmed that the elastic modulus is an intrinsic property of the material bulk.

In this work, not only the AFM again together with the nanoindenter have been taken from the "menu" of nanotechnological techniques. The analysis of surface composition was also addressed with advanced research tools of chemistry and physics. In fact, only the use of scanning electron microscopy (SEM) combined with chemical microanalysis by energy-dispersive spectroscopy (EDS) made it possible to identify the smear layer of polishing contamination products responsible for the apparent decrease in contact stiffness at the lowest indentation depths.

Another work across the morphological analysis of the surface and the mechanical elastic properties of the dental composites was carried out in Salerno et al. (2011). There, materials belonging to the class of flowable composites (Walter, 2013) were studied. The reference material for comparison was again VD, and the elastic properties were measured with the nanoindenter. The microscopic analysis by both AFM and SEM was the most important tool in this experiment, allowing to correlate the material properties to peculiar morphology of the matrix-filler distribution of the composites. In fact, we observed higher elastic modulus for the composites with higher filler particles loading, as expected. In particular, it was confirmed that AFM and SEM can be especially useful when used in combination. The access to the internal filler particles was obtained not by slicing the composites, which would probably damage the microtome diamond blade when crossing the matrix-filler modulus discontinuity of 2–70 GPa. Instead, some external material layers were removed by surface grinding. The main result was the observation that, as claimed for the latest generation flowable materials, at least in one case [Vertise Flow by Kerr (2014)] a flowable composite appeared enough stiff to be used also in bulk dental restorations.

The importance of SEM, in combination with AFM or as a faster alternative when no 3D surface topography is required, was confirmed in Salerno (2012b). This work focused on the standard restorative composite VD. With the support of EDS, we confirmed that the elasticity of the material was uniform inside the cured composite, same as qualitatively probed by AFM on a grinded section, due to material homogeneity. In fact, the elemental composition of the composite was uniform across the internal area exposed by grinding, which showed that the fillers, while well distinguished from the morphological point of view, were prepolymerized particles made of the same material surrounding them. Thus, the hybrid character of the composite (showing fillers of different size) was also confirmed in this study.

## ELASTIC PROPERTIES OF RESTORATIVE COMPOSITES

Whereas the elastic properties of dental restorative composites have been traditionally investigated by standard tensile and compressive tests, in the past decade, application of nanoindentation has been increasingly spreading. This technique makes it possible to provide a more comprehensive view of the material, for example, the ratio of elastic-to-plastic behavior may be obtained, as well as the time-dependency of creep, when staying at a given maximum load and monitoring the changes in indentation. In this respect, in the last decade, AFM has become the nanotechnology standard for elastic measurements at high accuracy and spatial resolution, by means of force-spectroscopy of force-volume modes (Cappella and Dietler, 1999; Salerno et al., 2012a,b). On the other hand, for time-dependent viscous response and thanks to the chance of changing the working temperature (see **Figures 4A,B**) dynamic mechanical analysis [DMA (Menard, 1999)] has been increasingly used in the field of dental materials (Ryou et al., 2011).

Atomic force microscope and DMA have been used jointly in Thorat et al. (2012) (**Figure 4C**), to investigate the elastic modulus of experimental restorative composites based on fillers of titania, nanosilica, and milled glass. These composites were studied in a simplified model version, without the use of a coupling agent bonding the fillers and the matrix. AFM had already been used in Salerno (2012b) to measure the contact stiffness (see **Figure 5**), but in this work (Salerno, 2012a) the authors moved on to quantitative evaluation of the elastic modulus, which is known to be a critical step (Salerno, 2012b). Concurrently, DMA was used for cross-checking the AFM values of modulus, and a good agreement was observed between the two techniques (**Figure 4C**). As a consequence, we decided to always rely on the combined use of AFM and DMA for future characterizations of the elastic properties of dental materials.

On the way to the development of coupling-agent-free novel dental restorative materials, in Thorat et al. (2013a,b), we passed on from vitreous filler materials to hard metal oxides such as titania and alumina. For materials characterization, AFM and DMA were used as established previously. In Thorat et al. (2012), the nanosilica filler performed better than microscale ball-milled glass. Similarly, in this work, the nanoscale alumina was found to perform better than microscale alumina. This confirmed that the nano-size fillers provide some advantage with respect to microscale fillers of similar materials, probably due to the higher interfacial area to the matrix, when properly dispersed.

In a more recent work, the authors tried to exploit the nanoscale structuring of the filler particles for both mechanically reinforcing the material and for replacing the missing coupling-agent. To this goal, nanoporous microparticles of alumina have been engineered (Thorat et al., 2014). The controlled nanoporosity in alumina is obtained by anodization of an aluminum foil according to optimized conditions of voltage, electrolyte acid concentration, and bath temperature. As a result, a self-organized coating material called anodic porous alumina is obtained (Shingubara, 2003; Salerno et al., 2009). This material can be set free in the form of a self-standing membrane by etching away the remaining metal substrate. The alumina membranes were further processed by ball-milling, resulting into microscale particles with similar width as the thickness of the original membrane, in the range of 5–10 $\mu$m. The elastic modulus was measured by both AFM and DMA, before and after accelerated aging obtained with an equivalent thermal treatment of the specimens. It was observed that the filler micro-particles of ball-milled porous alumina provided better elastic modulus stability than two commercial restorative materials used for comparison. Obviously, a mechanical interlocking effect occurred thanks to the nanopores, which were infiltrated by the resin matrix during the mixing assisted by sonication. The found procedures were used to file a patent, before publication in the scientific literature (Salerno et al., 2014).

## DENTAL IMPLANTS

Another important field of application of materials nanoscience and nanotechnology to dentistry is related to the characterization and development of dental implants. In our group, we have started to work with dental implants only few years ago, and so far we focused on the preliminary issue of the characterization of commercial implants. Given the well-assessed mechanical properties of the materials currently used, which are titanium (Ti) alloys and ceramics, presently the most required characterization work deals with advanced imaging of the implant surfaces and interfaces. The interfaces of interest in dental implants occur between the tooth ceramic crown on the top, the cement joining the crown to the metallic (Ti alloy) parts of the abutment, and the implant itself (i.e., the screw hidden in the maxillar bone).

### IMPLANT SURFACE MORPHOLOGY

In a recent work on dental implants (Cresti et al., 2013), we focused on 3D imaging of the interface between the abutment and the ceramic crown, namely the region of cement bonding. Because the recessed interface was generally not addressable by the comparatively short (15 $\mu$m) AFM probe tip due to too large $Z$-range, a stylus profilometer was used. A small dedicated software was written in Igor (Wavemetrics, USA; Salerno, 2013), allowing to reassemble into 2D maps the 1D profile scans typical of the instrument. Different from AFM, the profilometer presented intrinsic imaging asymmetry, since the pixel size was of the order of 5 $\mu$m between scan lines and 100 nm along the scan. Nevertheless, the rendering allowed one to identify reference marks made

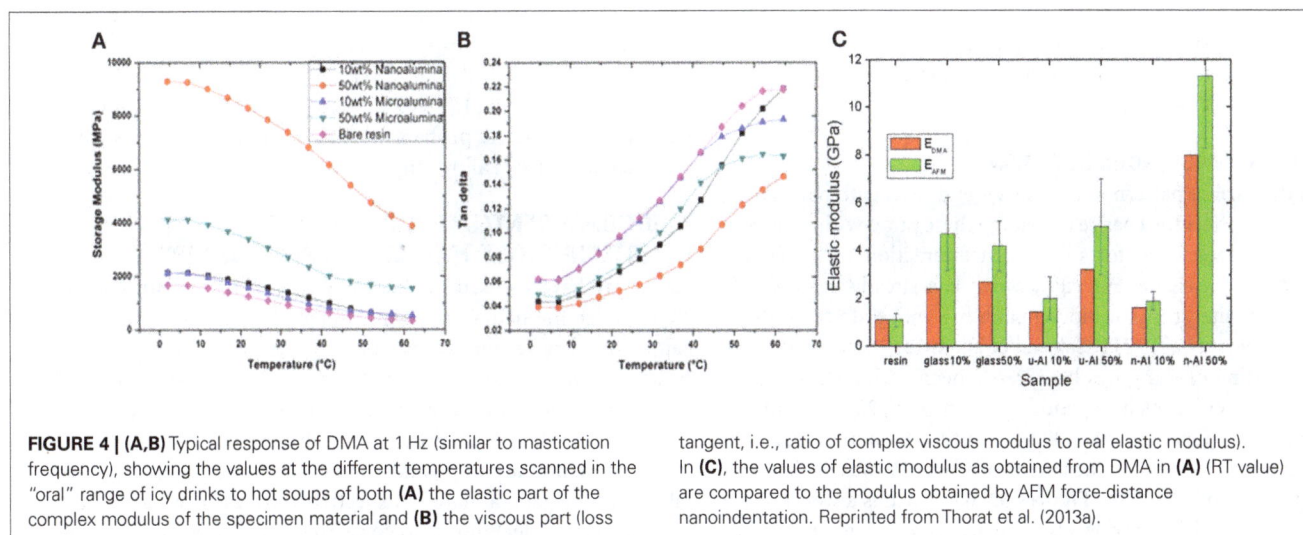

**FIGURE 4 | (A,B)** Typical response of DMA at 1 Hz (similar to mastication frequency), showing the values at the different temperatures scanned in the "oral" range of icy drinks to hot soups of both **(A)** the elastic part of the complex modulus of the specimen material and **(B)** the viscous part (loss tangent, i.e., ratio of complex viscous modulus to real elastic modulus). In **(C)**, the values of elastic modulus as obtained from DMA in **(A)** (RT value) are compared to the modulus obtained by AFM force-distance nanoindentation. Reprinted from Thorat et al. (2013a).

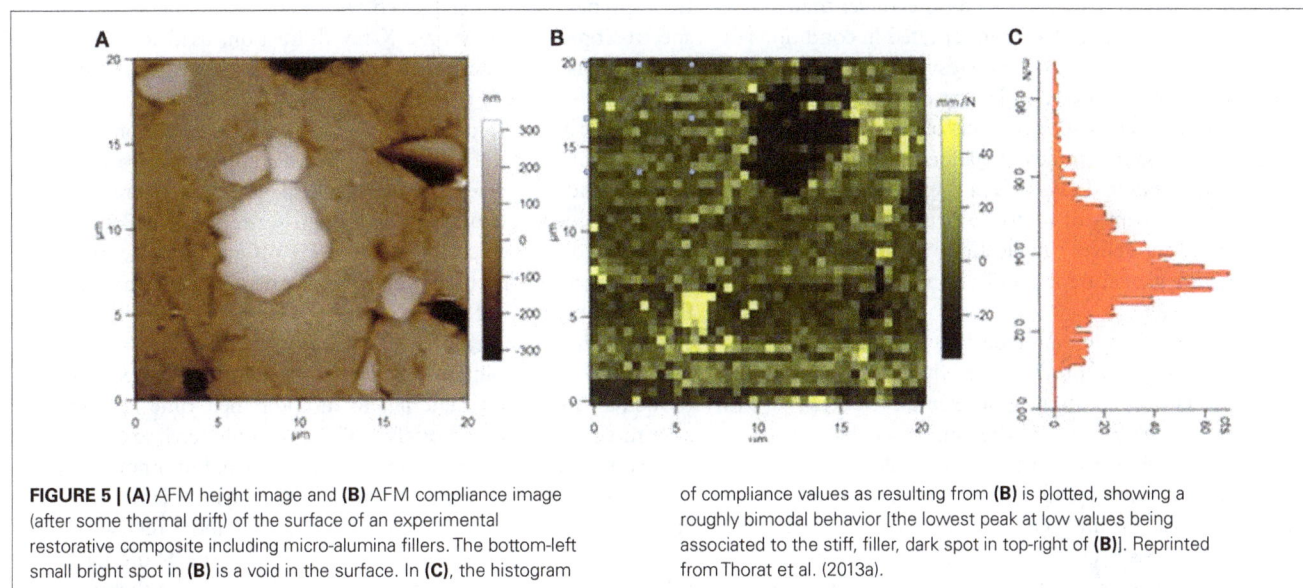

**FIGURE 5 | (A)** AFM height image and **(B)** AFM compliance image (after some thermal drift) of the surface of an experimental restorative composite including micro-alumina fillers. The bottom-left small bright spot in **(B)** is a void in the surface. In **(C)**, the histogram of compliance values as resulting from **(B)** is plotted, showing a roughly bimodal behavior [the lowest peak at low values being associated to the stiff, filler, dark spot in top-right of **(B)**]. Reprinted from Thorat et al. (2013a).

by indentation on the metal. Additionally, the 3D images obtained made it possible to use the standard techniques of AFM image processing to identify and isolate the cement region. To this goal, a 3 × 3 kernel filter based on a Sobel differential operator was used (Russ, 1998). Thus, we calculated the morphological parameters selected for description of the optimal interface, such as a step height at the materials edges, the width of cement (Cresti et al., 2013).

In a more recent work (Salerno et al., 2015), the possible morphological damage occurring on the surface of implant screws was evaluated, in terms of change of the surface after *in vitro* implantation tests in a model bone similar to the typical clinical bone [D1, Mish 90 classification (Misch, 1999)]. Within some limitations (only the top of the screw apex being accessible), AFM has been used in that case. The analysis of images pointed out relevant change in morphology and faced with the issue of its quantification. It should be stressed that having a change in the implant surface would not necessarily mean a decrease of performance, as in some instances the modified surface may even perform better than the original one. In any case, assessing the effect of insertion on the effective surface morphology and stability (i.e., possible delamination of coatings) is important not to misinterpret the success or failure of the implant.

Some researchers have recently experimented with implants based on metals other than Ti, such as zirconium and tantalum. Notably, a controlled roughness, similar to that usually made on medical grade Ti by means of either acid etching or sand-blasting, can also be obtained with the anodization treatment used to make anodic porous alumina. In fact, Ti is another so-called *valve metal* same as Al (Salerno, 2014). In particular, the surface of Ti, when anodized, tends to form adjacent nanotubes rather than nanopores joined by connected oxide walls (Uttiya et al., 2014). These porous titania nanotubes have been investigated so far mainly for applications in optics and photonics, thanks to the photocatalytic activity of titania (Mor et al., 2006). However, they could also be used as the surface modification of Ti implants in dentistry and orthopedics, similar to APA (Salerno, 2012b) and without the need for preliminary Al coating.

## IMPLANT SURFACE FUNCTIONALIZATION

When the surface pattern of implants is endowed with nanopores, it obviously offers the chance of loading these pores with nanoparticles or biomolecules, for subsequent interaction with the surrounding living tissue. For APA, which has already been widely tested in the literature for biocompatibility with different cellular types (Poinern et al., 2011; Salerno et al., 2013; Toccafondi et al., 2014), loading of the pores has already been demonstrated with bioactive materials such as peptides (Swan et al., 2005), remineralization ions (Okawa et al., 2009), or antibacterial nanoparticles (Thorat et al., 2013a,b). In the latter work, the APA was loaded with nanoparticles of silver (see **Figures 6A,B**), which is a well-known biocide. Limited loading levels were reached (up to maximum 6% volume), which would still exhibit a significant pharmaceutical effect, with sustained release during a time longer than 2 days (**Figure 6C**). This experiment has been repeated in conditions of APA sealed under a coating of the same resin base as that used in our novel restorative composite of Thorat et al. (2014). Also, in this case, an effective release of silver was observed in a solution of phosphate buffer saline. In fact, while decreasing the levels of released silver, the sealing delays and thus extends the elution to a significantly longer time, of up to ~5 days. This type of functionalization for drug-delivery applications (Gultepe et al., 2010) could be the future goal also for anodization patterned porous Ti of implants.

Concerning drug delivery and bioactivity of nanotechnological materials, a number of products already exist in the market for general medicine applications. For example, Acticoat™ and Algisite™ are patches for wound dressing based on antibacterial silver nanoparticles (Smith and Nephew, 2014). For products loaded instead with calcium phosphate or hydroxyapatite (HA) nanoparticles, Ostim™ [Osartis, Germany (Doessel and Schlegel, 2010)], Vitoss™ [Orthovita, USA (Kurien et al., 2013)], and NanOss™ [Angstrom Medica, USA (Roveri and Iafisco, 2010)]

can be mentioned, which have been launched in recent years as synthetic bone graft substitutes. For specific dental applications, some HA coatings of dental implants have also been introduced [e.g., Spline Twist MP-1™ by Zimmer (2014), USA], but generally without much success probably due to delamination effects at the interface with the Ti alloy implant core.

## ADDITIONAL DENTAL APPLICATIONS OF NANOTECHNOLOGY: MOLECULAR SPECTROSCOPY

So far, we have focused on the capability of nanotechnology to characterize the dental materials morphologically and mechanically – by ultra-resolution microscopy and advanced nanoprobe measurements – and to modify them – by nanostructuring and functionalization. In addition to this, for advanced characterization, a number of nanoscience techniques can also provide information on the composition and structure of materials, which fall under the general name of spectroscopy. A general advantage of spectroscopic techniques as compared to other testing methods is the non-invasiveness on the specimens. One example of spectroscopy has already been given with EDS. Additional common spectroscopic techniques are X-ray diffraction, used to identify the structure of crystalline inorganic nanoparticles (e.g., titania nanoparticles of HA) and also of biomolecules such as proteins and nucleic acids (Valdré et al., 1995); X-ray photoelectron spectroscopy, for chemical analysis of surface composition with part per thousand routine sensitivity (Kang et al., 2009); and vibrational molecular spectroscopy such as Raman scattering and Fourier-transformed infrared spectroscopy (FTIR) (Czichos et al., 2006). FTIR is already widely used in assessing the degree of conversion of the photopolymerized restorative resin composites, often in association with elastic measurements (Sideridou and Karabela, 2009; Thorat et al., 2012). Similarly, in a recent experiment, we used FTIR on dental impression materials to monitor their time response after mixing the base and catalyst. Thus, we could retrieve quantitative information on setting and working time of four impression materials, three belonging to the class of polyvinyl-siloxanes (VPS) and one being a conventional polyether (work under review). As compared to techniques such as shark fin test and Shore hardness durometer, FTIR can give a deeper insight into the chemistry

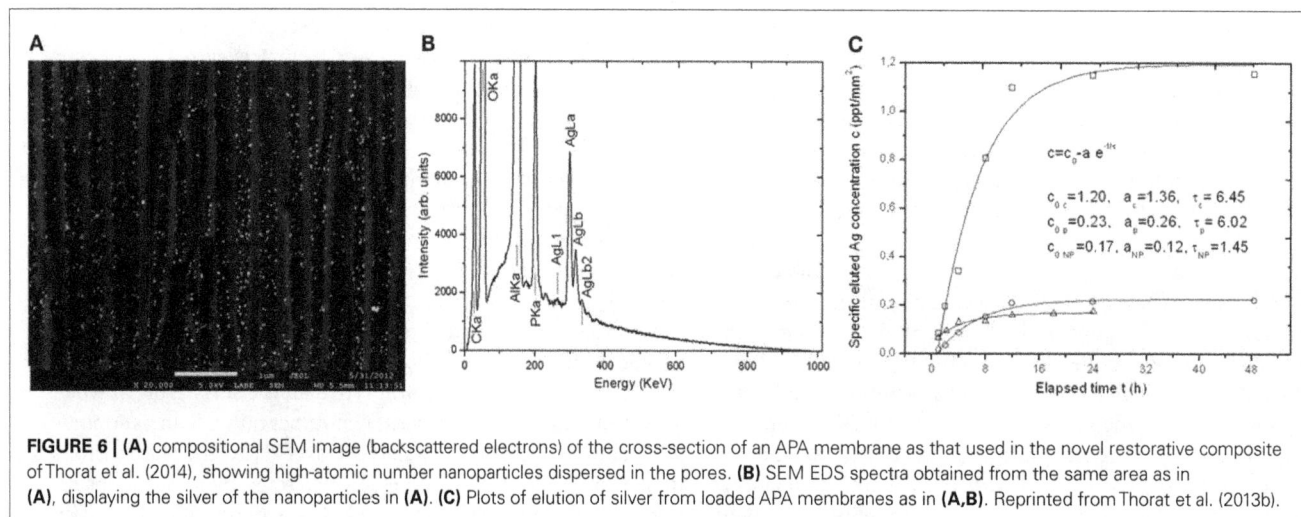

**FIGURE 6 | (A)** compositional SEM image (backscattered electrons) of the cross-section of an APA membrane as that used in the novel restorative composite of Thorat et al. (2014), showing high-atomic number nanoparticles dispersed in the pores. **(B)** SEM EDS spectra obtained from the same area as in **(A)**, displaying the silver of the nanoparticles in **(A)**. **(C)** Plots of elution of silver from loaded APA membranes as in **(A,B)**. Reprinted from Thorat et al. (2013b).

behind the decrease in viscosity and increase in stiffness of these elastomers during the setting. In fact, the spectral peaks describing the decrease in Si–H bonds upon increase of the Si–O ones during the setting have been identified in two materials, allowing to describe not only the setting kinetics but also their chemistry behind it. However, since the dental impression elastomers are complex materials that contain a number of components such as filler particles (to tune the rheology), color pigments, plasticizers, initiators, etc., the exact picture is difficult to obtain in the absence of detailed information available from the manufacturers.

Additional information may be obtained by complementary technique such as Raman scattering, since some molecular vibration modes are active in Raman and not (or not as sensitive) in FTIR, and vice-versa (Czichos et al., 2006). To be noted, both FTIR and Raman can today be coupled with optical microscope mapping, allowing spatial resolution on sample surfaces down to micrometer size spot (Mariani et al., 2010).

## CONCLUSION AND OUTLOOK

In this work, some results obtained during the past few years in the field of dental material science, both in the literature and particularly in our institute, have been reviewed. The background of our group in microscopy and microfabrication techniques has been employed to obtain the presented results, and we are still struggling to pursue the application of nanoscience and nanotechnology practices further in the field of dental sciences. We are confident that this attitude of increasing use of modern inspection and manufacturing techniques also in dentistry will proceed at a faster pace in the next few years. Especially, the use of scanning probe microscopy and advanced light irradiation techniques, and possibly combinations thereof, can be foreseen as a field of future development. Laser treatment has already entered the dental practice, both as drill replacement and for speed-up activation of peroxide bleaching solution during teeth whitening. The use of laser, in connection with photocatalytic materials such as titania, has also been proposed for disinfection (Riley et al., 2005). More recently, fluorescence probing has been introduced for detection of oral cancer (Shin et al., 2010). Similarly for AFM, endoscopic instruments imaging dental surfaces *in situ* in totally non-invasive manner have also been postulated (Stolz et al., 2007). In this application, the AFM could also be used as a nanoscalpel, or empowered with light irradiation to make real-time *in vivo* biosensors based on tip-enhanced Raman spectroscopy.

We find that there are no major technological barriers against this cross-contamination of research and technology between different areas of physics, engineering, and dentistry, but for the psychological barriers from different education of dental doctors. The point of view of dentists and major dental companies will have to change, and the latter in particular should open wider to this scenario and invest significantly in the field. A relatively small portion of the budget currently devoted to advertisement and promotion of dental products could actually foster the development of new practices and products of possibly great benefit to the clinical results. However, a shift is required in the mind of both dental practitioners and dental companies to take full advantage of nanoscience and nanotechnology, which for the dental area has so far remained mainly in the academia.

## REFERENCES

3M-ESPE. (2014). *Filtek Supreme*. Available at: http://www.3m.com/3M/en_US/Dental/Products/Catalog/~/Filtek-Supreme-Ultra-Universal-Restorative?N=5145652+3294736391&rt=rud

American Dental Association. (2014). *Glossary of Dental Clinical and Administrative Terms*. Available at: http://www.ada.org/glossaryforprofessionals.aspx

Zimmer. (2014). Anon Zimmer Dental Inc. 1900 Aston Avenue Carlsbad, CA 92008, USA (760) 929-471 4300. Available at: http://www.zimmerdental.com/Home/zimmerDental.aspx

Buzalaf, M. A., Barbosa, C. S., Leite Ade, L., Chang, S. R., Liu, J., Czajka-Jakubowska, A., et al. (2014). Enamel crystals of mice susceptible or resistant to dental fluorosis: an AFM study. *J. Appl. Oral Sci.* 22, 159–164. doi:10.1590/1678-775720130515

Cappella, B., and Dietler, G. (1999). Force-distance curves by atomic force microscopy. *Surf. Sci. Rep.* 34, 1–104. doi:10.1016/S0167-5729(99)00003-5

Cresti, S., Itri, A., Rebaudi, A., Diaspro, A., and Salerno, M. (2013). Microstructure of titanium-cement-lithium disilicate interface in CAD-CAM dental implant crowns: a three-dimensional profilometric analysis. *Clin. Implant Dent. Relat. Res.*, 4–6. Available at: http://www.ncbi.nlm.nih.gov/pubmed/23968260

Cricenti, A., Colonna, S., Girasole, M., Gori, P., Ronci, F., Longo, G., et al. (2011). Scanning probe microscopy in material science and biology. *J. Phys. D Appl. Phys.* 44, 464008. doi:10.1088/0022-3727/44/46/464008

Cripps, F., and Anthony, C. (2004). *Nanoindentation*. New York, NY: Springer.

Czichos, H., Saito, T., and Smith, L. (eds) (2006). *Springer Handbook of Materials Measurement Methods*. Springer.

Doessel, O., and Schlegel, W. C. (eds) (2010). "World congress on medical physics and biomedical engineering September 7-12, 2009 Munich, Germany," in *Biomaterials, Cellular and Tissue Engineering, Artificial Organs*, Vol. 25/X (Springer Science and Business Media).

Eliades, G., Vougiouklakis, G., and Palaghias, G. (1999). Effect of dentin primers on the morphology, molecular composition and collagen conformation of acid-demineralized dentin in situ. *Dent. Mater.* 15, 310–317. doi:10.1016/S0109-5641(99)00050-0

Ferracane, J. L. (2011). Resin composite – state of the art. *Dent Mater.* 27, 29–38. doi:10.1016/j.dental.2010.10.020

Folwaczny, M., Mehl, A., Haffner, C., and Hickel, R. (1998). Polishing and coating of dental ceramic materials with 308 nm XeCl excimer laser radiation. *Dent. Mater.* 14, 186–193. doi:10.1016/S0109-5641(98)00029-3

Freitas, R. (2000). Nanodentistry. *J. Am. Dent. Assoc.* 131, 1559–1565. doi:10.14219/jada.archive.2000.0084

Gatin, E., Luculescu, C., Iordache, S., and Patrascu, I. (2013). Morphological investigation by AFM of dental ceramics under thermal processing. *J. Optoelectron. Adv. Mater.* 15, 1136–1141.

Gerber, C., and Lang, H. P. (2006). How the doors to the nanoworld were opened. *Nat. Nanotechnol.* 1, 3–5. doi:10.1038/nnano.2006.70

Gultepe, E., Nagesha, D., Sridhar, S., and Amiji, M. (2010). Nanoporous inorganic membranes or coatings for sustained drug delivery in implantable devices. *Adv. Drug Deliv. Rev.* 62, 305–315. doi:10.1016/j.addr.2009.11.003

Hay, J. (2009). Introduction to instrumented indentation testing. *Exp. Tech.* 33, 66–72. doi:10.1063/1.2830028

Jandt, K. D. (1998). Structure of microfilled dental composite fractured surfaces. *Probe Microsc.* 1, 323–331.

Kang, B. S., Sul, Y. T., Oh, S. J., Lee, H. J., and Albrektsson, T. (2009). XPS, AES and SEM analysis of recent dental implants. *Acta Biomater.* 5, 2222–2229. doi:10.1016/j.actbio.2009.01.049

Kasas, S., and Thomson, N. (1997). Biological applications of the AFM: from single molecules to organs. *Int. J. Imaging Syst. Technol.* 8, 151–161. doi:10.1002/(SICI)1098-1098(1997)8:2<151::AID-IMA2>3.0.CO;2-9

Kerr. (2014). *Vertise Flow*. Available at: http://www.kerrdental.com/kerrdental-composites-vertiseflow-2

Kinney, J. H., Balooch, M., Marshall, S. J., Marshall, G. W. Jr., and Weihs, T. P. (1996). Atomic force microscope measurements of the hardness and elasticity of peritubular and intertubular human dentin. *J. Biomed. Eng.* 118, 133–135.

Kubinek, R., Zapletalova, Z., Vujtek, M., Novotný, R., Kolarova, H., and Chmelickova, H. (2007). Examination of dentin surface using AFM and SEM. *Dent. Mater.* 593–598. Available at: http://www.formatex.org/microscopy3/pdf/pp593-598.pdf

Kulzer, H. (2014). *Venus Diamond*. Available at: http://www.heraeus-venus.com/it/it/products_7/venusdiamond_4/venusdiamond_1.html

Kumar, S. R., and Vijayalakshmi, R. (2006). Nanotechnology in dentistry. *Indian J. Dent. Res.* 17, 62–65. doi:10.4103/0970-9290.29890

Kurien, T., Pearson, R. G., and Scammell, B. E. (2013). Bone graft substitutes currently available in orthopaedic practice: the evidence for their use. *Bone Joint J.* 95-B, 583–597. doi:10.1302/0301-620X.95B5.30286

Lombardini, M., Ceci, M., Colombo, M., Bianchi, S., and Poggio, C. (2014). Preventive effect of different toothpastes on enamel erosion: AFM and SEM studies. *Scanning* 36, 401–410. doi:10.1002/sca.21132

Luo, X. P., Silikas, N., Allaf, M., Wilson, N. H. F., and Watts, D. C. (2001). AFM and SEM study of the effects of etching on IPS-Empress 2 dental ceramic. *Surf. Sci.* 491, 388–394. doi:10.1016/S0039-6028(01)01301-2

Mariani, M. M., Day, P. J. R., and Deckert, V. (2010). Applications of modern micro-Raman spectroscopy for cell analyses. *Integr. Biol. (Camb)* 2, 94–101. doi:10.1039/b920572a

Marshall, G. W. Jr., Balooch, M., Tench, R. J., Kinney, J. H., and Marshall, S. J. (1993). Atomic force microscopy of acid effects on dentine. *Dent. Mater.* 9, 265–268. doi:10.1016/0109-5641(93)90072-X

Menard, K. P. (1999). *Dynamic Mechanical Analysis, A Practical Introduction.* Boca Raton, NY: CRC Press.

Misch, C. E. (1999). *Contemporary Implant Dentistry.* St. Louis, MO: Mosby.

Mor, G. K., Varghese, O. K., Paulose, M., Shankar, K., and Grimes, C. A. (2006). A review on highly ordered, vertically oriented TiO2 nanotube arrays: fabrication, material properties, and solar energy applications. *Sol. Energ. Mater. Sol. Cell.* 90, 2011–2075. doi:10.1016/j.solmat.2006.04.007

Okawa, S., Homma, K., Kanatani, M., and Watanabe, K. (2009). Characterization of calcium phosphate deposited on valve metal by anodic oxidation with polarity inversion. *Dent. Mater. J.* 28, 513–518. doi:10.4012/dmj.28.513

Oliver, W. C., and Pharr, G. M. (1992). An improved technique for determining hardness and elastic modulus using load and displacement sensing indentation experiments. *J. Mater. Res.* 7, 1564. doi:10.1557/JMR.1992.1564

Poggio, C., Lombardini, M., Dagna, A., Chiesa, M., and Bianchi, S. (2009). Protective effect on enamel demineralization of a CPP-ACP paste: an AFM in vitro study. *J. Dent.* 37, 949–954. doi:10.1016/j.jdent.2009.07.011

Poinern, G. E., Shackleton, R., Mamun, S. I., and Fawcett, D. (2011). Significance of novel bioinorganic anodic aluminum oxide nanoscaffolds for promoting cellular response. *Nanotechnol. Sci. Appl.* 4, 11–24. doi:10.2147/NSA.S13913

Pustan, M., and Belcin, O. (2009). Application of atomic force microscope for mechanical and tribological characterization of teeth and biomaterials. *Tribology in industry.* 31, 43–46.

Pyne, A., Marks, W., M Picco, L., G Dunton, P., Ulcinas, A., E Barbour, M., et al. (2009). High-speed atomic force microscopy of dental enamel dissolution in citric acid. *Arch. Histol. Cytol.* 72, 209–215. doi:10.1679/aohc.72.209

Riley, D. J., Bavastrello, V., Covani, U., Barone, A., and Nicolini, C. (2005). An in-vitro study of the sterilization of titanium dental implants using low intensity UV-radiation. *Dent. Mater.* 21, 756–760. doi:10.1016/j.dental.2005.01.010

Roveri, N., and Iafisco, M. (2010). Evolving application of biomimetic nanostructured hydroxyapatite. *Nanotechnol. Sci. Appl.* 3, 107–125. doi:10.2147/NSA.S9038

Russ, J. C. (1998). *The Image Processing Handbook*, 3rd Edn. CRC Press, Springer, IEEE Press.

Ryou, H., Niu, L. N., Dai, L., Pucci, C. R., Arola, D. D., Pashley, D. H., et al. (2011). Effect of biomimetic remineralization on the dynamic nanomechanical properties of dentin hybrid layers. *J. Dent. Res.* 90, 1122–1128. doi:10.1177/0022034511414059

Salerno, M. (2012a). Improved estimation of contact compliance via atomic force microscopy using a calibrated cantilever as a reference sample. *Measurement* 45, 2103–2113. doi:10.1016/j.measurement.2012.05.011

Salerno, M. (2012b). Improved estimation of contact compliance via atomic force microscopy using a calibrated cantilever as a reference sample. *Measurement* 45, 2103–2113. doi:10.1016/j.measurement.2012.05.011

Salerno, M. (2013). *Procedure for Assembling DAT ASCII Files of 1D Profiles from XP-2 (Ambios-USA) into 2D Maps of Height.* Available at: http://www.igorexchange.com/node/2699

Salerno, M. (2014). Introduction to the special issue on "nanostructures by valve metal anodization". *J. Mater. Sci. Nanotechnol.* 1(1):Se101. doi:10.15744/2348-9812.1.Se101

Salerno, M., Caneva-Soumetz, F., Pastorino, L., Patra, N., Diaspro, A., and Ruggiero, C. (2013). Adhesion and proliferation of osteoblast-like cells on anodic porous

alumina substrates with different morphology. *IEEE Trans. Nanobioscience* 12, 106–111. doi:10.1109/TNB.2013.2257835

Salerno, M., Derchi, G., Thorat, S., Ceseracciu, L., Ruffilli, R., and Barone, A. C. (2011). Surface morphology and mechanical properties of new-generation flowable resin composites for dental restoration. *Dent. Mater.* 27, 1221–1228. doi:10.1016/j.dental.2011.08.596

Salerno, M., Diaspro, A., and Thorat, S. (2014). *Combined Material Including Anodic Porous Alumina and a Polymer Matrix, and Its Use for the Dental Recondition.* Available at: http://worldwide.espacenet.com/publicationDetails/originalDocument?CC=WO&NR=2014053946A1&KC=A1&FT=D&ND=3&date=20140410&DB=EPODOC&locale=en_EP

Salerno, M., et al. (2015). Surface microstructure of dental implants pre- and post-insertion: an in vitro study by means of scanning probe microscopy. *Implant Dentistry.* (in press).

Salerno, M., Giacomelli, L., Derchi, G., Patra, N., and Diaspro, A. (2010a). Atomic force microscopy in vitro study of surface roughness and fractal character of a dental restoration composite after air-polishing. *Biomed. Eng. Online* 9, 59. doi:10.1186/1475-925X-9-59

Salerno, M., Giacomelli, L., and Larosa, C. (2010b). Biomaterials for the programming of cell growth in oral tissues: the possible role of APA bioinformation. *Bioinformation* 5, 291–293. doi:10.6026/97320630005291

Salerno, M., Patra, N., and Cingolani, R. (2009). Use of ionic liquid in fabrication, characterization, and processing of anodic porous alumina. *Nanoscale Res. Lett.* 4, 865–872. doi:10.1007/s11671-009-9337-3

Salerno, M., Patra, N., and Diaspro, A. (2012a). Atomic force microscopy nanoindentation of a dental restorative midifill composite. *Dent. Mater.* 28, 197–203. doi:10.1016/j.dental.2011.10.007

Salerno, M., Patra, N., Thorat, S., Derchi, G., and Diaspro, A. (2012b). Combined effect of polishing on surface morphology and elastic properties of a commercial dental restorative resin composite. *Sci. Adv. Mater.* 4, 126–134. doi:10.1166/sam.2012.1261

Shin, D., Vigneswaran, N., Gillenwater, A., and Richards-Kortum, R. (2010). Advances in fluorescence imaging techniques to detect oral cancer and its precursors. *Future Oncol.* 6, 1143–1154. doi:10.2217/fon.10.79

Shingubara, S. (2003). Fabrication of nanomaterials using porous alumina templates. *J. Nanopart. Res.* 5, 17–30. doi:10.1023/A:1024479827507

Sideridou, I. D., and Karabela, M. M. (2009). Effect of the amount of 3-methacyloxypropyltrimethoxysilane coupling agent on physical properties of dental resin nanocomposites. *Dent. Mater.* 25, 1315–1324. doi:10.1016/j.dental.2009.03.016

Silikas, N., Lennie, A. R., England, K., and Watts, D. C. (2001). AFM as a tool in dental research. *Microsc. Anal.* 19–21.

Silikas, N., Watts, D. C., England, K. E., and Jandt, K. D. (1999). Surface fine structure of treated dentine investigated with tapping mode atomic force microscopy (TMAFM). *J. Dent.* 27, 137–144. doi:10.1016/S0300-5712(98)00032-3

Smith and Nephew. (2014). *Acticoat Product Line.* Available at: http://www.smith-nephew.com/uk/products/wound_management/product-search/acticoat/

Stolz, M., Aebi, U., and Stoffler, D. (2007). Developing scanning probe-based nanodevices-stepping out of the laboratory into the clinic. *Nanomedicine* 3, 53–62. doi:10.1016/j.nano.2007.01.001

Swan, E. E. L., Popat, K. C., and Desai, T. A. (2005). Peptide-immobilized nanoporous alumina membranes for enhanced osteoblast adhesion. *Biomaterials* 26, 1969–1976. doi:10.1016/j.biomaterials.2004.07.001

Tălu, S, Stachb, S., Albc, Ş. F., and Salerno, M. (2015). Multifractal characterization of a dental restorative composite after air-polishing. *Chaos Solitons Fractals* 71, 7–13. doi:10.1016/j.chaos.2014.11.009

Thorat, S., Diaspro, A., and Salerno, M. (2013a). Effect of alumina reinforcing fillers in BisGMA-based resin composites for dental applications. *Adv. Mater. Lett.* 4, 15–21. doi:10.5185/amlett.2013.icnano.283

Thorat, S., Diaspro, A., Scarpellini, A., Povia, M., and Salerno, M. (2013b). Comparative study of loading of anodic porous alumina with silver nanoparticles using different methods. *Materials* 6, 206–216. doi:10.3390/ma6010206

Thorat, S., Patra, N., Ruffilli, R., Diaspro, A., and Salerno, M. (2012). Preparation and characterization of a BisGMA-resin dental restorative composites with glass, silica and titania fillers. *Dent. Mater. J.* 31, 635–644. doi:10.4012/dmj.2011-251

Thorat, S. B., Diaspro, A., and Salerno, M. (2014). In vitro investigation of coupling-agent-free dental restorative composite based on nano-porous alumina fillers. *J. Dent.* 42, 279–286. doi:10.1016/j.jdent.2013.12.001

Toccafondi, C., Thorat, S., La Rocca, R., Scarpellini, A., Salerno, M., Dante, S., et al. (2014). Multifunctional substrates of thin porous alumina for cell biosensors. *J. Mater. Sci. Mater. Med.* 25, 2411–2420. doi:10.1007/s10856-014-5178-4

Uttiya, S., Contarino, D., Prandi, S., Carnasciali, M. M., Gemme, G., Mattera, L., et al. (2014). Anodic oxidation of titanium in sulphuric acid and phosphoric acid electrolytes. *J. Mater. Sci. Nanotechnol.* 1, S106.

Vahabi, S., Nazemi Salman, B., and Javanmard, A. (2013). Atomic force microscopy application in biological research: a review study. *Iran. J. Med. Sci.* 38, 76–83.

Valdré, G., Mongiorgi, R., Monti, S., Corvo, G., Itro, A., Paroli, R., et al. (1995). [X-ray powder diffraction (XRD) in the study of biomaterials used in dentistry. 3]. *Minerva Stomatol.* 44, 21–32.

Walter, R. (2013). Bulk-fill flowable composite resins. *J. Esthet. Restor. Dent.* 25, 72–76. doi:10.1111/jerd.12011

**Conflict of Interest Statement:** The authors declare that the research was conducted in the absence of any commercial or financial relationships that could be construed as a potential conflict of interest.

# 23

# Covalently linked organic networks

*Matthew A. Addicoat[1]\* and Manuel Tsotsalas[2]\**

[1] *Engineering and Science, Jacobs University Bremen, Bremen, Germany*
[2] *Institute of Functional Interfaces (IFG), Karlsruhe Institute of Technology (KIT), Eggenstein-Leopoldshafen, Germany*

**Edited by:**
*Zhenyu Li, University of Science and Technology of China, China*

**Reviewed by:**
*François-Xavier Coudert, CNRS, France*
*Youyong Li, Soochow University, China*
*Wei-Qiao Deng, Dalian Institute of Chemical Physics, China*

**\*Correspondence:**
*Matthew A. Addicoat, Engineering and Science, Jacobs University Bremen, Campus Ring 1, 28759 Bremen, Germany*
*e-mail: m.addicoat@jacobs-university.de;*
*Manuel Tsotsalas, Institute of Functional Interfaces (IFG), Karlsruhe Institute of Technology (KIT), Hermann-von-Helmholtz-Platz 1, Eggenstein-Leopoldshafen 76344, Germany*
*e-mail: manuel.tsotsalas@kit.edu*

In this review, we intend to give an overview of the synthesis of well-defined covalently bound organic network materials such as covalent organic frameworks, conjugated microporous frameworks, and other "ideal polymer networks" and discuss the different approaches in their synthesis and their potential applications. In addition we will describe the common computational approaches and highlight recent achievements in the computational study of their structure and properties. For further information, the interested reader is referred to several excellent and more detailed reviews dealing with the synthesis (Dawson et al., 2012; Ding and Wang, 2013; Feng et al., 2012) and computational aspects (Han et al., 2009; Colón and Snurr, 2014) of the materials presented here.

**Keywords: covalent organic frameworks, conjugated microporous polymers, computational high-throughput screening, porous organic polymers, ideal network polymers**

## INTRODUCTION

Polymeric networks, with well-defined structures, can accommodate, interact with, and discriminate molecules, thereby leading to prominent applications. Indeed, due to the porous and robust nature of the frameworks as well as the chemical properties of the wall components, many of the materials have shown notable abilities for gas storage, gas separation, drug delivery, sensing, catalytic, and photovoltaic (Wan et al., 2008) applications (see **Figure 1** for possible applications of organic polymer networks) (McKeown and Budd, 2010; Xiang and Cao, 2013).

The high potential of covalently linked organic networks for these selected applications originates from the combination of large surface are, the high variability in the design of the organic building blocks and the tunable pore sizes. Another important feature of most covalent organic networks is their high physicochemical stability, which enables their application even at high temperatures and under humid conditions. In the following paragraphs, we will highlight the specific features for the different applications and refer to selected example from the literature.

Catalysis: the incorporation of catalytic sites within covalently linked organic frameworks showed to be comparable with the activities of the corresponding homogeneous catalysts, with the added potential of size selectivity, recyclability, and chirality (Kaur et al., 2011).

Gas separation: in gas separation vapors, gases, or liquids can be separated depending on their size (molecular sieving) and/or their affinity to the pore surface (selective adsorption). The organic networks can be either synthesized as freestanding membranes (Lindemann et al., 2014) or be used in pressure swing adsorption (Chang et al., 2013).

Gas storage: the main advantages of covalently linked organic networks for gas separation are their high surface area/low density due to their construction by lightweight elements in combination with their high versatility in terms of incorporation of functional groups (either within the building blocks or via post-synthetic functionalization). The high variability allows the storage of large amounts of gases with tailorable heat of adsorption within the covalent networks (Liebl and Senker, 2013; Arab et al., 2014).

Photovoltaics: in organic photovoltaics, the possibility to include functional building blocks into the organic networks and their organization in well-defined geometries promises large potential in order to optimize material properties toward high absorbance, efficient charge separation, and transfer at the donor–acceptor interface, fast diffusion of excitons, and efficient charge collection (Chen et al., 2010; Dogru and Bein, 2014).

To take full advantage of the possibilities of organic networks for applications is their macroscopic shaping in order to be able to incorporate the materials within functional devices. Recent examples include the interfacial synthesis (Colson and Dichtel, 2013) the layer-by-layer synthesis and the mechanochemical synthesis in order to create thin layers, freestanding membranes (Lindemann et al., 2014), or exfoliated layers (Biswal et al., 2013).

## SYNTHESIS OF COVALENTLY LINKED ORGANIC NETWORKS

Control over network topology and specific surface area of covalently linked organic materials can be achieved via three different

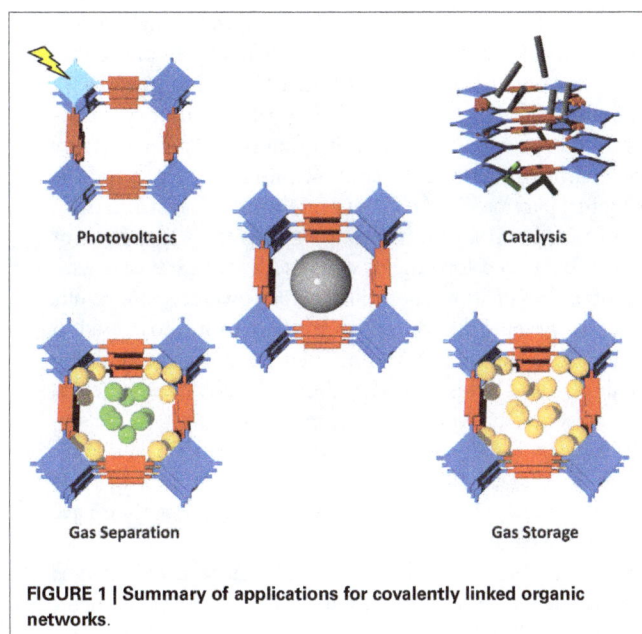

**FIGURE 1 | Summary of applications for covalently linked organic networks**.

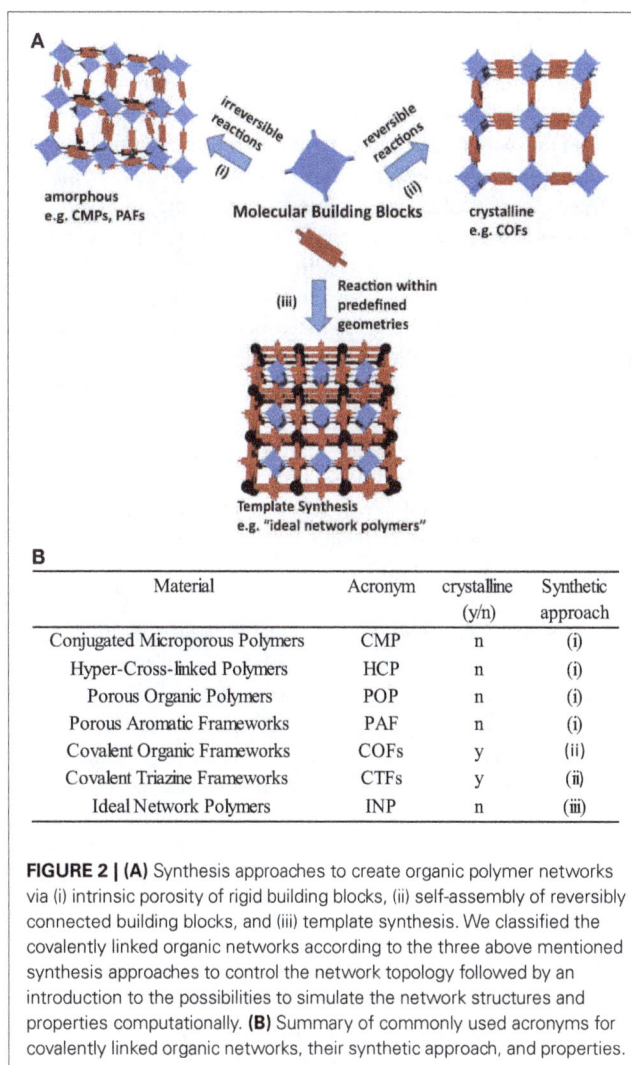

**FIGURE 2 | (A)** Synthesis approaches to create organic polymer networks via (i) intrinsic porosity of rigid building blocks, (ii) self-assembly of reversibly connected building blocks, and (iii) template synthesis. We classified the covalently linked organic networks according to the three above mentioned synthesis approaches to control the network topology followed by an introduction to the possibilities to simulate the network structures and properties computationally. **(B)** Summary of commonly used acronyms for covalently linked organic networks, their synthetic approach, and properties.

| Material | Acronym | crystalline (y/n) | Synthetic approach |
|---|---|---|---|
| Conjugated Microporous Polymers | CMP | n | (i) |
| Hyper-Cross-linked Polymers | HCP | n | (i) |
| Porous Organic Polymers | POP | n | (i) |
| Porous Aromatic Frameworks | PAF | n | (i) |
| Covalent Organic Frameworks | COFs | y | (ii) |
| Covalent Triazine Frameworks | CTFs | y | (ii) |
| Ideal Network Polymers | INP | n | (iii) |

methods: (i) The first method is to use rigid, sterically demanding molecular building blocks hold together via irreversible, but high yielding reactions, to create networks with large free volumes. (ii) The second method is to use reversible covalent reactions enabling self-assembly of the building block into networks of long-range order and crystallinity. (iii) The third method is to use a pre-synthesized template, which guides the structure and connectivity. **Figure 1** shows the three different approaches to create covalently linked organic networks with control over topology and degree of order – ranging to highly crystalline to amorphous – and with a wide range of mechanical properties – from soft gel-like to highly rigid (see **Figure 2A** for the three different synthesis methods to create organic polymer networks).

## COVALENT NETWORKS SYNTHESIZED VIA IRREVERSIBLE REACTIONS OF STERICALLY DEMANDING HIGHLY RIGID MOLECULAR BUILDING BLOCKS

Several different classes of organic polymer networks synthesized via high yielding reactions between rigid and sterically demanding molecular building blocks have been reported. The rigid organic building blocks are mostly based on aromatic subunits and the chemical reactions employed to connect them are mostly coupling reactions such as Ullmann, Yamamoto, or Click-Chemistry. Their common feature is permanent porosity upon removal of the solvent used in the synthesis and high physical and chemical stability. They typically lack long-range order and hence crystallinity, nevertheless often show narrow pore size distribution (PSD).

Naming of these compounds is somewhat arbitrary. Examples of such materials are conjugated microporous polymers (CMP), (Jiang et al., 2007) hyper-cross-linked polymers (HCP), (Tsyurupa and Davankov, 2002) porous organic polymers (POP), (Farha et al., 2009), and porous aromatic frameworks (PAF) (Ben et al., 2009). However, since a common feature of this class of materials is intrinsic porosity they can be regarded as a highly cross-linked sub-class of polymers of intrinsic microporosity (PIM).

The first reported porous poly(aryleneethynylene) CMPs networks are composed of aromatic halides and aromatic alkynes, which are connected using palladium-catalyzed Sonogashira–Hagihara cross-coupling reactions.

## POROUS POLYMERS SYNTHESIZED VIA REVERSIBLE COVALENT REACTIONS

This class of materials is known as covalent organic frameworks (COFs). COFs are composed of covalent building blocks made of boron, carbon, oxygen, and hydrogen, in many cases also including nitrogen or silicon, stitched together by organic subunits. The atoms are held together by strong covalent bonds. Depending on the selection of building blocks, the COFs may form 2D or 3D networks. Planar building blocks are the constituents of 2D COFs, whereas for the formation of 3D COFs, typically tetragonal building blocks are involved. High symmetric covalent linking, as it is perceived in reticular chemistry, was confirmed for the products.

Reversible reactions in the synthesis of COFs are usually condensation reactions where the reversability originates from the hydrolysis back reaction. Most COFs are synthesized either by boronic acid condensation forming boronic anhydrite or by boronic acid condensation with catechol (El-Kaderi et al., 2007). Other condensation reactions include imine (Uribe-Romo et al., 2009) or hydrozone (Uribe-Romo et al., 2011) formation through condensation of aldehyde and amine or aldehyde and hydrazide. Two reaction mechanism not based on condensation reaction have been reported to create crystalline frameworks. Covalent triazine frameworks (CTFs) produced by trimerization of dicyano compounds, however to generate reversibility the reaction has to be carried out under much harsher conditions (Kuhn et al., 2008). A recently reported reaction mechanism for the synthesis of crystalline, covalently connected organic networks is the dimerization of nitroso compound to azodioxides (Beaudoin et al., 2013). These compounds show excellent crystallinity and even enabled the first single crystal COF structures. However, due to the low stability, no permanent porosity could be achieved and upon removal of the solvent molecules the crystallinity was lost.

The first example of COFs (named COF-1) was prepared by self-condensation of benzene-1,4-diboronic acid, via the elimination of water (Côté et al., 2005).

### TEMPLATE SYNTHESIS OF COVALENT ORGANIC POLYMER NETWORKS

The synthesis of this class of materials is also based on irreversible covalent reactions; however, the synthesis is directed by a template. The template is usually formed via self-assembly using noncovalent interactions of the building blocks. In this class of materials, the network topology and overall morphology arises from the template, which is formed by self-assembly of molecular building blocks connected either via coordination bonds (e.g., a metalorganic framework or porous coordination polymer) or selfassembled via hydrogen bonding or van der Waals interactions.

These structures are, in a subsequent step, covalently cross-linked to enhance the thermal and mechanical stability. The templates employed for this process can be divided in completely preserved templates and sacrificial or semi-sacrificial templates.

The complete conversion of a template into a cross-linked structure involves the crystallization of a specifically designed photoreactive monomer into a layered structure and a photopolymerization step within the crystal (Kissel et al., 2012).

The (semi-)sacrificial conversion of a template structure was achieved by the use of metal-organic frameworks as template. In this approach, "ideal polymer networks" were created by covalent cross-linking of the organic linkers in the MOF structure followed by removal of the metal ions. The transformation of templates consisting of well-defined three-dimensional nanoporous network such as MOFs represents one of the most promising routes for creation of polymer networks with well-defined repeating units in the network structure, i.e., "ideal network polymers" (Ishiwata et al., 2013; Tsotsalas et al., 2014).

**Figure 2B** lists the commonly used acronyms for covalently linked networks of paragraphs 1.1–1.3 and their properties.

### MOLECULAR MODELING OF COVALENT NETWORKS

Covalent network chemistry is an area where experimentally and computationally available information is quite complementary and most studies contain some component of modeling. Typically, comparison of the refined experimental and calculated PXRD patterns is used to assign the structure, particularly in the case of

2D COFs, where it has not been possible to grow a single crystal suitable for XRD. Structural calculations generally employ either force field methods or Density Functional Tight Binding (DFTB), both of which are capable of calculating periodic unit cells containing hundreds or even thousands of atoms relatively cheaply and yield lattice parameters typically within a few percent of the experimental parameters (Addicoat et al., 2014b; Guo et al., 2013). Both force field and DFTB methods suffer from a limitation of scope. Parameterized force fields can readily compute cells with tens of thousands of atoms, but will generally produce poor results if applied to systems that differ too strongly from those used in the parameterization. DFTB can be used for systems containing hundreds of atoms but is limited by the availability of parameters, which are needed for every pair of atoms, X-Y, present in the system. Parameters for all combinations of H, C, N, O, S, and P are readily available, facilitating DFTB calculations of a majority of COFs, however, metal-doped or functionalized COFs often have no appropriate parameters available.

Calculations are often able to provide more detail than is available experimentally, in particular for the interlayer geometry of 2D networks. The experimentally measured PXRD of 2D COFs readily differentiate between eclipsed and staggered stacking (e.g., AA and AB stacking) but are unable to resolve the difference between fully eclipsed and slightly offset stackings. Computational studies on a variety of square and hexagonal 2D frameworks by Lukose et al. (2011), Spitler et al. (2011), and Zhou et al. (2010) all showed that slightly offset stackings were significantly lower in energy than the fully eclipsed structures, with only small effects on the PXRD pattern, and that this effect is primarily due to minimizing repulsive interactions between layers. Calculations are also able to elucidate the effects of framework flexibility. A straightforward approach may simply calculate several different framework structures, for example, rotating linkers. An elegant approach, developed recently for MOFs but which applies also to covalent frameworks, identifies the key flexible bonds and structural units in the framework as analogous to hinges and trusses, thus yielding an understanding of the mechanical flexibility of the framework (Sarkisov et al., 2014).

Beyond simply the determination of structural information, calculations are a means to access more detailed properties of the material. Of particular interest for the eventual application of these materials in gas storage and separation, are the surface area and void volume. Calculating these parameters may be done by a number of methods, including Voronoi decomposition and Delaunay tessellation. Both of these are geometric methods that work by partitioning the pore space into suitably small units. Another method, is to computationally mimic the experimental measurement by using a spherical probe, chosen as either an infinitely small point or given the radius of either the He atom or N2 molecule, and repeatedly inserting it into the structure at random locations. If the probe does not overlap with any framework atoms, then it must be in a void. Pore geometry, including the largest free sphere (which defines the Pore Limiting Diameter), the largest included sphere (which defines the Largest Cavity Diameter, i.e., the pore "size"), and pore connectivity (whether pores are connected in one, two, or three dimensions) can also be calculated similarly. Recording the largest probe that can fit at a given point, without overlapping any framework atoms yields the PSD, which shows how much of the void volume corresponds to particular pore sizes. The surface area accessible to a probe of a given size may be computed by effectively rolling the probe over the surface.

Moving beyond a static picture of a framework structure, calculations can also be used to describe and predict gas absorption properties. Molecular Dynamics (MD) calculations can be used to calculate the diffusion coefficients and transport properties of gas molecules through a framework. Grand Canonical Monte Carlo (CGMC) simulations yield enthalpies of absorption, absorption isotherms, and where a mixture of gases is simulated, selectivities. If the framework is quite rigid, the framework atoms may be fixed at their original (simulated or crystallographic) positions, however, in flexible frameworks or when interpenetrated frameworks may shift relative to each other, the motion of these atoms must be included in the simulation. Due to the expense of simulating large unit cells over a long time period, these calculations are usually based on classical mechanics, though several recent studies use quantum mechanical calculations to parametrize force fields, which are then used in the MD or GCMC simulations (Bureekaew et al., 2013).

Quantum mechanical, most often Density Functional Theory (DFT) calculations have had some use in directly investigating various periodic covalent network structures (Srepusharawoot et al., 2009). However, as mentioned above, this is prohibitively expensive and many studies, particularly those investigating binding of small molecules within framework structures, employ a co-called cluster model, whereby a section of the periodic structure is cut from the bulk. After saturating the excised bonds (e.g., with H atoms), the calculation may then proceed as a straightforward gas-phase calculation. Care must be taken that the size of the model is appropriate in order to avoid spurious results arising from interaction with the cut edges, and such studies may neglect spatial and electronic effects from neighboring framework atoms either through-bond or through-space. Despite these concerns, this approach has been extraordinarily fruitful (Klontzas et al., 2008; Assfour and Seifert, 2010; Choi et al., 2011).

The most significant recent development in the computational study of framework materials is in the high-throughput computational prediction of new covalent network structures. Several software packages capable of generating arbitrary network structures given a topology and the required building blocks were released (Martin and Haranczyk, 2014; Addicoat et al., 2014a; Gomez-Gualdron et al., 2014). In all cases, the topologies themselves are sourced from the Reticular Chemistry Structure Resource (RCSR) developed by O'Keeffe et al. (2008). For 3D covalent frameworks, the most common topologies are dia, ctn, bor, pto, and tbo (Bureekaew and Schmid, 2013). 2D frameworks are typically either hexagonal (layer symbol hca, hcb) or square planar (sql). This new software is now being employed in predictive high-throughput computational studies, such as the work of Martin et al. who generated an impressive 18,000 synthetically accessible covalent networks with diamond-like topology (RCSR symbol dia) (Martin et al., 2014) and identified promising candidates for methane storage. The predictive capability (pre-synthesis) offered by such software is fundamental to the development of frameworks with specific properties and represents a new way forward in the development of materials targeted to specific applications.

## ACKNOWLEDGMENTS

Matthew A. Addicoat gratefully acknowledges a Marie Curie Actions (MC-IIF: GA-MOF, Grant Agreement 327758) fellowship. We acknowledge support by Deutsche Forschungsgemeinschaft and Open Access Publishing Fund of Karlsruhe Institute of Technology.

## REFERENCES

Addicoat, M. A., Coupry, D. E., and Heine, T. (2014a). AuToGraFS: automatic topological generator for framework structures. *J. Phys. Chem. A* 118, 9607–9614. doi:10.1021/jp507643v

Addicoat, M. A., Vankova, N., Akter, I. F., and Heine, T. (2014b). Extension of the universal force field to metal-organic frameworks. *J. Chem. Theory Comput.* 10, 880–891. doi:10.1021/ct400952t

Arab, P., Rabbani, K. G., Sekizkardes, A. K., Islamoglu, T., and El-Kaderi, H. M. (2014). Copper(I)-catalyzed synthesis of nanoporous azo-linked polymers: impact of textural properties on gas storage and selective carbon dioxide capture. *Chem. Mater.* 26, 1385–1392. doi:10.1021/cm403161e

Assfour, B., and Seifert, G. (2010). Hydrogen adsorption sites and energies in 2D and 3D covalent organic frameworks. *Chem. Phys. Lett.* 489, 86–91. doi:10.1016/j.cplett.2010.02.046

Beaudoin, D., Maris, T., and Wuest, J. D. (2013). Constructing monocrystalline covalent organic networks by polymerization. *Nat. Chem.* 5, 830–834. doi:10.1038/nchem.1730

Ben, T., Ren, H., Ma, S., Cao, D., Lan, J., Jing, X., et al. (2009). Targeted synthesis of a porous aromatic framework with high stability and exceptionally high surface area. *Angew. Chem. Int. Ed. Engl.* 50, 9457–9460. doi:10.1002/anie.200904637

Biswal, B. P., Chandra, S., Kandambeth, S., Lukose, B., Heine, T., and Banerjee, R. (2013). Mechanochemical synthesis of chemically stable isoreticular covalent organic frameworks. *J. Am. Chem. Soc.* 135, 5328–5331. doi:10.1021/ja4017842

Bureekaew, S., Amirjalayer, S., Tafipolsky, M., Spickermann, C., Roy, T. K., and Schmid, R. (2013). MOF-FF – A flexible first-principles derived force field for metal-organic frameworks. *Phys. Status. Solidi B* 250, 1128–1141. doi:10.1002/pssb.201248460

Bureekaew, S., and Schmid, R. (2013). Hypothetical 3D-periodic covalent organic frameworks: exploring the possibilities by a first principles derived force field. *CrystEngComm* 15, 1551–1562. doi:10.1039/C2CE26473K

Chang, Z., Zhang, D.-A., Chen, Q., and Bu, X.-H. (2013). Microporous organic polymers for gas storage and separation applications. *Phys. Chem. Chem. Phys.* 15, 5430–5442. doi:10.1039/c3cp50517k

Chen, L., Honsho, Y., Seki, S., and Jiang, D. (2010). Light-harvesting conjugated microporous polymers: rapid and highly efficient flow of light energy with a porous polyphenylene framework as antenna. *J. Am. Chem. Soc.* 132, 6742–6748. doi:10.1021/ja100327h

Choi, Y. J., Choi, J. H., Choi, K. M., and Kang, J. K. (2011). Covalent organic frameworks for extremely high reversible CO2 uptake capacity: a theoretical approach. *J. Mater. Chem.* 21, 1073–1078. doi:10.1039/C0JM02891F

Colón, Y. J., and Snurr, R. Q. (2014). High-throughput computational screening of metal-organic frameworks. *Chem. Soc. Rev.* 43, 5735. doi:10.1039/c4cs00070f

Colson, J. W., and Dichtel, W. R. (2013). Rationally synthesized two-dimensional polymers. *Nat. Chem.* 5, 453–465. doi:10.1038/NCHEM.1628

Côté, A. P., Benin, A. I., Ockwig, N. W., O'Keeffe, M., Matzger, A. J., and Yaghi, O. M. (2005). Porous, crystalline, covalent organic frameworks. *Science* 310, 1166. doi:10.1126/science.1120411

Dawson, R., Cooper, A. I., and Adams, D. J. (2012). Nanoporous organic polymer networks. *Progress Polym. Sci.* 37, 530. doi:10.1016/j.progpolymsci.2011.09.002

Ding, S.-Y., and Wang, W. (2013). Covalent organic frameworks (COFs): from design to applications. *Chem. Soc. Rev.* 42, 548–568. doi:10.1039/C2CS35072F

Dogru, M., and Bein, T. (2014). On the road towards electroactive covalent organic frameworks. *Chem. Commun.* 50, 5531–5546. doi:10.1039/c3cc46767h

El-Kaderi, H. M., Hunt, J. R., Mendoza-Cortés, J. L., Côté, A. P., Taylor, R. E., O'Keeffe, M., et al. (2007). Designed synthesis of 3D covalent organic frameworks. *Science* 316, 268–272. doi:10.1126/science.1139915

Farha, O. K., Spokoyny, A. M., Hauser, B. G., Bae, Y.-S., Brown, S. E., Snurr, R. Q., et al. (2009). Synthesis, properties, and gas separation studies of a robust diimide-based microporous organic polymer. *Chem. Mater.* 21, 3033–3035. doi:10.1021/cm901280w

Feng, X., Ding, X., and Jiang, D. (2012). Covalent organic frameworks. *Chem. Soc. Rev.* 41, 6010–6022. doi:10.1039/C2CS35157A

Gomez-Gualdron, D. A., Gutov, O. V., Krungleviciute, V., Borah, B., Mondloch, J. E., Hupp, J. T., et al. (2014). Computational design of metal-organic frameworks based on stable zirconium building units for storage and delivery of methane. *Chem. Mater.* 26, 5632–5639. doi:10.1021/cm502304e

Guo, J., Xu, Y., Jin, S., Chen, L., Kaji, T., Honsho, Y., et al. (2013). Conjugated organic framework with three-dimensionally ordered stable structure and delocalized π clouds. *Nat. Commun.* 4, 2736. doi:10.1038/ncomms3736

Han, S. S., Mendoza-Cortés, J. L., and Goddard, W. A. III (2009). Recent advances on simulation and theory of hydrogen storage in metal-organic frameworks and covalent organic frameworks. *Chem. Soc. Rev.* 38, 1460–1476. doi:10.1039/B802430H

Ishiwata, T., Furukawa, Y., Sugikawa, K., Kokado, K., and Sada, K. (2013). Transformation of metal-organic framework to polymer gel by cross-linking the organic ligands preorganized in metal-organic framework. *J. Am. Chem. Soc.* 135, 5427–5432. doi:10.1021/ja3125614

Jiang, J.-X., Su, F., Trewin, A., Wood, C. D., Campbell, N. L., Niu, H., et al. (2007). Conjugated microporous poly(aryleneethynylene) networks. *Angew. Chem. Int. Ed.* 46, 8574–8578. doi:10.1002/anie.200701595

Kaur, P., Hupp, J. T., and Nguyen, S.-B. T. (2011). Porous organic polymers in catalysis: opportunities and challenges. *ACS Catal.* 1, 819–835. doi:10.1021/cs200131g

Kissel, P., Erni, R., Schweizer, W. B., Rossell, M. D., King, B. T., Bauer, T., et al. (2012). A two-dimensional polymer prepared by organic synthesis. *Nat. Chem.* 4, 287. doi:10.1038/nchem.1265

Klontzas, E., Tylianakis, E., and Froudakis, G. E. (2008). Hydrogen storage in 3D covalent organic frameworks. a multiscale theoretical investigation. *J. Phys. Chem. C* 112, 9095–9098. doi:10.1021/jp711326g

Kuhn, P., Antonietti, M., and Thomas, A. (2008). Porous, covalent triazine-based frameworks prepared by ionothermal synthesis. *Angew. Chem. Int. Ed.* 47, 3450. doi:10.1002/anie.200705710

Liebl, M. R., and Senker, J. (2013). Microporous functionalized triazine-based polyimides with high CO2 capture capacity. *Chem. Mater.* 25, 970–980. doi:10.1021/cm4000894

Lindemann, P., Tsotsalas, M., Shishatskiy, S., Abetz, V., Krolla-Sidenstein, P., Azucena, C., et al. (2014). Preparation of conjugated microporous polymer nanomembranes for gas separation. *Chem. Mater.* 26, 7189–7193. doi:10.1021/cm503924h

Lukose, B., Kuc, A., and Heine, T. (2011). The structure of layered covalent-organic frameworks. *Chemistry* 17, 2388–2392. doi:10.1002/chem.201001290

Martin, R. L., and Haranczyk, M. (2014). Construction and characterization of structure models of crystalline porous polymers. *Cryst. Growth Des.* 14, 2431–2440. doi:10.1021/cg500158c

Martin, R. L., Simon, C. M., Smit, B., and Haranczyk, M. (2014). In silico design of porous polymer networks: high-throughput screening for methane storage materials. *J. Am. Chem. Soc.* 136, 5006–5022. doi:10.1021/ja4123939

McKeown, N. B., and Budd, P. B. (2010). Exploitation of intrinsic microporosity in polymer-based materials. *Macromolecules* 43, 5163. doi:10.1021/ma1006396

O'Keeffe, M., Peskov, M. A., Ramsden, S. J., and Yaghi, O. M. (2008). The reticular chemistry structure resource (RCSR) database of, and symbols for, crystal nets. *Acc. Chem. Res.* 41, 1782–1789. doi:10.1021/ar800124u

Sarkisov, L., Martin, R. L., Haranczyk, M., and Smit, B. (2014). On the flexibility of metal-organic frameworks. *J. Am. Chem. Soc.* 136, 2228–2231. doi:10.1021/ja411673b

Spitler, E. L., Koo, B. T., Novotney, J. L., Colson, J. W., Uribe-Romo, F. J., Gutierrez, G. D., et al. (2011). A 2D covalent organic framework with 4.7-nm pores and insight into its interlayer stacking. *J. Am. Chem. Soc.* 133, 19416–19421. doi:10.1021/ja206242v

Srepusharawoot, P., Scheicher, R. H., Araújo, C. M., Blomqvist, A., Pinsook, U., and Ahuja, R. (2009). Ab initio study of molecular hydrogen adsorption in covalent organic framework-1. *J. Phys. Chem. C* 113, 8498–8504. doi:10.1021/jp809167b

Tsotsalas, M., Liu, J., Tettmann, B., Grosjean, S., Shahnas, A., Wang, Z., et al. (2014). Fabrication of highly uniform gel coatings by the conversion of surface-anchored metal-organic frameworks. *J. Am. Chem. Soc.* 136, 8–11. doi:10.1021/ja409205s

Tsyurupa, M. P., and Davankov, V. A. (2002). Hypercrosslinked polymers: basic principle of preparing the new class of polymeric materials. *React. Funct. Polym.* 53, 193–203. doi:10.1016/S1381-5148(02)00173-6

Uribe-Romo, F. J., Doonan, C. J., Furukawa, H., Oisaki, K., and Yaghi, O. M. (2011). Crystalline covalent organic frameworks with hydrazone linkages. *J. Am. Chem. Soc.* 133, 11478–11481. doi:10.1021/ja204728y

Uribe-Romo, F. J., Hunt, J. R., Furukawa, H., Klöck, C., O'Keeffe, M., and Yaghi, O. M. (2009). A crystalline imine-linked 3-D porous covalent organic framework. *J. Am. Chem. Soc.* 131, 4570–4571. doi:10.1021/ja8096256

Wan, S., Guo, J., Kim, J., Ihee, H., and Jiang, D. (2008). A belt-shaped, blue luminescent, and semiconducting covalent organic framework. *Angew. Chem. Int. Ed.* 47, 8826–8830. doi:10.1002/anie.200803826

Xiang, Z., and Cao, D. (2013). Porous covalent–organic materials: synthesis, clean energy application and design. *J. Mater. Chem. A* 1, 2691. doi:10.1039/c2ta00063f

Zhou, Z., Wu, H., and Yildirim, T. (2010). Structural stability and elastic properties of prototypical covalent organic frameworks. *Chem. Phys. Lett.* 499, 103–107. doi:10.1016/j.cplett.2010.09.032

**Conflict of Interest Statement:** The authors declare that the research was conducted in the absence of any commercial or financial relationships that could be construed as a potential conflict of interest.

# 24

# Physical properties of semiconducting/magnetic nanocomposites

*Petra Granitzer \* and Klemens Rumpf*

*Institute of Physics, Karl-Franzens-University Graz, Graz, Austria*

**Edited by:**
Samit K. Ray, Indian Institute of
Technology Kharagpur, India

**Reviewed by:**
Rabah Boukherroub, Centre National
de la Recherche Scientifique, France
Koji Yamada, Nippon Telegraph and
Telephone Corporation, Japan

**\*Correspondence:**
Petra Granitzer, Institute of Physics,
Karl-Franzens-University Graz,
Universitaetsplatz 5, A-8010 Graz,
Austria
e-mail: petra.granitzer@uni-graz.at

In this review, the fabrication of porous silicon/magnetic nanocomposite materials and their physical properties are elucidated. Especially, the investigation of the presented systems with respect to their magnetic properties is reported. Furthermore, the influence of the semiconducting matrix on the properties of the nanocomposites is highlighted. The main focus will be put on silicon used as template material. In general, the nanocomposite systems are fabricated in a two-step process, first by anodization of a silicon wafer to achieve porous silicon structures, and second by electrodeposition of a magnetic material into the pores. The morphology of the porous silicon template offers straight pores, grown perpendicular to the wafer surface. The magnetic nanostructures deposited within the pores lead to specific properties of the composite dependent on their size and shape. Due to their mutual arrangement, magnetic coupling between these structures can occur, whereas, coupling between adjacent pores depends on the porous silicon morphology. In the first section, different types of such template/metal systems are reviewed and second an experimental part follows implying the porous silicon formation as well as the subsequent metal deposition process. Third, the magnetic and optical properties of the systems are described. In the fourth chapter, the influence of the semiconducting matrix on these properties is elucidated and finally some prospects and conclusions are addressed.

**Keywords: nanostructured silicon, magnetic nanostructures, nanocomposite, electrodeposition**

## INTRODUCTION

Due to the miniaturization and integration of numerous microelectronic devices, low dimensional structures are under extensive investigation with respect to a simple and low-cost fabrication process but also concerning their specific properties. Beside lithographic bottom-up or top-down procedures, self-assembled arrangements of nanoparticles are also a key-topic in many of today's research fields. Self-organization of nanoparticles depends strongly on the interactions between them resulting in specific one-, two-, or three-dimensional arrangements. To avoid agglomeration of the particles and to stabilize them in general, they are coated with a surfactant. The kind of surfactant determines the inter-particle interactions and thus influences the resulting assembly. A further possibility of self-organization of particles is template guided. Porous materials are suitable candidates, whereas in most cases, the templates themselves are formed by self-organization. Beside, e.g., trench etched polymers (Chou St et al., 1995), porous alumina (Masuda and Fukuda, 1995; Masuda et al., 1997), and porous InP (Gerngross et al., 2013), porous silicon is a well-known and often-employed material. The pore formation is self-organized, nevertheless the morphology is tunable in a broad range and even quasi-regular arrangements can be achieved (Rumpf et al., 2010a).

Porous silicon, a versatile material, which has been discovered in the mid 1950s (Uhlir, 1956) and has been extensively investigated in the 1990s (Canham, 1990; Lehmann and Gösele, 1991; Koshida and Koyama, 1992; Zhang, 2001), is nowadays still an often-employed material in many research fields. After Canham showed in 1990 that microporous silicon emits light in the visible at room temperature due to quantum confinement, optoelectronic applications have been under intense discussion. The demonstration of electroluminescence by Koshida in 1995 pursued this research direction. Concurrent investigations of this material have been related to the tunable morphology and porosity (Föll et al., 2002; Lehmann, 2002), especially with respect to sensing of various molecules (Sailor, 1997; Buriak, 2006). After Canham found that porous silicon is biocompatible and biodegradable (Canham, 1995), this realm was booming (Anglin et al., 2008; Fernandez-Moure et al., 2014). Due to its tunable morphology (Föll et al., 2002) and high surface area (Buriak, 2006), it is applicable in various fields such as gas-sensors (Boarino et al., 2000), bio-sensors (Gupta et al., 2013), optics (Torres-Costa et al., 2005), and many more. The pore-diameters can be varied among the microporous regime (2–4 nm), mesoporous regime (5–50 nm), and macroporous regime (several tens of micrometers). Also its utilization as template material is of interest, e.g., for the deposition of various metals inside the pores (Fukami et al., 2008; Rumpf et al., 2012a; Gerngross et al., 2013). A further advantage to this material is the formation by self-organization (Kompan, 2003) and therefore expensive and time consuming nanopatterning by lithography can be avoided in many cases.

Magnetic nanostructures attract great attention since many years. They are under investigation for magnetic applications such as high density magnetic storage (Sellmyer and Skomski, 2006;

Shin et al., 2012) but also for biomedical applications (Pankhurst et al., 2003; Tartaj et al., 2003). The adjustability in their size, shape, and mutual arrangement is examined especially concerning the magnetic properties. Single domain (Goya et al., 2003) and superparamagnetic (Sinwani et al., 2014) particles are crucial in many realms. Their incorporation in a matrix material to determine the magnetic properties or for stabilization is also widespread. Such nanostructured magnetic materials offer completely different properties compared to their bulk materials. A big advantage of nanostructured materials is the tunability of their magnetic properties due to their size, shape, and mutual arrangement.

To fabricate nanostructures (wires), template-assisted methods are used to achieve three-dimensional arrangements of such nanostructures. Such arrays are investigated to get knowledge about the mutual interactions of the magnetic nanostructures, e.g., dipolar coupling (Vazquez et al., 2004) or magnetization reversal mechanisms (Uhlír et al., 2012). A further key-topic is the combination of nanostructured materials, nanocomposites, which are used to process new materials with unique physical properties (Wen and Krishnan, 2011; Wang and Gu, 2015). The intrinsic material characteristics are modified due to the reduced size or the material composition or due to interactions between the nanostructures. Magnetic semiconductors, which are in general magnetic ion-doped semiconductors, are under intense investigation since the last decade. The main goal of this research is the applicability in spintronics. A drawback of these systems is that so far their functionality at room temperature is limited (Dietl and Ohno, 2014). Nevertheless, there is a lot of progress in the choice of the used materials and thus the anticipation is high.

A further route to combine semiconductors and magnetic materials is the incorporation of magnetic nanostructures within a semiconducting template. This incorporation can occur in different ways, either, e.g., by deposition, infiltration, evaporation, or atomic layer deposition. The deposition can be performed electrochemically (Rumpf et al., 2014) or electroless (Nakamura and Adachi, 2012). For the infiltration of magnetic material into a porous structure, usually ready synthesized magnetic nanoparticles in solution (Granitzer et al., 2010) are used. In the following nanocomposite systems consisting of a porous silicon template with deposited magnetic nanostructures will be addressed. Thereby, the fabrication of the systems and the arising distinct properties will be emphasized.

### SELF-ORGANIZED POROUS TEMPLATES FOR METAL DEPOSITION
One attractive and often utilized template especially for the deposition of metal nanostructures is porous alumina (Masuda and Fukuda, 1995; Masuda et al., 1997). The pore-arrangement offers a honey-comb like structure and the pores are quite smooth. The pore-diameter is tunable in a regime from about 20 nm up to a few hundred nanometers. In the last decade, magnetic materials such as Ni, Co, and Fe have been deposited within this kind of templates and the resulting magnetic properties have been investigated intensely (Ramazani et al., 2012; Zhang et al., 2013). Recently, porous InP membranes have been employed for the deposition of magnetic metals whereas the deposition process has been examined by FFT-impedance spectroscopy (Gerngross et al., 2014).

A further self-organized template is porous silicon, which is also used for the deposition of magnetic or non-magnetic metal structures into the pores. Already in the 1990s, metal has been deposited within microporous silicon, on the one hand to improve the electric contact in the case of electroluminescence investigations (Ronkel et al., 1996) and on the other hand to influence the luminescence of the material (Huang, 1996; Herino, 1997). Mesoporous silicon with oriented pores of about 20 nm in diameter and thick pore-walls of about 50 nm has been employed as template for the deposition of Ni-wires, which show a magnetic anisotropy due to their shape (Gusev et al., 1994). Porous silicon formed by self-organization has been used with diameters between 25 and 100 nm and concomitant pore-distances between 60 and 40 nm for filling with different magnetic materials (Granitzer et al., 2012a). Depending on the pore-diameter and the concomitant pore-distance, the pore-arrangements offer more or less regularity. A quasi-regular pore-arrangement can be achieved with pore-diameters between 45 and 100 nm, whereas smaller diameters lead to higher irregularity (Rumpf et al., 2011). A further approach to fabricate regular porous silicon pore-arrangements is the pre-patterning of the silicon substrate by a porous alumina template (Zacharatos et al., 2008, 2009).

## EXPERIMENTS
### FORMATION OF NANOSTRUCTURED SILICON
In general, porous silicon can be fabricated by various procedures, wet and dry etching, or a further possibility is high power laser ablation (Laiho and Pavlov, 1995). One kind of a dry etching process is reactive ion etching (RIE) (Tserepi et al., 2003). These dry etching techniques are used in general to fabricate microporous, luminescent porous silicon, whereat a regular arrangement of the pores is not necessary. The most common techniques are wet etching processes such as anodization, stain etching, or metal assisted etching. In the following, these wet etching techniques are briefly addressed especially because the main advantage of these techniques is the tunable morphology and the fabrication of quasi-regular pore-arrangements by self-organization.

Stain etching is an electroless pore formation technique, which has been already described in 1960 (Turner, 1958) and (Archer, 1960). In this process supplementary to the HF solution, an oxidizing agent such as nitric acid solution is added. A lot of progress concerning the stain etching mechanism has been gained recently by Kolasinski (2010) who explained the self-limiting pore formation process and showed that hole injection into the silicon valence band initiates the etching and it is rate determining in the overall etch process. By using $V_2O_5$ as oxidant, Kolasinski and Barclay (2013) explained the stoichiometry of the reaction.

The findings are also important for metal-assisted etching for which metal nanoparticles (Ag, Au, Pd, Pt) are deposited on a silicon surface. This process also works electroless. It is mainly used to fabricate silicon nanowires or macropores. So far, literature deals with models in which holes are produced at the metal/silicon interface but which is inappropriate. An explanation of the pore formation has been recently given by Kolasinski (2014). The metal particles are charged with holes by the oxidizing agent resulting in an electric field, which enables the pore formation by anodization. In contrast to the common anodization process in the case of metal-assisted etching, the anodization is localized to the individual metal particles.

Pore formation by anodic dissolution of a silicon wafer leads to a tunable morphology in a broad range, whereas the pore-diameter, inter-pore spacing, and porosity depend on the kind of doping, doping density, electrolyte composition, and electrolyte concentration as well as on the applied current density (Föll et al., 2002). To achieve straight pores grown perpendicular to the surface with diameters <100 nm, one choice is to use highly $n$-doped silicon, which is anodized in a 10 wt% hydrofluoric acid solution. The bath temperature is at room temperature and the current density is kept constant at 100 mA/cm$^2$, which means the etching is performed under breakthrough conditions. The resulting pore-diameter is about 60 nm and the concomitant distance between the pores is around 40 nm. **Figure 1** shows a top-view scanning electron micrograph (SEM) of a typical porous silicon template.

A further parameter of the porous silicon morphology is the dendritic structure of the pores. In the investigated morphology regime, the pores offer such a dendritic growth (Rumpf et al., 2010b). Nevertheless, the occurring branches between adjacent pores are not connected, so the pores are separated from each other. The length of the dendrites increases with

decreasing pore-diameter and reaches a maximum of about twice the pore-diameter in the case of small pore-diameters of about 25–30 nm.

To reduce the dendritic structure of the pores, a method developed in the Koshida' laboratory, magnetic field-assisted etching, has been employed (Hippo et al., 2008). A magnetic field of 8 T applied perpendicular to the sample surface facilitates the directed growth of the pores due to a controlled motion of the holes, which are responsible for the pore formation. As a result, the pore-diameter is decreased and the dendritic pore-growth is drastically reduced.

Such different porous silicon templates are used for filling the pores with a ferromagnetic metal by electrodeposition. The deposition process is performed by a pulsed technique and as electrolyte an adequate metal salt solution is employed. On the one hand, the pulsed technique prevents the exhaustion of the electrolyte and on the other hand blocking of the pores at the pore-opening is avoided. By varying the electrochemical parameters, especially the applied current density and the pulse duration of the current, the shape and size of the deposited nanostructures can be adjusted (Rumpf et al., 2010c). For the deposition of Ni, either the so-called Watts electrolyte (0.2 M NiCl$_2$, 0.1 M NiSO$_4$) or a NiCl$_2$-solution (170 g/l NiCl$_2$) has been used (Rumpf et al., 2010b). For the deposition of Co-nanostructures within the pores, a CoSO$_4$-solution with a pH value of 4.5 has been employed. Considering the pulsed deposition of Ni, a modification of the pulse duration from 40 to 5 s results in an elongation of the deposited structures. By applying 40 s pulses, more or less spherical Ni-particles could be achieved (**Figure 2A**). A reduction of the pulse duration to 10 s leads to ellipsoidal structures of about 500 nm in length (**Figure 2B**) and by reducing the pulse duration further to 5 s nanowires up to 4 μm in length could be obtained (**Figure 2C**). A modification of the applied current density results in a change of the packing density of the deposits.

In using the equivalent deposition parameters in the case of Co, spherical and ellipsoidal structures with an aspect ratio of about five have been obtained (Rumpf et al., 2012a). The packing density within the pores can be modified similar as in the case of Ni-deposition and thus also the magnetic coupling between the Co-structures can be tuned.

## CHARACTERIZATION METHODS

For the structural characterization of such nanocomposite systems, the most powerful method is electron microscopy, whereas

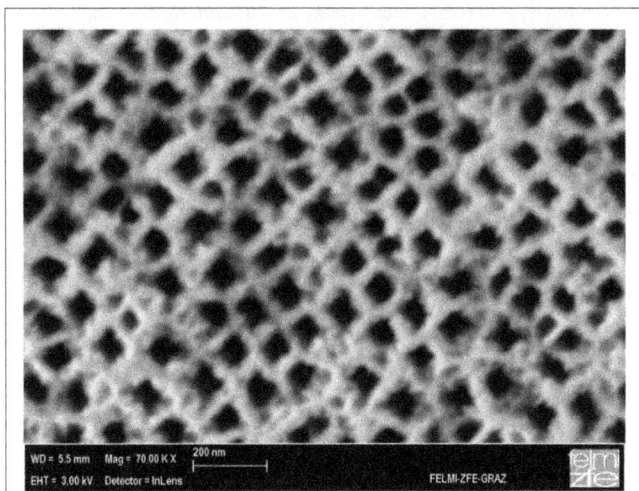

**FIGURE 1 | The top-view SEM image of a porous silicon sample shows pores with an average diameter of 60 nm and a mean distance between the pores of 40 nm.**

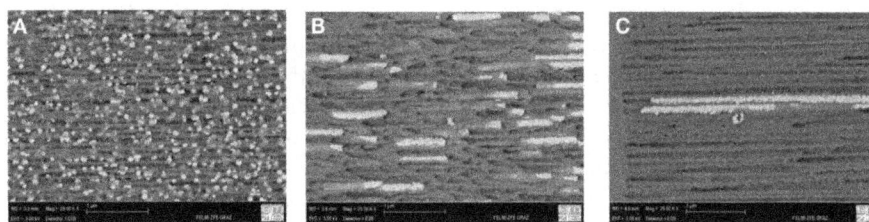

**FIGURE 2 | (A)** Deposited Ni-particles of about 60 nm in diameter. **(B)** Ellipsoidal Ni-structures of about 500 nm in length. **(C)** Ni-wires of about 4 μm in length. The diameter of all structures coincides with the pore-diameter (~60 nm).

scanning electron microscopy (SEM) is more appropriate in many cases as transmission electron microscopy (TEM). Images of the cross-section can also be used to estimate the filling factor of the porous silicon template. In this case, it has to be taken into account that SEM not only shows the top surface but also the electrons have a certain depth of penetration dependent on the energy of the primary electrons. In employing TEM, a difficile sample preparation is necessary, which is often a problem in the case of porous structures. Preparation side-effects can effect a slight modification of the pore-structure and also sputtering and re-deposition of silicon has to be considered. A further important feature is the utilization of back-scattered electrons (BSE) to achieve an element sensitive image. Energy dispersive X-ray (EDX) spectroscopy and mapping play a key role to get a survey of the elemental distribution within the pores (Rumpf et al., 2012b).

Fourier transform infrared spectroscopy (FTIR) has been used to figure out the oxidation status of the porous silicon template (Granitzer et al., 2009). Raman-spectroscopy is an appropriate tool to investigate stress caused by the porous silicon formation as well as by the metal filling within the pores (Granitzer et al., 2009).

Magnetic characterization of the samples has been performed by a superconducting quantum interference device (SQUID) and by a vibrating sample magnetometer (VSM). The magnetic field was adjustable between $\pm 6$ T in the case of the SQUID and $\pm 9$ T in the case of the VSM. The temperature can be varied between 4 and 300 K with both instruments. The magnetization has been measured in two directions, with the magnetic field applied parallel to the pores (easy axis) and perpendicular to the pores (hard axis), respectively.

## MAGNETIC AND OPTICAL PROPERTIES OF THE NANOCOMPOSITES

Due to the possibility to deposit magnetic nanostructures tunable in their size, shape, and also in their spatial arrangement within the pores, specific magnetic properties can be achieved (Rumpf et al., 2010c). A three-dimensional array of self-assembled ferromagnetic nanostructures renders possible especially to modify the coercivity, remanence, and magnetic anisotropy by tuning the electrochemical parameters and thus the mutual arrangement of the deposited nanostructures. The metal deposits can be modified in their spatial arrangement within the pores by the deposition parameters, especially the applied current density and the pulse duration of the current. A further parameter of modification is the morphology of the template, which allows to vary the wall-thickness and thus the magnetic interactions between nanostructures of adjacent pores. The coercivity increases with decreasing length of the nanostructures and also by decreasing magnetic coupling between metal deposits within adjacent pores (Granitzer et al., 2012b). **Figure 3** shows the dependence of the coercivity on the temperature for deposited Ni-wires and Ni-particles within a porous silicon template. The coercivity for deposited particles is always higher than for wires due to demagnetizing effects. Similar results concerning the coercivities and remanence have been reported from metal wires deposited within porous alumina templates (Bahiana et al., 2006).

Considering the magnetic properties of Ni-structures dependent on the shape, one sees that the coercivity as well as the

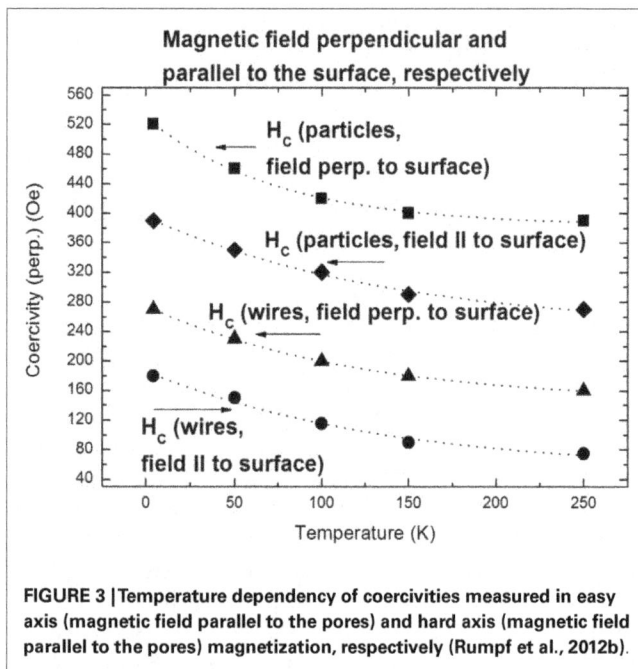

FIGURE 3 | Temperature dependency of coercivities measured in easy axis (magnetic field parallel to the pores) and hard axis (magnetic field parallel to the pores) magnetization, respectively (Rumpf et al., 2012b).

Table 1 | Coercivity and remanence in easy axis (magnetic field applied parallel to the pores) and hard axis (magnetic field perpendicular to the pores) magnetization in dependence on the Ni-structure length.

|  | $H_{C,\perp}$ (Oe) | $(M_R/M_S)_\perp$ | $H_{C,\parallel}$ (Oe) | $(M_R/M_S)_\parallel$ |
|---|---|---|---|---|
| Ni-particles | 520 | 57 | 400 | 50 |
| Ni-ellipsoides | 350 | 49 | 280 | 39 |
| Ni-wires | 280 | 41 | 190 | 32 |

*The diameter of the structures coincides in all cases with the pore-diameter.*

remanence increase with decreasing structure–length when dipolar coupling between the deposits can be neglected. In **Table 1**, a summary of the magnetic properties is given. In the case of elongated metal structures, a magnetic anisotropy between the two magnetization directions, easy axis and hard magnetization occurs. To investigate the crystalline structure of the deposited metal structures, XRD is a suitable tool. It has been reported that Ni-wires within mesoporous silicon are polycrystalline (Dolgiy et al., 2013) and also Co-wires deposited within an InP membrane offer a polycrystalline structure with very small grain size (Gerngross et al., 2014). From our preliminary electron back scatter diffraction (EBSD) experiments, we also found that Ni-deposits within porous silicon are polycrystalline. Due to the fact that the deposits are polycrystalline, the magnetocrystalline anisotropy can be neglected and thus the main anisotropy contribution can be attributed to the shape of the structures.

If the distance between the deposited Ni-particles within the pores decreases and a densely packed arrangement is achieved, magnetic coupling between the particles takes place resulting in the magnetic behavior of elongated Ni-structures (quasi-wires). **Figure 4** shows cross-sectional images of Ni-particles deposited

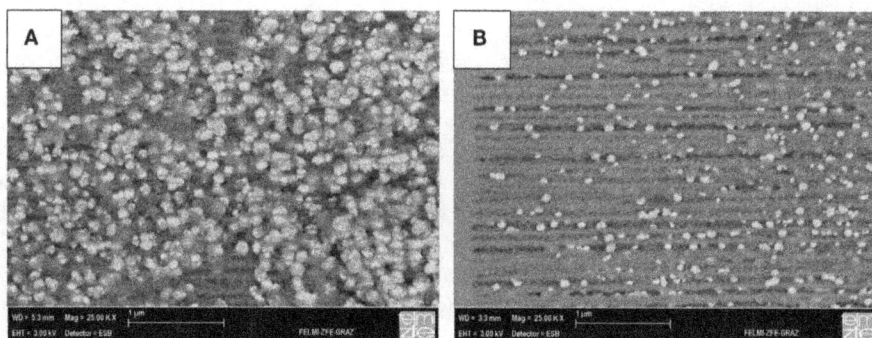

**FIGURE 4 | (A)** High filling fraction of Ni-particles within porous silicon and **(B)** low filling density of Ni-particles within the pores. The average particle size is 60 nm.

with a high and a low filling factor, respectively, within porous silicon.

The values of the coercivity and the remanence for interacting Ni-particles within the pores are comparable with the ones of Ni-wires deposited within porous silicon, $H_C$ (easy axis) = 290 Oe, $H_C$ (hard axis) = 210 Oe.

Optical methods such as FTIR and Raman-spectroscopy have been used to get knowledge especially about the porous silicon template, its surface, and the interface to the deposited metal structures. By FTIR, the oxidation status of the porous silicon has been figured out and Raman has been used to see if any stress occurs due to the pore formation and the subsequent metal deposition. As etched porous silicon offers a hydrogen terminated surface, which can be seen in FTIR-spectra exhibiting SiH-modes around 2087, 2115, and 2138 cm$^{-1}$. After aging, the sample in ambient air for 30 min oxide bands around 2250 cm$^{-1}$ occur, which indicate the formation of a native oxide layer (Granitzer et al., 2009). These results are also confirmed by TEM showing an oxide layer of about 2 nm on the porous silicon wall (**Figure 5**). After metal deposition within the pores, also oxide bands around 2250 cm$^{-1}$ appear accompanied by additional peaks around 2964 cm$^{-1}$, which are due to some SiO$_x$ modifications. These additional oxide bands occur during the metal deposition because the reduction of the metal ions is accompanied by an oxidation of the silicon matrix (Sasano et al., 2000).

Raman-spectra show a peak at 522 cm$^{-1}$ for bare silicon and a slight shift to 520 cm$^{-1}$ for porous silicon. Ni-deposition within the pores leads to a further shift of the peak to 505 cm$^{-1}$, which is due to compressive stress and shows the increasing mismatch between bulk silicon and the porous layer (Granitzer et al., 2009). **Figure 6** shows the peak shift of the Raman spectra between porous silicon and Ni-filled porous silicon.

Magneto-optical experiments, investigating the transverse magneto-optical Kerr-effect have been reported on Co deposited within porous silicon (Gan'shina et al., 2005). A correlation among the porosity of the template, the microstructure of the deposited Co, and the magnitude of the Kerr-effect has been found. An enhancement of the magneto-optical response of Ni-nanowires deposited within porous alumina with respect to bulk Ni has been observed (Melles et al., 2003).

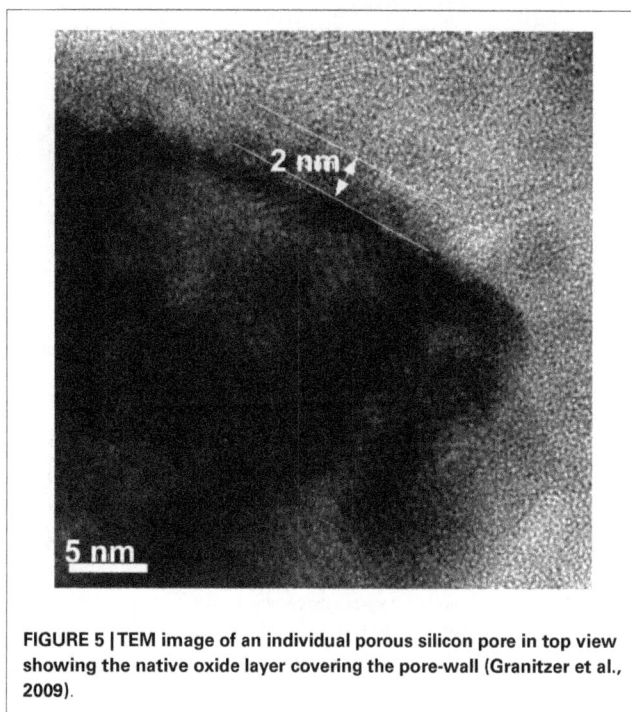

**FIGURE 5 | TEM** image of an individual porous silicon pore in top view showing the native oxide layer covering the pore-wall (Granitzer et al., 2009).

## INFLUENCE OF THE TEMPLATE ON THE PROPERTIES OF THE NANOCOMPOSITE

Due to the fact that the pores of the porous silicon templates offer a branched structure (**Figure 7**), the effective distance between the pores is in the range of 15 nm, which leads to an increase of the magnetic coupling between metal deposits. Concomitant to the porous structure, the metal deposits exhibit the same shape. For these investigations, Ni-wires of a few micrometers in length have been deposited. It could be shown that the magnetic coupling between metal wires of adjacent pores decreases with a decrease of the dendrite-length (Granitzer et al., 2012b). This behavior is accompanied by an increase of the coercivity with decreasing dendritic growth. Also the magnetic anisotropy between easy axis and hard axis magnetization is increased. In **Table 2**, the coercivities in dependence on the dendrites are presented.

The magnetic behavior of the nanocomposite is strongly influenced by the morphology of the template. As a result of the rough pore-walls, the deposited metal structures exhibit strong stray fields due to their branches, which also diminish the coercivity.

The best results concerning a reduction of the dendritic pore-growth could be achieved by magnetic field assisted etching. Average pore-diameters of about 25 nm with branches below 10 nm could be fabricated and thus also the stray fields effected by the dendrites are reduced, which leads to magnetic properties similar to an isolated nanowire. The decrease of the length of the dendrites has two effects, on the one hand an increase of the effective distance between the pores and on the other hand less magnetic stray fields caused by the branches of the metal deposits. This results in less magnetic coupling between metal structures of adjacent pores and an increase of the magnetization reversal field of the nanowires.

Considering such magnetic nanocomposite systems, one sees that the magnetic properties of the specimens can be tuned in various ways. First, the morphology of the template can be adjusted and thus the magnetic coupling between metal structures of adjacent pores can be tuned. Second, the size and shape of the metal deposits can be tuned, and a third parameter is the filling density and spatial distribution within the pores, which influences the magnetic coupling between deposits within the pores.

Such ferromagnetic/semiconducting composite materials are potential candidates for applications and they are compatible in today's microtechnology due to the silicon base material. It would be possible to integrate such structures on a chip by using lithography and produce porous silicon selective on localized regions (Hourdakis and Nassiopoulou, 2014).

## PERSPECTIVES AND CONCLUSION

Self-organized porous systems can be used as template for various materials. The deposition of metals allows to utilize the obtained nanocomposite for numerous applications such as sensor technology, optics, biomedicine, and many more. In incorporating magnetic materials within porous structures, three-dimensional arrays of nanomagnetic structures can be fabricated whereas the magnetic properties are variable through the interaction of the structures and their regularity and morphology of the template. So far, such nanocomposite systems are restricted to basic research but there is a high application potential. In using silicon as template material, the implementation of such nanostructured arrays into today's microtechnology becomes possible resulting in, e.g., miniaturized magnetic sensors integrated on a chip. Furthermore, the silicon/ferromagnet systems could be utilized in spintronics applications integrated on a chip. In considering spintronics, an advantage of silicon is the rather long spin lifetime, which is due to low spin orbit coupling, zero hyperfine coupling, and degenerate spin states. Furthermore, the spin lifetime in silicon is enhanced, e.g., by reducing electron–phonon scattering processes (Li et al., 2012) or by applying strain, which lengthens the spin coherence time and spin transport length (Tang et al., 2012).

Three-dimensional arrangements of self-organized magnetic nanostructures within a porous template material offer specific magnetic properties, which are tunable by the electrochemical parameters. On the one hand, the size and shape of the deposits can

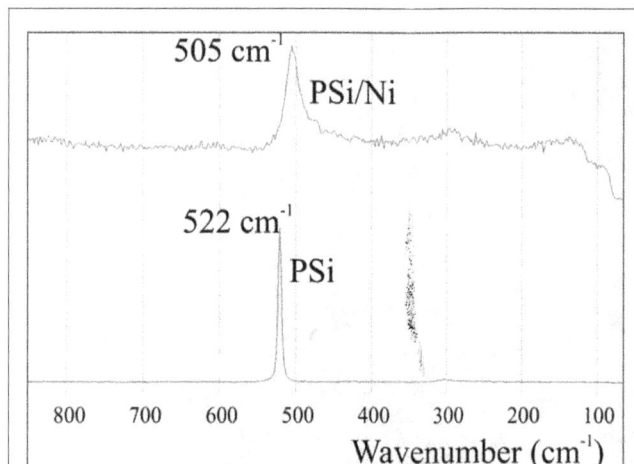

**FIGURE 6 | The Raman peak of porous silicon shifts from 522 to 505 cm⁻¹ for porous silicon with Ni filling (Granitzer et al., 2009).**

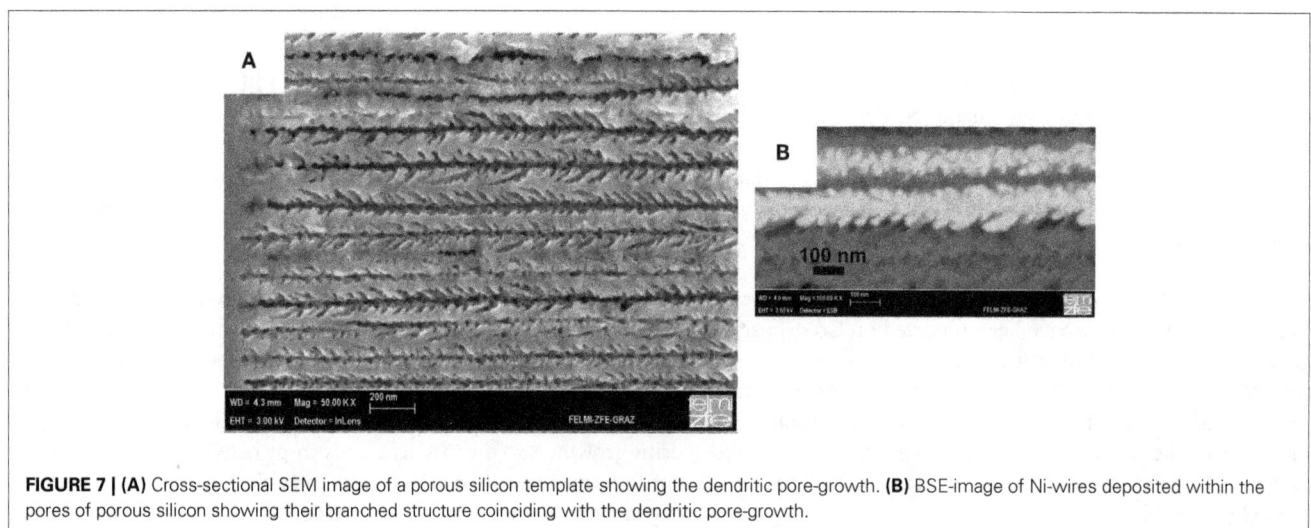

**FIGURE 7 | (A)** Cross-sectional SEM image of a porous silicon template showing the dendritic pore-growth. **(B)** BSE-image of Ni-wires deposited within the pores of porous silicon showing their branched structure coinciding with the dendritic pore-growth.

**Table 2 | Coercivity dependence on the length of the dendrites**.

| Dendrite length (nm) | $H_C$ (Oe) | $M_R/M_S$ |
|---|---|---|
| <10 (magnetic field assisted etched) | 650 | 0.85 |
| 20 | 355 | 0.42 |
| 30 | 320 | 0.35 |
| 50 | 270 | 0.28 |
| 60 | 100 | 0.21 |

*The deposited Ni-nanowires exhibit an aspect ratio of about 40.*

be adjusted by modifying the pulse duration and on the other hand their mutual arrangement within the pores can be varied by changing the current density. The porous silicon template can be used to tune the magnetic interactions between deposits in adjacent pores. Furthermore, the morphology of the template especially the more or less dendritic pore-growth influences the magnetic behavior. A branched pore-structure reduces the effective distance between the pores and therefore increases the magnetic coupling and also gives rise to stray fields and strong demagnetizing fields. In conclusion, one can say that template-assisted assemblies of nanostructures allow to tune the properties of nanocomposites in many ways to achieve desired specimens.

## ACKNOWLEDGMENTS

The authors thank Prof. N. Koshida from the Tokyo University of Agriculture and Technology for the preparation of magnetic field-assisted etched porous silicon samples and the Institute for Electron Microscopy (FELMI) at the University of Technology Graz for electron microscopy, FTIR, and Raman investigations. Furthermore, the authors are grateful for the possibility to perform magnetization measurements at the Institute of Solid State Physics at the Vienna University of Technology.

## REFERENCES

Anglin, E. J., Cheng, L., Freeman, W. R., and Sailor, M. J. (2008). Porous silicon in drug delivery devices and materials. *Adv. Drug Deliv. Rev.* 60, 1266–1277. doi:10.1016/j.addr.2008.03.017

Archer, R. (1960). Stain films on porous silicon. *J. Phys. Chem. Solids* 14, 104. doi:10.1016/0022-3697(60)90215-8

Bahiana, M., Amaral, F. S., Allende, S., and Altbir, D. (2006). Reversal modes in arrays of interacting magnetic Ni nanowires: Monte Carlo simulations and scaling technique. *Phys. Rev. B* 74, 174412. doi:10.1103/PhysRevB.74.174412

Boarino, L., Baratto, C., Geobaldo, F., Amato, G., Comini, E., Rossi, A. M., et al. (2000). NO2 monitoring at room temperature by a porous silicon gas sensor. *Mater. Sci. Eng. B* 6, 210–214. doi:10.1016/S0921-5107(99)00267-6

Buriak, J. M. (2006). High surface area silicon materials: fundamentals and new technology. *Phil. Trans. R. Soc. A* 364, 217. doi:10.1098/rsta.2005.1681

Canham, L. T. (1990). Silicon quantum wire array fabrication by electrochemical and chemical dissolution of wafers. *Appl. Phys. Lett.* 57, 1046–1048. doi:10.1063/1.103561

Canham, L. T. (1995). Bioactive silicon structure fabrication through nanoetching techniques. *Adv. Mater.* 7, 1033–1037. doi:10.1002/adma.19950071215

Chou St, Y., Krauss, P. R., and Renstrom, P. J. (1995). Imprint of sub 25 nm vias and trenches in polymers. *Appl. Phys. Lett.* 67, 3114. doi:10.1063/1.114851

Dietl, T., and Ohno, H. (2014). Dilute ferromagnetic semiconductors: physics and spintronic structures. *Rev. Mod. Phys.* 86, 187. doi:10.1103/RevModPhys.86.187

Dolgiy, A. L., Redko, S. V., Komissarov, I., Bondarenko, V. P., Yanushkevich, K. I., and Prischepa, S. L. (2013). Structural and magnetic properties of Ni nanowires grown in mesoporous silicon templates. *Thin Solid Films* 543, 133. doi:10.1016/j.tsf.2013.01.049

Fernandez-Moure, J. S., Evangelopoulos, M., Scaria, S., Martinez, J. O., Brown, B. S., Coronel, A. C., et al. (2014). "Multistage porous silicon for cancer therpy, chapter 16," in *Porous Silicon for Biomedical Applications*, ed. H. A. Santos (Cambridge, UK: Woodhead Publishing), 374–402.

Föll, H., Christophersen, M., Carstensen, J., and Hasse, G. (2002). Formation and application of porous silicon. *Mater. Sci. Eng. R* 39, 93–141. doi:10.1016/S0927-796X(02)00090-6

Fukami, K., Kobayashi, K., Matsumoto, T., Kawamura, Y. L., Sakka, T., and Ogata, Y. H. (2008). Electrodeposition of noble metals into ordered macropores in p-type silicon. *J. Electrochem. Soc.* 155, D443–D448. doi:10.1149/1.2898714

Gan'shina, E., Yu, A., Kochneva, M., Podgornyi, D. A., Shcherbak, P. N., Demidovich, G. B., et al. (2005). Structure and magneto-optical properties of "porous silicon-cobalt" granular nanocomposites. *Phys. Solid State* 47, 1383. doi:10.1134/1.1992622

Gerngross, M.-D., Carstensen, J., and Föll, H. (2014). Electrochemical growth of Co nanowires in ultra-high aspect ratio InP membranes: FFT-impedance spectroscopy of the growth process and magnetic properties. *Nanoscale Res. Lett.* 9, 316. doi:10.1186/1556-276X-9-316

Gerngross, M.-D., Chemnitz, S., Wagner, B., Carstensen, J., and Föll, H. (2013). Ultra-high aspect ratio Ni nanowires in single-crystalline InP membranes as multiferroic composite. *Phys. Status Solidi Rapid Res. Lett.* 7, 352. doi:10.1002/pssr.201307026

Goya, G. F., Fonseca, F. C., Jardim, R. F., Muccillo, R., Carreno, N. L. V., Longo, E., et al. (2003). Magnetic dynamics of single-domain Ni nanoparticles. *J. Appl. Phys.* 93, 6531. doi:10.1063/1.1540032

Granitzer, P., Rumpf, K., Ohta, T., Koshida, N., Poelt, P., and Reissner, M. (2012a). Porous silicon/Ni composites of high coercivity due to magnetic field-assisted etching. *Nanoscale Res. Lett.* 7, 384. doi:10.1186/1556-276X-7-384

Granitzer, P., Rumpf, K., Ohta, T., Koshida, N., Reissner, M., and Poelt, P. (2012b). Enhanced magnetic anisotropy of Ni nanowire arrays fabricated on nanostructured silicon templates. *Appl. Phys. Lett.* 101, 033110. doi:10.1063/1.4738780

Granitzer, P., Rumpf, K., Poelt, P., Albu, M., and Chernev, B. (2009). The interior interfaces of a semiconductor metal nanocomposite and their influence on its physical properties. *Phys. State Solid C* 6, 2222. doi:10.1002/pssc.200881730

Granitzer, P., Rumpf, K., Venkatesan, M., Roca, A. G., Cabrera, L., Morales, M. P., et al. (2010). Magnetic study of Fe3O4 nanoparticles incorporated within mesoporous silicon. *J. Electrochem. Soc.* 157, K145–K151. doi:10.1149/1.3425605

Gupta, B., Zhu, Y., Guan, B., Reece, P. J., and Gooding, J. J. (2013). Functionalized porous silicon as a biosensor: emphasis on monitoring cells in vivo and in vitro. *Analyst* 138, 3593. doi:10.1039/c3an00081h

Gusev, S. A., Korotkova, N. A., Rozenstein, D. B., and Fraerman, A. A. (1994). Ferromagnetic filaments fabrication in porous Si Matrix. *J. Appl. Phys.* 76, 6671.

Herino, R. (1997). "Impregnation of porous silicon," in *Properties of Porous Silicon*, ed. L. Canham (London: INSPEC), 66–76.

Hippo, D., Nakamine, Y., Urakawa, K., Tsuchiya, Y., Mizuta, H., Koshida, N., et al. (2008). Formation mechanism of 100-nm-scale periodic structures in silicon using magnetic-field-assisted anodization. *Jpn. J. Appl. Phys* 47, 7398. doi:10.1143/JJAP.47.7398

Hourdakis, E., and Nassiopoulou, A. G. (2014). Single photoresist masking for local porous Si formation. *J. Micromech. Microeng.* 24, 117002. doi:10.1088/0960-1317/24/11/117002

Huang, Y. M. (1996). Photoluminescence of copper-doped porous silicon. *Appl. Phys. Lett.* 69, 2855–2857. doi:10.1063/1.117341

Kolasinski, K. W. (2010). Charge transfer and nanostructure formation during electroless etching of silicon. *J. Phys. Chem. C* 114, 22098. doi:10.1021/jp108169b

Kolasinski, K. W. (2014). The mechanism of galvanic/metal-assisted etching of silicon. *Nanoscale Res. Lett.* 9, 432. doi:10.1186/1556-276X-9-432

Kolasinski, K. W., and Barclay, W. B. (2013). The stoichiometry of electroless silicon etching in solutions of V2O5 and HF. *Angew. Chem. Int. Ed.* 52, 6731. doi:10.1002/anie.201300755

Kompan, M. E. (2003). Mechanism of primary self-organization in porous silicon with regular structure. *Phys. Solid State* 45, 948. doi:10.1134/1.1575342

Koshida, N., and Koyama, H. (1992). Visible electroluminescence from porous silicon. *Appl. Phys. Lett.* 60, 347. doi:10.1063/1.106652

Laiho, R., and Pavlov, A. (1995). Preparation of porous silicon films by laser ablation. *Thin Solid Films* 255, 9. doi:10.1016/0040-6090(94)05621-J

Lehmann, V. (2002). *Electrochemistry of Silicon, Instrumentation, Science, Materials and Applications*. Weinheim: Wiley-VCH.

Lehmann, V., and Gösele, U. (1991). Porous silicon formation: a quantum wire effect. *Appl. Phys. Lett.* 58, 856. doi:10.1063/1.104512

Li, J., Qing, L., and Dery, H. I (2012). Appelbaum, field-induced negative differential spin lifetime in silicon. *Phys. Rev. Lett.* 108, 157201. doi:10.1103/PhysRevLett. 108.157201

Masuda, H., and Fukuda, K. (1995). Ordered metal nanohole arrays made by a two-step replication of honeycomb structures of anodic alumina. *Science* 268, 1466–1468. doi:10.1126/science.268.5216.1466

Masuda, H., Hasegawa, F., and Ono, S. (1997). Self-ordering of cell arrangement of anodic porous alumina formed in sulphuric acid solution. *J. Electrochem. Soc.* 144, L127–L130. doi:10.1149/1.1837634

Melles, S., Menendez, J. L., Armelles, G., Navas, D., Vazquez, M., Nielsch, K., et al. (2003). Magneto-optical properties of nickel nanowires arrays. *Appl. Phys. Lett.* 83, 4547. doi:10.1063/1.1630840

Nakamura, T., and Adachi, S. (2012). Properties of magnetic nickel/porous silicon composite powders. *AIP Adv.* 2, 032167. doi:10.1063/1.4754152

Pankhurst, Q. A., Connoly, J., Jones, S. K., and Dobson, J. (2003). Applications of magnetic nanoparticles in biomedicine. *J. Phys. D Appl. Phys.* 36, R167. doi:10.1088/0022-3727/36/13/201

Ramazani, A., Kashi, M. A., and Seyedi, G. (2012). Crystallinity and magnetic properties of electrodeposited Co nanowires in porous alumina. *J. Mag. Mag. Mater.* 324, 1826. doi:10.1016/j.jmmm.2012.01.009

Ronkel, F., Schultze, J. W., and Arens-Fischer, R. (1996). Electrical contact to porous silicon by electrodeposition of iron. *Thin Solid Films* 276, 40. doi:10.1016/0040-6090(95)08045-7

Rumpf, K., Granitzer, P., Hilscher, G., and Pölt, P. (2011). Interacting low dimensional nanostructures within a porous silicon template. *J. Phys. Conf. Ser.* 303, 012048. doi:10.1088/1742-6596/303/1/012048

Rumpf, K., Granitzer, P., Koshida, N., Poelt, P., and Reissner, M. (2014). Magnetic interactions between metal nanostructures within porous silicon. *Nanoscale Res. Lett.* 9, 412. doi:10.1186/1556-276X-9-412

Rumpf, K., Granitzer, P., and Pölt, P. (2010a). Influence of the electrochemical process parameters on the magnetic behaviour of a silicon/metal nanocomposite. *ECS Trans.* 25, 157. doi:10.1149/1.3422509

Rumpf, K., Granitzer, P., and Poelt, P. (2010b). Synthesis and magnetic characterization of metal-filled double-sided porous silicon. *Nanoscale Res. Lett.* 5, 379. doi:10.1007/s11671-009-9492-6

Rumpf, K., Granitzer, P., and Pölt, P. (2010c). Formation mechanism of 100-nm-scale periodic structures in silicon using magnetic-field-assisted anodization. *ECS Trans.* 25, 157. doi:10.1149/1.3422509

Rumpf, K., Granitzer, P., Reissner, M., Poelt, P., and Albu, M. (2012a). Investigation of Ni and Co deposition into porous silicon and the influence of the electrochemical parameters on the physical properties. *ECS Trans.* 41, 59. doi:10.1149/1.4718391

Rumpf, K., Granitzer, P., Hilscher, G., Albu, M., and Pölt, P. (2012b). Magnetically interacting low-dimensional Ni-nanostructures within porous silicon. *Microelectr. Eng.* 90, 83. doi:10.1016/j.mee.2011.05.016

Sailor, M. J. (1997). "Sensor applications of porous silicon," in *Properties of Porous Silicon*, ed. L. Canham (London: INSPEC), 364–370.

Sasano, J., Jorne, J., Yoshimi, N., Tsuboi, T., Sakka, T., and Ogata, Y. H. (2000). "Effects of chloride ion on copper deposition into porous silicon," in *Fundamental Aspects of Electrochemical Deposition and Dissolution.* Electrochemical Society Proceedings, 84–91.

Sellmyer, D. J., and Skomski, R. (eds) (2006). *Advanced Magnetic Nanostructures.* New York: Springer.

Shin, J., Goyal, A., Cantoni, C., Sinclair, J. W., and Thompson, J. R. (2012). Self-assembled ferromagnetic cobalt/yttria-stabilized zirconia nanocomposites for ultrohigh density storage applications. *Naotechnology* 23, 155602. doi:10.1088/0957-4484/23/15/155602

Sinwani, O., Reiner, J. W., and Klein, L. (2014). Monitoring superparamagnetic Langevin behavior of individual $SrRuO_3$ nanostructures. *Phys. Rev. B* 89, 020404. doi:10.1103/PhysRevB.89.020404

Tang, J. M., Collins, B. T., and Flatte, M. E. (2012). Electron spin-phonon interaction symmetries and tunable spin relaxation in silicon and germanium. *Phys. Rev. B* 85, 045202. doi:10.1103/PhysRevB.85.045202

Tartaj, P., Morales, M. P., Veintemillas-Verdaguer, S., Gonzalez-Carreno, T., and Serna, C. J. (2003). The preparation of magnetic nanoparticles for applications in biomedicine. *J. Phys. D Appl. Phys.* 36, R182. doi:10.1088/0022-3727/36/13/202

Torres-Costa, V., Agullo-Rueda, F., Martin-Palma, R. J., and Martinez-Duart, J. M. (2005). Porous silicon optical devices for sensing applications. *Opt. Mater.* 27, 1084. doi:10.1016/j.optmat.2004.08.068

Tserepi, A., Tsamis, C., Gogolides, E., and Nassiopoulou, A. G. (2003). Dry etching of porous silicon in high density plasma. *Phys. State Solid A* 197, 163. doi:10.1002/pssa.200306493

Turner, R. (1958). Electropolishing silicon in hydrofluoric acid solution. *J. Electochem. Soc.* 105, 402–408. doi:10.1149/1.2428873

Uhlir, A. Jr. (1956). Electrolytic shaping of Germanium and silicon. *Bell Syst. Tech. J.* 35, 333–347. doi:10.1002/j.1538-7305.1956.tb02385.x

Uhlír, V., Vogel, J., Rougemaille, N., Fruchart, O., Ishaque, Z., Cros, V., et al. (2012). Current-induced domain wall motion and magnetization dynamics in CoFeB/Cu/Co nanostripes. *J. Phys. Condens. Matter* 24, 024213. doi:10.1088/0953-8984/24/2/024213

Vazquez, M., Pirota, K., Hernandez-Velez, M., Prida, V. M., Navas, J., Sanz, R., et al. (2004). Magnetic properties of densely packed arrays of Ni nanowires as a function of their diameter and lattice parameter. *J. Appl. Phys.* 95, 6642. doi:10.1063/1.1687539

Wang, Y., and Gu, H. (2015). Core – shell-type magnetic mesoporous silica nanocomposites for bioimaging and therapeutic agent delivery. *Adv. Mater.* 27, 576–585. doi:10.1002/adma.201401124

Wen, T., and Krishnan, K. M. (2011). Cobalt-based magnetic nanocomposites: fabrication, fundamentals and applications. *J. Phys. D Appl. Phys.* 44, 393001. doi:10.1088/0022-3727/44/39/393001

Zacharatos, F., Gianneta, V., and Nassiopoulou, A. G. (2008). Highly ordered hexagonally arranged nanostructures on silicon through a self-assembled silicon-integrated porous anodic alumina masking layer. *Nanotechnology* 19, 495306. doi:10.1088/0957-4484/19/49/495306

Zacharatos, F., Gianneta, V., and Nassiopoulou, A. G. (2009). Highly ordered hexagonally arranged sub-200 nm diameter vertical cylindrical pores on p-type Si using non-lithographic pre-patterning of the Si substrate. *Phys. State Solid A Appl. Res.* 206, 1286–1289. doi:10.1002/pssa.200881111

Zhang, J. J., Li, Z. Y., Zhang, Z. J., Wu, T. S., and Sun, H. Y. (2013). Optical and magnetic properties of porous anodic alumina/Ni nanocomposite films. *J. Appl. Phys.* 113, 244305. doi:10.1063/1.4812466

Zhang, X. G. (2001). *Electrochemistry of Silicon and Ist Oxide.* New York: Kluwer Academic Plenum Publishers.

**Conflict of Interest Statement:** The authors declare that the research was conducted in the absence of any commercial or financial relationships that could be construed as a potential conflict of interest.

# Permissions

# List of Contributors

**Antonio Politano and Gennaro Chiarello**
Dipartimento di Fisica, University of Calabria, Cosenza, Italy

**Shinichi Saito, Frederic Yannick Gardes and Abdelrahman Zaher Al-Attili**
Faculty of Physical Sciences and Engineering, University of Southampton, Southampton, UK

**KazukiTani, Katsuya Oda, Yuji Suwa and Tatemi Ido**
Photonics Electronics Technology Research Association (PETRA), Tokyo, Japan
Institute for Photonics-Electronics Convergence System Technology (PECST), Tokyo, Japan
Central Research Laboratory, Hitachi Ltd., Tokyo, Japan

**Yasuhiko Ishikawa**
Department of Materials Engineering, Graduate School of Engineering, The University of Tokyo, Tokyo, Japan

**Satoshi Kako, Satoshi Iwamoto and Yasuhiko Arakawa**
Institute for Photonics-Electronics Convergence System Technology (PECST), Tokyo, Japan
Institute of Industrial Science, The University of Tokyo, Tokyo, Japan

**Eric Perim, Leonardo Dantas Machado and Douglas Soares Galvao**
Applied Physics Department, State University of Campinas, Campinas, Brazil

**Michele Brun**
Dipartimento di Ingegneria Meccanica, Chimica e dei Materiali, Università di Cagliari, Cagliari, Italy
Department of Mathematical Sciences, University of Liverpool, Liverpool, UK

**Gian Felice Giaccu**
Dipartimento di Architettura, Design e Urbanistica, Facoltà di Architettura, Università di Sassari, Alghero, Italy

**Alexander B. Movchan**
Department of Mathematical Sciences, University of Liverpool, Liverpool, UK

**Leonid I. Slepyan**
School of Mechanical Engineering, Tel Aviv University, Tel Aviv, Israel
Department of Mathematics and Physics, Aberystwyth University, Aberystwyth, UK

**Claudia Merlini , Guilherme Mariz de Oliveira Barra, Sílvia Daniela Araújo da Silva Ramôa and Giseli Contri**
Department of Mechanical Engineering, Universidade Federal de Santa Catarina, Florianópolis, Brazil

**Rosemeire dos Santos Almeida and Marcos Akira d'Ávila**
School of Mechanical Engineering, Universidade Estadual de Campinas (UNICAMP), Campinas, Brazil

**Bluma G. Soares**
Instituto de Macromoléculas, Universidade Federal do Rio de Janeiro, Rio de Janeiro, Brazi

**lInes Jimenez-Palomar**
School of Engineering and Materials Science, Queen Mary University of London, London, UK

**Anna Shipov and Ron Shahar**
Koret School of Veterinary Medicine, The Hebrew University of Jerusalem, Jerusalem, Israel

**Asa H. Barber**
School of Engineering and Materials Science, Queen Mary University of London, London, UK
School of Engineering, University of Portsmouth, Portsmouth, UK

**Miloslav Pekař**
Faculty of Chemistry, Brno University of Technology, Brno, Czech Republic

**Nael G.Yasri**
Department of Mechanical Engineering, University of Wisconsin-Milwaukee, Milwaukee, WI, USA
Department of Biological Systems Engineering, University of Wisconsin-Madison, Madison, WI, USA
Department of Chemistry, Faculty of Science, University of Aleppo, Aleppo, Syria

**Ashok K. Sundramoorthy**
Department of Biological Systems Engineering, University of Wisconsin-Madison, Madison, WI, USA

**Woo-Jin Chang**
Department of Mechanical Engineering, University of Wisconsin-Milwaukee, Milwaukee, WI, USA
Great Lakes WATER Institute, School of Freshwater Sciences, University of Wisconsin-Milwaukee, Milwaukee, WI, USA

**Sundaram Gunasekaran**
Department of Biological Systems Engineering, University of Wisconsin-Madison, Madison, WI, USA

**Maurício Chagas da Silva, Egon Campos dos Santos, Maicon Pierre Lourenço, Mateus Pereira Gouvea and Hélio Anderson Duarte**
Grupo de Pesquisa em Química Inorgânica Teórica, Departamento de Química, Instituto de Ciências Exatas (ICEx), Universidade Federal de Minas Gerais, Belo Horizonte, Brazil

**Antonios Kouloumpis, Konstantinos Dimos and Dimitrios Gournis**
Department of Materials Science and Engineering, University of Ioannina, Ioannina, Greece

**Konstantinos Spyrou**
Department of Materials Science and Engineering, University of Ioannina, Ioannina, Greece
Zernike Institute for Advanced Materials, University of Groningen, Groningen, Netherlands

**Vasilios Georgakilas**
Department of Materials Science, University of Patras, Rio, Greece

**Petra Rudolf**
Zernike Institute for Advanced Materials, University of Groningen, Groningen, Netherlands

**Pengfei Li, Caiyun Chen, Jie Zhang, Shaojuan Li, Baoquan Sun and Qiaoliang Bao**
Jiangsu Key Laboratory for Carbon-Based Functional Materials and Devices, Collaborative Innovation Center of Suzhou Nano Science and Technology, Institute of Functional Nano and Soft Materials (FUNSOM), Soochow University, Suzhou, China

**Aboozar Mosleh and Seyed Amir Ghetmiri**
Microelectronics-Photonics Graduate Program (mEP), University of Arkansas, Fayetteville, AR, USA
Department of Electrical Engineering, University of Arkansas, Fayetteville, AR, USA

**Murtadha A. Alher**
Department of Electrical Engineering, University of Arkansas, Fayetteville, AR, USA
Mechanical Engineering Department, University of Karbala, Karbala, Iraq

**Larry C. Cousar**
Microelectronics-Photonics Graduate Program (mEP), University of Arkansas, Fayetteville, AR, USA
Arktonics, LLC, Fayetteville, AR, USA

**Wei Du, Thach Pham, Hameed A. Naseem and Shui-QingYu**
Department of Electrical Engineering, University of Arkansas, Fayetteville, AR, USA

**Joshua M. Grant**
Engineering-Physics Department, Southern Arkansas University, Magnolia, AR, USA

**Greg Sun and Richard A. Soref**
Department of Engineering, University of Massachusetts Boston, Boston, MA, USA

**Baohua Li**
Arktonics, LLC, Fayetteville, AR, USA

**Vladimir S. Shiryaev**
G.G. Devyatykh Institute of Chemistry of High-Purity Substances of the Russian Academy of Sciences, Nizhny Novgorod, Russia
N.I. Lobachevski Nizhny Novgorod State University, Nizhny Novgorod, Russia

**Nicola Bordignon, Andrea Piccolroaz, Francesco Dal Corso and Davide Bigoni**
Department of Civil, Environmental and Mechanical Engineering (DICAM), University of Trento, Trento, Italy

**Xianshu Luo, Junfeng Song, Patrick Guo-Qiang Lo, Tsung-Yang Liow and MingbinYu**
Institute of Microelectronics, Agency for Science, Technology and Research (A*STAR), Singapore, Singapore

**Yulian Cao, Xiaonan Hu,Yuanbing Cheng, Chengming Li , Chongyang Liu, HongWang and Qi JieWang**
Photonics Center of Excellence (OPTIMUS), School of Electrical and Electronic Engineering, Nanyang Technological University, Singapore, Singapore

**Ram K. Gupta and J. Candler**
Department of Chemistry, Pittsburg State University, Pittsburg, KS, USA

**D. Kumar**
Department of Mechanical Engineering, North Carolina A&T State University, Greensboro, NC, USA

**Bipin K. Gupta**
National Physical Laboratory (CSIR), New Delhi, India

**Pawan K. Kahol**
Department of Physics, Pittsburg State University, Pittsburg, KS, USA

**Orkun Onal, Cemre Ozmenci and Demircan Canadinc**
Advanced Materials Group (AMG), Department of Mechanical Engineering, Koç University, Istanbul, Turkey

**Yaying Hong and WeijiaWen**
Department of Physics, Hong Kong University of Science and Technology, Hong Kong, China

**Alberto Leonardi and Paolo Scardi**
Department of Civil, Environmental and Mechanical Engineering, University of Trento, Trento, Italy

**Muhammad Rizwan Saleem**
Institute of Photonics, University of Eastern Finland, Joensuu, Finland
Center for Energy Systems (CES), USAID Center for Advance Studies, National University of Sciences and Technology (NUST), Islamabad, Pakistan

**Rizwan Ali , Seppo Honkanen and JariTurunen**
Institute of Photonics, University of Eastern Finland, Joensuu, Finland

**Mohammad Bilal Khan**
Center for Energy Systems (CES), USAID Center for Advance Studies, National University of Sciences and Technology (NUST), Islamabad, Pakistan

**Francesca Lionetto, Riccardo Dell'Anna, Francesco Montagna and Alfonso Maffezzoli**
Materials Science and Technology Group, Department of Engineering for Innovation, University of Salento, Lecce, Italy

**Marco Salerno  and Alberto Diaspro**
Department of Nanophysics, Istituto Italiano di Tecnologia, Genova, Italy

**Matthew A. Addicoat**
Engineering and Science, Jacobs University Bremen, Bremen, Germany

**Manuel Tsotsalas**
Institute of Functional Interfaces (IFG), Karlsruhe Institute of Technology (KIT), Eggenstein-Leopoldshafen, Germany

**Petra Granitzer and Klemens Rumpf**
Institute of Physics, Karl-Franzens-University Graz, Graz, Austria

www.ingramcontent.com/pod-product-compliance
Lightning Source LLC
Chambersburg PA
CBHW080521200326
41458CB00012B/4283